Supramolecular Assemblies Based on Electrostatic Interactions

M. Ali Aboudzadeh · Antonio Frontera
Editors

Supramolecular Assemblies Based on Electrostatic Interactions

Springer

Editors
M. Ali Aboudzadeh
CNRS, IPREM
University Pau and Pays Adour (UPPA)
Pau, France

Antonio Frontera
Universitat de Les Illes Balears
Palma de Mallorca (Baleares), Spain

ISBN 978-3-031-00659-3 ISBN 978-3-031-00657-9 (eBook)
https://doi.org/10.1007/978-3-031-00657-9

This Springer imprint is published by the registered company Springer Nature Switzerland AG
The registered company address is: Gewerbestrasse 11, 6330 Cham, Switzerland

To our families…

Preface

Ionic interactions play an essential role in the self-assembly of functional biological and synthetic systems. Ionic self-assembly, the coupling of two structurally distinct building blocks by electrostatic interactions, has recently become an effective approach to develop new nanostructured chemical objects and materials with interesting and tuneable properties. This ion pairing also plays an important role in the design of supramolecular assemblies, especially in an aqueous environment, where the presence of charges also guarantees adequate solubility. An appropriate molecular structure design of self-assembling building blocks is necessary to obtain tailored properties and even complex functions. Such functions have been already discussed in books and reviews on supramolecular chemistry, also with an emphasis on the underlying binding mechanism.

This volume presents a much-needed update of recent advances and current knowledge in the field of supramolecular assemblies based on electrostatic interactions. The flexibility and simplicity of constructing assemblies are explained via several examples, illustrations, figures, case studies, and historical perspectives. The first two chapters of the book have focused on synthesis aspects and properties of supramolecular ionic networks, including those prepared from small molecules or polymers (or a combination of both), and it is attempted to derive consistent relations between their structure, dynamics, and properties. Within the same context, Chap. 3 discusses the role of electrostatic interaction, purely or in combination with other non-covalent interactions, in the self-assembly of macroions (ranging from 0D to 3D supramolecular structures, e.g., metal-organic cages, dendrimers, biomacromolecules, etc.). Among all macroions, dendrimers have attracted great interest in recent times due to their unique self-assembling properties. Accordingly, a full chapter was included in this book which gives a comprehensive overview of functions, structures, and properties of dendrimers in bulk and in solution (Chap. 4). Moreover, recent developments suggest that electrostatic self-assembly is capable of yielding nano-objects that are well-defined in solution. This emerging field has been fully reviewed in Chap. 5. Layer-by-Layer (LbL) self-assembly is one of the versatile methods used to fabricate multilayered nano-objects, typically under aqueous assembly conditions which involve alternating deposition of multivalent compounds with complementary interactions. In this regard, Chap. 6 provides a broad perspective on the main physicochemical aspects of the fabrication of multilayered nanomaterials through LbL method.

The next two chapters discuss about σ-hole and π-hole interactions where electrostatic contributions are crucial. A deep understanding of noncovalent σ-hole interactions is necessary to advance in many fields, especially in crystal growth and crystal engineering, as illustrated in Chap. 7. Regium-bonding, a new player in supramolecular chemistry, is described in Chap. 8, including examples related to crystal engineering, biological systems, and surface absorption processes.

Nature has created the most beautiful and sophisticated examples of supramolecular assembled systems derived from amphiphilic molecules. Chapter 9 presents the most representative examples (including their design strategy) in electrostatic interaction-based photoresponsive molecular amphiphiles. In the next chapter, self-assembly behavior of amphiphilic salts in solution (mostly based on imidazolium and ammonium) was reviewed as a function of structural features of ionic tectons, emphasizing their implications on the different applications.

Finally, as there is a need for theory and modeling of interaction strength to help experimental studies, the last chapter of the book is dedicated to the topic "Modelling of supramolecular assemblies". It was shown in this chapter, how computational chemistry techniques are particularly helpful for comprehension at the molecular scale of the complex supramolecular polymerization process as well as the relevant properties that final self-assembled architectures could acquire.

This book aims to inspire and guide fellow scientists and students in this field. In this context, the book first merits broad and fast dissemination into the general scientific community especially for the chemists to examine this exciting branch of science to realize its full potential in the new century. Still have a long way to go for a complete understanding, but these volumes demonstrate that rapid and exciting progress is being made.

The Editors express their appreciation to all contributors from different parts of the world that have cooperated in the preparation of this volume. In this context, this international book gives the active reader different perspectives on the subject and encourages him/her to read the entire book.

Pau, France — M. Ali Aboudzadeh
Palma de Mallorca, Spain — Antonio Frontera

Contents

Contributors

M. Ali Aboudzadeh CNRS, University Pau & Pays Adour, E2S UPPA, Institut des Sciences Analytiques et de Physico-Chimie pour l'Environnement et les Matériaux, IPREM, UMR5254, 64000 Pau, France

Juan Aragó Instituto de Ciencia Molecular (ICMol), Universidad de Valencia, Paterna, Spain

Antonio Bauzá Universitat de Les Illes Balears, Palma de Mallorca (Baleares), Spain

Joaquín Calbo Instituto de Ciencia Molecular (ICMol), Universidad de Valencia, Paterna, Spain

Alberto Concellón Department of Chemistry, Massachusetts Institute of Technology, Cambridge, MA, USA

Azahara Doncel-Giménez Instituto de Ciencia Molecular (ICMol), Universidad de Valencia, Paterna, Spain

Francesca D'Anna Dipartimento di Scienze e Tecnologie Biologiche, Chimiche, Farmaceutiche (STEBICEF), Università degli Studi di Palermo, Palermo, Italy

Antonio Frontera Universitat de Les Illes Balears, Palma de Mallorca (Baleares), Spain

Franziska Gröhn Department of Chemistry and Pharmacy and Interdisciplinary Center for Molecular Materials (ICMM), Bavarian Polymer Institute (BPI), Friedrich-Alexander University (FAU) Erlangen-Nürnberg, Erlangen, Germany

Eduardo Guzmán Departamento de Química Física, Universidad Complutense de Madrid, Madrid, Spain;
Instituto Pluridisciplinar, Universidad Complutense de Madrid, Madrid, Spain

Shaghayegh Hamzehlou POLYMAT, Department of Applied Chemistry, Faculty of Chemistry, University of the Basque Country UPV/EHU, Joxe Mari Korta Zentroa, Donostia, Spain

Verónica Iguarbe Departamento de Química Orgánica, Universidad de Zaragoza, Zaragoza, Spain

Anja Krieger Department of Chemistry and Pharmacy and Interdisciplinary Center for Molecular Materials (ICMM), Bavarian Polymer Institute (BPI), Friedrich-Alexander University (FAU) Erlangen-Nürnberg, Erlangen, Germany

Franco King-Chi Leung State Key Laboratory of Chemical Biology and Drug Discovery, Department of Applied Biology and Chemical Technology, The Hong Kong Polytechnic University, Kowloon, Hong Kong SAR, China

Tianbo Liu The School of Polymer Science and Polymer Engineering, The University of Akron, Akron, OH, USA

Salvatore Marullo Dipartimento di Scienze e Tecnologie Biologiche, Chimiche, Farmaceutiche (STEBICEF), Università degli Studi di Palermo, Palermo, Italy

Ana Mateos-Maroto Max Planck Institute for Polymer Research, Mainz, Germany

Francisco Ortega Departamento de Química Física, Universidad Complutense de Madrid, Madrid, Spain;
Instituto Pluridisciplinar, Universidad Complutense de Madrid, Madrid, Spain

Enrique Ortí Instituto de Ciencia Molecular (ICMol), Universidad de Valencia, Paterna, Spain

Ehsan Raee The School of Polymer Science and Polymer Engineering, The University of Akron, Akron, OH, USA

Carla Rizzo Dipartimento di Scienze e Tecnologie Biologiche, Chimiche, Farmaceutiche (STEBICEF), Università degli Studi di Palermo, Palermo, Italy

Ramón G. Rubio Departamento de Química Física, Universidad Complutense de Madrid, Madrid, Spain;
Instituto Pluridisciplinar, Universidad Complutense de Madrid, Madrid, Spain

Yuqing Yang The School of Polymer Science and Polymer Engineering, The University of Akron, Akron, OH, USA

Yifan Zhou The School of Polymer Science and Polymer Engineering, The University of Akron, Akron, OH, USA

Alexander Zika Department of Chemistry and Pharmacy and Interdisciplinary Center for Molecular Materials (ICMM), Bavarian Polymer Institute (BPI), Friedrich-Alexander University (FAU) Erlangen-Nürnberg, Erlangen, Germany

Abbreviations

1D	One-Dimensional
1N36S	1-naphthol-3,6-disulfonate
3D	Three-Dimensional
AA	All-atom
AA	Amino Acid
AA-FF	All-atom force field
ABnOHRA	Disodium 4-((3-(hydroxymethyl)phenyl)diazenyl)-3-hydroxynaphthalene-2,7-disulfonate
ACQ	Aggregation caused quenching
AEE	Aggregation enhanced emission
AFM	Atomic force microscopy
AIE	Aggregation induced emission
Ala	L-alaninate
ANAC	Abscisic acid-responsive NAC
Ar26	Acid red 26
Ar27	Acid red 27
ASAP	Sodium polyacrylate
ASAXS	Anomalous small-angle X-ray scattering
AT	Anthracene
ATR	Attenuated total reflectance
AuNPs	Gold nanoparticles
Ay38	Acid yellow 38
BETI	Bis[(pentafluoroethyl)sulfonyl]imide
Bis-MPA	2,2-Bis(hydroxymethyl)propionic acid
BS	Benzenesulfonate
BTA	Benzene-1,3,5-tricarboxamide
BTF	Bis-(trifluoromethanesulfonimide)
C_{14}DMAO	Tetradecyldimethylamine oxide
CB7	Cucurbit[7]uril
CBS	Complete basis set
CCSD(T)	Coupled Cluster with single double (triple)

CD	Circular dichroism
CG	Coarse-grained
CG-FF	Coarse-grained force field
CG-MD	Coarse-Grained molecular Dynamic
Ch	Chalcogen atom
ChB	Chalcogen bond
CHI	Chitosan
CiB	Coinage or Regium bond
CLSM	Confocal laser scanning microscopy
CM	Continuum models
CMC	Critical micellization concentration
CNDO	Complete neglect of differential overlap
CNS	Charged clay nanosheets
Col_h	Hexagonal columnar mesophase
Col_r	Rectangular columnar mesophase
COSMO	Conductor-like screening solvation model
Cryo-TEM	Cryogenic transmission electron microscopy
CSD	Cambridge structural database
CTAB	Cetyltrimethylammonium bromide
CTAT	Cetyl trimethylammonium tosylate
CTC	Chlorotetracycline
Cub_I	Micellar cubic mesophase
Cub_V	Bicontinuous cubic mesophase
Cys	Cysteine
DFT	Density functional theory
DFT-SAPT	DFT-symmetry-adapted intermolecular perturbation theory
DiCOOH-PEG	Carboxylic acid-telechelic poly(ethylene glycol)s
DLS	Dynamic light scattering
DLVO theory	Derjaguin-Landau-Verwey-Overbeek theory
DMF	N,N-dimethylformamide
DMP[5]A	Dimethoxypillar[5]arene
DMPC	2-dimyristoyl-sn-glycero-3-phosphocholine
DMSO	Dimethylsulfoxide
DNA	Deoxyribonucleic Acid
DOPC	1,2-dioleoyl-*sn*-glycero-3-phosphocholine
DPPC	Dipalmitoyilphosphocoline
DSC	Differential scanning calorimetry
EDTA	Ethylenediaminetetraacetate
ee	Enantiomeric excess
EHT	Extended Hückel theory
ETAB	4-ethyl-4′-(trimethylaminohexyloxy) azobenzene bromide
EWG	Electron withdrawing groups
FeNP	Iron nanoparticles
FF	Force fields
Fmoc	Fluorenylmethyloxycarbonyl

FRET	Fluorescence resonance energy transfer
FTIR	Fourier transform infrared
GAFF	General amber force field
GB	Generalized born
GBSA	Generalized Born and surface area continuum salvation
GFN	Geometries, vibrational Frequencies, and Noncovalent interactions
GGA	Generalized gradient approximation
GLU	Gluconate
$GV_3A_3E_3$	Glycine-Valine-Valine-Valine-Alaine-Alaine-Alaine-Glutamine-Glutamine-Glutamine
Ha	Halogen atom
HA	Hyaluronic acid
HaB	Halogen bond
HB	Hydrogen bonding
HBC	Hexaperi-hexabenzocoronenes
HBS	4-hydroxybenzenesulfonate
HC_1im	1-methylimidazolium
HEWL	Hen egg-white lysozyme
HF	Hartree–Fock
HFB	Hexafluorobenzene
HFIP	Hexafluoroisopropanol
His	Histidine
HMDA	Hexamethylene diamine
HNC	3-hydroxy-2-naphthoate
HOMO	Highest occupied molecular orbital
Ile	Isoleucine
INDO	Intermediate neglect of differential overlap
ITC	Isothermal titration calorimetry
IUPAC	International union of pure and applied chemistry
KS	Kohn–Sham
L-Ala	L-alanine
LbL	Layer-by-Layer
LC	Liquid crystal
LCD	Liquid crystal display
LDA	Local-density approximation
Leu	Leucine
LiTFSI	Lithium bis(trifluoromethylsulphonyl)imide
LJ	Lennard–Jones potential
LLS	Laser light scattering
LP	Lone pair
L-Phe	L-phenylalanine
L-Pro	L-proline
LUMO	Lowest unoccupied molecular orbital
Mb	Myoglobin

MCH	Methylcyclohexane
MCH^+	Merocyanine form
MD	Molecular dynamics
MEP	Molecular electrostatic potential
Met	Methionine
MFRM	Melamine-formaldehyde resin microspheres
MM/MD	Classical molecular mechanics and molecular dynamics
MM	Classical molecular mechanics
MOC	Metal organic cage
MOM	Metal organic macrocycle
MOPV	Oligophenylenevinylene
MP2	Møller-Plesset 2
MR	Majority rules
MRI	Magnetic resonance imaging
NCI	Non Covalent Interactions
N_D	Nematic Discotic Mesophase
NDC	Naphthalene dicarboxylic acid
NDDO	Neglect of differential diatomic overlap
NDI	Naphthalene diimide
NgB	Noble gas bond
NIR	Near-infrared
NMR	Nuclear magnetic resonance
NOESY	Nuclear overhauser effect spectroscopy
NPA	p-nitrophenylacetate
N-PBIs	*N*-annulated perylene bisimides
NR	Neutron reflectometry
NSA	Naphthalenesulfonate
NT	Naphtalene
NTf_2	Bis[(trifluoromethyl)sulfonyl]imide
OPE-TA	Oligo(phenylene ethynylene) tricarboxamide
PA	Poly(acrylate)
PAA	Poly(acrylic acid)
PAH	Poly(allylamine-hydrochloride)
PAMAM	Poly(amidoamine)
PB	Poisson−Boltzmann
PB	Polybutadiene
PBE	Perdew–Burke–Ernzerhof functional
PCM	Polarizable continuum model
PDADMAC	Poly(diallyldimethylammonium chloride)
PDB	Protein data bank
PDI	Perylene-bis(dicarboximide)
PDMAEMA	Poly[2-(N,N-diethylamino)ethyl methacrylate]
PDMS	Poly(dimethylsiloxane)
PEG	Polyethylene glycol
PEI	Hyperbranched polyethyleneimine

PEIMe	Fully methylated hyperbranched polyethyleneimine
PEMs	Polyelectrolyte Multilayers
PFAT	Perfluoroanthracene
PFG-NMR	Pulsed field gradient NMR
PFNT	Perfluoronaphtalene
PGA	Poly(glutamic acid)
PGSE-NMR	Pulsed gradient spin-echo NMR
Phe	Phenylalanine
PLL	Poly(L-lysine)
PM2VP	Poly(N-methyl-2-vinyl pyridinium chloride)
Pn	Pnictogen atom
PnB	Pnictogen bond
POM	Polarized optical microscopy
POM	Polyoxometalate
POPC	Palmitoyl oleyl phosphatidylcholine
POSS	Polyhedral oligomeric silsesquioxane
PPI	Poly(propylene imine)
Pro	L-prolinate
PSA	Polar surface area
PSS	Photostationary state
PSS	Poly(4-styrenesulfonate of sodium)
PTS	p-toluenesulfonate
QM/MM	Quantum mechanics/Molecular mechanics
QM	Quantum mechanics
QTAIM	Quantum theory of atoms in molecules
Rg	Regium atom
RgB	Regium bond
rhB	Rhodamine B
RVFV	Rift valley fever virus
SAIL	Surface active ionic liquids
Sal	Salicilate
SANS:	Small-angle neutron scattering
SaS	Sergeants and soldiers
SASA	Solvent-accessible surface area
SAXS	Small-angle X-ray scattering
SDS	Sodium dodecyl sulfate
SEM	Scanning Electron Microscopy
SFTSV	Severe fever with thrombocytopenia syndrome virus
SLS	Static light scattering
SmA	Smectic A mesophase
SmA^+	Frustrated smectic A mesophase
SmC	Smectic C mesophase
SmCP	Polar smectic C mesophas
SMD	Solvation model based on density
SP	Spiropyrans form

SpB	Spodium bond
SPs	Supramolecular polymers
SQM	Semiempirical quantum mechanical
SSCIS	Spray coating of interacting species
TAEA	Tris(2-aminoethyl)amine
TAPP	Meso-tetrakis-(4-(trimethyl-ammonium) phenyl) porphyrin
TEM	Transmission electron microscopy
TFB	Trifluorobenzene
TGA	Thermogravimetric analysis
THF	Tetrahydrofuran
THI	Thermal helix inversion
TMPyP	Meso-tetrakis(4-Nmethyl-pyridinium)porphyrin
TrB	Triel bond
Trp	Tryptophan
TR-SANS	Time-resolved small-angle neutron scattering
TR-SAXS	Time-resolved small-angle X-ray scattering
TS	Tkatchenko–Scheffler approximation
Tt	Tetrel
TtB	Tetrel bond
Tyr	Tyrosine
UPy	2-ureido-4[1H]-pyrimidinone
ur-Py	Ureido-pyrimidone
UV	Ultra-violet
VFT	Vogel-Fulcher-Tamman
WF	Wavefunction
XC	Exchange-correlation
XDM	Becke–Johnson exchange–hole dipole model
XPS	X-ray photoelectron spectroscopy
XRR	X-Ray reflectometry
xTB	Extended tight-binding
ZDO	Zero differential overlap

Nomenclature and Symbols

AC_{60}	Lactic acid-functionalized chiral fullerene
B	Benzene
c	Concentration
$C_1C_{14}im$	1-methyl-3-tetradecylimidazolium
$C_1C_{16}im$	1-methyl-3-hexadecylimidazolium
d	Diameter
D_H	Hydrodynamic Radius
E_a	Activation energy
F	Phenylalanine
G	Dendrimer generation
G4	4th generation
G5	5th generation
G7.5	7.5th generation
G'	Storage moduli
G''	Loss viscous moduli
G_e	Equilibrium elasticity parameter
G_p	Plateau modulus
H_2im	1H-imidazolium
K	Equilibrium constant
M_i	Molar weight between elastically effective interactions
M_n	Number average molecular weight
M_w	Weight average molecular weight
N	Aggregation number
N	Number of bilayers
N	Nematic mesophase
P	Packing parameter
p	PO_4^{2-}
pKa	Acid dissociation constant
R_C	Compensation ratio
r	Radius
R_h	Hydrodynamic radius

R_g	Radius of gyration
$-SO_2-N-SO_2-CF_3$	(trifluoromethane-sulfonyl)imide
$-SO_2-C(CN)_2$	(propylsulfonyl)methanide
$-SO_2-N-CN$	(cyano-propylsulfonyl)imide
T	Temperature
T_g	Glass transition temperature
T_{ln}	Liquid–network transition
T_m	Melting temperature
T_{nl}	Network–liquid transition
Y	Tyrosine
ΔG	Free energy
ΔH	Enthalpy change
ΔS	Entropy change
ε'	Dielectric permittivities
γ_c	Crosslinking density
τ	Relaxation time
ω	Frequency

Chapter 1
Supramolecular Ionic Networks: Design and Synthesis

M. Ali Aboudzadeh

Abstract Supramolecular polymer networks are chains of low molecular mass monomers held together by reversible non-covalent interactions, such as hydrogen bonds, metal–ligand bonds, hydrophobic or $\pi-\pi$ stacking interactions. The reversibility and low energy bonding bring about additional features compared to conventional covalent polymers, which potentially lead to new properties such as improved processing, self-healing behavior, and stimuli-responsiveness. Whereas the use of (multiple) hydrogen bonds is leading the discoveries in this area, the emerging ionic chemistry has also been translated to the development of supramolecular assemblies based on ionic interactions. This approach provides exciting opportunities for synthesizing new supramolecular materials via manipulation of the type and strength of the ion pair as well as the number of interactions. In this chapter, the most relevant advances and current knowledge in design and synthesis of supramolecular ionic networks, including those prepared from low molecular weight molecules, polymers, or a combination of the two are briefly reviewed. Their flexible and simple construction is depicted via several examples and case studies. Finally, the important concerns and possible opportunities are explained to inspire critical discussions and boost further findings.

1.1 Introduction

Conventional polymers are long-chain molecules made up of repeating structural units linked through covalent bonds. They have been employed widely in ordinary life and advanced technologies for more than a half-century. In the last thirty years, the universality of reversible non-covalent interactions has been perceived with the evolution of supramolecular chemistry, defined by Lehn as the chemistry beyond molecules, who firstly reported the application of hydrogen bonds to create polymer

M. A. Aboudzadeh (✉)
CNRS, University Pau & Pays Adour, E2S UPPA, Institut des Sciences Analytiques et de Physico-Chimie pour l'Environnement et les Matériaux, IPREM, UMR5254, 64000 Pau, France
e-mail: m.aboudzadeh-barihi@univ-pau.fr

M. A. Aboudzadeh and A. Frontera (eds.), *Supramolecular Assemblies Based on Electrostatic Interactions*,
https://doi.org/10.1007/978-3-031-00657-9_1

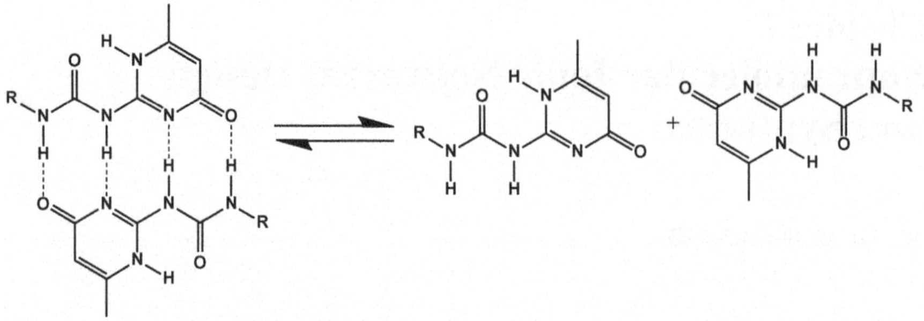

Fig. 1.1 Network structure of the monomer (UPy) units connected by reversible hydrogen bonds. The UPy units form four hydrogen bonds to each other (dotted lines)

structures [1]. One important aspect of supramolecular chemistry is the application of clear-cut molecules or macromolecules and intermolecular forces to generate larger, more intricate chemical species with novel and extraordinary properties [2–4]. This discovery encouraged scientists to build up reversible alternatives for covalent polymers.

Nevertheless, in the beginning, it was not simple how to incorporate highly directional, amply strong, but still reversible interactions to direct small molecules to be assembled into polymeric patterns. Particular breakthrough was not reported until the pioneering work by Meijer et al. [5]. They assembled 2-ureido-4[1H]-pyrimidinone (UPy) units into extended chains by means of quadruple cooperative hydrogen bonds in an array and developed products with similar mechanical properties that until then could only be achieved with covalent polymers (Fig. 1.1). In such systems, by increasing the temperature the viscosity of the supramolecular polymers could decrease, indicating an exceptional thermo-responsiveness coming from reversible hydrogen bond interactions. This property allowed supramolecular polymeric materials to be easily processed.

Following this achievement and in recent years, many scientists have been studying and exploring the ability to utilize non-covalent intermolecular forces to build controlled supramolecular structures and tailor their properties. Among various supramolecular assemblies known, polymer networks are specifically interesting as their properties can be completely different from the properties of their covalently bonded counterparts or the individual macromers [6–8]. In particular, reversible binding through cooperative hydrogen bonds [9–11], ionic interactions [12, 13] and metal–ligand complexes [14, 15], etc. can lead to aggregation, gelation, or sudden viscosity changes that are triggered by changes in molecular concentration, pH, or temperature. Networks established through this approach have particular advantages over conventional polymer networks based on covalent bonding because they merge the features of traditional polymers with the reversibility of bonds that hold monomer segments. Nevertheless, when these reversible holding interactions are removed, for instance by heating, their elastic or rigid nature can be converted into

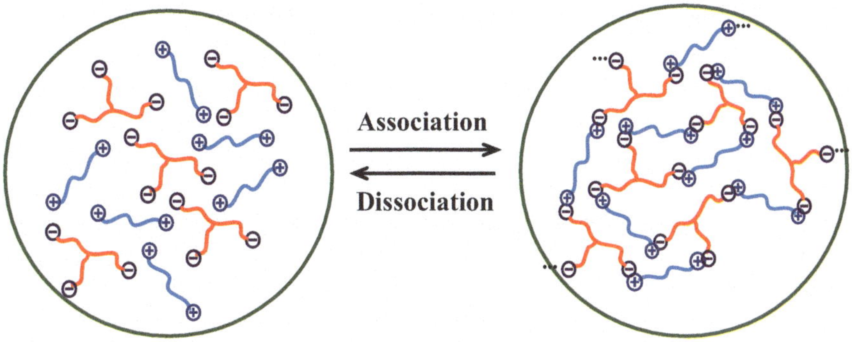

Scheme 1.1 Schematic of supramolecular polymer networks generated from non-covalent ionic interactions. The major difference from traditional polymer is the equilibrium between association and dissociation of the interactions

low viscosity small molecules [16]. These unique characteristics facilitate a new category of stimuli-responsive and functional polymers named supramolecular polymer networks [17].

Scheme 1.1 graphically shows a supramolecular polymer network made up of end-to-end chaining of low molecular mass oligomers. Similar to conventional polymers, supramolecular networks present analogous macromolecular properties in solutions as well as in bulk. However, because the connecting forces for their structures are reversible interactions, supramolecular networks still behave like small molecules when their reversible interactions dissociate with changing conditions such as heating or solvent switching.

One undesirable disadvantage of conventional polymers is their high melt viscosities as a result of their chain entanglements, which make them difficult to process. In contrast, supramolecular polymer networks generally exhibit a strongly temperature-dependent melt viscosity, which improves their processability in a less viscous state at temperatures only moderately higher than their melting or glass transition temperatures. The reversible properties of supramolecular polymers make them capable of self-repair or healing after disruption of the interaction, which opened a quite new research field in the last two decades. To obtain these novel features, the most challenging task is to design appropriate building block functionalities with synthetic accessibility and high stability. It is the aim of this chapter to present an overall view of the synthetic methods to create these new networks. Particularly, we start by presenting different types of supramolecular assemblies, and then we address the advances in the field of supramolecular ionic assemblies, including those composed of complementarily charged polymers, low molecular weight molecules, or a combination of the two. Finally, the motivation for further discoveries in this field and future perspectives are discussed.

1.2 Different Types of Supramolecular Assemblies

Supramolecular assemblies can be categorized on the basis of three different principles: (1) the physical nature of the non-covalent force that lies at the origin of the reversible interaction, (2) the type of structural monomer(s) used, for example, supramolecular polymerization of an A_2 monomer or an A-B type monomer, and (3) the Gibbs free thermodynamics which describes different mechanisms of supramolecular polymerizations and shows how the conversion is based on temperature, concentration, etc. Each classification has its own scientific merits [18]. In this chapter, we classify different supramolecular assemblies on the basis of the physical nature of the various types of interactions that can behave as driving forces for the design of large supramolecular assemblies. Important non-covalent interactions include hydrogen bonding, electrostatic interactions, metal–ligand complex, and π–π stacking. Some examples of supramolecular assemblies via different non-covalent interactions will be discussed. The electrostatic interactions as the main synthetic strategy in this chapter will be discussed in much more detail in the next section. There are still other non-covalent interactions that could lead to supramolecular structures, such as hydrophobic forces [19, 20]. Furthermore, some supramolecular networks may include more than one type of non-covalent interaction. Table 1.1 summarizes different noncovalent interactions involved in the formation of supramolecular polymeric networks [21, 22].

1.2.1 Hydrogen Bond-Assisted Supramolecular Assemblies

Hydrogen bonding is the most investigated interaction among all types of reversible bonds. Due to their excellent directional selectivity, hydrogen bonds are ideal for molecular engineering of desired polymer networks. The strength of hydrogen

Table 1.1 Bond strengths of the different non-covalent interactions used in construction of supramolecular assemblies in comparison to the one of covalent bonds

Reversibility	Bond Strength		Type of interaction
Irreversible	Strong	> 60 kcal/mol	Covalent bond
Reversible	Medium	20–60 kcal/mol	Reversible covalent bond (e.g., –S–S–)
			metal–ligand coordination
			ionic interaction
			multiple hydrogen bonds
	Weak	0–20 kcal/mol	Hydrogen bond
			π–π stacking
			Hydrophobic interaction

Reused with permission from Ref. [22] Copyright (2013) (Wiley)

bonding depends on temperature, pressure, solvent, bond angle, and environment. It is worth remarking that the strength of a hydrogen-bonding motif is governed by the number of individuals involved hydrogen bonds. A higher number of hydrogen bonds generally signify stronger hydrogen bonding. Supramolecular assemblies based on hydrogen bonding gave rise to an array of novel materials. Following some examples are reviewed.

Meijer et al. reported thermal reversible polymers through developing ureidopyrimidinone (UPy) group with a bonding constant K of $6 \times 10^7\ M^{-1}$. They observed that the building blocks are held together by UPy hydrogen bonds at room temperature resulting in a flexible polymer. While by increasing the temperature, UPy hydrogen bonds break, the building blocks lose their connections and the material behaves like a viscous liquid. This dramatic phase and property transitions induced by the dynamic nature of hydrogen bonds bring about a unique behavior for UPy groups [23].

Long et al. were capable of pairing UPy side-groups into linear poly(butyl acrylate)s to achieve a novel thermoplastic elastomer material. Mechanical studies revealed that melts of UPy consisted of linear poly(butyl acrylate) chains that act as a rubbery elastomer at room temperature, bearing a classical elastomer Young's modulus at about hundreds of kPa. Whereas at elevated high temperatures (80 °C), the Young's modulus decreased considerably, behaving like a viscous polymer melt. The authors attributed this novel effect to the dynamic nature of UPy hydrogen bonds and they concluded that at low temperature, UPy groups are attached together by means of hydrogen bonding serving as crosslinking junctions for the polymer network. At high temperatures, UPy hydrogen bonds dissociate, the linear poly(butyl acrylate) chains lose their connectivity and consequently flow again like a melt [24].

In another interesting example, a self-healing supramolecular elastomer based on multiple hydrogen-bonding interactions was developed by Leibler et al. They initiated their approach with vegetable-based fatty diacids and triacids, then the acid groups were condensed with a controlled excess of diethylenetriamine. In the end, the obtained product was reacted with urea leading to various oligomers with multiple self-complementary hydrogen bonding sites. The new resulting plastic-like material showed a glass transition temperature (T_g) at about 28 °C. Above this temperature, the material displayed typical characteristics of elastomers, i.e., it was deformed by applying stress and recovered its shape when the force is removed. At even higher temperatures (> 160 °C), the material could flow like a viscous liquid. In contrast to classical rubbers, it exhibited excellent self-healing abilities as depicted in Fig. 1.2. Once the cut parts were put in contact together, the material was capable of self-healing with time through rebinding the fractured hydrogen bonds [25–27].

Similar to this study, Odriozola et al. reported a covalently cured poly(urea–urethane) elastomeric network with self-healing ability at room temperature which was achieved through combination of dynamic covalent bonds with hydrogen bonds with superior mechanical strength [29]. The significant self-healing ability of this system could be associated with two structural features: metathesis reaction of aromatic disulfide which is in constant exchange at room temperature [30, 31] and two existing urea groups, capable of forming a quadruple hydrogen bond (Fig. 1.3).

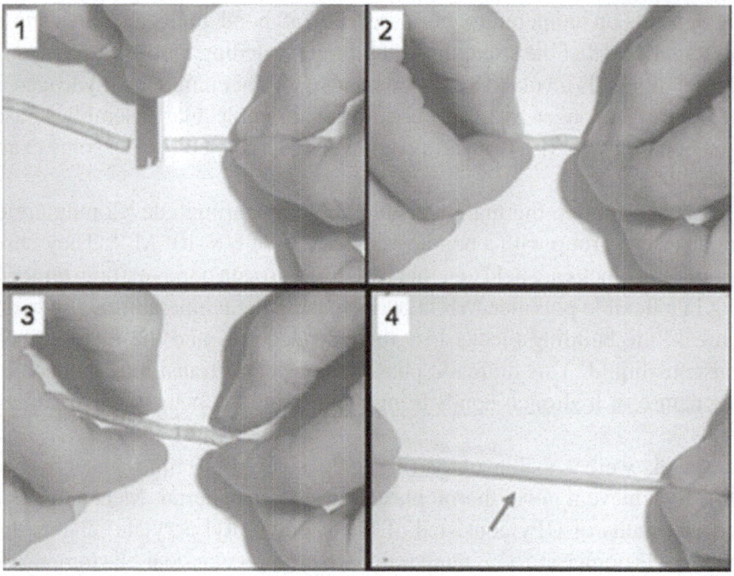

Fig. 1.2 Self-healing supramolecular elastomer developed by Leibler et al. Reproduced with permission from the Ref. [28] Copyright (2015) (Wiley)

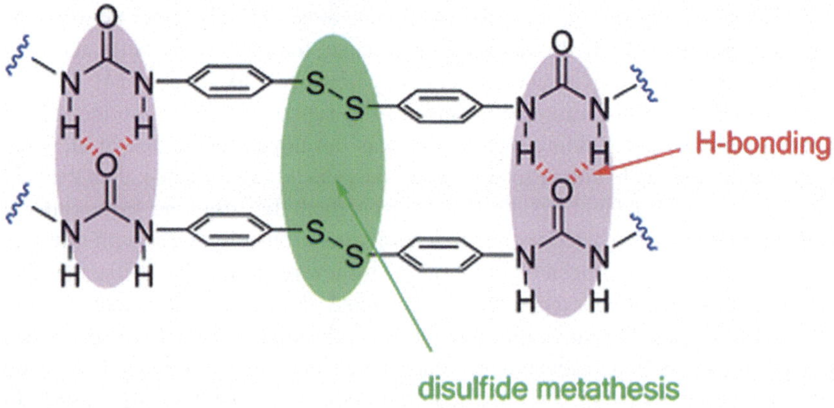

Fig. 1.3 Proposed bonds involved in the self-healing material developed by Odriozola et al. Reproduced with permission from the Ref. [28] Copyright (2014) (Royal Society of Chemistry)

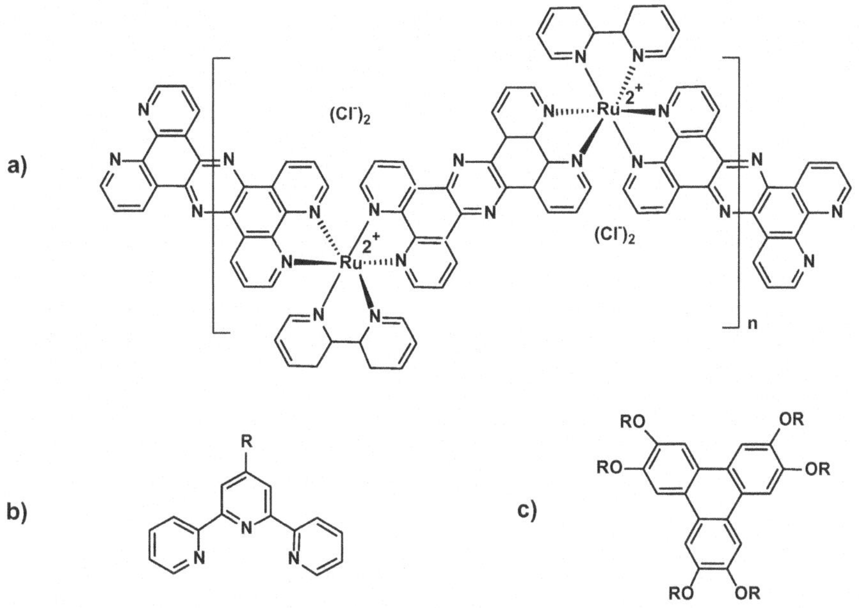

Fig. 1.4 **a** An example of Ruthenium(II) coordination polymer complex, **b** a model structure of terpyridine ligand and **c** molecular structure of triphenylene

1.2.2 Metal–Ligand Induced Supramolecular Assemblies

Reversible metal–ligand complexes are extensively employed in molecular self-assembly and host–guest recognition applications [32–34]. Ruthenium (II) coordination polymers are one of the first examples of metal–ligand induced supramolecular assemblies [35, 36]. The Ruthenium (II) pyridine type ligand complex is well known and relatively stable. By designing the appropriate monomer structure, Ruthenium (II) coordination polymers can be assembled (Fig. 1.4a). Terpyridine is another important class of ligands used in metal–ligand-induced supramolecular assemblies [37–40]. A model structure of this ligand is presented in Fig. 1.4b. Similar to the structure shown in Fig. 1.4a, terpyridine is able to bind various metals such as Fe^{2+}, Zn^{2+}, Ru^{2+}, Co^{2+}, etc. to form bisterpy metal complexes [16].

1.2.3 Supramolecular Assemblies Based on π–π Stacking Interaction

Many conjugated or aromatic molecular structures can undergo π–π stacking secondary interactions. This type of interaction mostly induces discotic stacking which results in crystalline or liquid crystalline state. Supramolecular assemblies can be constructed by means of π–π stacking. In this regard, triphenylenes are the most investigated disc-shaped molecules to form supramolecular structures [16]. The general structure of triphenylene is presented in Fig. 1.4c, in which changing the pendant R groups into appropriate side chains allows triphenylenes to generate aggregated polymeric columns in either water phase or organic solutions [41–44]. Burattini et al. developed an original healable polymer network formed by π–π stacking interactions between pyrenyl end-groups and chain folded polyimides [45–47].

Moreover, π–π stacking interaction is very prevalent in copolymers or conjugated polymers. The cooperative π–π interactions inside some polymers lead to generation of new supramolecular complexes which could have promising applications in the fields such as transistor electronics, light-emitting cells, and photovoltaic devices [48–50].

1.3 Supramolecular Assemblies Based on Electrostatic Interactions

Non-covalent electrostatic interactions via ion pairing play a major role in the design of supramolecular assemblies, especially in an aqueous environment wherein due to the water-solubility of the charged groups, they could form strong electrostatic interactions. Rise of ionomers and polyelectrolytes in the early twentieth century led to the concept development and application of electrostatic interactions in polymer structures. Ionomers are copolymers in which an ionic group (*e.g.,* carboxylate) is loaded into a polymer at a small quantity (generally < 15%) as a means for altering polymer properties. These components were commercially available in the 1950s. Mixtures of ionomers with randomly distributed positively and negatively charged groups can lead to formation of materials held together by electrostatic interactions. For example, van der Zwaag and Varley have reported different classes of ionomers and their self-healing behavior [51–53].

Polyelectrolytes are similar to ionomers but they have higher number of ionic groups (high charge density) and most of them are soluble in aqueous solution [54]. Polyelectrolytes are divided into three types: polyanions, polycations, and polyampholytes. Polyanions and polycations have negatively and positively charged groups, respectively and they possess extended chain conformations due to repulsive interactions between like-charged groups. Polyampholytes are ionic polymers having both positive and negatively charged groups and they have compact conformations because of attractive interactions between the unlike charges. The high quantity of charged groups on the backbones of polyelectrolytes (polycations or polyanions) attracts many counterions to its nearby neighborhood (counterion condensation) [55–57].

Ionomers and polyelectrolytes, with their ionic recognition groups, are extensively used to generate supramolecular assemblies. In spite of their long history, interesting functionality, and diverse structures available to use, electrostatic interactions have not been as widely studied as metal–ligand coordination and hydrogen bonding in the field of supramolecular materials. Compared to other non-covalent analogs, ionic interactions have particular characteristics as follows. First, they are stronger and less directional [21]; second, they may develop larger agglomerated structures dependent on the steric surroundings of the ion pair [58, 59]; third, coulombic interactions are asymmetric and sensitive to the local constant of the environment they are in [60]; and fourth, ionic interactions are simply adjustable via the combination of different cationic (e.g., primary or secondary amine, quaternary ammonium) and anionic structures (e.g., sulfonate vs. carboxylate) that are available through straightforward chemistry. In particular, advantages of working with electrostatic bonds alone or in combination with other non-covalent interactions are that a considerable number of molecules and macromolecules are ready for use and that the preparation methods are simple and appear generalizable [22].

In the following subsections, advances in supramolecular ionic networks from synthetic aspects are reviewed. These networks are classified if they were prepared from low molecular weight molecules, polymers, or a combination of the two. This type of classification was first introduced in a review paper published on this topic in 2013 [22]. Then in the next chapter, we will review the dramatic impact of molecular design on how these ionic interactions could control the bulk properties.

1.3.1 Supramolecular Ionic Assemblies from Low Molecular Weight Molecules

Low molecular weight molecules can attach to each other through non-covalent interactions to form linear or polymer network systems. The challenges related to applying ionic interactions between low molecular weight molecules for the balanced engineering of supramolecular assemblies include first, the relative deficiency of specificity between counterparts [21] and second, the isotropy of Coulombic potentials between discrete charges, which compromises even the validity of considering stoichiometric 1:1 pairwise interactions between oppositely charged partners [61].

Contrary to supramolecular assemblies formed by polymers, low molecular weights molecules can induce faster equilibration and the obtained assemblies may exhibit a higher degree of mesoscopic order [22]. In one of early studies, Hosseini et al. synthesized a self-assembled structure by combining ion-pairing electrostatic interaction and hydrogen bonding. In this study, a phthalate dianion was reacted with a dicationic compound composed of two cyclic amidinium groups with four hydrogen bonding donor sites to form a linear polymer chain (Fig. 1.5a). X-ray study confirmed the proposed linear structure if the anion was terephthalate or isophthalate [62]. The same research group showed that by adding two carboxylic acids to the anion (using pyromellitate dianion), a self-assembled crystalline 2D molecular network can be achieved (Fig. 1.5b) in addition to the linear polymer chains [63].

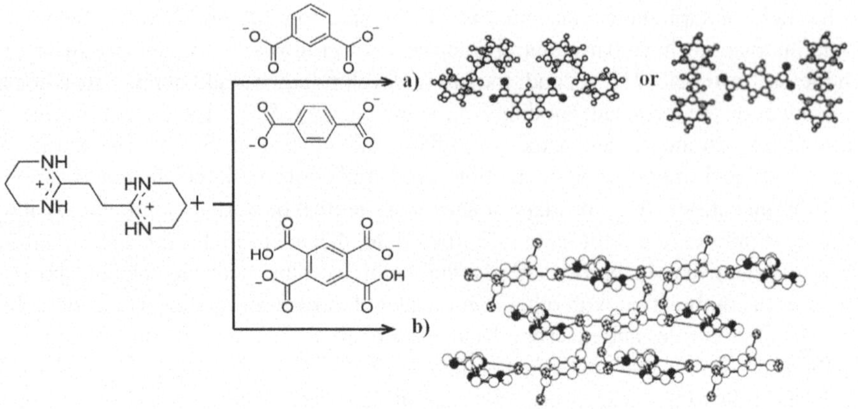

Fig. 1.5 Chemical structures of dicationic cyclic amidinium, phthalate dianions, pyromellitate dianion and X-ray structures of the resulting **a** linear or **b** 2-D network assemblies. Adapted with permissions from the references [61, 62] Copyright (1994, 2001) (Royal Society of Chemistry, Elsevier), respectively

In 2008, Grinstaff et al. suggested a distinct, practical, and complementary strategy for the construction of ionic networks from entirely non-coordinating ionic pairs, particularly those discovered from ionic liquids chemistry. These scientists combined a tetraanion, ethylenediaminetetraacetate ($EDTA^{4-}$), and a dication, consisting of two covalently linked tetraalkyl phosphonium components to assemble an ionic liquid. Since there were adequate units of cationic and anionic groups on each species and assuming that the Coulombic forces are governed by two-by-two interactions among individual cationic and anionic groups, the expected complex to be formed was effectively a supramolecular ionic network (Fig. 1.6). To check the generalizability of the approach, the authors replaced EDTA with *para*-tetracarboxy-5,10,15,20-tetraphenyl-21*H*,23*H*-porphine (H_2TPP^{4-}) aiming at developing an ionic porphyrin assembly. As result, they observed that porphyrin sustained its fluorescence suggesting that ionic networks having particular functional building blocks maintain their original features [64]. Combining the reversibility of the disulfide bond and the use of ionic interactions, this research group designed a crosslinked ionic network based on disulfide diphosphonium ionic liquid and tetraanion ($EDTA^{4-}$). By applying external stimuli such as temperature or inducing a mild redox reaction, the resulting ionic network underwent a reversible transition from a network to a non-network state [65].

Whereas this pioneering study made use of intricate ionic liquid units like alkyl phosphonium dications, Aboudzadeh et al. presented a simple strategy to synthesize supramolecular ionic networks through employing commercially accessible di- or trifunctional amines and carboxylic acids (Fig. 1.7a). The basic chemical reaction that they used was proton transfer, a reaction mechanism step applied in constructing protic ionic liquids, also recognized as acid–base complexation. Protic ionic liquids are readily synthesized through the neutralization and subsequent proton transfer

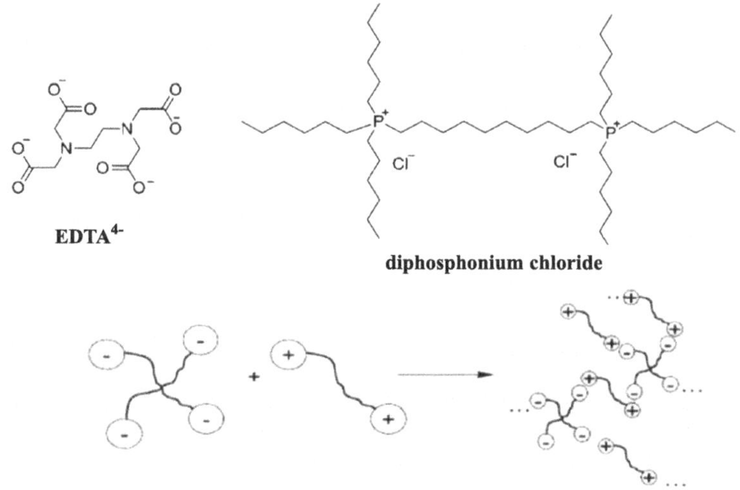

Fig. 1.6 Formation of a supramolecular ionic network reported by Grinstaff et al. (up): chemical structures involved; (down): schematic diagram of network formation. Reproduced with permission from Ref. [22] Copyright (2013) (Wiley)

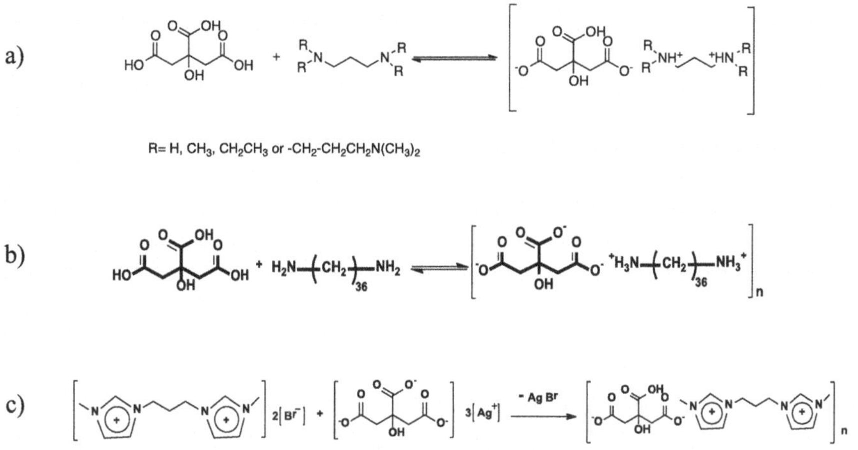

Fig. 1.7 Proposed reaction pathway for the synthesis of supramolecular ionic networks based on citric acid and **a** commercially available di- or trifunctional amines **b** a fully renewable C36 biobased diamine (Priamine 1074) and **c** geminal dicationic ionic liquids. Reproduced with permissions from the references [70–72] Copyright (2012, 2014, 2013) (American Chemical Society, Wiley, Royal Society of Chemistry), respectively

between a base and a Brønsted acid [66–68]. A group of (di-/tri-) functional primary- and tertiary amine bases and (di-/tri-) functional carboxylic acids were tested by these authors. In all selected combinations, the reactants were added all at once in an equimolar quantity and stirred together at room temperature in the presence of a solvent, such as water or methanol, to eliminate the issues of the exothermic neutralization reactions. The reactions were quick and after solvent removal, a series of products were synthesized effortlessly. Characterizing the products using nuclear magnetic resonance (NMR) and Fourier transform infrared (FTIR) spectroscopies, the authors obtained a direct evidence of proton transfer reaction between the carboxylic acid and the amine. The important advantage of this synthetic method was that it did not require a previous ionic-monomer synthetic stage [69]. Citric acid, among all examined carboxylic acid molecules, displayed a promising ability leading to the formation of supramolecular ionic networks with remarkable ionic conductivity and self-healing characteristics. The probability of involving hydrogen bonding in these systems was studied by the authors as well. They performed a control test in which instead of citric acid, the equivalent triol (glycerol) was reacted with a diamine molecule (1,3-propanediamine in this case) at mole ratio 1:1. As the product of this reaction, a liquid with very low viscosity was achieved, suggesting the small impact of the hydrogen bonding in comparison to the ionic interactions [70].

Moreover, the same authors via the same synthetic strategy developed supramolecular ionic networks fully based on chemicals coming from renewable feedstock (Fig. 1.7b). In this study, Priamine 1074 a biobased fatty diamine molecule (as the dication source) was reacted separately with a series of naturally occurring carboxylic acids such as malonic acid, citric acid, tartaric acid, and 2,5-furandicarboxylic acid [71]. These efforts allowed incorporation of natural products such as citric acid in the design of supramolecular networks. Following these reports, these scientists also achieved another type of supramolecular ionic network based on citrate anions and the most important types of dicationic ionic liquids such as di-imidazolium, di-pyridinium, or di-pyrrolidonium (Fig. 1.7c) [72]. The properties (rheological, conductivity, self-healing, etc.) of these three categories (shown in Fig. 1.7) will be fully discussed in chapter two. However, these ionic networks possessed certain intrinsic issues related to the constituent carboxylate compounds such as water sensitivity, poor thermal stability temperature, and low ionic conductivity. Therefore, in order to overcome these weak points, the same research group developed another approach to synthesize supramolecular ionic networks. This approach involved first the preparation of ionic monomers which contained highly delocalized attached (trifluoromethane-sulfonyl)imide $-SO_2-N-SO_2-CF_3$, or (propylsulfonyl)methanide $-SO_2-C(CN)_2$, or (cyano-propylsulfonyl)imide $-SO_2-N-CN$ anions (Fig. 1.8 left). The route to synthesize these compounds was similar to the reported procedure in the case of methacrylic anionic monomers [73]. In the second step, these dianions were combined with geminal di-imidazolium dications as the building blocks for preparation of the targeted supramolecular ionic networks. Although most of the synthesized networks were semi-crystalline (Fig. 1.8 right, **a**, **b** and **d**), amorphous

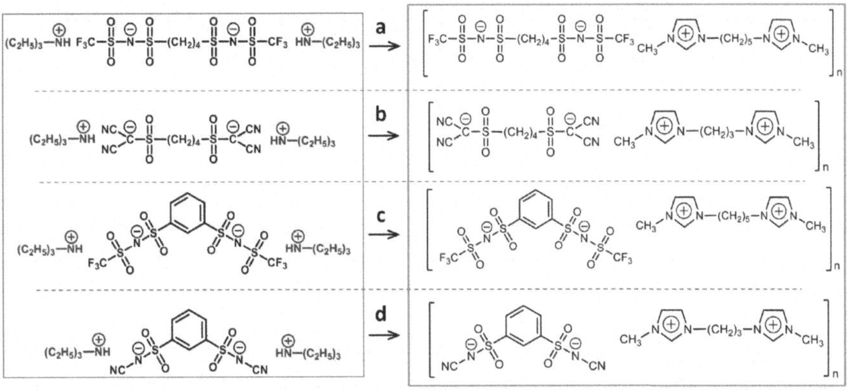

Fig. 1.8 Left) Ionic monomers containing highly delocalized attached $–SO_2–N–SO_2–CF_3$, $–SO_2–C(CN)_2$, and $–SO_2–N–CN$ anions. Right) Structures of supramolecular ionic networks based on highly delocalized dianions. Adapted with permission from the reference [74] Copyright (2015) (Royal Society of Chemistry)

networks were also obtained using aromatic asymmetric dianions (Fig. 1.8 right, **c**) [74].

In another study, supramolecular porous ionic networks were designed and prepared through one-pot procedure, involving the quaternization of triimidazole triazine with cyanuric chloride followed by hydrolysis and in-situ assembly (Fig. 1.9). The authors declared that the process of ionic crosslinking formation forces their chain conformation to be fixed to some extent, and this conformation fixation by ionic crosslinking is one of the reasons to create porosity [75].

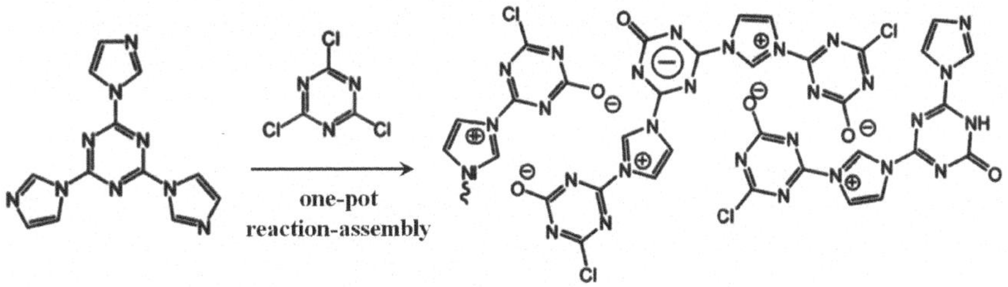

Fig. 1.9 Reaction pathway for the synthesis of supramolecular porous ionic network developed by Huang et al. Adopted with permission from the reference [75] Copyright (2016) (Royal Society of Chemistry)

1.3.2 Supramolecular Ionic Assemblies from Low Molecular Weight Molecules and Polymers

Appropriate selection of a low molecular weight molecule (*e.g.*, lipids, peptides, dyes, or surfactants) to be paired with a polymer via electrostatic interactions can provide an approach to developing specific structures with excellent functions. Polyelectrolyte–surfactant complexes are one of the early investigated assemblies in this category. These complexes combine the properties of polymers (mechanical strength and viscoelasticity) and surfactants (formation of highly hierarchical mesophases) [76, 77]. In these complexes, the ionic interaction between the ion pairs, as well as the hydrophobic interaction between the hydrocarbon tails of the surfactant, bring about thermodynamically stable ordered phases or aggregates depending on the preparation method [78].

In one of pioneering studies, Antonietti et al. synthesized and characterized a self-assembled complex formed between poly(styrenesulfonate) and alkyltrimethylammonium cations with different alkyl tails by common precipitation in water. Similarly, copolymers of poly(alkylacrylamide) and poly(acrylic acid) were assembled with surfactants to form ordered structures. Considering the simplicity of the "synthesis" and the availability of the starting materials, the authors predicted promising applications for this new class of products [79, 80]. The complexation of natural lipids such as diverse soybean-lecithins with a cationic polyelectrolyte, poly(dimethyldiallylammonium chloride) (PDADMAC), was carried out by the same research group too. The authors considered such systems as "Plastic Membranes" to mimic the structure of the cell membrane [81–83]. In another study, Thünemann et al. investigated the mixture of PDADMAC and surfactants with a pendant trimethylsilyl moiety complex and found smectic A-like lamellar mesophases and low energy surfaces (20 and 36 mN/m) for them. Due to the low surface energy and the high mechanical deformability of the complexes, the authors introduced them as flexible water-repellent coatings [84]. This scientist in another structural study using X-ray scattering showed that poly(ethylenimine) and n-alkyl carboxylic acids produce lamellar structures which are dependent on composition of the coordinating acid as well as the ratio of the two components [85].

Cationic dendrimers and oppositely charged small divalent organic ions in solution can give rise to stable functional supramolecular architectures. For example, Gröhn et al. introduced a simple route to prepare nanoparticles with cylindrical or spherical shapes through electrostatic self-assembly between poly(amidoamine) (PAMAM) dendrimer ions and the divalent 1,4- or 2,3-naphthalene dicarboxylic acid in methanolic solution (Fig. 1.10) [86, 87]. Use of dendritic macromolecules to create macroscopic ionic networks was reported in another study by Hvilsted et al. too. In this study, the networks were prepared by simple mixing of stoichiometric amounts of carboxylic acid-telechelic poly(ethylene glycol)s (DiCOOH-PEG) at different molecular mass (M_n ~ 250, M_n ~ 600, and M_n ~ 4800) with poly(propylene imine) (PPI) dendrimers or simpler amines such as tris(2-aminoethyl)amine (TAEA) or hexamethylene diamine (HMDA). These structures are shown in Fig. 1.11.

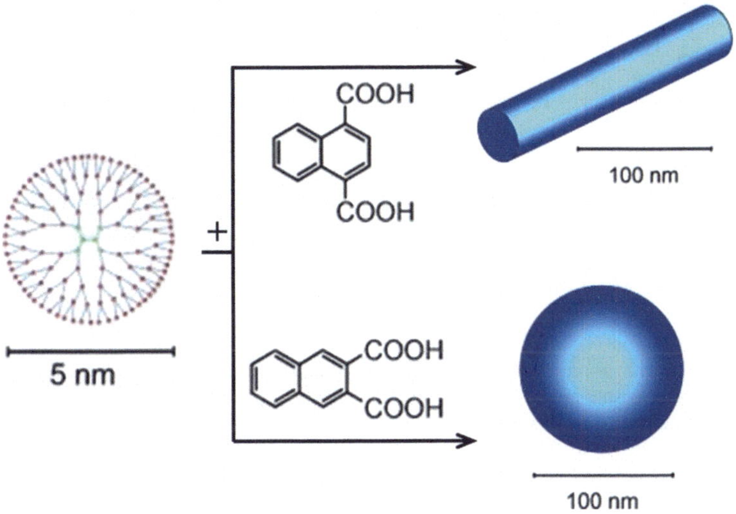

Fig. 1.10 Schematic representation of a PAMAM dendrimer, formulas of the counterions, and different shapes of the resulting nanoparticles. Adopted with permission from the Ref. [86] Copyright (2008) (Wiley)

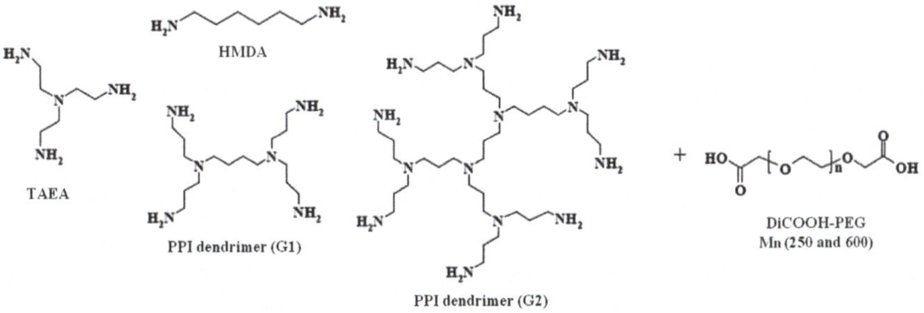

Fig. 1.11 Structures of PPI dendrimers, amines, and carboxylic acid end-functionalized PEGs DiCOOH-PEG that have been used to create the supramolecular ionic networks by Hvilsted et al. Reproduced with permission from the Ref. [88] Open access under a CC BY 4.0 license, https://creativecommons.org/licenses/by/4.0/

Performing a comprehensive ^{1}H NMR analysis, it was revealed that only the four primary amines of the PPI dendrimer (with molecular formula $C_{16}N_6H_{40}$) residing at the periphery take part in the ionic network formation [88].

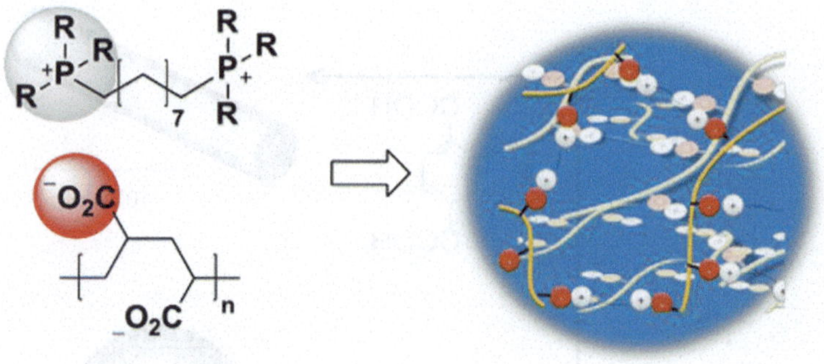

Fig. 1.12 Schematic representation of supramolecular polymer networks formed through ion pairing between phosphonium dications and the multianion, poly(acrylic acid). R = butyl, hexyl, or octyl. Reproduced with permission from the Ref. [90] Copyright (2012) (American Chemical Society)

Building upon these results, Grinstaff et al. investigated a self-organizing polyelectrolyte, a polymerized phosphonium ionic liquid, and mixed this polymer with different carboxylic acid-containing surfactants to form ordered structures. Depending on the number of ionic interactions as well as the nature of the anionic surfactant, the complexes exhibited completely different features including a brittle material, a rubbery ball that bounces, or a sticky fiber [89]. In most of the aforementioned studies, the charge stoichiometry is held at one to one (*i.e.*, each charge on the polymer is paired with a single charge species). Substituting the mono anion (or cation) with a multivalent anion (or cation) produces a polymeric network assembly. In this context, the same authors synthesized network structures by the complexation of various phosphonium dications with the multianion, poly(acrylic acid). In this study, the alkyl chains around the phosphonium dication were altered from butyl, hexyl, to octyl in order to explore the effect of sterics and ion pairing on the resulting macroscopic properties of the assemblies (Fig. 1.12). This study revealed insight into the structure–property relationship for the design of ionic supramolecular assemblies with desired properties [90].

Utilization of short peptides with a sequence-ordered chemical structure can bring about supramolecular networks a special functionality [91]. Among these short peptides, those containing fluorenylmethyloxycarbonyl (Fmoc) functional groups and bearing phenylalanine amino acids are prone to self-assemble via π–π interactions between aromatic groups. In a recent study, Boulmedais et al. presented an approach to trigger and regulate the self-assembly of Fmoc-FFpY peptides (F: phenylalanine; Y: tyrosine; p: PO_4^{2-}), through direct electrostatic interactions with a polycation (polyallylamine hydrochloride), which yielded hydrogels without dephosphorylation of the peptides. The authors suggested that electrostatic interactions between the charged amine groups of the polycations and the phosphate groups of the peptides lead to the generation of core–shell cylindrical structures in which flexible polycation chains decorate the micellar aggregates of Fmoc-FFpY peptides [92].

1.3.3 Supramolecular Ionic Assemblies from Polymers

Oppositely charged polymers can be used to develop different classes of self-assembled architectures governed by ionic interactions. The resulting supramolecular structures are commonly regarded as polyion assemblies. Their self-assembly capabilities bring about ordered mesoscopic scaffolds such as membranes, capsules, flocculants, and micelles by simple blending processes. A pioneering work on construction of supramolecular ionic assemblies from polymeric building blocks was carried out by mixing poly(vinyl-N-butylpyridinium bromide) and sodium poly(styrenesulfonate) together and studying the properties of the resultant polyanion-polycation complex [93]. Another early example reported by Miekka et al., in which sodium poly(styrenesulfonate) was also reacted with poly(vinylbenzyltrimethylammonium chloride) in dilute aqueous solution. As the product of the reaction, a water-insoluble extremely thin film was formed at the two-solution interface which contained equivalent quantities of each polyelectrolyte. The film displayed high diffusivity against salts such as sodium chloride, indicating a potential application as semipermeable membranes or as a solid electrolytes [94]. Depending on the concentration of the polyion solutions, the molecular weight of the polymers, pH, and the ionic strength, such complexes could be soluble in water or they could precipitate due to charge neutralization. Based on this fact, Kataoka and Harada developed polymeric micelles through spontaneous formation between a PEG-polycation block copolymer and a PEG-polyanion block copolymer in an aqueous solution. The authors chose biodegradable poly(Llysine) and poly(aspartic acid) as the polycation and polyanion blocks in the copolymer, respectively with the intention of making these polymer micelles as a promising vehicle for the delivery of various charged compounds to the body [95].

Different charged compounds can be used for the formation of supramolecular assemblies such as DNA, small interfering RNA, charged proteins, dendrimers, etc. [96–99]. In this context, these assemblies could be excellent candidates for the transport and delivery of biological agents since molecules such as DNA and proteins can be easily incorporated via ionic interactions. One of the earliest studies was the complexation between polyethyleneimine (PEI) and DNA reported by Boussif et al. The authors stated that every third atom of PEI is a protonable amino nitrogen atom, which makes the polymeric network an effective "proton sponge" at virtually any pH. This fact plus its low cytotoxicity makes PEI a promising vector for in vitro and in vivo gene therapy and delivery [100].

Viral gene vectors are considered the leading vehicles for delivering new genetic code to cells with outstanding results in multiple clinical tests [101]. To address the safety and manufacturing issues of viral vectors, a wide range of cationic polymers (associated with DNA via electrostatic interactions), have been studied as alternatives due to their relatively low cost and tunability [102–107]. In this regard, the reader is referred to some comprehensive reviews on this topic [108–110].

Ionic dendrimers are the classical examples of rigid-rod and globular-shaped polyelectrolytes. Unlike linear polyelectrolytes, their ionic side chains decrease the degrees of freedom. Charge neutralization during the formation of supramolecular assembly shrinks the hydrodynamic volume, but the original globular or rod shape of each polymer is maintained. Aida et al. presented the first study on electrostatic self-assembly between negatively and positively charged dendrimers that contain within their frameworks a fluorescence probe. They observed that only globular-shaped particles or precipitates are formed by mixing two oppositely charged dendrimers [111]. Aida et al. also reported another interesting study, in which environmentally friendly 3D network hydrogels based on electrostatic interaction were formed. These gels were prepared using a PEG chain with cationic G3-dendrons on each end (G3-binder, 3 = generation number), mixed with negatively charged clay nanosheets (CNSs) that were dispersed by sodium polyacrylate (ASAP). G3-binder had two dendron units, which were decorated with multiple guanidinium ion pendants on their peripheries (Fig. 1.13). Multiple dendrons bind to each clay nanosheet, forming a free-standing gel that can self-heal and be remolded [112] (Fig. 1.13). In a follow-up study, the same research group prepared similar supramolecular hydrogels by using linear ABA triblock copolyethers carrying guanidinium (Gu^+) groups at the side-chain termini of A blocks, while comprising a flexible PEG linker as the central B block. These hydrogels were as tough as the dendrimer-based gels, but the linear binders could be obtained by much less complicated syntheses from starting materials that were readily available [113].

Preparation of supramolecular ionic assemblies via acid–base complexation between functionalized polymers has been reported in some studies. For example, Matsushita et al. reported polymer blends made from carboxylic acid-terminated poly(dimethylsiloxane) (PDMS) and hyperbranched PEI. The parent polymers were both liquids and almost colorless at room temperature, while the blend samples had a yellow color originating from the redshift of the absorption band in a UV–vis region and looked more viscous due to acid–base complexation [114]. Furusho et al. reported a material developed by blending telechelic carboxylic acid-terminated polybutadiene and a linear polyamidine having N, N-di-substituted acetamidine group in the main chain. Charge interaction between the carboxylic acid and amidine groups led to the formation of a solid ionic network, which was not seen when the polyamidine was exchanged for PEI. Considering the high stability of amidinium-carboxylate salt bridge, the authors introduced it as an efficient and attractive non-covalent crosslinking concept [115]. In another study, Xu et al. designed a self-healing and thermoreversible rubber network by self-assembly of complementary polybutadiene oligomers bearing amine and carboxylic acid functionalities through reversible ionic interactions involving in acid–base reaction. Afterward, they further covalently crosslinked the rubber via the thiol-ene reaction by adding a small amount of trifunctional thiol. By varying the degree of covalent crosslinking, the authors could tune the mechanical properties and the stimuli-responsiveness of the resulting rubber [116]. Recently, Suriano et al. obtained supramolecular networks through

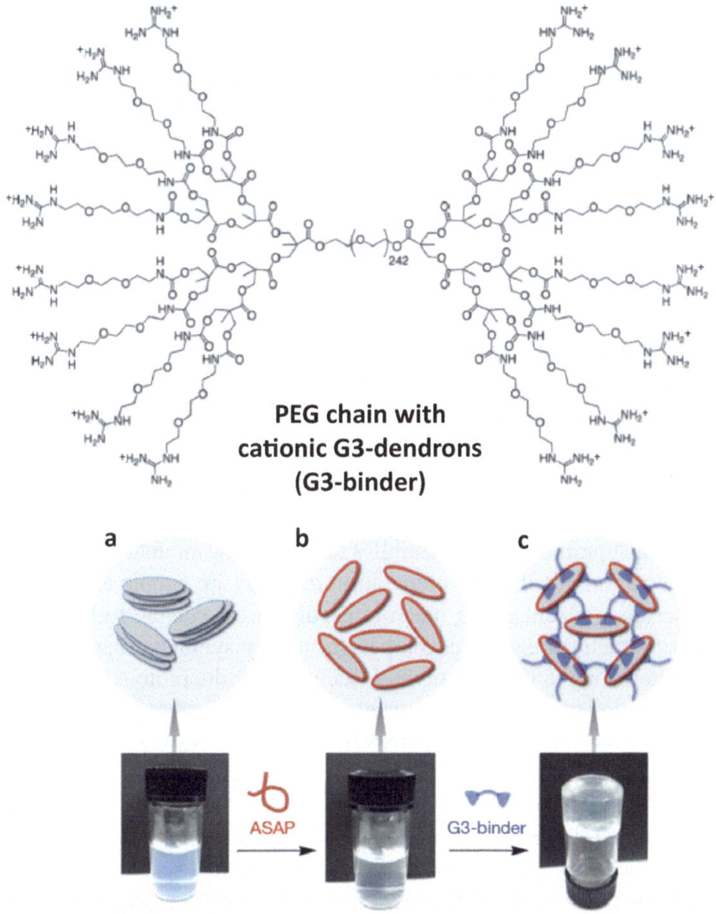

Fig. 1.13 Top: structure of G3-binder. Bottom: proposed mechanism for hydrogelation: **a** CNSs entangled with one another, **b** dispersing CNSs with ASAP, **c** addition of G3-binder to form electrostatic supramolecular crosslinks. Adopted with permission from the Ref. [112] Copyright (2016) (Royal Society of Chemistry)

implementing acid–base interactions between very viscous carboxylic-terminated and amino-terminated telechelic PDMS oligomers, with different molecular masses (ranging from 900 to 27,000 g/mol). Through analyzing the number of ionic interactions by FTIR analyses, higher values of ionic interactions were achieved in the case of samples with lower molecular mass. A similar same trend was detected for the apparent crosslinking density [117].

The electrostatic interaction between charged polymers also allowed scientists to develop a technique named layer-by-layer (LbL) for designing multilayer films through alternating deposition of polycations and polyanions [118]. In this approach, repeating the electrostatic attachment of polyelectrolytes and washing process leads to the formation of LbL films onto the solid substrate under precise control. Layers' width, surface charge, morphology, and composition can be adjusted by changing the reaction conditions, the nature of polyelectrolytes, and the number of washing steps [119–122]. LbL approach is specifically appropriate for the immobilization and delivery of biomolecules, such as DNA, antibodies, enzymes, and polysaccharides, as well as bioactive drug molecules [123–126]. For comprehensive understanding of the LbL technique, the reader is directed to some references [127–129] and chapter six of this volume.

1.4 Summary and Future Perspectives

In the recent years, supramolecular ionic networks are broadening the chemistry, physicochemical properties, and applications of conventional polymers. Their self-ordering, guided by electrostatic interactions, provides a simple pathway to highly complex structures. The first advantage of applying electrostatic interactions in the construction of supramolecular assemblies (alone or in combination with other non-covalent interactions) is that the preparation methods are simple and in the second place, numerous molecules and macromolecules are available to be used. In this regard, these structures can be developed from only synthetic materials or from a synthetic material and a natural polymer, such as peptide, protein, and DNA.

The introduction of new cations and anions is extending the synthetic strategies of these networks. As reviewed in this chapter, in terms of synthesis a large number of new supramolecular ionic assemblies have been already designed. However, there is still a high potential for developing innovative ones due to the new types of ions which are being explored driven by the ionic liquid chemistry and the recent progress in precise/controlled polymer chemistry. Placing more emphasis on stimuli-responsive co-assembled networks and nanostructures is identical of great importance and establishes future routes for the design of advanced functional smart nano-objects. In addition, there are still some open questions and challenges remaining. For example, analyzing the ionic assemblies' features and structure at molecular level still is a challenge and additional fundamental research is necessary to check the existing short- and long-range order in the assembly. Also, in order to understand better experimental studies, it is essential to carry out theoretical analyses and modeling of the interaction strength. Once measuring numerous ionic bond strengths and ionic assembly stabilities have been done, then preparing a directory containing value of various ionic interaction strengths would guide scientists to design new assemblies.

From the synthetic perspective, supramolecular ionic assemblies of complex structures (branched, cylindrical, etc.) and well-defined morphologies (nanoparticles, micelles, capsules, etc.) will surely draw growing attention in the field of polymer science. At the same time, material researchers should take into account the costs and sustainability of supramolecular ionic assemblies in their future research directions. Particularly, the assemblies that are developed for different applications should be entirely biocompatible, if not fully biodegradable, and new synthetic routes for designing polarizable, soft, organic cations and anions may be based on low cost naturally occurring materials, such as amino acids or other functional organic acids and amines. Successively, this strategy will open up the potential commercialization of supramolecular ionic products for everyday life applications.

References

1. Lehn J (1995) Supramolecular chemistry: concepts and PerspectiVes. Wiley-VCH Verlag GmbH & Co, KGaA, Weinheim
2. Atwood JL, Davies JED, MacNicol DD et al (1996) Comprehensive supramolecular chemistry. Pergamon, Oxford
3. Whitesides GM, Simanek EE, Mathias JP et al (1995) Noncovalent synthesis: using physical-organic chemistry to make aggregates. Acc Chem Res 28:37–44. https://doi.org/10.1021/ar00049a006
4. Nelson A, Cantrill SJ, Turnbull WB, Stoddart JF (2005) Multivalency and cooperativity in supramolecular. Chemistry 38:723–732
5. Sijbesma RP, Beijer FH, Brunsveld L et al (1997) Reversible polymers formed from self-complementary monomers using quadruple hydrogen bonding Published by : American Association for the advancement of science stable. http://www.jstor.org/stable/2894863 Your use of the JSTOR archive indicates your a. 278:1601–1604
6. Brunsveld L, Folmer BJB, Meijer EW, Sijbesma RP (2001) Supramolecular polymers 5–8
7. Serpe MJ, Craig SL, Uni V, Carolina N (2007) Physical organic chemistry of supramolecular polymers. Langmuir 23:1626–1634. https://doi.org/10.1021/la0621416
8. Lehn J (2007) From supramolecular chemistry towards constitutional dynamic chemistry and adaptive chemistry. 151–160. https://doi.org/10.1039/b616752g
9. Yamauchi K, Lizotte JR, Hercules DM et al (2002) Combinations of microphase separation and terminal multiple hydrogen bonding in novel macromolecules. J Am Chem Soc 124:8599–8604
10. Lehn BJ (1990) Perspectives in supramolecular chemistry-from molecular recognition towards molecular information processing and self-organization. Angew Chem Int Ed Engl 29:1304–1319
11. Feldman KE, Kade MJ, de Greef TFA et al (2008) Polymers with multiple hydrogen-bonded end groups and their blends. Macromolecules 4694–4700
12. Broze G, Jerome R, Teyssie P (1981) Halato-telechelic polymers. 1. Gel formation and its dependence on the ionic content. Macromolecules 14:224–225. https://doi.org/10.1021/ma50002a052
13. Broze G, Jerome R, Teyssi P (1982) Halato-telechelic polymers. 4 Synthesis and dilute-solution behavior. Macromolecules 15:920–927
14. Yount WC, Juwarker H, Craig SL (2003) Orthogonal control of dissociation dynamics relative to thermodynamics in a main-chain reversible polymer. J Am Chem Soc 125:15302–15303
15. Yount WC, Loveless DM, Craig SL (2005) Strong means slow: dynamic contributions to the bulk mechanical properties of supramolecular networks**. 2746–2748. https://doi.org/10.1002/anie.200500026
16. Li J (2011) Polymer networks containing reversibly associating side-groups. University of Rochester, New York
17. Voorhaar L, Hoogenboom R (2016) Supramolecular polymer networks: hydrogels and bulk materials. Chem Soc Rev 45:4013–4031. https://doi.org/10.1039/c6cs00130k
18. De GTFA, Smulders MMJ, Wolffs M et al (2009) Supramolecular polymerization. Chem Rev 109:5687–5754
19. Wang J, Jiang M (2006) Polymeric self-assembly into micelles and hollow spheres with multiscale cavities driven by inclusion complexation. J Am Chem Soc 128:3703–3708

20. Zhong C, Luo P (2007) Characterization, solution properties, and morphologies of a hydrophobically associating cationic terpolymer. J Polym Sci Part B Polym Phys 45:826–839. https://doi.org/10.1002/polb
21. Faul CFJ, Antonietti M (2003) Ionic self-assembly: facile synthesis of supramolecular materials. Adv Mater 15:673–683. https://doi.org/10.1002/adma.200300379
22. Lin X, Grinstaff MW (2013) Ionic supramolecular assemblies. Isr J Chem 53:498–510. https://doi.org/10.1002/ijch.201300034
23. Lange RFM, Gurp MVAN, Meijer EW (1999) Hydrogen-bonded supramolecular polymer networks. J Polym Sci Part A Polym Chem 37:3657–3670
24. Yamauchi K, Lizotte JR, Long TE (2003) Thermoreversible poly(alkyl acrylates consisting of self-complementary multiple hydrogen bonding. 1083–1088
25. Cordier P, Tournilhac F, Soulié-Ziakovic C, Leibler L (2008) Self-healing and thermoreversible rubber from supramolecular assembly. Nature 451:977–980. https://doi.org/10.1038/nature06669
26. Montarnal D, Tournilhac F, Hidalgo M et al (2009) Versatile one-pot synthesis of supramolecular plastics and self-healing rubbers. J Am Chem Soc 131:7966–7967
27. Montarnal D, Cordier P, Soulié-Ziakovic C et al (2008) Synthesis of self-healing supramolecular rubbers from fatty acid derivatives, diethylene triamine, and urea. J Polym Sci Part A Polym Chem 46:7925–7936. https://doi.org/10.1002/pola
28. Diesendruck CE, Sottos NR, Moore JS, White SR (2015) Biomimetic self-healing. Angew Chem Int Ed 54:10428–10447. https://doi.org/10.1002/anie.201500484
29. Rekondo A, Martin R, De LAR et al (2014) Catalyst-free room-temperature self-healing elastomers based on aromatic disulfide metathesis. Mater Horiz 1:237–240. https://doi.org/10.1039/c3mh00061c
30. Sarma RJ, Otto S, Nitschke JR (2007) Disulfides, imines, and metal coordination within a single system: interplay between three dynamic equilibria. Chem Eur J 13:9542–9546. https://doi.org/10.1002/chem.200701228
31. Belenguer AM, Friščić T, Day GM, Sanders JKM (2011) Solid-state dynamic combinatorial chemistry: reversibility and thermodynamic product selection in covalent mechanosynthesis. Chem Sci 2:696–700. https://doi.org/10.1039/c0sc00533a
32. Lehn J (1988) Supramolecular chemistry-scope and perspectives molecules, supermolecules, and molecular devices (Nobel lecture). Angew Chem Inr Ed 27:89–112
33. Lehn J (1993) Supramolecular chemistry. Science (80)260:1762–1763
34. Piguet C, Bernardinelli G, Hopfgartner G (1997) Helicates as versatile supramolecular complexes. Chem Rev 97:2005–2062
35. Kelch S, Rehahn M (1997) High-molecular-weight Ruthenium(ii) coordination polymers: synthesis and solution properties. Macromolecules 30:6185–6193
36. Knapp R, Schott A, Rehahn M (1996) A novel synthetic strategy toward soluble, well-defined Ruthenium(II) coordination polymers. Macromolecules 29:478–480
37. Lohmeijer BGG, Schubert US (2002) Supramolecular engineering with macromolecules: an alternative concept for block copolymers. Angew Chem Int Ed 41:3825–3829
38. Schmatloch S, González MF, Schubert US (2002) Metallo-supramolecular diethylene glycol: water-soluble reversible polymers. Macromol Rapid Commun 23:957–961
39. Hofmeier H, Schubert US (2003) supramolecular branching and crosslinking of terpyridine-modified copolymers: complexation and decomplexation studies in diluted solution. Macromol Chem Phys 204:1391–1397. https://doi.org/10.1002/macp.200350003
40. Dobrawa R, Würthner F (2005) Metallosupramolecular approach toward functional. J Polym Sci Part A Polym Chem 43:4981–4995. https://doi.org/10.1002/pola.20997
41. Markovitsi D, Bengs H, Ringsdorf H (1992) Charge-transfer absorption in doped columnar liquid crystals. J Chem Soc Faraday Trans 88:1275–1279
42. Ringsdorf H, Schlarb B, Venzmer J (1988) Molekulare Architektur und Funktion von polymeren orientierten Systemen – Modelle für das Studium von Organisation, Oberflächenerkennung und Dynamik bei Biomembranen. Angew Chem Int Ed 100:117–162

43. Henderson JR (2000) Discotic amphiphiles. J Chem Phys 113:5965–5970. https://doi.org/10.1063/1.1308105
44. Percec V, Imam MR, Peterca M et al (2009) Self-assembly of dendronized triphenylenes into helical pyramidal columns and chiral spheres. J Am Chem Soc 131:7662–7677
45. Burattini S, Greenland BW, Merino DH et al (2010) A healable supramolecular polymer blend based on aromatic π–π stacking and hydrogen-bonding interactions. J Am Chem Soc 132:12051–12058
46. Burattini S, Greenland BW, Hayes W et al (2011) A supramolecular polymer based on tweezer-type π–π stacking interactions: molecular design for healability and enhanced toughness. Chem Mater 23:6–8. https://doi.org/10.1021/cm102963k
47. Burattini S, Colquhoun HM, Fox JD et al (2009) A self-repairing, supramolecular polymer system: healability as a consequence of donor-acceptor π–π stacking interactions. Chem Commun 6717–6719. https://doi.org/10.1039/b910648k
48. Romaner BL, Pogantsch A, Scandiucci De Freitas P et al (2003) The origin of green emission in polyfluorene-based conjugated polymers: on-chain defect fluorescence. Adv Funct Mater 13:597–601. https://doi.org/10.1002/adfm.200304360
49. Hoeben FJM, Jonkheijm P, Meijer EW, Schenning APHJ (2005) About supramolecular assemblies of π-conjugated systems. Chem Rev 105
50. Russell DM, Arias AC, Friend RH et al (2002) Efficient light harvesting in a photovoltaic diode composed of a semiconductor conjugated copolymer blend. Appl Phys Lett 80:2204–2206. https://doi.org/10.1063/1.1464226
51. Varley RJ, van der Zwaag S (2008) Towards an understanding of thermally activated self-healing of an ionomer system during ballistic penetration. Acta Mater 56:5737–5750. https://doi.org/10.1016/j.actamat.2008.08.008
52. Varley RJ, van der Zwaag S (2010) Autonomous damage initiated healing in a thermo-responsive ionomer. Polym Int 59:1031–1038. https://doi.org/10.1002/pi.2841
53. Kalista SJ, Pflug JR, Varley RJ (2013) Effect of ionic content on ballistic self-healing in EMAA copolymers and ionomers. Polym Chem 4:4910–4926. https://doi.org/10.1039/c3py00095h
54. Masanori H (1993) Polyelectrolytes, science and technology. Marcel Dekker, New York
55. Manning GS (1969) Limiting laws and counterion condensation in polyelectrolyte solutions I. Colligative properties. J Chem Phys 51:924–933. https://doi.org/10.1063/1.1672157
56. Manning GS (1976) The application of polyelectrolyte limiting laws to the helix–coil transition of DNA. VI. The numerical value of the axial phosphate spacing for the coil form. Biopolymers 15:2385–2390. https://doi.org/10.1002/bip.1976.360151206
57. Anderson CF, Record MT (1982) Polyelectrolyte theories and their applications to DNA. Annu Rev Phys Chem 33:191–222. https://doi.org/10.1146/annurev.pc.33.100182.001203
58. Consorti CS, Suarez PAZ, de Souza RF et al (2005) Identification of 1, 3-dialkylimidazolium salt supramolecular aggregates in solution. J Phys Chem B 109:4341–4349
59. Paulechka YU, Kabo GJ, Blokhin AV et al (2009) IR and X-ray study of polymorphism in 1-Alkyl-3-methylimidazolium bis (trifluoromethanesulfonyl) imides. J Phys Chem B 113:9538–9546
60. Hoogenboom R, Fournier D, Schubert US (2008) Asymmetrical supramolecular interactions as basis for complex responsive macromolecular architectures. Chem Commun 155–162. https://doi.org/10.1039/b706855g
61. Craig SL (2009) From ionic liquids to supramolecular polymers. Angew Chem Int Ed 48:2645–2647. https://doi.org/10.1002/anie.200805603
62. Hosseini MW, Ruppert R, Schaeffer P et al (1994) A molecular approach to solid-state synthesis: prediction and synthesis of self-assembled infinite rods. J Chem Soc Chem Commun 2135–2136
63. Felix O, Hosseini MW, De Cian A (2001) Design of 2-D hydrogen bonded molecular networks using pyromellitate dianion and cyclic bisamidinium dication as complementary tectons. Solid State Sci 3:789–793

64. Wathier M, Grinstaff MW (2008) Synthesis and properties of supramolecular ionic networks. J Am Chem Soc 130:9648–9649
65. Lin X, Godeau G, Grinstaff MW (2014) A reversible supramolecular assembly containing ionic interactions and disulfide linkages. New J Chem 38:5186–5189. https://doi.org/10.1039/c4nj00895b
66. Greaves TL, Drummond CJ (2008) Protic ionic liquids: properties and applications. Chem Rev 108:206–237. https://doi.org/10.1021/cr068040u
67. Burrell GL, Burgar IM, Separovic F, Dunlop NF (2010) Preparation of protic ionic liquids with minimal water content and 15 N NMR study of proton transfer. Phys Chem Chem Phys 12:1571–1577. https://doi.org/10.1039/b921432a
68. Stoimenovski J, Izgorodina EI, Macfarlane DR (2010) Ionicity and proton transfer in protic ionic liquids. Phys Chem Chem Phys 12:10341–10347. https://doi.org/10.1039/c0cp00239a
69. Aboudzadeh MA, Muñoz ME, Santamaría A et al (2012) Facile synthesis of supramolecular ionic polymers that combine unique rheological, ionic conductivity, and self-healing properties. Macromol Rapid Commun 33:314–318. https://doi.org/10.1002/marc.201100728
70. Aboudzadeh MA, Muñoz ME, Santamaría A et al (2012) Synthesis and rheological behavior of supramolecular ionic networks based on citric acid and aliphatic diamines. Macromolecules 45:7599–7606. https://doi.org/10.1021/ma300966m
71. Aboudzadeh A, Fernandez M, Muñoz ME et al (2014) Ionic supramolecular networks fully based on chemicals coming from renewable sources. Macromol Rapid Commun 35:460–465. https://doi.org/10.1002/marc.201300732
72. Aboudzadeh MA, Muñoz ME, Santamaría A, Mecerreyes D (2013) New supramolecular ionic networks based on citric acid and geminal dicationic ionic liquids. RSC Adv 3:8677–8682. https://doi.org/10.1039/c3ra40629f
73. Shaplov AS, Vlasov PS, Armand M et al (2011) Design and synthesis of new anionic "polymeric ionic liquids" with high charge delocalization. Polym Chem 2:2609–2618. https://doi.org/10.1039/c1py00282a
74. Aboudzadeh MA, Shaplov AS, Hernandez G et al (2015) Supramolecular ionic networks with superior thermal and transport properties based on novel delocalized di-anionic compounds. J Mater Chem A 3:2338–2343. https://doi.org/10.1039/c4ta05792a
75. Hei ZH, Song GL, Zhao CY et al (2016) Supramolecular porous ionic network based on triazinonide and imidazolium: a template-free synthesis of meso-/macroporous organic materials: via a one-pot reaction-assembly procedure. RSC Adv 6:92443–92448. https://doi.org/10.1039/c6ra20590a
76. Thünemann AF (2002) Polyelectrolyte-surfactant complexes (synthesis, structure and materials aspects). Prog Polym Sci 27:1473–1572. https://doi.org/10.1016/S0079-6700(02)00017-5
77. Ober CK, Wegner G (1997) Polyelectrolyte-surfactant complexes in the solid state: facile building blocks for self-organizing materials. Adv Mater 9:17–31. https://doi.org/10.1002/adma.19970090104
78. Goddard ED (1986) Polymer-surfactant interaction: part II: polymer and surfactant of opposite charge. Colloids Surf 19:301–329. https://doi.org/10.1016/0166-6622(86)80341-9
79. Antonietti M, Conrad J, Thuenemann A (1994) Polyelectrolyte-surfactant complexes: a new type of solid, mesomorphous material. Macromolecules 27:6007–6011. https://doi.org/10.1021/ma00099a011
80. Antonietti M, Maskos M (1996) Fine-tuning of phase structures and thermoplasticity of polyelectrolyte-surfactant complexes: copolymers of ionic monomers with *N*-alkylacrylamides. Macromolecules 29:4199–4205. https://doi.org/10.1021/ma9518870
81. Antonietti M, Thünemann A (1996) Polyelectrolyte-lipid complexes as membrane mimetic systems. Curr Opin Colloid Interface Sci 1:667–671. https://doi.org/10.1016/s1359-0294(96)80106-3
82. Antonietti M, Kaul A, Thunemann A (1995) Complexation of lecithin with cationic polyelectrolytes: "Plastic membranes" as models for the structure of the cell membrane? Langmuir 11:2633–2638. https://doi.org/10.1021/la00007a050

83. Antonietti M, Wenzel A, Thünemann A (1996) The "Egg-carton" phase: a new morphology of complexes of polyelectrolytes with natural lipid mixtures. Langmuir 12:2111–2114. https://doi.org/10.1021/la950620r
84. Thünemann AF, Lochhaas KH (1998) Self-assembly of solid polyelectrolyte—silicon—surfactant complexes. Langmuir 14:6220–6225. https://doi.org/10.1021/la980229g
85. Thünemann AF, General S (2000) Poly(ethylene imine) *n*-alkyl carboxylate complexes. Langmuir 16:9634–9638. https://doi.org/10.1021/la000991u
86. Gröhn F, Klein K, Brand S (2008) Facile route to supramolecular structures: self-assembly of dendrimers and naphthalene dicarboxylic acids. Chem A Eur J 14:6866–6869. https://doi.org/10.1002/chem.200800650
87. Gröhn F (2008) Electrostatic self-assembly as route to supramolecular structures. Macromol Chem Phys 209:2295–2301. https://doi.org/10.1002/macp.200800290
88. González L, Ladegaard Skov A, Hvilsted S (2013) Ionic networks derived from the protonation of dendritic amines with carboxylic acid end-functionalized PEGs. J Polym Sci Part A Polym Chem 51:1359–1371. https://doi.org/10.1002/pola.26503
89. Godeau G, Navailles L, Nallet F et al (2012) From brittle to pliant viscoelastic materials with solid state linear polyphosphonium-carboxylate assemblies. Macromolecules 45:2509–2513. https://doi.org/10.1021/ma3002092
90. Lin X, Navailles L, Nallet F, Grinstaff MW (2012) Influence of phosphonium alkyl substituents on the rheological and thermal properties of phosphonium-PAA-based supramolecular polymeric assemblies. Macromolecules 45:9500–9506. https://doi.org/10.1021/ma3019624
91. Makam P, Gazit E (2018) Minimalistic peptide supramolecular co-assembly: expanding the conformational space for nanotechnology. Chem Soc Rev 47:3406–3420. https://doi.org/10.1039/c7cs00827a
92. Criado-Gonzalez M, Wagner D, Rodon Fores J et al (2020) Supramolecular hydrogel induced by electrostatic interactions between polycation and phosphorylated-fmoc-tripeptide´. Chem Mater 32:1946–1956. https://doi.org/10.1021/acs.chemmater.9b04823
93. Fuoss RM, Sadek H (1949) Mutual interaction of polyelectrolytes. Science (80)110:552–554. https://doi.org/10.1126/science.110.2865.552
94. Michaels AS, Miekka RG (1961) Polycation-polyanion complexes: preparation and properties of poly-(vinylbenzyltrimethylammonium) poly-(styrenesulfonate). J Phys Chem 65:1765–1773
95. Harada A, Kataoka K (1995) Formation of polyion complex micelles in an aqueous milieu from a pair of oppositely-charged block copolymers with poly(ethylene glycol) segments. Macromolecules 28:5294–5299
96. Itaka K, Kanayama N, Nishiyama N et al (2004) Supramolecular nanocarrier of siRNA from PEG-based block catiomer carrying diamine side chain with distinctive pK a directed to enhance intracellular gene silencing. J Am Chem Soc 126:13612–13613. https://doi.org/10.1021/ja047174r
97. Kabanov AV, Kabanov VA (1995) DNA complexes with polycations for the delivery of genetic material into cells. Bioconjugate Chem 6:7–20. https://doi.org/10.1021/bc00031a002
98. Kataoka K, Togawa H, Harada A et al (1996) Spontaneous formation of polyion complex micelles with narrow distribution from antisense oligonucleotide and cationic block copolymer in physiological saline. Macromolecules 29:8556–8557. https://doi.org/10.1021/ma961217+
99. Vinogradov S, Batrakova E, Li S, Kabanov A (1999) Polyion complex micelles with protein-modified corona for receptor-mediated delivery of oligonucleotides into cells. 10:851–860. https://doi.org/10.1021/bc990037c
100. Boussif O, Lezoualc'h F, Zanta MA et al (1995) A versatile vector for gene and oligonucleotide transfer into cells in culture and in vivo: Polyethylenimine. PNAS 92:7297–7301. https://doi.org/10.1073/pnas.92.16.7297
101. Verma IM, Somia N (1997) Gene therapy—promises, problems and prospects. Nature 389:239–242

102. Jeong JH, Kim SW, Park TG (2003) Novel intracellular delivery system of antisense oligonucleotide by self-assembled hybrid micelles composed of DNA/PEG conjugate and cationic fusogenic peptide. Bioconjugate Chem 14:473–479
103. Jiang X, Dai H, Ke C et al (2007) PEG-b-PPA/DNA micelles improve transgene expression in rat liver through intrabiliary infusion. J Control Release 122:297–304. https://doi.org/10.1016/j.jconrel.2007.06.014
104. Oishi M, Sasaki S, Nagasaki Y, Kataoka K (2003) pH-responsive Oligodeoxynucleotide (ODN)—Poly (Ethylene Glycol) conjugate through acid-labile—thiopropionate linkage: preparation and polyion complex micelle formation. Biomacromol 4:1426–1432
105. Takae S, Miyata K, Oba M et al (2008) PEG-detachable polyplex micelles based on disulfide-linked block catiomers as bioresponsive nonviral gene vectors. J Am Chem Soc 130:6001–6009
106. Wakebayashi D, Nishiyama N, Itaka K et al (2004) Polyion complex micelles of pDNA with acetal-poly(ethylene glycol)-poly(2-(dimethylamino)ethyl methacrylate) block copolymer as the gene carrier system: physicochemical properties of micelles relevant to gene transfection efficacy. Biomacromol 5:2128–2136
107. Jewell CM, Zhang J, Fredin NJ, Lynn DM (2005) Multilayered polyelectrolyte films promote the direct and localized delivery of DNA to cells. J Control Release 106:214–223. https://doi.org/10.1016/j.jconrel.2005.04.014
108. Wong SY, Pelet JM, Putnam D (2007) Polymer systems for gene delivery-past, present, and future. Prog Polym Sci 32:799–837. https://doi.org/10.1016/j.progpolymsci.2007.05.007
109. Park TG, Jeong JH, Kim SW (2006) Current status of polymeric gene delivery systems. Adv Drug Deliv Rev 58:467–486. https://doi.org/10.1016/j.addr.2006.03.007
110. Pack DW, Hoffman AS, Pun S, Stayton PS (2005) Design and development of polymers for gene delivery. Nat Rev Drug Discov 4:581–593. https://doi.org/10.1038/nrd1775
111. Tomioka N, Takasu D, Takahashi T, Aida T (1998) Electrostatic assembly of dendrimer electrolytes: negatively and positively charged dendrimer porphyrins. Angew Chem Int Ed 37:1531–1534. https://doi.org/10.1002/(SICI)1521-3773(19980619)37:11%3c1531::AID-ANIE1531%3e3.0.CO;2-T
112. Wang Q, Mynar JL, Yoshida M et al (2010) High-water-content mouldable hydrogels by mixing clay and a dendritic molecular binder. Nature 463:339–343. https://doi.org/10.1038/nature08693
113. Tamesue S, Ohtani M, Yamada K et al (2013) Linear versus dendritic molecular binders for hydrogel network formation with clay nanosheets: studies with ABA triblock copolyethers carrying guanidinium ion pendants. J Am Chem Soc 135:15650–15655. https://doi.org/10.1021/ja408547g
114. Noro A, Ishihara K, Matsushita Y (2011) Nanophase-separated supramolecular assemblies of two functionalized polymers via acid-base complexation. Macromolecules 44:6241–6244. https://doi.org/10.1021/ma201440v
115. Furusho Y, Endo T (2014) Supramolecular polymer gels formed from carboxy-terminated telechelic polybutadiene and polyamidine through amidinium—carboxylate salt bridge. J Polym Sci, Part A Polym Chem 52:1815–1824. https://doi.org/10.1002/pola.27187
116. Wang D, Guo J, Zhang H et al (2015) Intelligent rubber with tailored properties for self-healing and shape memory. J Mater Chem A 3:12864–12872. https://doi.org/10.1039/c5ta01915j
117. Suriano R, Boumezgane O, Tonelli C, Turri S (2020) Viscoelastic properties and self-healing behavior in a family of supramolecular ionic blends from silicone functional oligomers. Polym Adv Technol 31:3247–3257. https://doi.org/10.1002/pat.5049
118. Decher G (1997) Fuzzy nanoassemblies: toward layered polymeric multicomposites. Science (80) 227:1232–1237. https://doi.org/10.1126/science.277.5330.1232
119. Wang Y, Angelatos AS, Caruso F (2008) Template synthesis of nanostructured materials via layer-by-layer assembly. Chem Mater 20:848–858. https://doi.org/10.1021/cm7024813
120. Anzai J, Kobayashi Y, Nakamura N et al (1999) Layer-by-layer construction of multilayer thin films composed of avidin and biotin-labeled poly (amine)s. Langmuir 15:221–226

121. Cheung JH, Stockton WB, Rubner MF (1997) Molecular-level processing of conjugated polymers. 3. Layer-by-layer manipulation of polyaniline via electrostatic interactions. Macromolecules 30:2712–2716
122. Lvov Y, Ariga K, Onda M et al (1997) Alternate assembly of ordered multilayers of SiO_2 and other nanoparticles and polyions. Langmuir 13:6195–6203. https://doi.org/10.1021/la970517x
123. Borden MA, Caskey CF, Little E et al (2007) DNA and polylysine adsorption and multilayer construction onto cationic lipid-coated microbubbles. Langmuir 23:9401–9408. https://doi.org/10.1021/la7009034
124. Lvov Y, Munge B, Giraldo O et al (2000) Films of manganese oxide nanoparticles with polycations or myoglobin from alternate-layer adsorption. Langmuir 16:8850–8857
125. Lynn DM (2006) Layers of opportunity: nanostructured polymer assemblies for the delivery of macromolecular therapeutics. Soft Matter 2:269–273. https://doi.org/10.1039/b517860f
126. Choi J, Konno T, Takai M, Ishihara K (2009) Smart controlled preparation of multilayered hydrogel for releasing bioactive molecules. Curr Appl Phys 9:e259–e262. https://doi.org/10.1016/j.cap.2009.06.054
127. Hammond PT (2012) Building biomedical materials layer-by-layer. Mater Today 15:196–206. https://doi.org/10.1016/S1369-7021(12)70090-1
128. Hammond BPT (2004) Form and function in multilayer assembly: new applications at the nanoscale. Avd mater 16:1271–1293. https://doi.org/10.1002/adma.200400760
129. Alkekhia D, Hammond PT, Shukla A (2020) Layer-by-layer biomaterials for drug delivery. Annu Rev Biomed Eng 22:1–24

Chapter 2
Supramolecular Ionic Networks: Properties

M. Ali Aboudzadeh and Shaghayegh Hamzehlou

Abstract Non-covalent ionic interactions are present in all elements of nature and can be used adequately to generate a wide variety of supramolecular assemblies. In recent years, a lot of scientists have been studying and exploring the ability to utilize ionic interactions to build controlled polymeric networks and to tailor desirable properties. The reversibility and low energy bonding of these interactions bring about additional features compared to conventional covalent polymers, which potentially lead to new properties such as improved processing, self-healing behaviour and stimuli responsiveness. In spite of these advantages, a basic physical–chemical characterization of supramolecular ionic networks is still far from complete. This is because obtaining an overall picture of the structure, the dynamics, and the properties of these networks is a key step to their use as high-performance materials. This chapter summarizes the current knowledge on different characterizations (e.g. morphology, thermal, electrical, rheological, and self-healing) of supramolecular ionic networks and attempts to derive consistent relations between their structure, dynamics, and properties. Various case studies are discussed in this chapter, using them as examples in order to elucidate their structure–property relation.

2.1 Introduction

Combination of supramolecular chemistry and polymer science has led to design of smart materials named supramolecular networks that integrate both macromolecular architecture and dynamic flexibility. These "smart" assemblies are foreseen to present unique and even superior performances than their covalent analogs in everyday

M. A. Aboudzadeh (✉)
CNRS, University Pau & Pays Adour, E2S UPPA, Institut des Sciences Analytiques et de Physico-Chimie pour l'Environnement et les Matériaux, IPREM, UMR5254, 64000 Pau, France
e-mail: m.aboudzadeh-barihi@univ-pau.fr

S. Hamzehlou
POLYMAT, Department of Applied Chemistry, Faculty of Chemistry, University of the Basque Country UPV/EHU, Joxe Mari Korta Zentroa, Tolosa Hiribidea 72, 20018 Donostia, Spain

M. A. Aboudzadeh and A. Frontera (eds.), *Supramolecular Assemblies Based on Electrostatic Interactions*,
https://doi.org/10.1007/978-3-031-00657-9_2

applications, such as biomaterials, energy transfer and storage, printing, coatings, adhesives, and cosmetics. Choosing the specific building blocks and the type of non-covalent interactions determine the structure and the macroscopic properties of the resulting networks. Each non-covalent interaction has its particular bond strength and features, which bring about different properties such as the networks' strength (moduli), its viscosity and flow, as well as the intrinsic ordering of the formed chains. For example, the mechanical properties of quadruple hydrogen-bonded ureidopyrimidinone (Upy) end-functionalized monomers are superior compared to those of supramolecular assemblies formed by weaker single or triple hydrogen bonding [1]. Moreover, the reversibility of these interactions brings additional features compared to usual covalent polymer networks, which potentially lead to new properties, such as improved processability, self-healing behaviour and stimuli responsiveness. Finally, the supramolecular interactions are generally dynamic and transient, allowing conformational transitions and molecular relaxations on time scales not otherwise available to covalent structures.

The use of ionic interactions for construction of supramolecular assemblies is less investigated but presents an interesting complementary approach with advantages of easy preparation and wide diversity [2]. Long before the concept of supramolecular ionic polymer came along, ion containing polymers like ionomers and polyelectrolytes have already been important materials and have been studied extensively [3, 4]. Then by the introduction of new cations and anions (coming from ionic liquids chemistry), the properties and classical applications of polyelectrolytes were extended to other fields [5], such as gas membranes [6] or stimuli-responsive materials [7]. These new polyelectrolytes are being named polymeric ionic liquids or poly(ionic liquids) in analogy to their monomeric constituents. Since ionic liquids generally show unique characteristics such as non-volatility and non-flammability, preparation of polymer homologue of ionic liquids is a promising approach to design novel class of polymer materials possessing such properties. The charged nature of these polyelectrolytes also provides properties such as high hydrophilicity and ion conductivity, enabling a variety of potential applications in the field of ionic devices such as solid-state conducting polymers in Li-ion battery and fuel cells [8–11]. This emerging ionic chemistry has been translated to the development of supramolecular assemblies based on ionic interactions amongst positively and negatively charged species and, thus, always involves asymmetrical self-assembly. The strategy based on using multicationic and multianionic molecules provides an opportunity to prepare new supramolecular ionic networks with properties not attainable with current ionic liquids. Moreover, these self-assembled networks can be easily prepared from a diverse set of readily available starting componentss, facilitating the design of new ionic materials with a high level of molecular control.

The aim of this chapter is to focus on different physicochemical properties and technological applications of supramolecular ionic networks and to relate these properties to the ionic bond dynamics and responsiveness. The range of properties (*e.g.* morphological, thermal, electrical, rheological, and self-healing) they exhibit depend on their molecular design. Consequently, it is possible to tune the properties and to add functionality through chemical design. In this context, various examples are discussed in this chapter, where macroscopic properties of these networks are clearly demonstrated.

2.2 Different Properties of Supramolecular Ionic Networks

Networks made through supramolecular chemistry have been reported based on all major types of non-covalent interactions. The use of electrostatic interactions between groups of monomers can be used for preparing rather strong networks with tunable properties. The following subsections describe the most reported characteristics of supramolecular ionic networks, particularly emphasizing their physicochemical features with regard to the molecular architecture and number of interactions between monomers.

2.2.1 Morphology

Based on the number of ionic interactions as well as the nature of the involving anionic or cationic molecules, the bulk assemblies may display dramatically different properties. In general, crystallization of low molar mass molecules is favoured in the absence of any solvent. Thus, reducing crystallinity through a proper molecular design will lead to achieve polymer like properties. As example in a study, wide range of products from crystalline solids, viscous liquids or non-crystalline solids were likely to obtain when commercially accessible di- or trifunctional amines and carboxylic acids (as shown by numbers in Scheme 2.1) were used to form supramolecular ionic networks via proton transfer reaction. The authors observed that products containing regular and stiff dicarboxylates, like malonate and terephthalate dicarboxylates (compounds 6 and 8), were prone to crystallization, refraining the formation of 3D ionic networks. Contrarily, when ductile dicarboxylates such as itaconate (compound 7), or other long alkyl carboxylates were utilized, sticky liquids that could flow at room temperatures were achieved. Identical results were detected with the flexible octane diammonium compound (5). The third class of products were obtained through combination of natural (di-/tri-) carboxylates such as acotinates (compound 9), or citrates (compound 10) with nearly all of the diammonium molecules (compounds 1–4). These samples were completely solid and transparent, which as it will be demonstrated later in this chapter, they displayed features analogous to other supramolecular polymer networks. The authors stated that the formation of 3D networks was also facilitated by other functional groups (besides the dicarboxylates) existent in citrates and acotinates, such as hydroxyl or pendant carboxylic acids, as these groups may additionally interact with themselves and the diammonium molecules via hydrogen bonding [12]. The effect of carboxylic acid type on the morphology of the final supramolecular ionic network (obtained via proton transfer reaction) was reported by these authors in another study too [13].

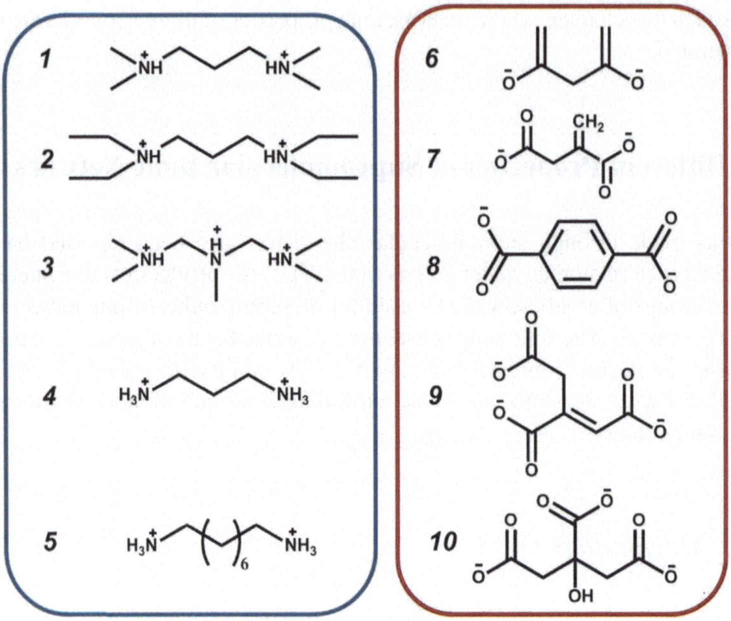

Scheme 2.1. Chemical structures of diammonium and dicarboxylate molecules investigated to form supramolecular ionic networks through proton transfer reaction. Adapted with permission from the reference [12] Copyright (2012) (John Wiley and Sons)

In another study, Grinstaff et al. reported that a linear polystyrenylphosphonium polymer (a self-organizing polyelectrolyte) was able to form diverse supramolecular assemblies with mono- and dicarboxylates. Based on the nature of the anion, the properties of these ionic materials were found to be quite different, including (a) brittle, glass-like materials, (b) rubbery balls that bounce, and (c) sticky fibres that flow upon drawing (Fig. 2.1) [14]

The stoichiometry of polyelectrolyte charge is also an important factor to control the morphology of the self-assembled systems [15]. Combining oppositely charged polyelectrolytes at 1:1 charge ratio normally generates aggregates or precipitates in solution. Nevertheless, if one polyelectrolyte is in large excess, it envelops the counter polyelectrolyte and the assembly remains soluble. Like-charged polyion complexes will also not aggregate because of electrostatic repulsion. In this context by varying the mixing ratio, cylindrical, spherical, lamellar, and macrophase-segregated morphologies can all be easily obtained [16, 17]. The original shapes and stiffness of a positively charged polymer can affect the morphologies of the resulting combinations; for example in complexation with negatively charged DNA, it causes DNA to be tightly condensed or packed into a nanometer-scale structure and protects the DNA from external hydrolytic and enzymatic digestion. Also, it was shown that the addition of polycations to DNA induces phase transitions to twisted ribbons,

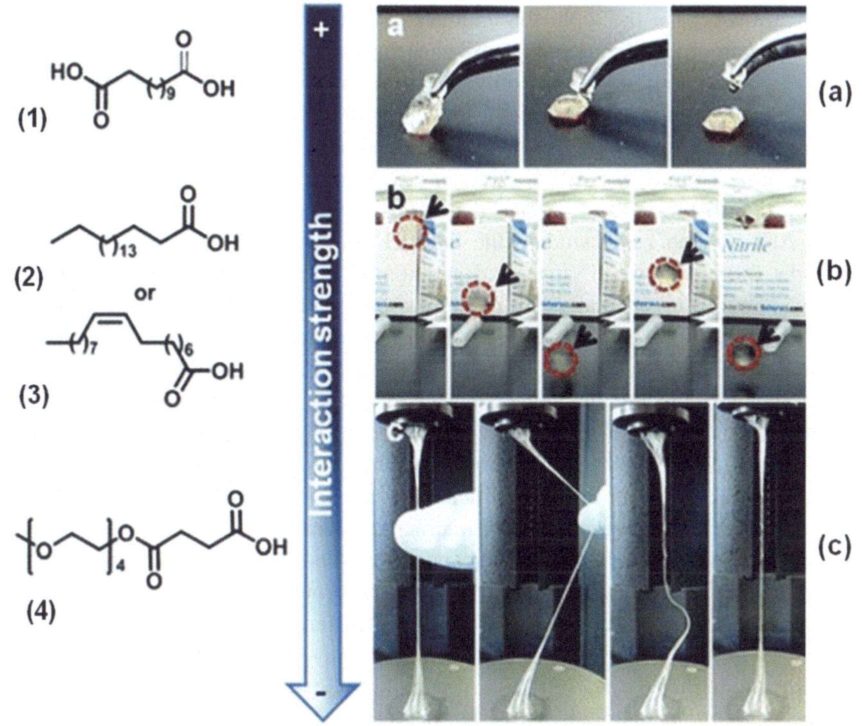

Fig. 2.1 **a** Photograph of a brittle ionic material formed between polystyrenylphosphonium and dodecandioic acid (1), **b** Photograph of a supramolecular ionic network formed between polystyrenylphosphonium and a mono fatty carboxylate counteranion such as stearic acid (2) or oleic acid (3), which bounces. (c) Photograph of a sticky elastic ionic material formed between polystyrenylphosphonium and tetraethylenglycol carboxylate (4) as the counteranion. Adapted with permission from the reference [14] Copyright (2012) (American Chemical Society)

donut-like structures, and compact particles because of the rigidity of the polycation backbone [18].

2.2.2 *Thermal Properties*

The effect of the molecular architecture on the thermal properties of supramolecular ionic networks including the phase behaviours and the degradation temperature was investigated in literature using differential scanning calorimetry (DSC) thermogravimetric analysis (TGA), respectively. For example, Grinstaff and coworkers investigated the thermal properties of supramolecular networks formed via ionic interactions between poly(acrylic acid) (PAA) and a diphosphonium ionic liquid crosslinker and they found out that these properties were highly dependent on the alkyl chain length of the diphosphonium crosslinker (-*R* group shown in Fig. 1.12 of Chap. 1). In order to examine the effect of sterics and ion pairing on the resulting macroscopic properties of the assemblies, these authors varied systematically the alkyl chains connected to phosphonium cation from butyl, hexyl, to octyl. All of the assemblies exhibited glass transition temperatures (T_g) lower than room temperature ranging from 8 to −40 °C. Network having the shortest alkyl chain (butyl) length afforded the highest T_g and decomposition temperature [19]. In a similar study, Aboudzadeh et al. studied the thermal behaviour of supramolecular ionic networks formed between citric acid and four different geminal dicationic ionic liquids. They observed that T_g in these networks followed the same trend as the equilibrium elasticity parameter (G_e) which itself was influenced by the chemical composition of the dicationic molecule. The type of the anion being the same (citrate) and taking into account the same alkyl chain length used (three carbons) in the structure of the all dications, two factors affected G_e: the nature of the dication (*e.g.* imidazolium versus pyrrolidinium) and the substituents on the dication. The authors related the effect of the chemical structure on T_g to a reduction of the free volume, as G_e (and consequently ionic interactions density) augments [20]. The same researchers characterized the thermal properties of some fully bio-based supramolecular ionic networks, which were prepared by the proton transfer reaction between a fatty diamine molecule (priamine 1074) and different carboxylic acids coming from renewable sources. As a general tendency, they observed for these combinations that the higher the T_m of the starting carboxylic acid, the higher the T_g of the obtained supramolecular ionic network. The highest decomposition temperature and T_g (≈10 °C) were obtained for the system based on 2,5-furandicarboxylic acid, which was due to the presence of a heterocyclic group in this carboxylic acid that rigidified the system [13]. Furthermore, these scientists evaluated the thermal behaviour of another family of supramolecular ionic networks based on highly delocalized dianions having (trifluoromethane-sulfonyl) imide, (propylsulfonyl)methanide or (cyano-propylsulfonyl)imide groups (structures shown in Fig. 1.9 of Chap. 1). As the main findings, most of the synthesized compounds were semi-crystalline possessing melting temperature (T_m) ≈ 100 °C; however, amorphous networks having T_g ≈ −20 °C were also obtained using aromatic asymmetric dianions. Also, networks containing aromatic dianion having (trifluoromethane-sulfonyl)imide group (-SO_2-N-SO_2-CF_3) showed higher onset decomposition temperature (≈295 °C) than the previously synthesized networks based on carboxylates (≈135 °C). Using the same type of dianions, the transfer from an aromatic spacer to an aliphatic one resulted in a decrease of thermal stability of the final network by 40 °C [21]. Upon addition of lithium salt (10 mol%) to a semi-crystalline network containing -SO_2-N-SO_2-CF_3 group, both T_m and T_g decreased (from 100 to 96 °C and from −19 to −23 °C, respectively), the melting peak also broadened significantly and the enthalpy change (ΔH) slightly decreased (≈3%) (Fig. 2.2). These changes led to a slight reduction in the crystallinity of the network [22].

In another interesting report, Hvilsted et al. investigated the thermal behaviour of poly(propylene imine) (PPI) dendrimers that were ionically crosslinked into hydrogels using dicarboxylic acid functionalized poly(ethylene glycol)s (DiCOOH-PEGs)

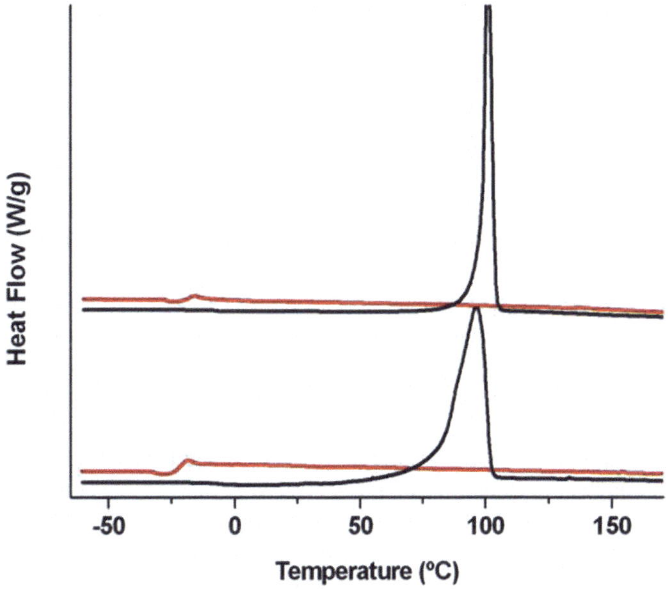

Fig. 2.2 DSC profile of neat network **a** (upper) and after addition of 10 mol% LiTFSI (lower), first heating cycle (black) and second heating cycle (red). Reproduced with permission from the reference [22] Copyright (2015) (Elsevier)

(structures are shown in Fig. 1.12 of Chap. 1). It was observed that decreasing molecular mass of the DiCOOH-PEG in all the series of dendritic ionic networks led to a significant increase in T_g. This revealed that more brittle (harder or more glassy) ionic networks generate when using a low molecular mass DiCOOH-PEG (M_n 250), which can be due to higher crosslinking density consequent from the use of short DiCOOH-PEG. Moreover, adding an excess of DiCOOH-PEG during synthesis led to a lower T_g of the resulting ionic network compared to the one formed with equimolar amounts of PPI dendrimer and DiCOOH-PEG. This indicated that an excess of carboxylic acid induces a decrease in the ionic crosslinking density of the ionic network. The authors concluded that network components and composition determine both T_g and the thermal stability of the synthesized materials which are also indirectly related to the crosslink density [23].

2.2.3 Electrical Characteristics

Electrical characteristics of supramolecular polymer networks have been hardly investigated. The dynamics of supramolecular networks can be studied by dielectric spectroscopy technique. Considering the dynamics on the molecular level play an important role in the conversion of chemical structure into macroscopic properties, dielectric spectroscopy has been applied to explain this relation [24–27]. In this context, dielectric properties of polyelectrolyte-surfactant supramolecular complexes have been analysed in pioneering works of Antonietti et al. Performing dielectric relaxation measurements, they discovered that both liquid-like and solid-like complexes showed a β-relaxation mode, which was associated to the movements of the surfactant chains grafted into the polyelectrolytes because the β-relaxation mode exhibited Arrhenius temperature dependence regardless of the polyelectrolyte-surfactant complex species. Moreover, depending on phase-morphology and volume fraction of alkyl-tails of surfactants, electrical conductivities at room temperature of up to 10^{-4} S/cm were found for these complexes. This conductivity for solid-like complexes was found to be Arrhenius temperature dependent, whilst liquid-like complexes displayed Vogel–Fulcher–Tamman (VFT) temperature dependence [28, 29].

The first insight on dielectric properties of supramolecular ionic networks created from stoichiometric proton transfer reaction was presented by Hvilsted et al. As stated before, in their study the ionic crosslinker molecules such as PPI dendrimers or multifunctional amines were, respectively, reacted with DiCOOH-PEG of two different molecular mass (M_n 250 and M_n 600). All the prepared networks exhibited very high relative dielectric permittivities (ε') ($\approx 10^2$–10^6) at low frequencies, and significantly lower values (from 1 up to 26) at high frequencies. The networks that displayed higher relative permittivities over the whole range of measured frequencies were formed by mixing PPI dendrimers (PPI G1 and PPI G2) with DiCOOH-PEG 600. In order to obtain a structure-dielectric property relationship, the authors calculated the crosslinking density (γ_c) for all the supramolecular structures according to the following equation by assuming volume conservation as no considerable volume changes could be observed upon mixing of the two reactants.

$$\gamma_c = \frac{f_1 \times \left(\frac{m_1}{M_1}\right)}{m_1 + m_2} \tag{2.1}$$

where f_1, m_1 and M_1 are the functionality, mass and molecular weight of reactant 1, respectively, and m_2 is the mass of reactant 2. However, it was not concluded any clear relationship between the dielectric permittivity and the crosslinking density of different networks, as there are a number of other factors that affect the permittivity of these materials such agglomeration of the ionic groups and the length of the reactants [30].

In the studies of polyelectrolyte-surfactant complexes of conducting polymers described above it has been detected that conductivity jumps sharply when the mixture is heated above a certain temperature [31]. Correspondingly, supramolecular networks formed through ionic interactions between citric acid and a di-amine exhibited sharp changes in its conductivity property (determined by electrochemical impedance spectroscopy) by four orders of magnitude in the temperature range between 30 and 80°C. Based on Walden rule, the authors correlated this jump in ionic conductivity to the sharp decrease in viscosity [12]. The ionic conductivity of these supramolecular ionic structures was still low (10^{-5} S cm^{-1}) at high temperature (80 °C) to be used as electrolytes in conventional devices. Nonetheless, due to this big jump in conductivity, they potentially can be used in temperature-dependent electrochemical applications such as temperature-sensitive electrochromic tags or frozen junctions in light-emitting electrochemical cells [32]. Similarly, Grinstaff et al. designed a redox active ionic supramolecular assembly based on a disulfide diphosphonium ionic liquid and ethylenediaminetetraacetic acid (EDTA) and they determined the relationship between temperature, viscosity, and conductivity. The assembled network was viscous (≈4000 Pa s) at 25 °C and displayed a low conductivity (<1 μS cm^{-1}). However, by increasing temperature up to 90 °C, its conductivity value increased (100 μS cm^{-1}), because at this temperature the disulfide bonds are cleaved or the ionic interactions are disrupted, the resulting network is lost presenting a freely flowing composition with lower viscosity and a higher conductivity [33]. In another study carried out by Aboudzadeh et al., ionic conductivity of supramolecular ionic networks based on highly delocalized dianions containing hydrophobic (trifluoromethane-sulfonyl) imide group were investigated (structures **a** and **c** shown in Fig. 2.3). In these structures, the change of aliphatic dianion to its aromatic analogue led to an increase in the conductivity of nearly three orders of magnitude (from 10^{-6} S cm^{-1} to 10^{-3} S cm^{-1} at 100 °C). These scientists also investigated the temperature dependence of conductivity for these networks. The crystalline network (structure **a**) at temperatures below 70 °C, displayed an almost

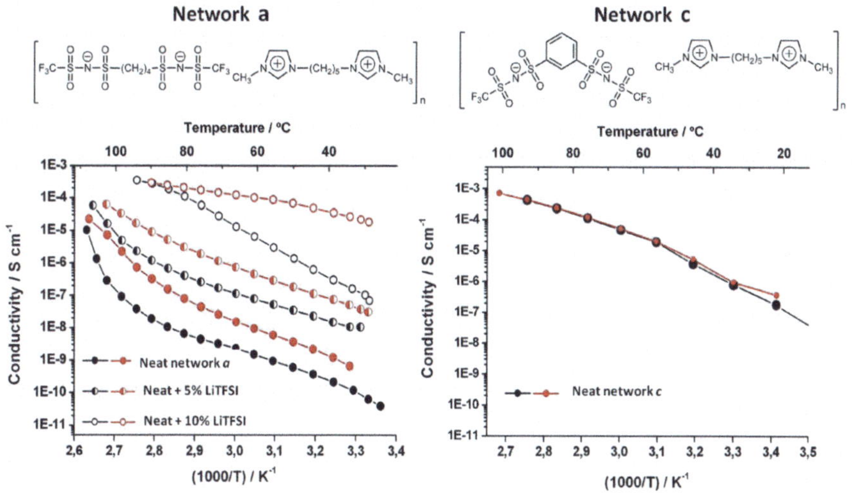

Fig. 2.3 (left) Temperature dependence of ionic conductivity for network **a** and after addition of 5 or 10 mol% LiTFSI. (right) Temperature dependence of ionic conductivity for neat network c. Heating and cooling cycles are shown in black and red colours, respectively. Chemical structure of each network is shown above each graph. Adapted with permission from the reference [22] Copyright (2015) (Elsevier)

linear behaviour (Arrhenius behaviour) with the conductivity increasing from 1.3 $\times 10^{-10}$ to 4.4 $\times 10^{-9}$ S cm^{-1} upon heating. However, at temperatures close to T_m ($\approx$ 100 °C), a faster increase in ionic conductivity, equal to three orders of magnitude, was observed leading to a conductivity of 10^{-5} S cm^{-1}. Similarly, during the cooling phase this network showed a high conductivity at its melted state, which was decreased at temperature below T_m. Nevertheless, it exhibited a hysteresis of more than one order of magnitude between the cooling and heating cycles, which could be directly related to the crystalline content of the structure. Upon doping this semi-crystalline network with lithium bis(trifluoromethylsulphonyl)imide (LiTFSI), the conductivity reached values 10^{-3} S cm^{-1}, which was due to the enhanced mobility of the network confirmed by solid-state static NMR studies (Fig. 2.3 left). On the other hand, the amorphous network (structure **c**) demonstrated a different temperature dependent ionic conductivity performance, which was predominantly linear with highest values of ionic conductivity (between 10^{-6} and 10^{-3} S cm^{-1}) for the entire temperature range (30–100 °C) and without any apparent conductivity hysteresis between heating and cooling cycling (Fig. 2.3 right) [21, 22].

2.2.4 *Rheological Properties*

One undesirable disadvantage of conventional polymers is their high melt viscosities as a result of their chain entanglements, which make them difficult to process. In contrast, supramolecular polymer networks generally demonstrate a strongly temperature-dependent melt viscosity, which enhances their processability in a less viscous state at temperatures only slightly higher than their melting or glass transition temperatures. Apart from the increase in viscosity, polymer solutions and melts can present viscoelastic properties i.e. showing both liquid-like (viscous) and solid-like (elastic) behaviour, which depend on the experimental time scale. Reversible supramolecular polymers possess viscoelastic behaviour as well, and this feature is mostly studied by rheology. In rheology science, flow and deformation of materials are studied by measuring their response of the material to an oscillating stress or strain.

The history of measuring the viscosity of supramolecular ionic assemblies dates back to 1980s towards polyelectrolyte-surfactant complexes [34, 35]. Some of these structures directly influence the macroscopic viscosity of such solutions [36–39], whilst others show scarcely any changes or even a decrease [40, 41]. The reason for these contradictions is not very clear, and extensive characterization of the viscosity behaviour of these complexes is still needed. Whereas in these studies, the viscosity of structures was measured in the presence of a solvent, the first insight on determining the viscosity of supramolecular ionic networks after removing all the volatile components was done by Grinstaff et al. In their study, the addition of EDTA to the phosphonium dication resulted in formation of a network with a 12-fold increase in viscosity, which was not observed when using structural analogs of the compounds with fewer charges [42]. In another work performed by the same authors, EDTA

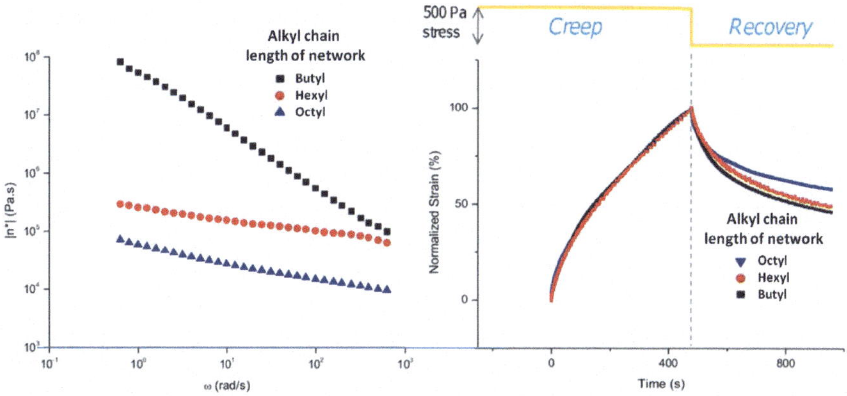

Fig. 2.4 (left) Frequency sweep and (right) creep-recovery curves of three supramolecular ionic networks based on PAA and phosphonium dications with different alkyl chain lengths (■ butyl, ●hexyl and ▲octyl) at 25 °C. Adapted with permission from the reference [19] Copyright (2012) (American Chemical Society)

was replaced with PAA and the new networks showed creep-recovery behaviour (confirming viscoelasticity) that were highly reliant on the alkyl chain length of the diphosphonium cation (*–R* group shown in Fig. 1.12 of Chap. 1). When the alkyl chain was the shortest one (butyl group), the resulting network exhibited the highest viscosity and storage and loss moduli amongst the others. In creep-recovery experiments, this network also presented the highest percent of strain recovery after removing the stress (Fig. 2.4). In addition, by performing reversibility test in frequency and temperature sweep rheological tests, the authors proved that the ionic networks could fully reassemble within a short time period after disassembly of the network due to heat or shear without dropping the mechanical properties [19, 43].

Aida et al. investigated the rheological properties of high-water-content mouldable hydrogels that were obtained simply by mixing clay, a dendritic molecular binder and sodium poly(acrylate) at room temperature (structure shown in Fig. 1.14 of Chap. 1). The hydrogels exhibited high mechanical properties, for instance when G3-binder (3 = generation number) was used to formulate them, G' and G" were achieved as 10^4 Pa and 5×10^3 Pa, respectively. The hydrogels that were formed without incorporating sodium poly(acrylate) salt, displayed weaker mechanical properties [44].

Aboudzadeh et al. through various studies deepened more on the viscoelastic properties of supramolecular networks constructed by ionic interactions between small molecules. In one of their synthetic approaches, citric acid in combination with commercially accessible di- or trifunctional amines was used to design supramolecular ionic networks. The resulting networks displayed unique rheological properties and they studied the importance of the chemical nature of the di- or triamine and the strength of ionic bonds on the rheological characteristics of these networks [12,

45]. These rheological features were analysed through performing frequency and temperature sweeps in small-amplitude oscillatory flow tests. It was found that at high frequencies and/or reduced temperatures, the ionic interactions generated an elastic gel or network, which disappeared at reduced frequencies and/or high temperatures and became a viscoelastic liquid. As an example, the dynamic viscoelastic functions G′ (storage moduli) and G″ (loss moduli) of an ionic network, made up of citrate and 1,3-diammoniumpropane, at two different temperatures are shown in Fig. 2.5. At reduced temperature (20 °C), the mechanical span is distinguished by almost frequency-independent moduli, with a G′ approximately two orders of magnitude greater than G″ (Fig. 2.5 left). This behaviour has been reported for polymer and biopolymer gels too [46] and verified the construction of a supramolecular ionic network that created a flexible solid with an elastic modulus approximately 10^7 Pa. However, at 60 °C this network or gel behaviour vanished completely (Fig. 2.5 right), since the mechanical spectrum conformed the G′ and G″ dependency on frequency and the typical response of viscoelastic liquids was observed.

Observing this viscoelastic behaviour, the authors assumed the existence of a transitory network derived from the involved ionic interactions. They obtained the transition temperature from the network to the liquid form, T_{nl} (network – liquid

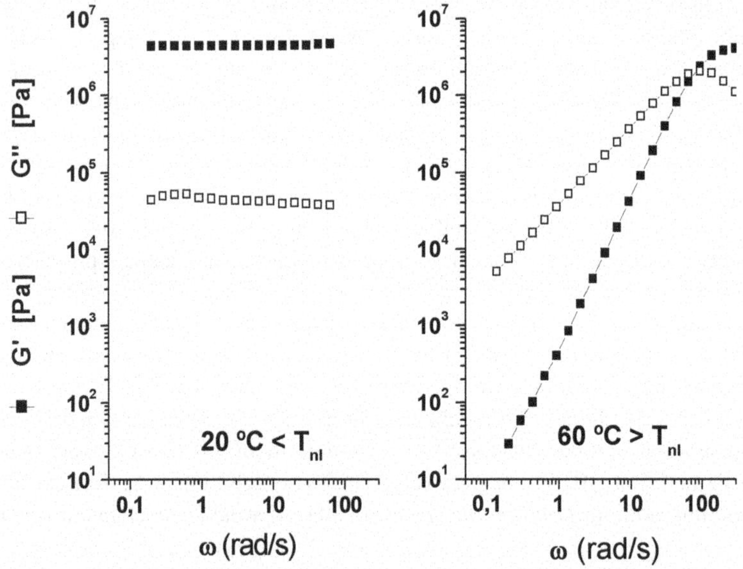

Fig. 2.5 Dynamic moduli G′ (filled symbols) and G″ (empty symbols) as a function of frequency, (left) below and (right) above the network—liquid transition (T_{nl}) for supramolecular ionic network based on citrate and 1,3-diammoniumpropane. At low temperatures (left) the presence of an elastic network or gel, characterized by no effect of frequency on both moduli and by G′ > G″, was observed, whereas at high temperatures (right) the typical response of viscoelastic liquids was detected. Reproduced with permission from the reference [45] Copyright (2012) (American Chemical Society)

transition), by alteration of the dynamic moduli as a function of temperature, taken at a fixed frequency (1 Hz in this case). This transition temperature, T_{nl}, was determined by the highest value of G″, which was corresponded also with the temperature at the crossing point G′ = G″ as displayed in Fig. 2.6. As can be seen in this figure, T_{nl} changed from −1 °C for the network assembled through reaction between citrate and ethyl-substituted tertiary diamine such as tetraethyl-1,3-propanediamonium to 32 °C for the one based on citrate and methyl-substituted diamine counterpart. The maximum transition temperature was detected for the ionic network formed between citrate and primary 1,3-diammoniumpropane.

According to these rheological results, the authors concluded that the dynamic viscoelastic functions of these ionic networks at high temperatures are very well fitted to the simplest model of the linear viscoelasticity, which is the Maxwell model with a single relaxation time and a well-defined plateau at high frequencies [47]. From this model, the following equations are inferred:

$$G'(\omega) = \frac{G_p(\omega\tau)^2}{1 + (\omega\tau)^2} \quad (2.2)$$

$$G''(\omega) = \frac{G_p\omega\tau}{1 + (\omega\tau)^2} \quad (2.3)$$

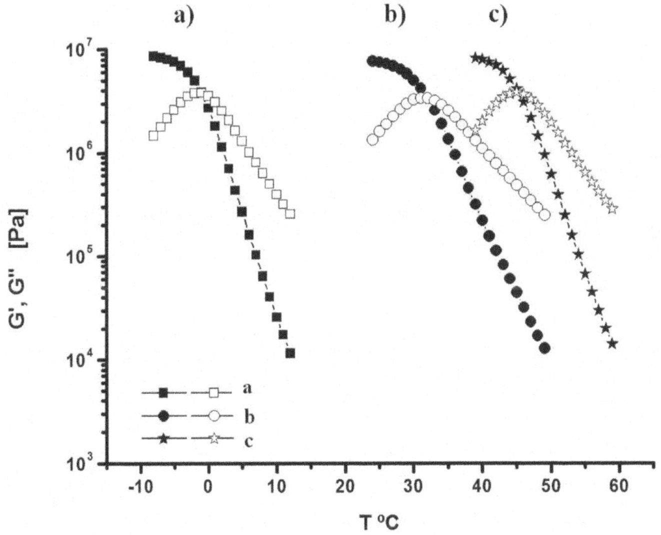

Fig. 2.6 Evolution of the dynamic moduli G′ (filled symbols) and G″ (empty symbols) with increasing temperature at a constant frequency of 1 Hz for supramolecular ionic networks based on citrate and **a** N,N,N′,N′-tetraethyl-1,3-propanediammonium, **b** N,N,N′,N′-tetramethyl-1,3-propanediammonium and **c** 1,3-propanediammonium. Reproduced with permission from the reference [45] Copyright (2012) (American Chemical Society)

where ω is the frequency, τ is the relaxation time, and G_p is the plateau modulus. Based on this curve fitting, the relaxation time (τ) for each ionic network was calculated and was plotted against temperature leading to an Arrhenius-like dependency ($\tau = A \exp E_a/RT$), in which E_a is the activation energy for each system.

In addition to the above-mentioned general viscoelastic behaviour, these networks demonstrated rapid thermal reversibility (Fig. 2.7), meaning that they could reversibly break and re-form rapidly through non-covalent interactions. This is one of the basic characteristics of supramolecular polymers (also named as reversible or equilibrium polymers) [48]. This result along with the fact that there was only one single relaxation time in the Maxwell model (associated with a single exponential stress) indicated that the formation and breakage of the ionic interactions control the dynamics of these supramolecular ionic networks.

Two different characteristic time scales control the common dynamics of supramolecular polymer networks. The first one is the time scale of formation and rupture of supramolecular associations, and the second one is the time scale for the relaxation of chains by reptation due to entanglements [49]. However, the rheological results obtained by Aboudzadeh et al. could not confirm the presence of two relaxation procedures. The existence of two relaxation times would generate two plateaus in the elastic modulus: the classical plateau attributed to the entanglement network and a secondary plateau owing to the additional reversible ionic bonds. For instance, Craig et al. have reported a double plateau for metallopincer-crosslinked

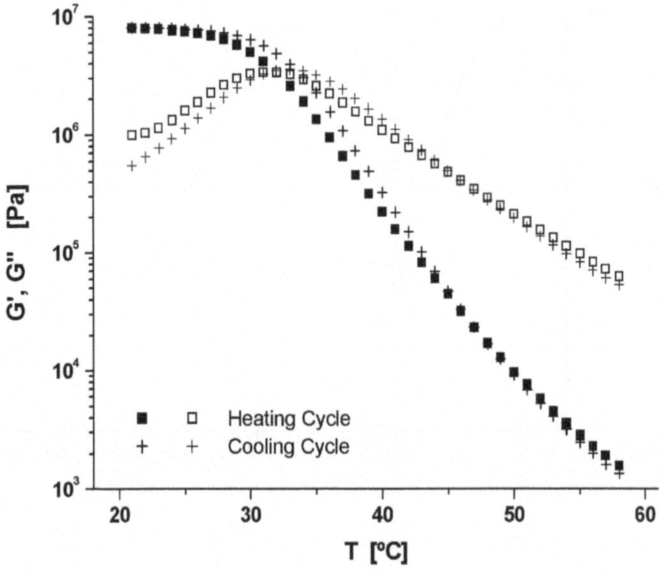

Fig. 2.7 G′ (filled symbols) and G″ (empty symbols) data obtained on continuous cooling and heating cycles, showing the thermal reversibility for an ionic network based on citrate and N,N,N,N′-tetramethyl-1,3-propanediammonium, measured at $\omega = 1$ Hz and a strain of 0.5%. Reproduced with permission from the reference [45] Copyright (2012) (American Chemical Society)

poly(4-vinylpyridine) [1]. Comparing the molecular nature of the building blocks of these ionic networks with the polymeric ones can discard the presence of any entanglement. In addition, considering that the elasticity is governed by the entropy correlated with the conformational distribution between the ionic interactions, the authors applied the rubber elasticity equation (Eq. 2.4) to their G_p data. This equation describes the correlation between the plateau modulus (G_p) and the molar weight between elastically effective interactions (M_i).

$$G_p = \rho RT/M_i \tag{2.4}$$

As a result, they obtained the values of M_i around $\approx$550 g/mol for all the samples. This value, was lower than the smallest value of the entanglement molecular weight M_i reported for conventional polymers [50]. Therefore, Aboudzadeh et al. concluded that entangled physical interactions were not the reason for the detected plateau modulus at high frequencies.

Furthermore, these authors have concluded some initial relationships between the physical properties and the chemical composition of the constituent molecules. First, at high frequencies all the systems brought about the same value (5×10^6 Pa) as plateau modulus (G_p), which suggested that the density of effective interactions or transient crosslinks was the same and, thus, independent of the amine used to synthesize the supramolecular ionic networks. Second, the plots of the relaxation time versus temperature (not shown here) represented a remarkable difference between the network obtained using tetraethyl-1,3-propanediamine and the rest of the samples. This network shifted the relaxation times to lower temperatures, which was compatible with its lower T_{nl} (taken at $\omega = 1$ Hz), in comparison with the other samples (Fig. 2.6). This implied that weaker interactions are involved in the case of this sample, as the required energy to break them, inducing network–liquid transition, depends on their strength. The authors presented a possible reason for this case; the length of the units attached to nitrogen for tetraethyl-1,3-propanediamine is longer than for the other samples (i.e. ethyl units comparing to methyl units), which can induce more strain upon the temporary crosslinks, facilitating its breakage and decreasing the relaxation time and the T_{nl}. Third, the highest T_{nl} was observed for the system that contained 1,3-diaminopropane. The authors observed by FTIR that the carboxylate and ammounium bands appeared at lower wavelengths, which mean that the ionic bonds involving primary amines are stronger than ionic bonds with tertiary amines. Finally, in the liquid form, at the same comparable temperature to the network–liquid transition, $T = T_{nl} + 20$, all the networks demonstrated a similar relaxation time value $\tau = 0.02$ s. Therefore, the authors concluded that all the studied ionic networks displayed an identical behaviour in their respective solid and liquid forms [45].

Based on the similar synthetic approach, the same scientists developed fully bio-based supramolecular ionic networks composed of an aliphatic fatty diamine (derived from natural oils) and series of naturally occurring carboxylic acids including malonic acid, citric acid, tartaric acid, and 2,5-furandicarboxylic acid. Likewise, the resulting networks showed a viscoelastic behaviour identified by G′ > G″ at low temperature

(Fig. 2.8). As it can be seen in Fig. 2.8a–c by increasing temperature, a transition from an elastic network (with G′ > G″) to a viscoelastic liquid with viscous modulus superior to the elastic modulus (G″ > G′) was detected. Also, the reversed trend (from liquid to network) was observed for these networks in the cooling cycle (Fig. 2.8a–c). For combinations that included malonic acid and citric acid, the transition temperature from the network to the liquid phase, T_{nl} (on heating cycle) coincided with the transition from the liquid to the solid phase, T_{ln} (on cooling cycle). However, in the case of network composed of 2,5-furandicarboxylic acid, T_{ln} was higher than T_{nl}, revealing that on cooling the ionic interactions were formed at a higher temperature than that required to break them on heating.

For the semi-crystalline network that contained tartaric acid, the transitions were governed by the melting of the crystals on heating and the crystallization process on cooling. The correlation between T_g and T_m (obtained by DSC) and dynamic viscoelastic results are illustrated in Fig. 2.8d, T_{nl} and T_{ln} corresponded with melting and crystallization temperatures, respectively. The authors proposed that the nature of

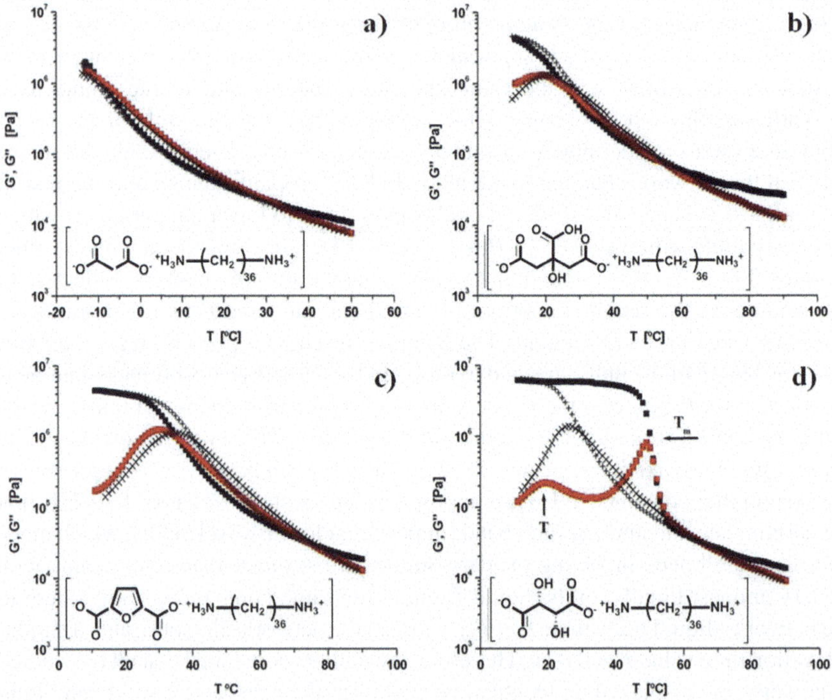

Fig. 2.8 Evolution of the dynamic moduli G′ (heating ■ and cooling +) and G″ (heating ■ and cooling ×) with increasing temperature at a constant frequency of 6.28 Rad/s for supramolecular ionic networks based on an aliphatic fatty diamine (Priamine 1074) and **a** malonic acid, **b** citric acid, **c** 2,5-furandicarboxylic acid and **d** tartaric acid. Arrows in the figure indicate the melting (T_m) and crystallization temperatures (T_c). Reproduced with permission from the reference [13] Copyright (2013) (John Wiley and Sons)

the ionic network in fatty diamine–tartaric acid combination led to the crystallization of the system.

The former category of ionic networks developed by these scientists displayed continuously increasing difference between the viscous and the elastic modulus (with $G'' > G'$) as the temperature increased (Fig. 2.6). Whilst for this category (as can be noticed in Fig. 2.8), at high temperatures the elastic modulus went beyond the viscous modulus, which entailed the suppression of the flow behaviour expected at high temperatures, as the ionic network has been demolished. The authors related this behaviour to the high molecular weight (540 g mol $^{-1}$) and chemical structure of fatty diamine that could build up a secondary network [13].

Moreover, in a follow-up study, Aboudzadeh et al. examined the viscoelastic characteristics of supramolecular ionic networks synthesized using citrate anions and the most relevant classes of dicationic ionic liquids e.g. di-imidazolium, di-pyridinium or di-pyrrolidonium. Likewise, at low temperatures the resulting materials exhibited an elastic network which turned into a viscoelastic liquid at high temperatures. The authors analysed the effect of frequency on the rheological results obtained in temperature sweep tests. For a sample within this category, the change of the dynamic moduli with temperature, measured at different frequencies is shown in Fig. 2.9. As can be seen in this figure, on the one hand the value of the elastic modulus G_e calculated at reduced temperatures was not influenced by frequency and on the other hand, by increasing the frequency, T_{nl} was shifted to higher temperatures. This shift can be explained by the characteristic feature of polymer viscoelasticity, temperature-frequency equivalence, that could bring about a solid-like response at reduced temperatures and/or high frequencies and vice versa. Therefore, the solid response should extend to higher temperatures as frequency is increased, causing the shift of T_{nl} [20].

The discovered T_{nl} for all these supramolecular ionic networks (developed by Aboudzadeh et al.) facilitated their use in self-healing applications. This rheological property could be translated into self-healing by a small temperature variation before and after T_{nl} transition. Below T_{nl}, these materials display solid film like properties; when a crack is provoked, the material heals itself in some minutes even without applying heat, and recovers its original shape/aspect. In this context, it is also possible to evaluate the self-healing feature of supramolecular materials by rheological measurements. For example, Wan et al. reported self-healable hydrogels based on the electrostatic interaction between macroanionic polyoxometalates and cationic-neutral-cationic ABA triblock copolymers. The self-healing performance of these hydrogels was examined through rheological measurements. Upon applying a large-amplitude force ($\gamma = 50\%$, $\omega = 2$ rad s^{-1}) on these materials, the G' value dropped off from 30 kPa to 0.4 kPa and the tan δ (tan $\delta = G''|G'$) value increased from 0.15 to 4.0, suggesting the collapse of the hydrogel to a quasi-liquid state. By removing the strain, G' fully recovered to its primary value (30 kPa) within 20 s and returned to its original gel state. Furthermore, the recovery behaviour was repeated three times without any apparent degradation (Fig. 2.10) [51].

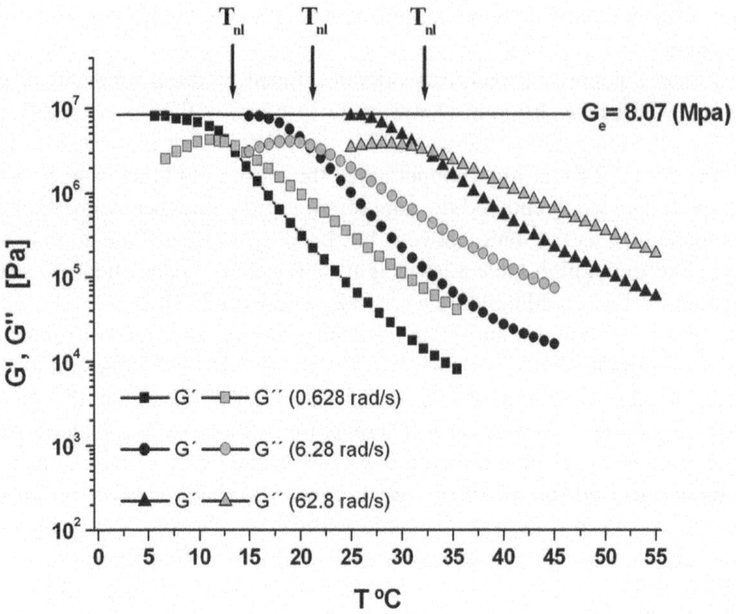

Fig. 2.9 Dynamic elastic modulus G′ (black colour) and viscous modulus G″ (grey colour) as function of temperature at different constant frequencies for supramolecular ionic networks based on citrate and 1,3-propanedimethylimidazolium. The horizontal line shows the value of the G_e modulus. The arrows indicate the T_{nl} transition which is shifted to higher temperatures as frequency is increased. Reproduced with permission from the reference [20] Copyright (2013) (Royal Society of Chemistry)

2.2.5 *Self-Healing Properties*

The self-healing property is always advantageous in polymer science and engineering fields as a lot of polymer materials have limited lifespan because of irrevocable damages. Microcrack formation and crack propagation are common causes of material damage. The reversible properties of supramolecular assemblies make them capable of self-repair or healing after disruption of the interaction, which opened a moderately new research field in the last two decades. The healing can either occur independently or upon exposure to an external stimulus such as light, pressure, heat, or mechanical stress [52].

Supramolecular ionic networks are closely associated to two polymer families, recognized for their self-healing properties, which are ionomers and supramolecular polymers [53, 54]. On the one hand, ionomers present ionic clusters that can serve as reversible networking points and bring about self-healing behaviour. For example, different types of ionomers and their self-healing behaviour were reported by van der Zwaag and Varley [55–57]. On the other hand, supramolecular polymers are based on

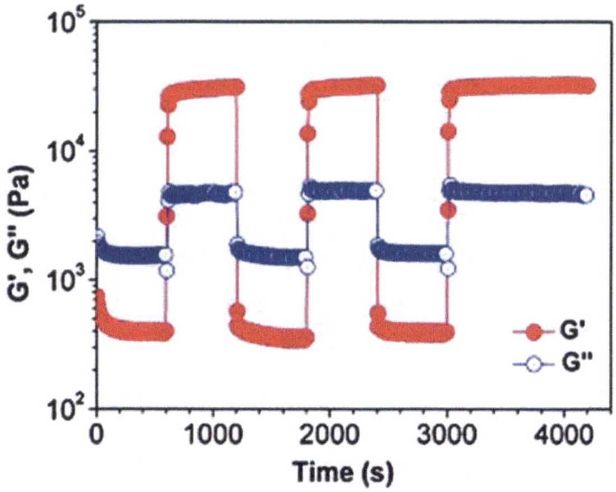

Fig. 2.10 Evaluation of self-healing behaviour for a hydrogel by rheological measurements. The graph shows G′ and G″ values of the hydrogel in continuous step strain measurements. (ω = 2 rad s^{-1}, γ = 50% (600 s) → 0.8% (600 s) → 50% (600 s) → 0.8% (600 s) → 50% (600 s) → 0.8% (1200 s)). Reproduced with permission from the reference [51] Copyright (2014) (Royal Society of Chemistry)

reversible interactions and exhibit interesting temperature transitions between solid and liquid states that can be used to develop self-healing materials.

As stated earlier Aida et al. created a transparent self-healable hydrogels consist of water, clay nanosheets, a dendritic macromolecule, and sodium polyacrylate. These authors pushed together several freshly cut surfaces of these hydrogels so that these surfaces came into contact with each other and they observed that the resulting bridge-like structure was able to hold firmly when hung up horizontally, between two supports, or vertically, from a needle (Fig. 2.11a). In addition, it was found out that the fusion between surfaces occurs only if they are freshly cut and if the blocks are cut but left for more than a minute alone, they do not attach to each other. This is because the freshly cut blocks were hydrophilic as a deposited water droplet spread over a cut surface immediately but beaded on an uncut surface [44]. This time-reliant self-healing property is associated to surface rebuilding resulting from the amphiphilic nature of these self-healing materials. In order to obtain time-independent self-healing, this surface rebuilding must be prevented. In this context, Jiang et al. developed hydrogels by free radical polymerization of carboxybetaine acrylamide that could spontaneously be healed without the need for additional healing reagents or the input of external energy, driven by a mechanism called "zwitterionic fusion". Electrostatic interactions between the zwitterionic groups formed physical crosslinks in the materials, which could self-heal repeatedly independent of time after the damage under physiological conditions (Fig. 2.11b) [58].

In another similar study, Xu et al. generated a thermal-reversible rubber built by thiol-ene functionalized polybutadiene (PB) via a combination of dynamic ionic bonds and covalent crosslinks. By adjusting the covalent crosslinking density, the

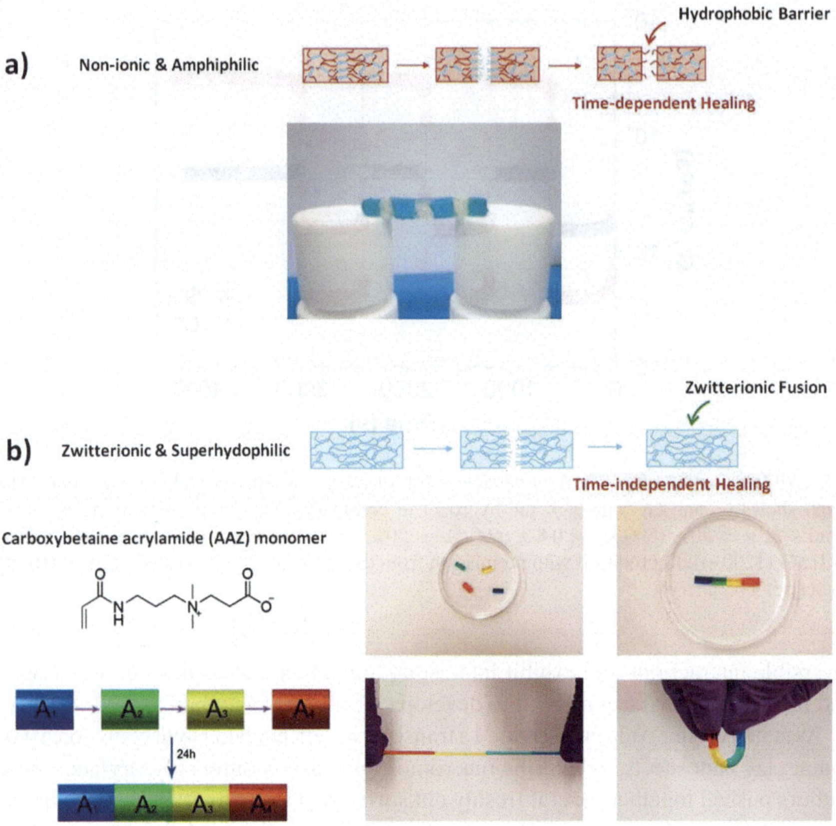

Fig. 2.11 Schematic description of a) time-dependent self-healing behaviour of hydrogels based on amphiphilic polyethylene glycol (PEG) dendrimers; a bridge constructed by connecting together seven hydrogel blocks can be suspended horizontally and b) time-independent self-healing behaviour based on zwitterionic fusion mechanism; chemical structure of zwitterionic individual AAZ hydrogel building block and the pictorial construction process of a block hydrogel from different AAZ hydrogels are shown. Individual hydrogel building blocks dyed with different colours for easy inspection. The constructed block hydrogel can withstand a rather large deformation and bending without tearing apart from the joined connections. Figure excerpts are adapted with permission from the references [58, 59] Copyright (2013, 2014) (John Wiley and Sons, Elsevier), respectively

resulting rubbers exhibited unique self-healing properties at room temperature as a result of sufficient dynamic ionic bonds that existed in the networks. The authors cut the rubber at different areas with a razor blade, and they brought into contact the fracture surfaces. The cut sample could spontaneously heal itself without any external stimulus at room temperature. In addition, after 1 h healing, the recovered sample could be elongated from 2 to 6 cm without breaking, (Fig. 2.12a). Studying the healed surface by optical microscopy, the authors observed that the cracks between the fracture surfaces were almost invisible after healing for 1 h (Fig. 2.12b), indicating

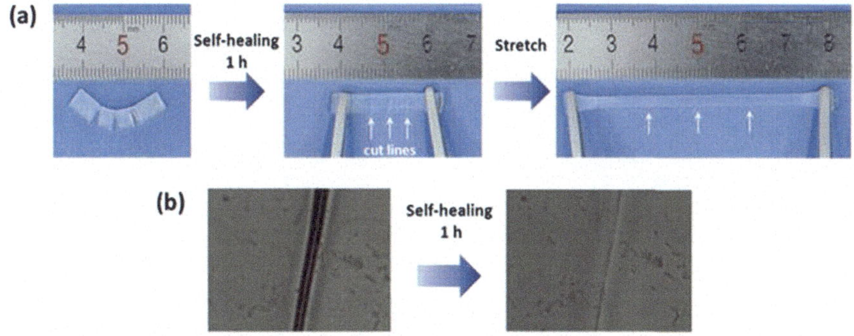

Fig. 2.12 **a** Images displaying the self-healing behaviour of the supramolecular rubber crosslinked with 2 wt% tri-thiol. **b** Optical microscopy images of cut interfaces before and after healing. Reproduced with permission from the reference [60] Copyright (2015) (Royal Society of Chemistry)

the automatic fusion of the fracture surface as well as effective self-healing [60]. Related to this study, bromobutyl rubber functionalized with imidazolium groups was reported by Heinrich et al. Converting bromine functionalities in this rubber into ionic imidazolium bromide groups led to the creation of reversible ionic interactions that showed physical crosslinking capability. Consequently, higher elastic modulus, higher stretchability, and tensile strength was achieved for the ionically crosslinked rubber due to its higher degree of crosslinking compared to the typical sulfur-cured bromobutyl rubber. At high temperatures (above 130 °C) the ionic crosslinks were broken and cut samples could be healed over several days at room temperature. However, increasing the temperature to 100 °C during the first ten minutes of the healing decreased the overall healing time [61].

An oligomer-based self-healing supramolecular ionic networks was created by Aboudzadeh et al. These materials were formed by mixing low molecular weight (di-/tri) carboxylic acids and (di-/tri-) alkyl amines, holding together by a combination of ionic interactions and hydrogen bonds. These viscoelastic gel-like materials could quickly be healed by heating to 50 °C or slowly at room temperature (Fig. 2.13) [12]. Nevertheless, the self-healing behaviour of these materials was closely related to their nature, for example in the case of network formed between tartaric acid and an aliphatic fatty diamine (priamine 1074), the authors did not observe any self-healing behaviour, due to semi-crystalline nature of this sample. In addition, the self-healing test was not successful for malonic acid—priamine and 2,5-furandicarboxylic acid—priamine combinations, because the first one was very soft (quite above from its T_{nl}) at room temperature and the second one was very rigid and brittle (below its T_{nl} in room temperature) [13].

It is worth mentioning that despite the potential dynamic nature of the supramolecular ionic interactions in providing self-healable materials, sometimes kinetically trapped materials are produced unexpectedly, either because of too strong interactions or due to vitrification of the supramolecular materials. To achieve soft or rubbery materials possessing self-healing property, multiple weak interactions between low

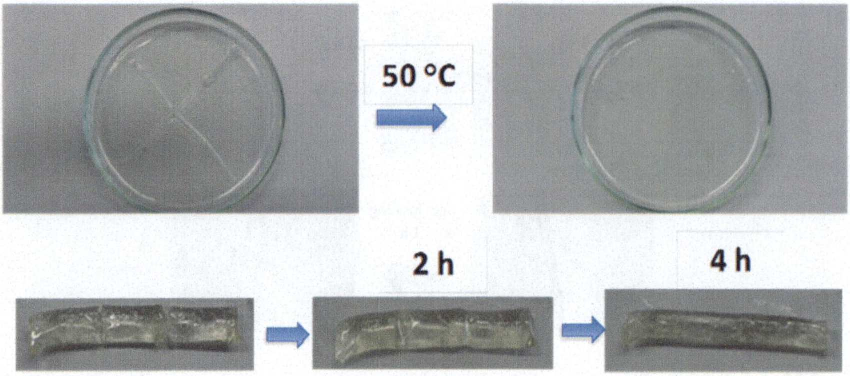

Fig. 2.13 Images of scratched supramolecular ionic network film before and after heating at 50 °C (upper side) and supramolecular ionic bulk cylinder before and after healing by standing at 25 °C under nitrogen atmosphere for 2 h and 4 h respectively (lower side). Reproduced with permission from the reference [12] Copyright (2012) (John Wiley and Sons)

T_g polymers or oligomers can be applied as a standard approach. For example, rigid bulk supramolecular networks that at room temperature are below their T_g, do not show any dynamic nature. To acquire sufficient chain mobility for self-healing, they should be heated up above the T_g either thermally or by photothermal treatment. This also directs us to the principal challenge of bulk supramolecular network materials, which is developing a stiff substance with enough chain dynamics capable of self-healing at ambient temperature [62]. Another challenge is the rate of self-healing which is important for both practical applications and intrinsic interests. Currently how fast the material self-heals is remained mostly undetermined and the parameters associated to these rates have not been studied comprehensively. Achieving effective healing over a short time period still is a crucial need in this regard.

2.3 Conclusions and Future Aspects

Ionic interactions have been used extensively to generate physical and/or chemical crosslinks in network materials. Supramolecular networks based on ionic interactions constitute a very versatile family of compounds with interesting properties. Because of the reversibility of the interactions involved, these networks are under thermodynamic equilibrium. Consequently, their properties can be adjusted beforehand by a careful structural design. In this chapter, we presented an overview of the main properties of supramolecular ionic networks (e.g. morphology, thermal, electrical, rheological, and self-healing) and we tried to describe how the structure of these networks conditions their properties. A wide variety of applications is foreseen to be feasible for these materials. Particularly as their synthetic strategy is accomplished just by mixing monomers, it makes hybrids between blocks of oligomers, macromolecules, and supramolecular polymers easy-to-achieve. Therefore, novel self-healing rubbers, drug and gene delivery systems, electrochemical devices such as batteries are within reach.

Although a lot of research has been devoted to the field and many interesting achievements have been made, establishing a logical general description on physical–chemical properties of supramolecular ionic networks is far from complete yet. The existing literature has mainly focused on the synthesis of supramolecular polymers and networks (mostly mentioned in Chap. 1), and only very few recent researches have been attempted to determine rational and quantitative relationships between their structure, dynamics, and properties. Like for conventional polymerization methods, an effective programmable mixing of multiple building blocks and thereby coherent modulation of properties is an important unmet challenge. It is therefore highly advantageous to carry out more systematic studies with an especial focus on the relation of the structure and dynamics of the supramolecular polymer networks simultaneously. This point could be interesting for polymer theorists and new polymer theories can be derived for the thermodynamic nature of the supramolecular networks, as the kinetic constraints that are usually involved in the study of conventional polymers are not present for these materials.

References

1. Serpe MJ, Craig SL, Uni V, Carolina N (2007) Physical organic chemistry of supramolecular polymers. Langmuir 23:1626–1634. https://doi.org/10.1021/la0621416
2. Faul CFJ, Antonietti M (2003) Ionic self-assembly: facile synthesis of supramolecular materials. Adv Mater 15:673–683. https://doi.org/10.1002/adma.200300379
3. MacKnight WJ, Earnest TR (1981) The structure and properties of ionomers. J Polym Sci Macromol Rev 16:41–122. https://doi.org/10.1002/pol.1981.230160102
4. Masanori H (1993) Polyelectrolytes, science and technology. Marcel Dekker, New York
5. Mecerreyes D (2011) Progress in polymer science polymeric ionic liquids: broadening the properties and applications of polyelectrolytes. Prog Polym Sci 36:1629–1648. https://doi.org/10.1016/j.progpolymsci.2011.05.007
6. Bara JE, Camper DE, Gin DL, Noble RD (2010) Room-temperature ionic liquids and composite materials: platform technologies for CO 2 capture. Acc Chem Res 43:152–159
7. Vijayakrishna K, Jewrajka SK, Ruiz A et al (2008) Synthesis by RAFT and ionic responsiveness of double hydrophilic block copolymers based on ionic liquid monomer units. Macromolecules 41:6299–6308
8. Appetecchi GB, Kim G, Montanino M et al (2010) Ternary polymer electrolytes containing pyrrolidinium-based polymeric ionic liquids for lithium batteries. J Power Sources 195:3668–3675. https://doi.org/10.1016/j.jpowsour.2009.11.146
9. Lutkenhaus JL, Hammond PT (2007) Electrochemically enabled polyelectrolyte multilayer devices: from fuel cells to sensors. Soft Matter 3:804–816. https://doi.org/10.1039/b701203a
10. Philipp B, Dautzenberg H, Linow K-J et al (1989) Polyelectrolyte complexes—recent developments and open problems. Prog Polym Sci 14:91–172
11. Meyer WH (1998) Polymer electrolytes for lithium-ion batteries. Adv Mater 10:439–448
12. Aboudzadeh MA, Muñoz ME, Santamaría A et al (2012) Facile synthesis of supramolecular ionic polymers that combine unique rheological, ionic conductivity, and self-healing properties. Macromol Rapid Commun 33:314–318. https://doi.org/10.1002/marc.201100728

13. Aboudzadeh A, Fernandez M, Muñoz ME et al (2014) Ionic supramolecular networks fully based on chemicals coming from renewable sources. Macromol Rapid Commun 35:460–465. https://doi.org/10.1002/marc.201300732
14. Godeau G, Navailles L, Nallet F et al (2012) From brittle to pliant viscoelastic materials with solid state linear polyphosphonium-carboxylate assemblies. Macromolecules 45:2509–2513. https://doi.org/10.1021/ma3002092
15. Piculell L, Norrman J, Svensson AV et al (2009) Ionic surfactants with polymeric counterions. Adv Colloid Interface Sci 147–148:228–236. https://doi.org/10.1016/j.cis.2008.09.009
16. Hansson P (2009) Phase behavior of aqueous polyion-surfactant ion complex salts: a theoretical analysis. J Colloid Interface Sci 332:183–193. https://doi.org/10.1016/j.jcis.2008.12.011
17. Oskolkov NN, Potemkin II (2007) Complexation in asymmetric solutions of oppositely charged polyelectrolytes: phase diagram. Macromolecules 40:8423–8429. https://doi.org/10.1021/ma0709304
18. Osada K, Oshima H, Kobayashi D et al (2010) Quantized folding of plasmid DNA condensed with block catiomer into characteristic rod structures promoting transgene efficacy. J Am Chem Soc 132:12343–12348
19. Lin X, Navailles L, Nallet F, Grinstaff MW (2012) Influence of phosphonium alkyl substituents on the rheological and thermal properties of phosphonium-PAA-based supramolecular polymeric assemblies. Macromolecules 45:9500–9506. https://doi.org/10.1021/ma3019624
20. Aboudzadeh MA, Muñoz ME, Santamaría A, Mecerreyes D (2013) New supramolecular ionic networks based on citric acid and geminal dicationic ionic liquids. RSC Adv 3:8677–8682. https://doi.org/10.1039/c3ra40629f
21. Aboudzadeh MA, Shaplov AS, Hernandez G et al (2015) Supramolecular ionic networks with superior thermal and transport properties based on novel delocalized di-anionic compounds. J Mater Chem A 3:2338–2343. https://doi.org/10.1039/c4ta05792a
22. Aboudzadeh MA, Zhu H, Pozo-Gonzalo C et al (2015) Ionic conductivity and molecular dynamic behavior in supramolecular ionic networks; The effect of lithium salt addition. Electrochim Acta 175:74–79. https://doi.org/10.1016/j.electacta.2015.02.064
23. González L, Ladegaard Skov A, Hvilsted S (2013) Ionic networks derived from the protonation of dendritic amines with carboxylic acid end-functionalized PEGs. J Polym Sci Part A Polym Chem 51:1359–1371. https://doi.org/10.1002/pola.26503
24. Gold BJ, Hövelmann CH, Lühmann N et al (2017) Importance of compact random walks for the rheology of transient networks. ACS Macro Lett 6:73–77. https://doi.org/10.1021/acsmacrolett.6b00880
25. Shabbir A, Javakhishvili I, Cerveny S et al (2016) Linear viscoelastic and dielectric relaxation response of unentangled UPy-based supramolecular networks. Macromolecules 49:3899–3910. https://doi.org/10.1021/acs.macromol.6b00122
26. Xing K, Tress M, Cao P et al (2018) Hydrogen-bond strength changes network dynamics in associating telechelic PDMS. Soft Matter 14:1235–1246. https://doi.org/10.1039/C7SM01805C
27. Tress M, Xing K, Ge S et al (2019) What dielectric spectroscopy can tell us about supramolecular. Eur Phys J E 42.https://doi.org/10.1140/epje/i2019-11897-4
28. Antonietti M, Maskos M, Kremer F, Blum G (1996) Dielectric relaxation and conductivity in polyelectrolyte-surfactant complexes. Acta Polym 47:460–465. https://doi.org/10.1002/actp.1996.010471007
29. Antonietti M, Neese M, Blum G, Kremer F (1996) Dielectric and mechanic relaxation in polyelectrolyte-supported bilayer stacks: a model for the dynamics of membranes? Langmuir 12:4436–4441. https://doi.org/10.1021/la960221b
30. González L, Yu L, Hvilsted S, Skov AL (2014) Dielectric properties of supramolecular ionic structures obtained from multifunctional carboxylic acids and amines. RSC Adv 4:36117–36124. https://doi.org/10.1039/c4ra06195k
31. Ober CK, Wegner G (1997) Polyelectrolyte-Surfactant Complexes in the Solid State: Facile building blocks for self-organizing materials. Adv Mater 9:17–31. https://doi.org/10.1002/adma.19970090104

32. Rosseinsky DR, Mortimer RJ (2001) Electrochromic systems and the prospects for devices. Adv Mater 13:783–793. https://doi.org/10.1002/1521-4095(200106)13:11%3c783::AID-ADMA783%3e3.0.CO;2-D
33. Lin X, Godeau G, Grinstaff MW (2014) A reversible supramolecular assembly containing ionic interactions and disulfide linkages. New J Chem 38:5186–5189. https://doi.org/10.1039/c4nj00895b
34. Abuin EB, Scaiano JC (1984) Exploratory study of the effect of polyelectrolyte-surfactant aggregates on photochemical behavior. J Am Chem Soc 106:6274–6283. https://doi.org/10.1021/ja00333a028
35. Bekturov EA, Kudaibergenov SE, Kanapyanov GS (1984) Interaction of synthetic polyampholytes with anionic and cationic detergents in aqueous solution. Polym Bull 11:551–555
36. Beheshti N, Kjøniksen A, Zhu K et al (2010) Viscosification in polymer— surfactant mixtures at low temperatures. J Phys Chem B 114:6273–6280. https://doi.org/10.1021/jp100333f
37. Kästner U, Hoffmann H, Dönges R, Ehrler R (1996) Interactions between modified hydroxyethyl cellulose (HEC) and surfactants. Colloids Surf A Physicochem Eng Asp 112:209–225. https://doi.org/10.1016/0927-7757(96)03557-1
38. Marques EF, Regev O, Khan A et al (1999) Interactions between catanionic vesicles and oppositely charged polyelectrolytes s phase behavior and phase structure. Macromolecules 32:6626–6637
39. Patruyo LG, Müller AJ, Sáez AE (2002) Shear and extensional rheology of solutions of modified hydroxyethyl celluloses and sodium dodecyl sulfate. Polymer (Guildf) 43:6481–6493. https://doi.org/10.1016/S0032-3861(02)00598-0
40. Guillot S, McLoughlin D, Jain N et al (2003) Polyelectrolyte—surfactant complexes at interfaces and in bulk polyelectrolyte—surfactant complexes at interfaces and. J Phys Condens Matter 15:S219–S224
41. Jain N, Trabelsi S, Guillot S et al (2004) Critical aggregation concentration in mixed solutions of anionic polyelectrolytes and cationic surfactants. Langmuir 20:8496–8503. https://doi.org/10.1021/la0489918
42. Wathier M, Grinstaff MW (2008) Synthesis and properties of supramolecular ionic networks. J Am Chem Soc 130:9648–9649
43. Wathier M, Grinstaff MW (2010) Synthesis and creep-recovery behavior of a neat viscoelastic polymeric network formed through electrostatic interactions. Macromolecules 43:9529–9533. https://doi.org/10.1021/ma101506p
44. Wang Q, Mynar JL, Yoshida M et al (2010) High-water-content mouldable hydrogels by mixing clay and a dendritic molecular binder. Nature 463:339–343. https://doi.org/10.1038/nature08693
45. Aboudzadeh MA, Muñoz ME, Santamaría A et al (2012) Synthesis and rheological behavior of supramolecular ionic networks based on citric acid and aliphatic diamines. Macromolecules 45:7599–7606. https://doi.org/10.1021/ma300966m
46. Graessley WW (2003) Polymeric liquids & networks structure and properties. Taylor & Francis Inc., New York
47. Ferry JD (1980) Viscoelastic properties of polymers, 3rd edn. Wiley, New York
48. Vermonden T, Van SMJ, Besseling NAM et al (2004) Linear rheology of water-soluble reversible neodymium (III) coordination polymers. J Am Chem Soc 126:15802–15808. https://doi.org/10.1021/ja0458928
49. Seiffert S, Sprakel J (2012) Physical chemistry of supramolecular polymer networks. Chem Soc Rev 41:909–930. https://doi.org/10.1039/c1cs15191f
50. Graessley WW (2004) Polymeric liquids & networks dynamics and rheology. Taylor & Francis Inc., London
51. Wei H, Du S, Liu Y et al (2014) Tunable, luminescent, and self-healing hybrid hydrogels of polyoxometalates and triblock copolymers based on electrostatic assembly. Chem Commun 50:1447–1450. https://doi.org/10.1039/c3cc48732f
52. Hart LR, Harries JL, Greenland BW et al (2013) Healable supramolecular polymers. Polym Chem 4:4860–4870. https://doi.org/10.1039/C3PY00081H

53. Syrett JA, Becer CR, Haddleton DM (2010) Self-healing and self-mendable polymers. Polym Chem 1:978–987. https://doi.org/10.1039/c0py00104j
54. Burattini S, Colquhoun HM, Fox JD et al (2009) A self-repairing, supramolecular polymer system: healability as a consequence of donor-acceptor π–π stacking interactions. Chem Commun 6717–6719https://doi.org/10.1039/b910648k
55. Vega JM, Grande AM, Van Der Zwaag S, Garcia SJ (2014) On the role of free carboxylic groups and cluster conformation on the surface scratch healing behaviour of ionomers. Eur Polym J 57:121–126. https://doi.org/10.1016/j.eurpolymj.2014.05.005
56. Varley RJ, van der Zwaag S (2010) Autonomous damage initiated healing in a thermo-responsive ionomer. Polym Int 59:1031–1038. https://doi.org/10.1002/pi.2841
57. Varley RJ, van der Zwaag S (2008) Towards an understanding of thermally activated self-healing of an ionomer system during ballistic penetration. Acta Mater 56:5737–5750. https://doi.org/10.1016/j.actamat.2008.08.008
58. Bai T, Liu S, Sun F et al (2014) Zwitterionic fusion in hydrogels and spontaneous and time-independent self-healing under physiological conditions. Biomaterials 35:3926–3933. https://doi.org/10.1016/j.biomaterials.2014.01.077
59. Lin X, Grinstaff MW (2013) Ionic Supramolecular Assemblies. Isr J Chem 53:498–510. https://doi.org/10.1002/ijch.201300034
60. Wang D, Guo J, Zhang H et al (2015) Intelligent rubber with tailored properties for self-healing and shape memory. J Mater Chem A 3:12864–12872. https://doi.org/10.1039/c5ta01915j
61. Das A, Sallat A, Böhme F et al (2015) Ionic modi fi cation turns commercial rubber into a self-healing material. ACS Appl Mater Interfaces 7:20623–20630. https://doi.org/10.1021/acsami.5b05041
62. Voorhaar L, Hoogenboom R (2016) Supramolecular polymer networks: hydrogels and bulk materials. Chem Soc Rev 45:4013–4031. https://doi.org/10.1039/c6cs00130k

Chapter 3
The Role of Electrostatic Interaction in the Self-assembly of Macroions

Yuqing Yang, Ehsan Raee, Yifan Zhou, and Tianbo Liu

Abstract With sizes (1–5 nm) between simple ions and large colloids, solution behaviors of macroions cannot be described either by Debye-Hückel limiting theory or DLVO theory. In addition, the large size disparity between macroions and small counterions makes their self-assembly process even more complicated. With charges carried by macroions, electrostatic interaction usually plays a critical role during self-assembly. A well-known feature of these structurally well-defined macroions with moderate charges is the spontaneous formation of hollow, spherical, single-layered blackberry structures, whose size can be accurately controlled via pH, solvent polarity, and salt concentration, based on counterion-mediated attraction. These blackberry structures show some unique properties, e.g., unique kinetic properties similar to the virus capsid formation, self-recognition, chiral-recognition, and permeation of small counterions through their membranes. Based on the complex structures of macroions, various interactions, such as hydrophobic interaction, van der Waals forces, hydrogen bonding, and cation-π interactions, can be involved to compete or cooperate with electrostatic interaction to tune their self-assembly behaviors.

Y. Yang · E. Raee · Y. Zhou · T. Liu (✉)
The School of Polymer Science and Polymer Engineering, The University of Akron, Akron, OH 44325-3909, USA
e-mail: tliu@uakron.edu

Y. Yang
e-mail: yy43@uakron.edu

E. Raee
e-mail: er60@uakron.edu

Y. Zhou
e-mail: yz99@uakron.edu

M. A. Aboudzadeh and A. Frontera (eds.), *Supramolecular Assemblies Based on Electrostatic Interactions*,
https://doi.org/10.1007/978-3-031-00657-9_3

3.1 Introduction

Among the intermolecular forces, the electrostatic interaction has the unique feature of longer effective range than many other physical forces, such as hydrophobic interaction, hydrogen bonding, van der Waals forces, cation-π attraction and π-π stacking. It is very essential in charged species and sometimes can regulate the molecular self-assembly processes, with or without other physical forces.

A typical example to demonstrate the role of electrostatic interaction on noncovalent molecular self-assembly is hydrophilic macroions. For people in different areas, the word "macroion" has different meanings, herein, the Macroions (Fig. 3.1) represent a large group of charged macromolecules with sizes (1–5 nm) filling the gap between simple ions and large colloids, e.g., polyoxometalate (POM) molecular clusters, metal organic cages (MOCs), polyhedral oligomeric silsesquioxane (POSS), dendrimers, functionalized fullerenes, organic-inorganic hybrids, cyclodextrins and biomacromolecules. These species exhibit promising applications in a wide range of areas, such as drug delivery, water purification, catalysis and so on so forth [1, 2]. Also, they show unique solution behaviors due to their nano-scaled sizes. Various self-assembled structures including blackberry structures[3], nanosheets [4, 5], nanofibers [6], vesicles[7, 8], reverse vesicles [9, 10] nanrods [11, 12], nanoflowers [13] and nanoribbons [6], which formed by macroions are reported.

Some macroions, such as some polyoxometalates (POMs) and uranyl peroxide molecular clusters, possess fully hydrophilic surface with inherent or surface charges (could be either localized or delocalized) and intact molecular structure in dilute solution. Such macroions are ideal models for exploring the role of electrostatic interaction as other intermolecular interactions are absent or negligible.

For functionalized macroions, various interactions, including hydrophobic interactions (a type of special van der Waals force), π-π stacking, van der Waals forces, hydrogen bonding, host-guest, and cation-π interactions, can participate in their self-assembly behaviors, leading to all sorts of self-assembled structures. The strength and effective distance of these interactions are listed below in Table 3.1. The van der Waals

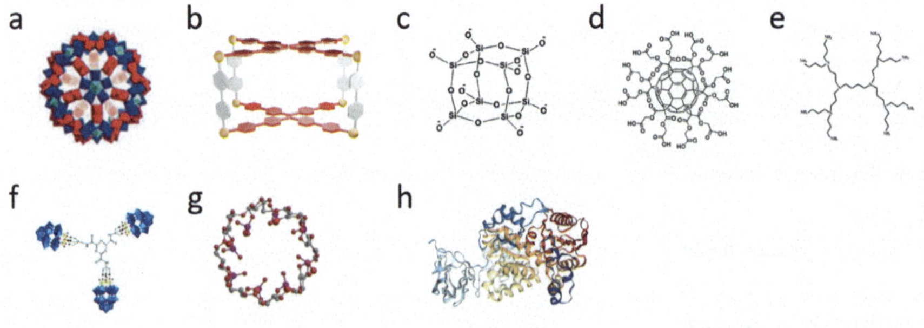

Fig. 3.1 Examples of different macroions, including **a** polyoxometalate molecular clusters, **b** metal-organic cages, **c** polyhedral oligomeric silsesquioxane, **d** functionalized fullerenes, **e** dendrimers, **f** polyoxometalate-organic hybrid, **g** cyclodextrins and **h** biomacromolecules

Table 3.1 Features of attractive forces in macroionic solutions [14]

	Strength (kJ/mol)	Effective distance (A)
Electrostatic interaction	4–80	Long
Hydrogen bonding	8–40	<3
Van der waals forces	0.4–4	<6
Cation-pi interaction	20–160	<6
Pi-pi stacking	<40	<4
Hydrophobic interaction	<40	<10

forces consist of dipole-dipole, dipole-induced dipole, and induced dipole-induced dipole interactions. In addition, hydrophobic interaction is a type of van der Waals forces, that is, induced dipole-induced dipole interaction, which is much stronger in solution than in air due to the solvation effects in solution. Cation-π interaction shares some features with electrostatic interactions since it bears the electrostatic nature. To have these interactions involved, the macroions must contain some certain domains, for example, electron rich and electron poor components to trigger host-guest interactions, hydrogen bond accepting and donating sites for hydrogen bonding, large π area for cation-π interactions or π-π stacking, or hydrophobic components for hydrophobic interactions. Among all these physical interactions, electrostatic interaction will always play a role since all the macroions carry certain amount of charges. On the other hand, the long-range nature of electrostatic interaction also benefits the self-assembly processes, drawing dispersed macroions coming close to each other.

In this chapter, the supramolecular structures purely based on electrostatic interactions and the ones which driven by the competition or cooperation of electrostatic interaction and other interactions (for example, hydrogen bonding, hydrophobic interactions, π-π stacking, van der Waals forces, host–guest, and cation-π interactions) will be discussed.

3.2 Theoretical Challenges of Nanoscale Macroions—Between Simple Ions and Colloids

The most well-known theory to describe a simple dilute ionic solution is the Debye-Hückel limiting theory, in which an ionic atmosphere model is assumed, and the ions are considered as homogenously distributed in solution [15]. The Debye-Hückel limiting theory gives several conclusions: 1. the solution is overall neutral; 2. oppositely charged ions attract each other and lower the overall free energy of the solution; 3. each ion is surrounded more closely by oppositely charged ions than by ions with like-charges. However, the Debye-Hückel limiting theory is only valid for very dilute ionic solution, in which the ionic size is negligible. For macroions and colloidal particles, this theory is no longer valid as they cannot be treated as point charges.

One the other hand, the large, charged colloids behave differently from simple ions. Colloidal particles are thermodynamically unstable, which is not a homogeneous system comparing with simple ionic solutions, and form suspension when dissolved in solvent. The Derjaguin-Landau-Verwey-Overbeek theory (DLVO theory), developed in 1940s, is widely applied for charged colloids [16, 17]. In DLVO theory, colloids repel each other based on purely repulsive Coulomb forces between like-charge molecules; and the only source of attraction forces between two colloidal particles is believed to be van der Waals forces. The competition between these two interactions determines the stability of colloidal particles in solution. One thing that needs to be clarified here is that the approximation that applied in Debye-Hückel theory to describe the electrical potential of a point charge can also be applied here for colloidal as their counterion distribution can be described by Boltzmann distribution like simple ions. But in the colloidal systems, we have to consider the size of the colloidal particles.

However, with the unique size range of macroions (between simple ions and colloids), the description of electrostatic interaction between macroions in solution is a challenge. The macroions are too big to be treated as charged points due to their surface charge density and surface charge distribution. In this case, the well understood Debye-Hückel theory for simple ions cannot be applied for macroions. In the meanwhile, DLVO theory is not suitable to describe solution behaviors of macroions either as van der Waals forces are negligible for macroions which still form real solutions. What's more, the large size disparity between macroions and their counterions is significant but not dominant, which leads to moderate counterion-association around macroions and contributes to the complex solution behaviors of these species.

3.3 Self-assembly of Ideal Macroions Regulated by Electrostatic Interaction

A well-reported feature of macroions with moderate charges is the spontaneous formation of hollow, spherical, single-layered blackberry structures based on counterion-mediated attraction[7]. The formation/disassociation and size of blackberry structures can be accurately controlled via solvent polarity, addition of counterions, or macroionic charge density [18, 19]. The driven force for constructing blackberry structures is not hydrophobic effect (which is dominant for double-layered vesicular structures) or van der Waals forces; instead, electrostatic interactions among counterions and macroions play critical roles, which will be discussed in detail in this section.

3.3.1 The Discovery of Blackberry Structures

Among various macroions, self-assembly behaviors of POMs based on electrostatic interactions have been studied widely [20]. The unique properties of POM clusters, e.g., well-defined molecular structure, uniform shape, tunable charges, high stability and monodispersity in solution, as well as no intra-molecular charge interactions, make them perfect models for the macroion solution study. Interestingly, these POM clusters tend to come close to each other and slowly form large structures in aqueous solution instead of staying as single clusters even though the solubility of POMs are quite high [21, 22].

Laser light scattering (LLS), including both static and dynamic light scattering (SLS and DLS), can be used to characterize large structures in solution. Weight average molecular weight (M_w), second virial coefficient and radius of gyration (R_g) of the particles can be obtained by SLS via measuring the scattered intensity at different scattering angles [22]. One can study the nature of the interaction between solvent and solute particles through these information determined by SLS, e.g., identifying whether it's attractive or repulsive interaction by second virial coefficient. By analyzing the intensity-intensity time correlation function measured by DLS via CONTIN method [23], the hydrodynamic radius (R_h) and particle size distribution can be obtained. Taking the combination of SLS and DLS, the shape of the large particles (e.g., spheres, sheets, or fibers) can be indicated.

In 2003, T. Liu first found that a dilute aqueous solution of $\{Mo_{154}\}$ (0.010 mg/mL) at pH = 3.0 showed Tyndall effect which revealed the large assembly formation in the system [22]. The spherical assemblies with relatively uniform size can be noticed from transmission electron microscopy (TEM) images (Fig. 3.2a). Moreover, as indicated by atomic force microscopy (AFM) study, these assemblies suddenly burst when shifted from normal mode to vacuum model, revealing their hollow structures. The spontaneously formation of these large structures is unexpected based on the fact that the solubility of $\{Mo_{154}\}$ clusters in aqueous solution is extremely high (~100 mg/mL). To further dig out the nature of these assemblies, DLS and SLS measurements were conducted. The average R_h of large assemblies was determined to be 45 nm with a narrow size distribution (Fig. 3.2b) using CONTIN analysis of the DLS studies. The average R_h is consistent with the size of spherical assemblies in the TEM images (Fig. 3.2a). Analyzing SLS study by Zimm plot suggested that the R_g of assemblies is around 45 nm and an average M_W of these assembled structures is determined to be about $(2.54 \pm 0.25) \times 10^7$ g/mol, that is 1150 $\{Mo_{154}\}$ clusters per assembly. The fact that R_g is equal to the R_h obtained from DLS indicated the hollow nature of these spherical assemblies, since $R_h = R_g$ is a characteristic of hollow spherical structures in which all the mass of the particle is distributed on its surface ($R_g/R_h \sim 0.77$ if they are solid). All $\{Mo_{154}\}$ clusters are expected to be on the surface based on the hollow nature of these assemblies. With the average radius (45 nm) and average M_W ($(2.54 \pm 0.25) \times 10^7$ g/mol) of the hollow spheres, the intermolecular distance was calculated to be 0.9 nm and the neighboring clusters are not touching each other. A model was proposed as shown in Fig. 3.2c. The

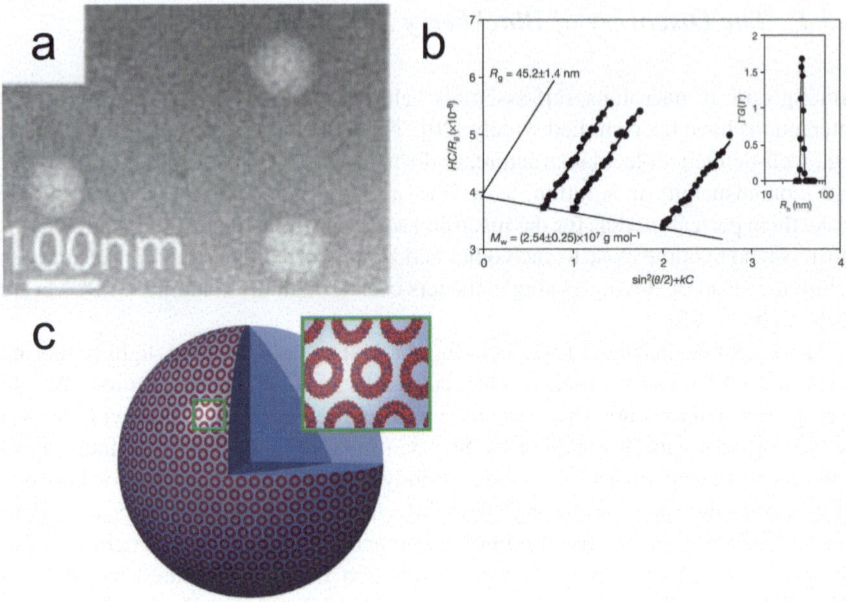

Fig. 3.2 **a** TEM image of {Mo_{154}} aqueous solution showing spherical assemblies (R ~ 45 nm). **b** Zimm plot based on the SLS study of the {Mo_{154}} aqueous solutions at pH 3; (inset) CONTIN analysis on the DLS study of the same solution, showing R_h ~ 45 nm. **c** Schematic plot showing the supramolecular blackberry structure formed by {Mo_{154}} macroions in water. [22] Reprinted with permission from Ref.. Copyright 2003, Macmillan Magazines Ltd

single-layered, hollow structures with a nickname "blackberry" are self-assembled by highly soluble {Mo_{154}}. Although the spherical morphology of blackberry structures is similar to vesicles formed by amphiphilic surfactants, these two structures are fundamentally different: the single-layered blackberry structures are formed by fully hydrophilic molecules, while amphiphilic surfactants are involved in bilayer vesicle formation. The driving force for vesicle formation is hydrophobic effect, however blackberry structures are formed based on a different driving force due to their highly hydrophilic feature.

3.3.2 *Driving Force for Blackberry Structure Formation*

Afterwards, blackberry structures were observed for several POM-based macroions, e.g., {$Mo_{72}Fe_{30}$} [19], {$Mo_{72}Cr_{30}$} [24], and {Mo_{132}} [18]. Although all these POMs are purely hydrophilic, they self-assembled into blackberry structures instead of staying as single molecules. The most essential issue for the blackberry formation is the driving force to bring these like-charged, purely hydrophilic POMs together.

Many efforts have been made to investigate the effect of van der Waals forces in blackberry formation. Van der Waals forces are short-ranged attractive forces and they are dominant in colloidal systems since colloids are large particles composed of many nuclei [25]. For POM solutions, the effect of van der Waals forces is much weaker than that of colloids based on their smaller sizes. A study with $\{Mo_{132}\}$ in acetone/water mixtures can help to figure out its role [18]. $\{Mo_{132}\}$ is a type of molybdenum brown "Keplerate" anions with diameter of around 2.9 nm and 42 negative charges [26]. $\{Mo_{132}\}$ is highly soluble in water, in which case, only discrete $\{Mo_{132}\}$ macroions are observed in water due to strong electrostatic repulsion based on −42 charges on their surfaces. However, when $\{Mo_{132}\}$ are dissolved in acetone/water mixed solvents containing 3–70 vol% acetone, blackberries can be found. Moreover, the blackberry sizes increase from 45 to 100 nm with increasing acetone content, while no blackberries in pure acetone, as shown in Fig. 3.3.

When dissolved in an acetone-rich solvent, the strong counterion association balances the charges on $\{Mo_{132}\}$ clusters and makes them almost neutral. In such a case, they existed as single clusters. When increasing water content, the strength of van der Waals attractive forces remain unchanged since the size of the cluster does not vary; in the meanwhile, the clusters are expected to carry more charges and enlarge

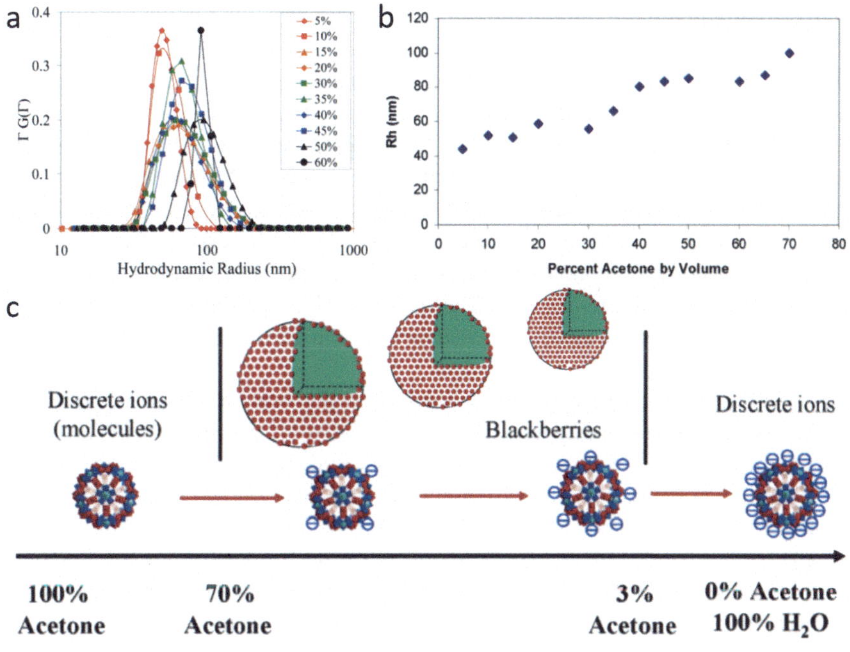

Fig. 3.3 **a** CONTIN analysis of DLS study on $\{Mo_{132}\}$ aqueous solutions of 5–60 vol % acetone. **b** Average R_h of the $\{Mo_{132}\}$ blackberries in water/acetone mixed solvents. **c** Schematic plot showing transition from discrete macroions to blackberries, then to discrete macroions due to the change of solvent content. [18] Reprinted with permission from Ref.. Copyright 2007 American Chemical Society

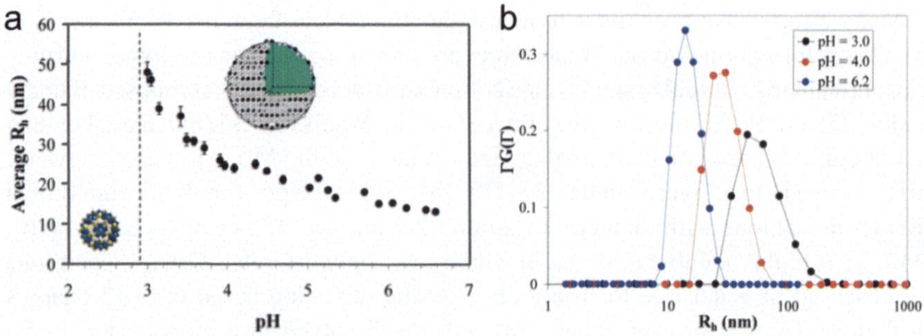

Fig. 3.4 **a** Average R_h of blackberries formed in 0.5 mg/mL aqueous solutions of $\{Mo_{72}Fe_{30}\}$ at different pH. **b** CONTIN analysis of DLS measurements on $\{Mo_{72}Fe_{30}\}$ solution at different pH. Reprinted with permission from Ref. [19]. Copyright © 2006, American Chemical Society

the repulsive forces between macroions. However, blackberry formation is achieved at high water content. In addition, at high acetone content van der Waals forces should be dominant resulting in obvious aggregation since the electrostatic repulsive forces are weakened. These facts rule out the possibility of van der Waals forces as the major driving force for blackberry assembly and indicate the self-assembly process is charge regulated.

The study of $\{Mo_{72}Fe_{30}\}$ blackberry structures further confirms the inapplicability of van der Waals forces for blackberry formation [19]. The charge density of $\{Mo_{72}Fe_{30}\}$, a weak acid-type POM with 30 potential deprotonation sites [27], can be controlled by changing solution pH. At low pH, $\{Mo_{72}Fe_{30}\}$ stays as discrete clusters due to its low charge density. With increasing pH, $\{Mo_{72}Fe_{30}\}$ repels each other strongly based on the larger number of charges carried due to enhanced deprotonation. However, when pH of solution reaches 3.0, $\{Mo_{72}Fe_{30}\}$ starts to attract each other and form blackberry structures. Moreover, the R_h of blackberry structures decreases linearly with increasing pH, as shown in Fig. 3.4. These evidence suggest the importance of charge effect in blackberry formation rather than van der Waals forces.

For POMs with many strongly associated water and oxo-ligands, hydrogen bonding is expected to be important in their self-assembly process. Two POMs with same shape and size, $\{Mo_{72}Fe_{30}\}$ and $\{Mo_{72}Cr_{30}\}$, are tested for blackberry formation. Blackberry structures with different sizes are observed even when the charge density of $\{Mo_{72}Fe_{30}\}$ and $\{Mo_{72}Cr_{30}\}$ are controlled to be the same by tuning solution pH (Fig. 3.5) [24]. This can be due to the water bridged hydrogen bonding. But these clusters exist as single macroions in solution at low pH, proving that the hydrogen bonding can play roles in blackberry formation process, but it's not a primary driving force [24].

As shown above, the fact that no blackberry structures are observed when the charge density of POM clusters are too high or too low indicates that electrostatic interaction plays a critical role in the blackberry formation process. When POMs carry very large amount of charges, the counterion-mediated attraction is not strong

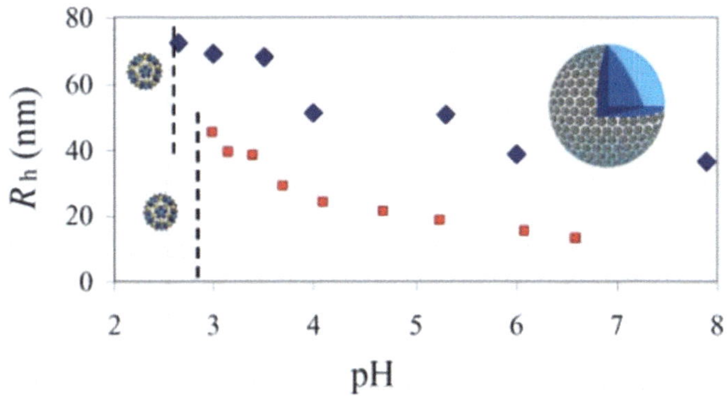

Fig. 3.5 Average R_h of the blackberry structures formed in 0.5 mg/mL aqueous solutions of $\{Mo_{72}Cr_{30}\}$ (blue diamonds) and $\{Mo_{72}Fe_{30}\}$ (red squares) under different pH conditions showing a decrease of R_h with increasing degree of deprotonation. No self-assembly occurs under very low pH conditions. Reprinted with permission from Ref. [24]. Copyright © 2009 Royal Society of Chemistry

enough to overcome the repulsive forces among POMs, while POMs with very low charge density are not able to form ion-pairs to gain enough attractive forces [3]. Only when proper amount of charges are carried by POMs, blackberry structures can be observed.

Similar cases among counterions and polyelectrolyte or biomacromolecular solutions have been reported [28]. Sogami and Ise first proposed the weak counterion-mediated attraction between like-charged colloids [29], nevertheless, this weak interaction is not treated as a significant force due to large size disparity between colloids and their counterions. For macroions, which are smaller than colloids, they can strongly associate with their counterions, forming ion-pairs, in which case, the counterion-mediated attraction is expected to be remarkable for blackberry formation.

3.3.3 Macroion-Counterion Interaction

To fully understand the role of counterion-mediated attraction in self-assembly, the interaction between macroions and counterions need to be addressed which can be explored by small angle X-ray scattering (SAXS) and anomalous small angle X-ray scattering (ASAXS). Based on the giant size and well-defined structures of POMs, they are ideal models for such study [30–32]. Either association among POMs and counterions or the self-assembly of POMs can be accurately detected from SAXS by calculating the R_g value and the distance pair distribution functions [30, 31]. While ASAXS is a powerful technique to directly measure the counterion distribution[33–34].

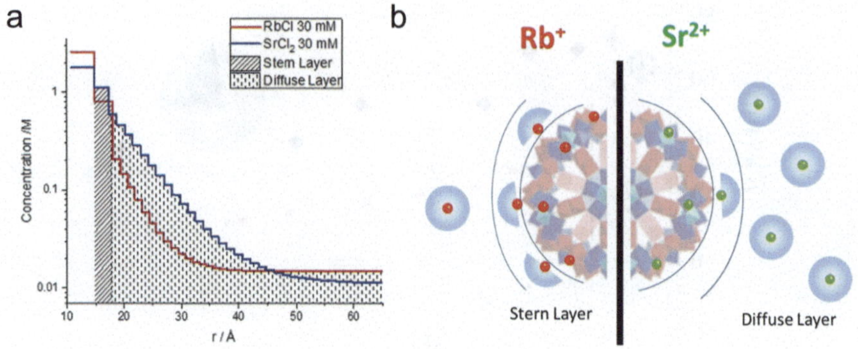

Fig. 3.6 **a** Radial concentration profiles of 30 mM Rb^+ and Sr^{2+} in $\{Mo_{132}\}$ solution obtained from ASAXS fitting using multi-layered spherical shell model. r: distance of the counterion to $\{Mo_{132}\}$ center. **b** Graphical representation of Rb^+ (red) and Sr^{2+} (green) hydration (semi-transparent bubble) and distribution around $\{Mo_{132}\}$. Most of Rb^+ ions closely associate with $\{Mo_{132}\}$, while a large portion of Sr^{2+} ions only loosely associate with $\{Mo_{132}\}$. Reprinted with permission from Ref. [32]. Copyright © 2020 Wiley-VCH GmbH

3.3.3.1 Counterion Distribution Around Macroions

For dilute $\{Mo_{72}V_{30}\}$ aqueous solution, the R_g was calculated (from SAXS curve) to be 10.8 Å suggesting the single molecule state of $\{Mo_{72}V_{30}\}$ macroions in solution [30]. Moreover, the distance pair distribution p(r), which is the probability of finding the vector length r in a molecule, showed a peak correspond to single $\{Mo_{72}V_{30}\}$ macroions while a new peak appeared when acetone content was increased, which indicated that the $\{Mo_{72}V_{30}\}$ clusters still stay as discrete ions while the counterion association is enhanced [30]. In the meantime, the peak appearance was consistent with the appearance of the blackberry structure in solution, suggesting the strong connection between counterion association and blackberry formation. In 2020, more accurate distribution of counterions around $\{Mo_{132}\}$ was directly measured using ASAXS [32]. As shown in Fig. 3.6, Rb^+ ions tend to stay near the skeleton of $\{Mo_{132}\}$ or in the Stern layer, showing closely associated with $\{Mo_{132}\}$; while Sr^{2+} ions prefer a looser association with $\{Mo_{132}\}$ by staying in the diffuse layer. The stronger interaction between Rb^+ ions and $\{Mo_{132}\}$ than that of Sr^{2+} ions explains the anomalous lower critical coagulation concentration of $\{Mo_{132}\}$ with Rb^+ compared to Sr^{2+}.

3.3.3.2 Counterion Exchange Around Macroions

With additional counterions presented in POM solution, different counterions can be distinguished and selected by macroions based on their hydrated sizes and valence. Counterions with smaller hydrated size and higher valence are preferred by macroions and the exchange of counterions associated by macroions can be achieved. The

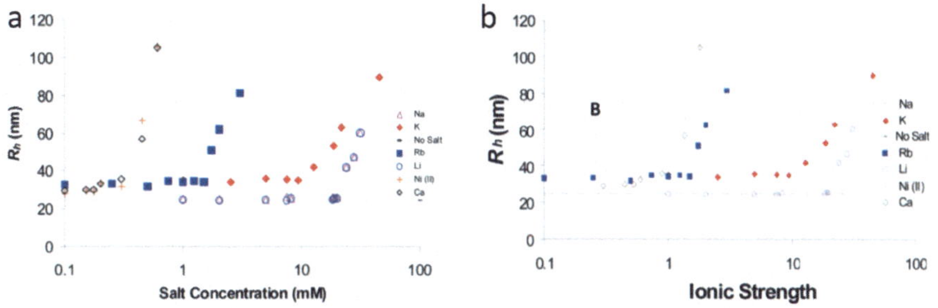

Fig. 3.7 Change of blackberry size (R_h) with **a** added chloride salt concentration and **b** total ionic strength for 0.5 mg/mL $\{Mo_{72}Fe_{30}\}$ solutions. Reprinted with permission from Ref. [31]. Copyright © 2010 American Chemical Society

change of blackberry size with different counterions added into POM solution was observed (Fig. 3.7) [31]. $\{Mo_{72}Fe_{30}\}$, with protons as original counterions, can form blackberry structures in aqueous solution, whose size remains unchanged with addition of 1–20 mM LiCl or NaCl. However, the larger blackberries were observed when adding 0.1–10 mM KCl or RbCl. The enlarged blackberry structures result from the replacement of original protons by K^+ or Rb^+, which decreased the surface charge density of $\{Mo_{72}Fe_{30}\}$ and increased the attraction forces between POMs. Highly hydrated ions, Li^+ and Na^+, cannot replace the protons which would lead to no obvious change of blackberry sizes.

This explanation was further confirmed by isothermal titration calorimetry (ITC) studies. No measurable binding of Na^+ to $\{Mo_{72}Fe_{30}\}$ can be detected from ITC. While the ITC results demonstrated that the binding between $\{Mo_{72}Fe_{30}\}$ and K^+ or Rb^+ are present, additionally, the Rb^+ can bind to $\{Mo_{72}Fe_{30}\}$ much stronger than K^+. From ITC study, the binding strength between $\{Mo_{72}Fe_{30}\}$ and counterions follows an order of (Li^+, Na^+) < H_3O^+ < K^+ < Rb^+ < Cs^+, which is fully consistent with the blackberry size changes with different counterions. All these connection between counterion-macroion interaction and blackberry formation prove the importance of counterion-mediated attraction in constructing blackberry structures.

3.3.4 *Features and Applications of Blackberry Structures*

The blackberry structure has a lot of fascinating features, which are broadly explored in the past decades, such as the special kinetic properties, cation transportation through blackberry membrane, self-recognition behaviors, chiral recognition, and so on. Due to these features, the applications of the blackberry structure are also well-studied.

3.3.4.1 The Kinetic Properties and Lag Phase of Blackberry Formation

Blackberries are thermodynamically stable [35–37], which distinguishes them from colloidal particles. Comparing with fast self-assembly of amphiphilic molecules, the blackberry formation processes take a long period of time (even several months) [38, 39] which enables the detailed exploration of the mechanism of the self-assembly. As shown in Fig. 3.8 [39], the macroions first associate into oligomers, which is the rate-determining step. Once there are enough amount of oligomers, they quickly assemble into blackberries [38, 39].

With presence of no or small amount of extra salts (NaCl, NaBr, NaI, and Na_2SO_4), the time-resolved intensity of $\{Mo_{72}Fe_{30}\}$ blackberry formation process showed a linear relationship with time, which is similar to a first order reaction [38]. In addition, there is a lag phase observed at the beginning of the assembly. As shown in Fig. 3.9,

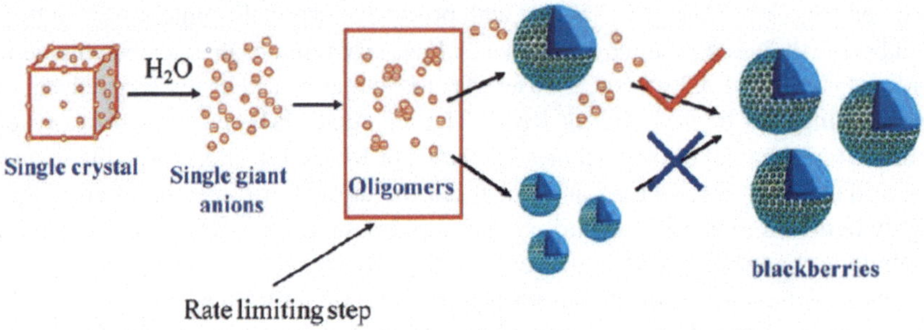

Fig. 3.8 Possible mechanisms of $\{Mo_{72}Fe_{30}\}$ blackberry formation in dilute aqueous solution. Reprinted with permission from Ref. [39]. Copyright © 2006, American Chemical Society

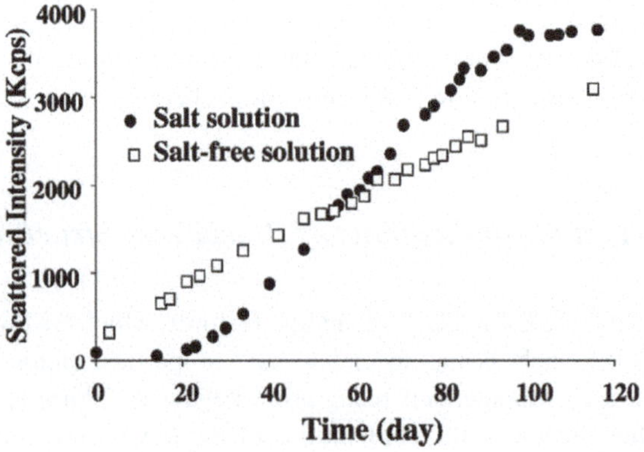

Fig. 3.9 Comparison of scattered intensity increment of $\{Mo_{72}Fe_{30}\}$ solution with or without NaCl. Reprinted with permission from Ref. [40]. Copyright © 2009, American Chemical Society

the lag phase became significant when introducing larger amount of salt. The overall intensity growth is shown as a sigmoidal curve [40].

The formation process (sigmoidal curve) and assembly structure (single-layer, hollow, spherical structures) of blackberries are very similar to the virus capsid formation [41], indicating that they may share similar mechanisms [3, 7]. Although the hydrophobic effect is believed to be the driving force for virus capsid formation, it definitely does not play a role in blackberry formation as there is no any hydrophobic moiety for POMs. Whether the electrostatic interaction plays a role in virus shell formation is still a mystery. POMs are promising to be a simplified model to explore the solution behaviors of complicated biomacromolecular systems [3, 7].

3.3.4.2 Cation Transportation Through Blackberry Membrane

With the same type of charge carried by macroions, they do not touch each other on the surface of blackberries. This contributes to a soft and "porous" membrane which can serve as channel for simple ions to pass through. The fluorophores, which are specifically sensitive to one or two types of ions, e.g., chlorotetracycline (CTC) for Ca^{2+} and Mg^{2+}; 6-methoxyquinoline (6-MQ) for Cl^- and Coumarin 1 for Br^-, were used to track the transportation of small ions through the membrane of $\{Mo_{72}Fe_{30}\}$ blackberries [42]. The fluorophores were partially captured inside the blackberries, which was confirmed by an 18-nm shift of the fluorescent signals. After completing the formation of blackberry, specific ions which are sensitive to the fluorophores were added into the solution. As shown in Fig. 3.10, the immediate intensity increment with addition of Ca^{2+} and Mg^{2+} suggested that the CTCs outside blackberries rapidly saturated after dominant amounts of Ca^{2+}/Mg^{2+}. In addition, there was a continuous and slow increase of fluorescence signal after initial process, which indicated a slow

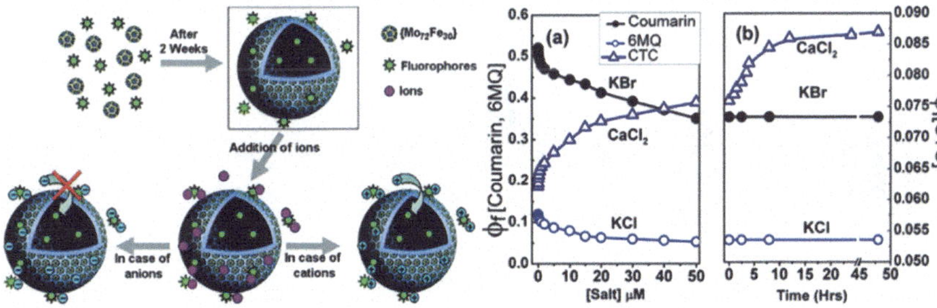

Fig. 3.10 (Left) Formation of fluorophore-containing $\{Mo_{72}Fe_{30}\}$ blackberries in solution. The additional cations first interact with fluorophores in bulk solution and on blackberry surfaces, subsequently enter into the blackberries, and interact with the fluorophores inside. While the anions could not pass through the membrane. (Right) Change in fluorescence quantum yield of Coumarin 1, 6-MQ, and CTC with addition of KBr, KCl, and $CaCl_2$, respectively; **a** instantaneous change occurs with the addition of salts; **b** change in fluorescence quantum yield with time. Reprinted with permission from Ref. [42]. Copyright 2008 American Chemical Society

and continuous binding between Ca^{2+}/Mg^{2+} and CTCs. This can be interpreted by the transportation of Ca^{2+}/Mg^{2+} through blackberry membrane and the cations can interact with CTC inside blackberry. However, anions (such as Cl^- and Br^-) cannot pass through the membrane, since the highly negatively charged blackberries will repel the anions.

3.3.4.3 Self-recognition Behaviors During Self-assembly

Different from vesicles, which formed by short-range hydrophobic interactions, the blackberries, controlled by long-range electrostatic interactions, have chance to show self-recognition during formation.

The mixed aqueous solutions of two macroions, $\{Mo_{72}Fe_{30}\}$ and $\{Mo_{72}Cr_{30}\}$, were studied and showed the presence of two types of blackberries with different sizes [43]. The sizes of blackberries in mixture are corresponding to those of individual solutions which contains either $\{Mo_{72}Cr_{30}\}$ or $\{Mo_{72}Fe_{30}\}$. Moreover, the self-recognition feature is also confirmed by analyzing the elements of the two assemblies which were separated by filtration from mixed solutions. During the self-assembly, the macroions can self-recognize each other in a mixed solution, forming two homogeneous blackberry structures instead of heterogeneous ones (Fig. 3.11).

Another study of two similar rod-shape clusters $((C_4H_9)_4\ N)_7{}^-[Mo_6O_{18}NC(OCH_2)_3XMo_6O_{18}(OCH_2)_3\ CNMo_6O_{18}]$ (X = Mn^{III} or Fe^{III})) [44] with identical morphology but different central metal atoms also showed self-recognition feature, demonstrating the two types of blackberries in their mixture. The self-recognition behaviors of these macroions are attributed to the different charge distribution of the macroions, as confirmed by DFT calculations. With more positively charged Mn atom and more negative planar molecular unit, the larger charge distribution of Mn-containing macroions influences the interaction

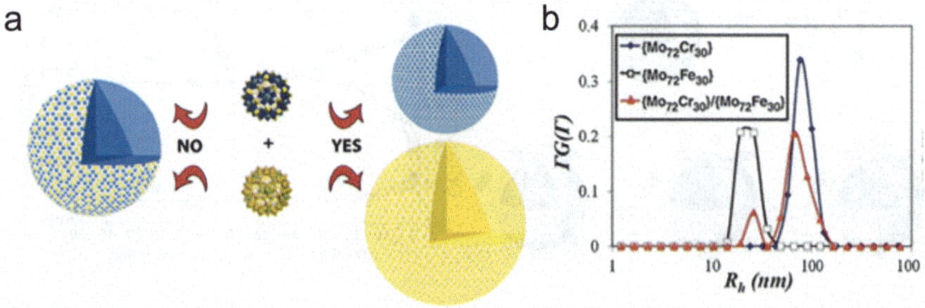

Fig. 3.11 **a** In mixed dilute aqueous solutions, the clusters $\{Mo_{72}Fe_{30}\}$ (top) and $\{Mo_{72}Cr_{30}\}$ (bottom) self-assembled into individual blackberry structures of the Cr_{30} (yellow) and Fe_{30} type (blue) and do not form mixed species. **b** CONTIN analysis of the DLS studies at 90° scattering angle measured for aqueous solutions containing $\{Mo_{72}Cr_{30}\}$ or $\{Mo_{72}Fe_{30}\}$ (0.1 mg/ml each), as well as for a solution containing both species (0.1 mg/ml of each) (pH = 4.2). Reprinted with permission from ref. [43]. Copyright © 2011, American Association for the Advancement of Science

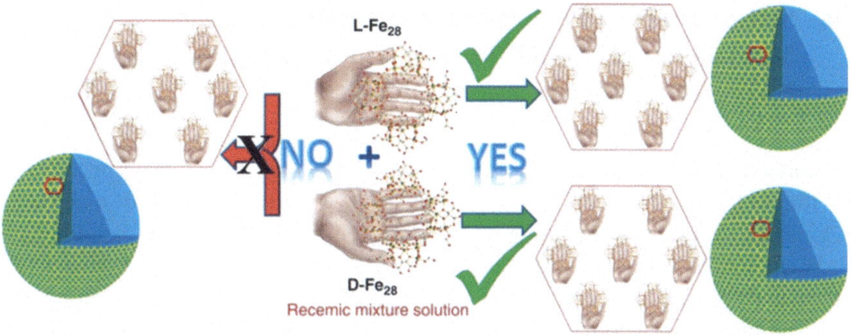

Fig. 3.12 Graphical representation of the chiral recognition behavior. Reprinted with permission from Ref. [45]. Copyright © 2015, Nature Publishing Group, a division of Macmillan Publishers Limited

between macroion and counterions, and consequently their self-assembly behaviors to show self-recognition. Not only difference of charge density, but also charge distribution of macroions will affect their self-assembly behaviors.

3.3.4.4 Chiral Recognition and Chiral Selection

During self-assembly process, the chiral macroions show enantioselective behavior toward chiral small organic molecules [45–47]. Two enantiomeric wheel-shaped macroions, $[Fe_{28}(\mu_3\text{-}O)_8(Tart)_{16}(HCOO)_{24}]^{20-}$ (Tart = D- or L-tartaric acid tetra-anion) were studied, showing the formation of their individual assemblies in racemic mixture solution as shown in the Fig. 3.12 [45]. With addition of chiral co-ions, the self-assembly process of the enantiomeric macroions could be selectively suppressed. The energy barrier for homo-oligomers formation is slightly lower than that of hybrid oligomers during lag phase of self-assembly process, leading to the formation of homogeneous blackberries rather than heterogeneous ones. Besides, the chiral selection is achieved during lag phase as well as based on the compatibility between the chiral microenvironment from the macroions and the external added small chiral co-ions.

The effect of chiral counterions has been studied in systems containing lactic acid-functionalized chiral fullerene [47] (AC_{60}) or chiral metal organic cages [46]. Chiral counterions affect the self-assembly process of chiral macroions, by either interacting differently with chiral macroions or inhibiting the interaction between macroions and original counterions at different level. Obvious chiral discriminations between assemblies from enantiomeric macroions observed in these systems suggest the chiral recognition and selection could be a common feature during self-assembly of macroions with electrostatic interactions as driving forces.

3.4 Self-assembly of Complex or Functionalized Macroions—Competition and Cooperation Among Different Attractive Forces

Since macroions are charged molecules, electrostatic interaction always plays a critical role in their self-assembly process. However, with diverse types of macroions, all kinds of interactions including hydrophobic interaction, hydrogen bonding, van der Waals forces, cation-π attraction and π-π stacking, can be involved during their self-assembly process to construct a wide range of 0D-3D supramolecular structures. How to control shape and size of these assembly structures is of great importance. The competition or cooperation among electrostatic interactions with other interactions and also geometrical restrictions lead to wonderful availability of supramolecular structures, which will be discussed in this section.

3.4.1 Electrostatic Interaction Versus Hydrophobic Interaction

Considering the hydrophilic nature of charged macroions such as POMs, presence of hydrophobic moieties introduces amphiphilic nature to macroions which will influence their self-assembly behaviors [7, 8, 48–52] Different parameters including counterion effect, large charged polar heads, molecular morphology, hydrophobicity, solvent polarity, and concentration should be taken into consideration when studying their solution behaviors, which makes the case more complicated than simple macroions and amphiphilic surfactants.

With small organic ligands grafted on macroions, their self-assembly behavior is still dominated by counterion-mediated attraction as macroions have large and rigid structures [14]. While hydrophobic effect from the organic ligands can contribute but it cannot regulate it. Two similar molecules, Pd_{13} and its hybrid $Pd_{13}(Ph)_8$ (Fig. 3.13) [53, 54], self-assembled into blackberries in aqueous solution. They share similar chemical structures and same counterions, while $Pd_{13}(Ph)_8$ contains eight phenyl groups on the surface and carries slightly less charges than that of Pd_{13}^{8-}. Pd_{13} shows a slow self-assembly process (>40 days), however, $Pd_{13}(Ph)_8$ can assemble rapidly which takes only several hours. With less charges, the $Pd_{13}(Ph)_8^{6-}$ exhibits weaker counterion-mediated attraction, in which case, the faster assembly is the result of another attractive force, that is, hydrophobic interaction from phenyl groups on its surface. Moreover, as shown in Fig. 3.13, larger $Pd_{13}(Ph)_8$ blackberries are formed in more polar solvents in water/DMSO mixture, which is caused by the competition of counterion-mediated attraction and hydrophobic effect. While Pd_{13} blackberries become smaller in more polar solvents which is a common behavior of blackberry formation solely by counterion-mediated attraction. In $Pd_{13}(Ph)_8$ self-assembly process, as solvent polarity increases, the counterion-mediated attraction

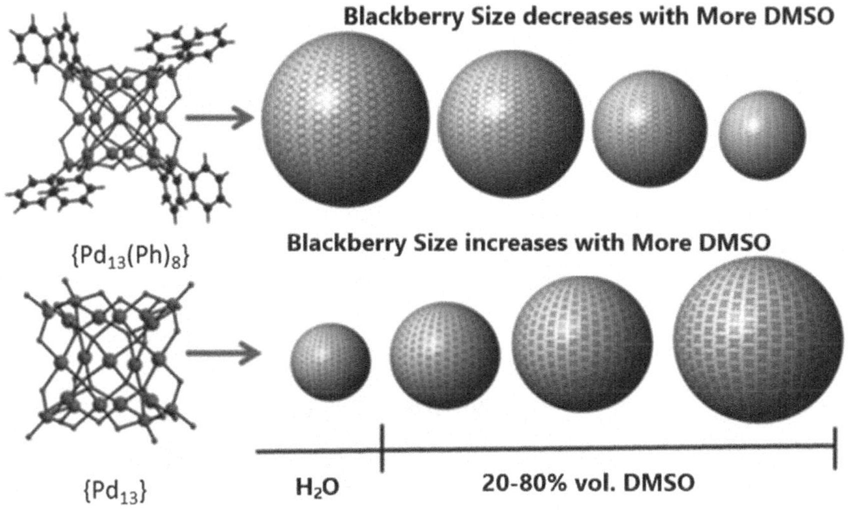

Fig. 3.13 A model depicting the blackberry formation process in solutions of $\{Pd_{13}(Ph)_8\}$ and $\{Pd_{13}\}$ in water and DMSO/water mixed solvents. Reprinted with permission from Ref. [53]. Copyright © 2015 WILEY-VCH Verlag GmbH & Co. KGaA, Weinheim

is weaker, while the hydrophobic effect becomes much stronger. The overall attractive interaction is stronger in more polar solvent due to the enhanced hydrophobic interaction, and consequently lead to larger blackberries.

When large hydrophobic domains are introduced to macroions, the amphiphilic nature of these hybrids affect their self-assembly behaviors, making them act more like a surfactant since the hydrophobic effect becomes more critical. Our group first studied the solution behavior of a C_{16}-POM-C_{16} amphiphile (Fig. 3.14a) [8]. The hydrophobic-hydrophilic-hydrophobic structures formed double-layered vesicles in acetonitrile/water mixed solvents with hydrophobic interaction acting as the major driving force. The two flexible hydrophobic alkyl tails were packed on one side and assembled into vesicles. A similar hybrid with shorter alkyl chains, C_6-POM-C_6, arranged themselves into reverse vesicles in non-polar solvent [52]. In this case, the hydrophobic tails tended to face outside with the hydrophilic POM heads staying inside the vesicular structures to avoid contacting with the non-polar solvent. When one alkyl chain (here is C_{15}) is grafted on POM, forming hydrophilic-hydrophobic structures, the POM-C_{15} exhibited relative fast vesicle formation as no bending of tails needed [55]. The length and number of the alkyl chains influence the solvophobic interaction, thus controlling the self-assembly process and final assembled structures.

The hydrophobic moieties could be even larger by using long polymer chains, and this will significantly decrease solubility of corresponding hybrid in water. Solution behavior of a POM-polymer hybrid as shown in Fig. 3.14b was studied in non-polar solvent [10]. Therein, reverse vesicles were formed driving by solvophobic interaction, as POMs act as solvophobic parts, while the polymer chains are solvophilic.

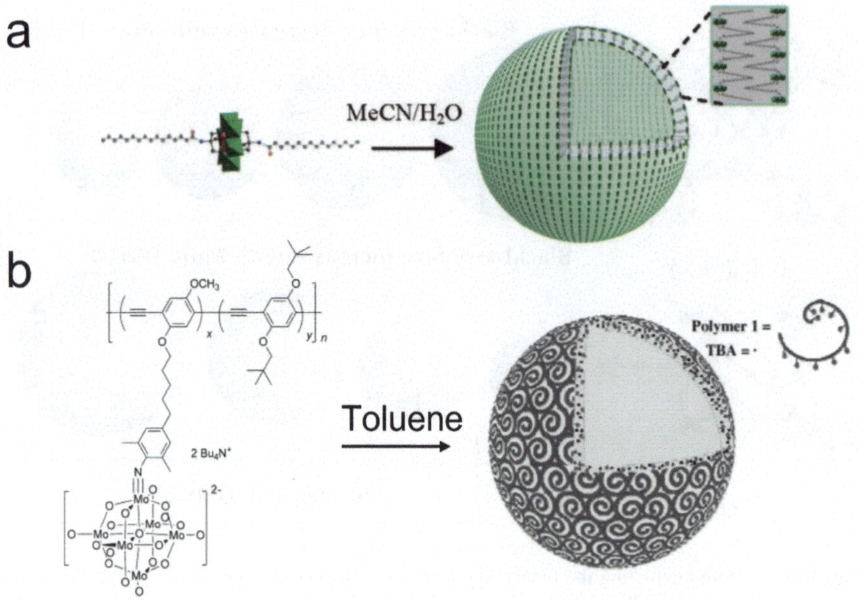

Fig. 3.14 Scheme of vesicle formation of **a** C_{16}-POM-C_{16} hybrid and **b** polymer hybrids. [8][10] Adapted with permission from Refs. and. Copyright 2008 American Chemical Society and 2012 WILEY-VCH

Polymer chains formed circles around POMs and kept them in between to avoid contact with the organic solvent.

3.4.2 Electrostatic Interaction Versus Hydrogen Bonding

Hydrogen bonding is a special type of dipole-dipole interaction, usually resulting from interaction between H and electronegative atoms such as N, O, or F. With the oxo and water ligands of $\{Mn_{40}P_{32}W_{224}\}$, which consists of four Dawson trimers connected to one P_8W_{48} wheel, hydrogen bonding can have an effect on supramolecular structure formation [11]. As shown in Fig. 3.15a, rod-like structures were formed in acetone/water mixed solvent due to the directional growth based on anisotropic charge distribution of the POM. However, in methanol/water mixtures, hydrogen bonding effect become stronger. The isotropic hydrogen bonding formed by POMs and solvent weakens the orientation effect caused by the counterion-mediated attraction, thus, making it behave like regular POMs and forming blackberry structures. When dissolved in deuterated solvent, rod-like structures were observed as in acetone/water mixture, which confirmed the significance of hydrogen bonding in macroionic self-assembly.

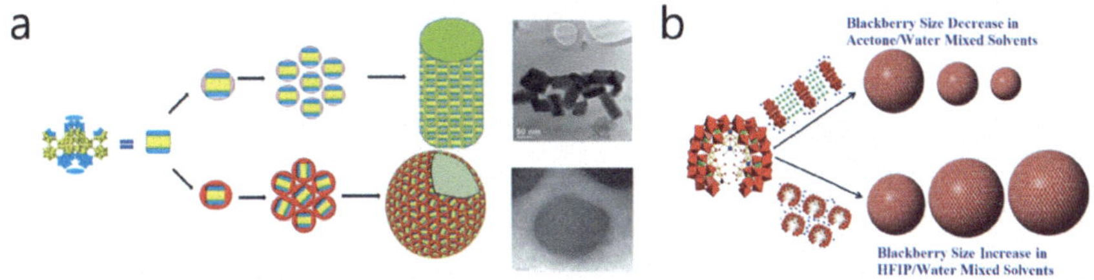

Fig. 3.15 **a** Scheme of anisotropic and isotropic self-assembly of $\{Mn_{40}P_{32}W_{224}\}$. **b** Graphic showing blackberry size change of hydrogen bonding functionalized POM in acetone/water and HFIP/water mixed solvent. [11][56] Adapted with permission from Refs. and. Copyright 2013 Royal Society of Chemistry and 2016 WILEY-VCH

Presence of hydroxo and aqua moieties in the core of POM framework led to cooperation of hydrogen bonding and electrostatic interaction during the self-assembly process of POM, as shown in Fig. 3.15b [56]. Relatively rapid formation of blackberry structures can be achieved compared to typical ones, which suggests the importance of hydrogen bonding. In addition, stronger hydrogen bonding and weaker counterion-mediated attraction in more polar solvent lead to larger blackberry structures. When introducing strong hydrogen bond-donor solvent, such as hexafluoroisopropanol (HFIP), to the solution, the opposite trend can be observed as dominated by the counterion-mediated attraction in HFIP/ water mixed solvents as a result of the cutting hydrogen bonds among hybrid molecules.

In addition to hydrogen bonding acceptors and donors in macroionic structure, presence of different numbers of hydrogen bonding sites in the structure of counterions can significantly affect the supramolecular structure formation process of macroions. In the self-assembly of $Na_9[EuW_{10}O_{36}]\cdot 32H_2O$ (EuW_{10}) POM, different alkylamines with different numbers of amine groups were used as counterions [13]. Alkylamines play an important role as a "glue" and bind primary assembled nanospheres to form nanopetals and finally obtain nanoflowers. Alkylamines with four and five amine groups (TETA and TEPA in Fig. 3.16) were able to act as binding agents and form secondary nanostructures; however, DETA with three amine groups only led to the formation of nanospheres and no growth toward nanopetals and nanoflowers was observed. Here, amines play a dual role both as electrostatic interaction sites (in case of protonation) and hydrogen bonding sites. Therefore, higher number of amine groups in alkylamines contributes to obtain nanopetals and nanoflowers. This indicates that the number of hydrogen bonding sites of counterions plays a critical role in the self-assembly process as well.

3.4.3 Electrostatic Interaction Versus Cation-π Interaction

Along with the interactions mentioned above, cation-π interaction is another important interaction, which becomes more significant when aromatic functional groups

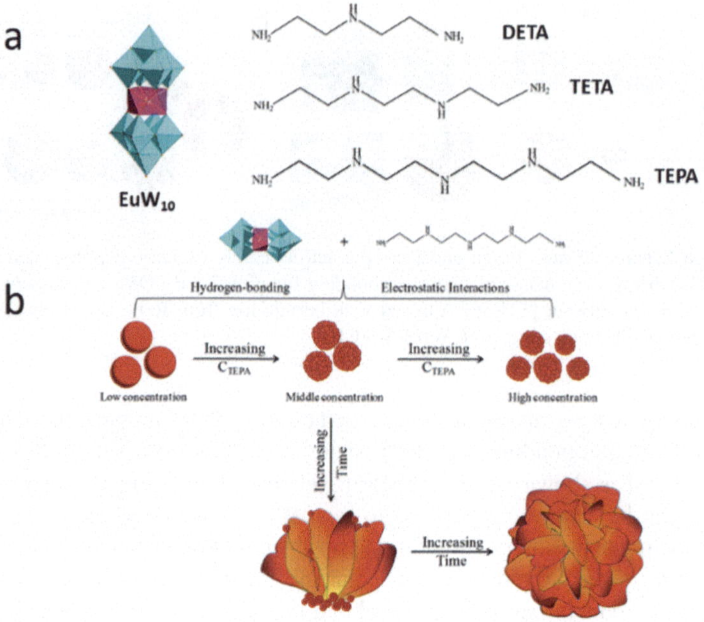

Fig. 3.16 **a** Structure of EuW_{10} POM and alkylamine counterions diethylenetriamine (DETA), tetraethylenetetramine (TETA), and tetraethylenepentamine (TEPA). **b** Self-assembly of POMs into nanospheres, nanopetals, and finally nanoflowers. [13] Reprinted with permission from Ref.. Copyright 2017 American Chemical Society

are attached to macroions[57]. For such aromatic functionalized macro-anions, the counter-cations will interact both with the macro-anions due to electrostatic interaction and aromatic groups through cation-π interaction. The competition between them will influence the self-assembly process and regulate the supramolecular structures obtained. A series of phenyl group functionalized rod-shape POM-based hybrids have been synthesized to explore the importance of cation-π interaction [58]. The POMs (with or without phenyl groups) formed blackberries at high cation concentrations driven by counterion-mediated attraction. However, the size trends in assemblies for these POMs (with or without phenyl groups) are opposite: smaller size at higher cation concentrations for the phenyl group modified POMs while larger size at higher cation concentrations for the regular POMs (Fig. 3.17). The phenyl groups can draw cations close to the hybrids and significantly enhance the cation-hybrid binding through cation-π interaction, changing the cation distribution around the hybrids and leading to the opposite trend toward salt concentration. What's more, self-recognition between two similar phenyl groups decorated POMs could be achieved due to cation−π interaction (Fig. 3.17).

Not only in anionic macroionic systems, cation-π interaction also plays a role in positively charged macroionic solutions. Strong cation-π interaction was noticed in macrocycle solutions [59]. Although the phenyl-ligands-linked macrocycles carry

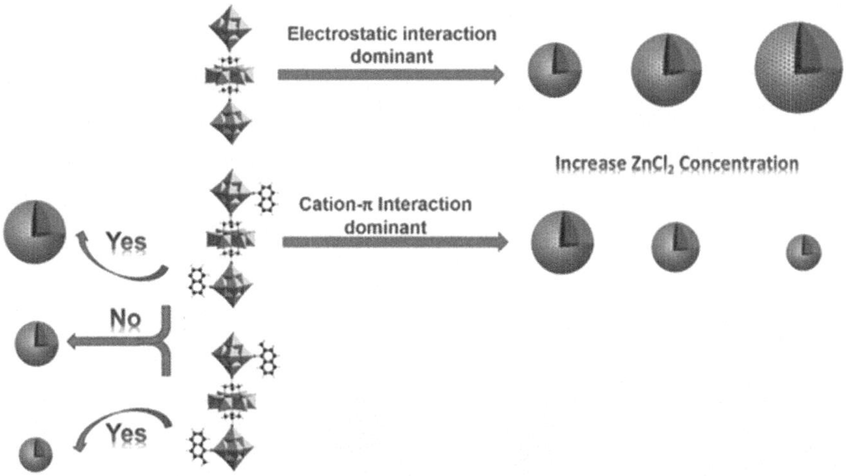

Fig. 3.17 A model showing the effect of cation-π interaction on the self-assembly and self-recognition processes of different macro-anions with and without π-rich groups. [58] Reprinted from Ref.. Copyright © 2018 Wiley-VCH

positive charges, their self-assembly process still can be regulated by co-ions via cation-π interaction.

3.4.4 Electrostatic Interaction vs. Van Der Waals Forces

In order to show the importance of secondary interactions alongside with electrostatic interactions, a steroid-Anderson POM-steroid hybrid can be considered in which cholic acid (POM-Ca), dehydrocholic acid (POM-Dhca), and cholesterol (POM-Chol) are three different steroids used (Fig. 3.18a) [6]. In this system, electrostatic interaction between POMs and van der Waals interactions between steroids are cooperating to trigger the self-assembly of supramolecular structures. Depending on the solvent composition, hybrids based on cholesterol can form various well-defined assemblies due to the rigid and planar skeleton of cholesterol which strengthens the van der Waals interactions (Fig. 3.18b). However, when it comes to the hybrids based on the other two steroids, rigid and twisty structure of the steroids reduces van der Waals interactions and no stable assemblies are observed for such systems. This indicates the importance of a secondary interaction, such as van der Waals, on the self-assembly process of macroions acting synergistically with electrostatic interactions.

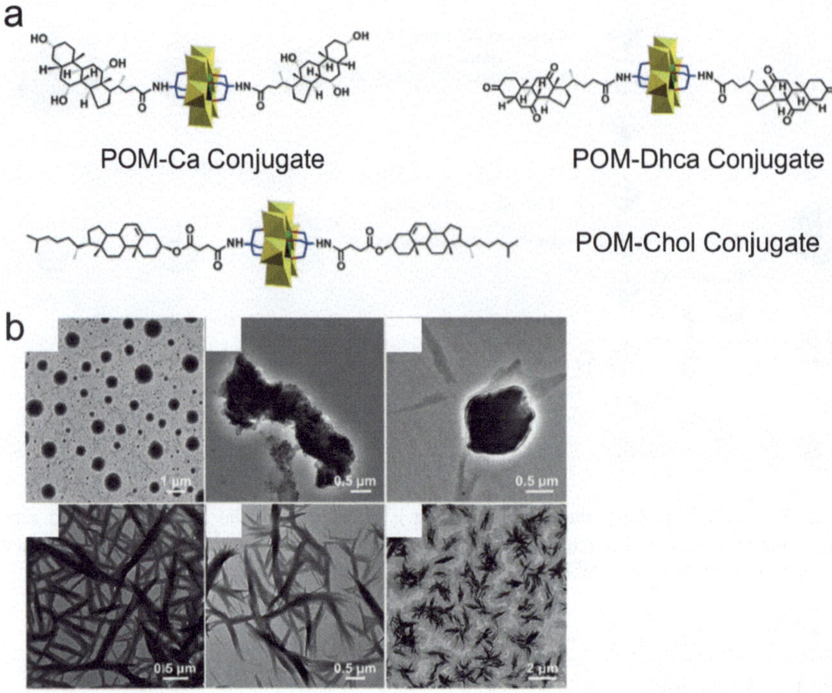

Fig. 3.18 **a** Structure of three different steroid-POM-steroid hybrids. **b** Supramolecular structures based on POM-Chol in different solvent compositions. [6] Reprinted with permission from Ref.. Copyright 2016 Royal Society of Chemistry

3.4.5 Multiple Interactions

With complex structure of macroions, multiple interactions are usually involved at the same time. Three different MOCs based on TPE ligands, dicarboxylate ligands, and $Pt(PEt_3)_2$ metal ions (MOCs 1, 2, and 3 in Fig. 3.19a) were synthesized to study their self-assembly process into 2D nanosheets with the addition of different salts via counterion-mediated attraction.[4] It was shown that presence of longer dicarboxylic ligands with a length of 1.52 Å (MOC 1) provides more chance for $\sigma - \pi$ interaction between metal ions and dicarboxylic ligands stabilizing 2D nanosheets. MOCs with shorter dicarboxylic ligands (MOC 2) and MOC 3 only formed precipitates indicating the importance of $\sigma - \pi$ interaction between the edges and corners is significant for stabilizing the nanosheets. Therefore, any minor or major variation in the structure of macroions such as changing the flexibility of macroions or anchored ligands to macroions, and changing size and geometry of macroion's skeleton, can impose geometrical restrictions and/or induce secondary intermolecular interactions leading to variations in shape and size of the supramolecular structures formed based on electrostatic interactions.

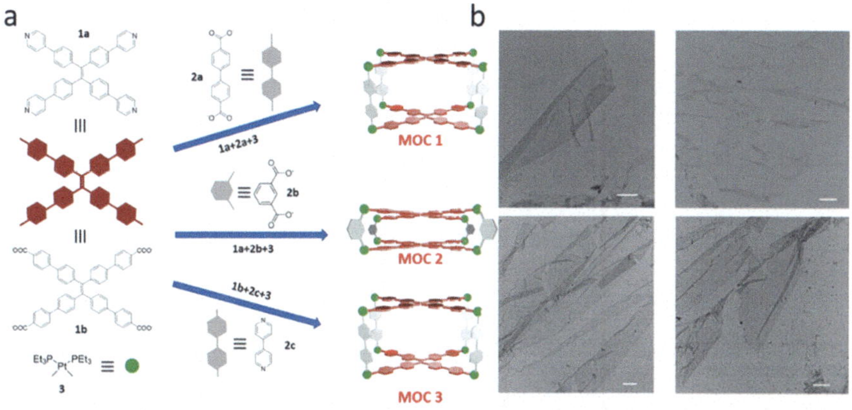

Fig. 3.19 **a** Structure of MOCs used to self-assemble into 2D nanosheets. **b** TEM images of nanosheets assembled by MOC 1. [4] Reprinted with permission from Ref.. Copyright 2020 American Chemical Society

Another example of self-assembly process with multi-interaction cooperated is based on a long-alkyl-chain-functionalized POM hybrid in aqueous solution [60]. The hexavanadate-cluster-based surfactant quickly formed micelles via hydrophobic effect which induced by the long alkyl chains attached on POM. With addition of less hydrated ions (K^+ and NH_4^+), the binding of cations to the anionic micellar surface, decreasing the stability of the micelles, leads to the further coagulation to nanobelts. The electrostatic interaction, cooperated with directional hydrogen bonding between the polar head groups, leads to these 1D anisotropic assemblies.

3.4.6 Electrostatic Interaction vs. Geometrical Restrictions

Geometrical restrictions and features of macroions and counterions can interplay with the strength of electrostatic interaction, regulate the self-assembly process, and lead to totally different supramolecular structures. One example is the self-assembly of different generations of poly (amido amine) (PAMAM) dendrimers (G2-G8), ranging from isotropic spheres to elliptical cylinders, using organic dye counterions with different valency [61]. Higher generations of poly (amido amine) (PAMAM) dendrimers have a larger size and consequently less flexibility. While it is possible to obtain isotropic assemblies for G4, only anisotropic ones were seen for G8. Besides, at a constant ratio of azo dye (Acid Red 26) to dendrimer (1/8), higher generations of dendrimers lead to assemblies with more anisotropy, *e.g.,* ellipsoidal structures are obtained for G2, elliptical cylinders for G4 and G5, and flexible elliptical cylinders for G7 and G8 (Fig. 3.20). Moreover, it was observed that trivalent organic counterions form isotropic assemblies; while, divalent ones can only form anisotropic structures (Fig. 3.20). Trivalent dyes bind to a greater extent to protonated amine

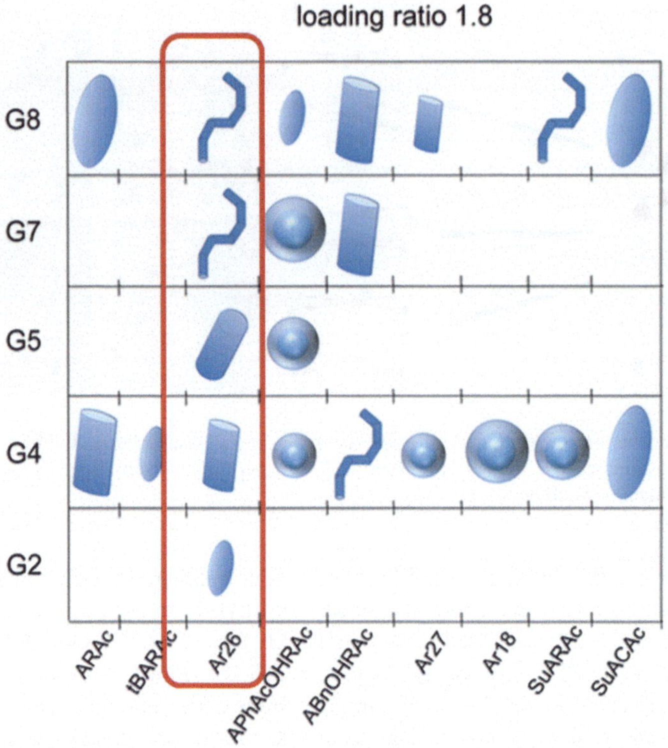

Fig. 3.20 Representation of assemblies obtained by different generations of PAMAM dendrimers and different azo dyes. Red rectangle indicates supramolecular structures obtained by Ar26 azo dye and G2-G8 PAMAM dendrimers [61] Reprinted with permission from Ref.. Copyright 2016 American Chemical Society

groups of dendrimers due to their additional charge and this restricts the available geometric conformations for the assemblies [62]. In addition, the change of molar ratio between dendrimers and counterions also affect the self-assembly structures [61, 63]. Therefore, increasing size, decreasing flexibility of macroions, and the extent of binding between macroion and counterion imposes geometrical restrictions during the self-assembly process and leads to different nano-structures.

Another example is the self-assembly of PAMAM dendrimer and organic counterions that demonstrated different supramolecular structures when slightly changing the position of carboxylic acid groups from 1,4 to 2,3 in divalent counterions (naphthalene dicarboxylic acid (NDC)) [12]. As can be seen in Fig. 3.21, 2,3-NDC triggers the self-assembly of solid spheres by G5; while, 1,4-NDC results in cylindrical assemblies. This may result from a delicate interplay of interactions, that is, electrostatic contributions, π–π interactions, and geometric constraints. Thus, a slight change in the structure of organic counterions can impose different macroion-counterion binding and geometrical restrictions during the self-assembly process.

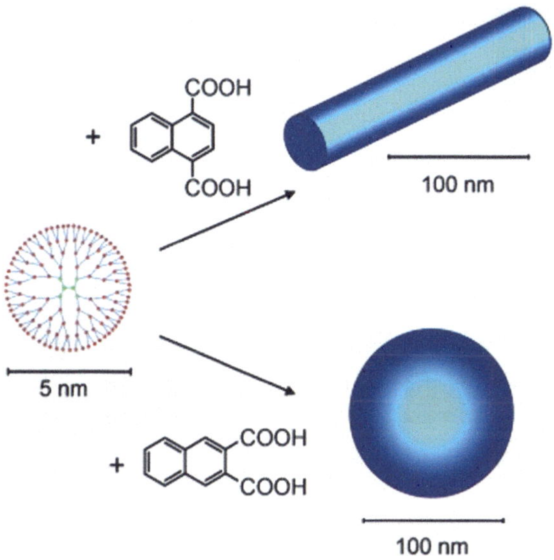

Fig. 3.21 Self-assembly of G5 dendrimer when 1,4- and 2,3-NDC counterions are present. [12] Reprinted with permission from Ref.. Copyright 2008 Wiley–VCH

Moreover, self-assembly of a G4 PAMAM dendrimer at the presence of photoisomerizable diazo dye, Acid Yellow 38, indicated that UV irradiation and consequent isomerization of the dye significantly impacted assembly structures [64]. Azo dye is primarily in a *trans* isomeric state leading to nanospheres with a size of 65 nm. After 20 min UV irradiation dyes were partially converted to *cis* isomers and size of the assemblies was increased to 330 nm (Fig. 3.22). Zeta potential measurements indicated fewer negative charges on the assemblies which is due to the dissociation of dyes after isomerization. Therefore, geometric effects let macroions to accept more dyes and form closer molecular packing in *trans* state leading to much smaller assemblies than in *cis* state.

Flexibility or rigidity is a determining factor not only in macroions, but also in organic counterions, and it can impose different geometrical restrictions and affect the self-assembly process to a great extent. For example, flexible ether linkage-enriched ammonium counterions adopted a cis conformation when binding to Keggin-type POM during its self-assembly and formed 2D nanosheets, replacing the counterion with rigid alkylammoniums leads to the self-assembly of 3D microparticles.[5]

Metal-organic cages (MOCs) and metal-organic macrocycles (MOMs) are another interesting type of macroions in which structural and functional diversity play a critical role in their self-assembly process [65]. Another example of the importance of geometrical features of macroions can be observed in the self-assembly process of a positively charged hexagonal MOM with negatively charged polymer, heparin, into entangled pearl-necklace networks (Fig. 3.23a) [66]. Simulation results indicated that heparin tends to pass through the cavity of MOM due to the strong electrostatic

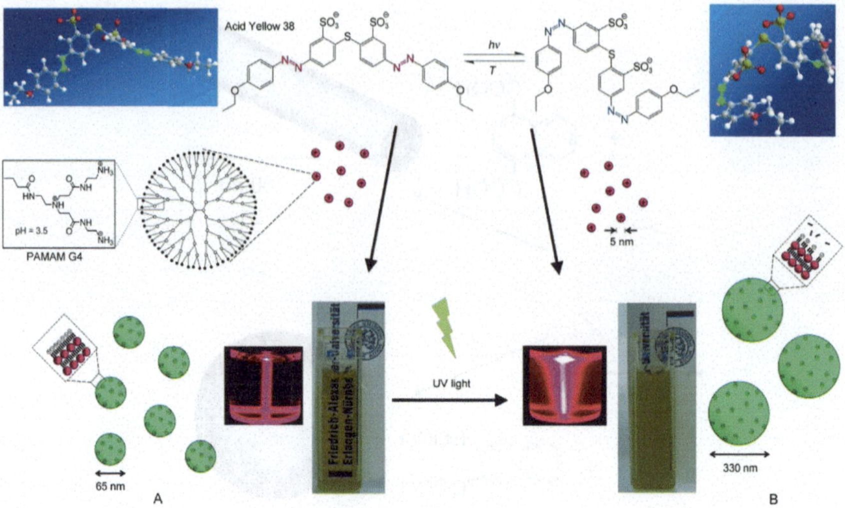

Fig. 3.22 Effect of dye isomerization by UV irradiation on the self-assembly process of G4 PAMAM dendrimers. [64] Reprinted with permission from Ref.. Copyright 2010 Wiley–VCH

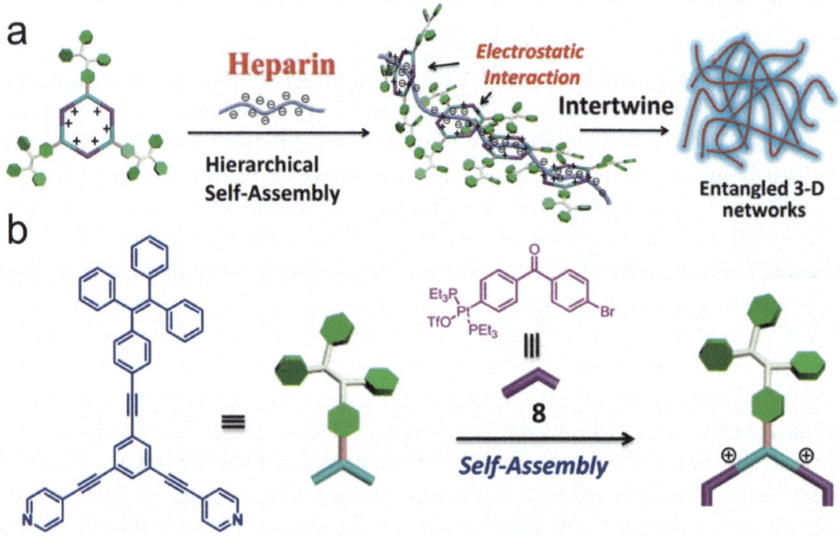

Fig. 3.23 **a** Self-assembly of hexagonal MOM macroion and heparin into entangled pearl-necklace networks. b) Structure of an open-frame organometallic structure which its self-assembly with heparin leads to short-rod structure. [66] Reprinted with permission from Ref.. Copyright 2015 American Chemical Society

interaction and the size of the cavity which is larger than heparin's diameter. In order to prove the importance of skeleton of MOM in the self-assembly process, another positively charged organometallic (Fig. 3.23b) was used with heparin. It was shown that heparin forms short rods with this organometallic due to its open skeleton which cannot confine heparin. Consequently, geometrical shape of macroions can play an extremely important role in the self-assembly process of macroions via electrostatic interactions such as changing the shape and size of the assemblies.

3.5 Conclusion

Contributed from counterion-mediated attraction, the fully hydrophilic macroions come close to each other and form the spherical, hollow, single-layer blackberry structures, which can be tuned by solvent polarity, macroion charges, pH, and counterions. With different functional groups cooperated in macroions, diverse self-assembled structures, ranging from 0 to 3D supramolecular structures, can be observed with competition or cooperation with various interactions, such as hydrophobic interaction, van der Waals forces, hydrogen bonding, and cation-π interactions. Geometrical restrictions can also change macroion-macroion and macroion-counterion binding behavior and lead to different self-assembly behaviors. The study on their self-assembly processes makes it possible for us to gain further understanding on the solution behavior of macroions.

Acknowledgements T.L. acknowledges support from the National Science Foundation (NSF CHE1904397) and The University of Akron.

References

1. Long D-L, Burkholder E, Cronin L (2007) Polyoxometalate clusters, nanostructures and materials: from self assembly to designer materials and devices. Chem Soc Rev 36(1):105–121
2. Webber MJ, Pashuck ET (2021) (Macro)molecular self-assembly for hydrogel drug delivery. Adv Drug Deliv Rev 172:275–295
3. Liu T (2010) Hydrophilic macroionic solutions: what happens when soluble ions reach the size of nanometer scale? Langmuir 26(12):9202–9213
4. Yang Y, Rehak P, Xie T-Z, Feng Y, Sun X, Chen J, Li H, Král P, Liu T (2020) Nanosheets and hydrogels formed by 2 nm metal-organic cages with electrostatic interaction. ACS Appl Mater Inter 12(50):56310–56318
5. Ishiba K, Noguchi T, Iguchi H, Morikawa M-A, Kaneko K, Kimizuka N (2017) Photoresponsive nanosheets of polyoxometalates formed by controlled self-assembly pathways. Angew Chem Int Ed 56(11):2974–2978
6. Yang H-K (2016) Structure- and solvent-triggered influences in the self-assembly of polyoxometalate–steroid conjugates. RSC Adv 6(71):66431–66437
7. Yin P, Li D, Liu T (2012) Solution behaviors and self-assembly of polyoxometalates as models of macroions and amphiphilic polyoxometalate–organic hybrids as novel surfactants. Chem Soc Rev 41(22):7368–7383

8. Zhang J, Song Y-F, Cronin L, Liu T (2008) Self-assembly of organic–inorganic hybrid amphiphilic surfactants with large polyoxometalates as polar head groups. J Am Chem Soc 130(44):14408–14409
9. Zhang J, Song YF, Cronin L, Liu T (2010) Reverse-vesicle formation of organic–inorganic polyoxometalate-containing hybrid surfactants with tunable sizes. Chem A Europ J 16(37):11320–11324
10. Yin P, Jin L, Li D, Cheng P, Vezenov DV, Bitterlich E, Wu X, Peng Z, Liu T (2012) Supramolecular assembly of conjugated polymers containing polyoxometalate terminal side chains in polar and nonpolar solvents. Chem Europ J 18(22):6754–6758
11. Haso F, Fang X, Yin P, Li D, Ross JL, Liu T (2013) The self-assembly of a macroion with anisotropic surface charge density distribution. Chem Commun 49(6):609–611
12. Gröhn F, Klein K, Brand S (2008) Facile route to supramolecular structures: self-assembly of dendrimers and naphthalene dicarboxylic acids. Chem Eur J 14(23):6866–6869
13. Xia C, Wang Z, Sun D, Jiang B, Xin X (2017) Hierarchical nanostructures self-assembled by polyoxometalate and alkylamine for photocatalytic degradation of dye. Langmuir 33(46):13242–13251
14. Luo J, Liu T (2019) Competition and cooperation among different attractive forces in solutions of inorganic-organic hybrids containing macroionic clusters. Langmuir 35(24):7603–7616
15. Debye P, Bewilegua L, Ehrhardt F (1923) Physikalische Zeitschrift. Physikalische Zeitschrift 24(9)
16. Russel WB (1989) Formulation and processing of colloidal dispersions. MRS Online Proc Libr (OPL), 177
17. Verwey EJW (1947) Theory of the stability of lyophobic colloids. J Phys Colloid Chem 51(3), 631–636
18. Kistler ML, Bhatt A, Liu G, Casa D, Liu T (2007) A Complete macroion–"Blackberry" assembly–macroion transition with continuously adjustable assembly sizes in mo132 water/acetone systems. J Am Chem Soc 129(20):6453–6460
19. Liu T, Imber B, Diemann E, Liu G, Cokleski K, Li H, Chen Z, Müller A (2006) Deprotonations and charges of well-defined mo72fe30 nanoacids simply stepwise tuned by pH allow control/variation of related self-assembly processes. J Am Chem Soc 128(49):15914–15920
20. Yin P, Li D, Liu T (2011) Counterion interaction and association in metal-oxide cluster macroanionic solutions and the consequent self-assembly. Isr J Chem 51(2):191–204
21. Müller A, Diemann E, Kuhlmann C, Eimer W, Serain C, Tak T, Knöchel A, Pranzas PK (2001) Hierarchic patterning: architectures beyond 'giant molecular wheels.' ChemComm 19:1928–1929
22. Liu T, Diemann E, Li H, Dress AWM, Müller A (2003) Self-assembly in aqueous solution of wheel-shaped Mo154 oxide clusters into vesicles. Nature 426(6962):59–62
23. Provencher SW (1982) CONTIN: a general purpose constrained regularization program for inverting noisy linear algebraic and integral equations. Comput Phys Commun 27(3):229–242
24. Kistler ML, Liu T, Gouzerh P, Todea AM, Müller A (2009) Molybdenum-oxide based unique polyprotic nanoacids showing different deprotonations and related assembly processes in solution. Dalton Trans 26:5094–5100
25. Israelachvili JN (2011) Intermolecular and surface forces. Academic press
26. Müller A, Krickemeyer E, Bögge H, Schmidtmann M, Peters F (1998) Organizational forms of matter: an inorganic super fullerene and keplerate based on molybdenum oxide. Angew Chem Int Ed 37(24):3359–3363
27. Müller A, Sarkar S, Shah SQN, Bögge H, Schmidtmann M, Sarkar S, Kögerler P, Hauptfleisch B, Trautwein AX, Schünemann V (1999) Archimedean synthesis and magic numbers: "Sizing" giant molybdenum-oxide-based molecular spheres of the keplerate type. Angew Chem Int Ed 38(21):3238–3241
28. Stevens MJ (2001) Simple simulations of DNA condensation. Biophys J 80(1):130–139
29. Sogami I, Ise N (1984) On the electrostatic interaction in macroionic solutions. J Chem Phys 81(12):6320–6332

30. Pigga JM, Kistler ML, Shew C-Y, Antonio MR, Liu T (2009) Counterion distribution around hydrophilic molecular macroanions: the source of the attractive force in self-assembly. Angew Chem Int Ed 48(35):6538–6542
31. Pigga JM, Teprovich JA, Flowers RA, Antonio MR, Liu T (2010) Selective monovalent cation association and exchange around keplerate polyoxometalate macroanions in dilute aqueous solutions. Langmuir 26(12):9449–9456
32. Chen J, Bera MK, Li H, Yang Y, Sun X, Luo J, Baughman J, Liu C, Yao X, Chuang SSC, Liu T (2021) Accurate determination of the quantity and spatial distribution of counterions around a spherical macroion. Angew Chem Int Ed 60(11):5833–5837
33. Schweins R, Goerigk G, Huber K (2006) Shrinking of anionic polyacrylate coils induced by Ca2+, Sr2+ and Ba2+: a combined light scattering and ASAXS study. Europ Phys J E 21(2):99–110
34. Goerigk G, Schweins R, Huber K, Ballauff M (2004) The distribution of Sr2+ counterions around polyacrylate chains analyzed by anomalous small-angle X-ray scattering. EPL (Europhysics Letters) 66(3):331
35. Liu T (2003) An unusually slow self-assembly of inorganic ions in dilute aqueous solution. J Am Chem Soc 125(2):312–313
36. Liu G, Cai Y, Liu T (2004) Automatic and subsequent dissolution and precipitation process in inorganic macroionic solutions. J Am Chem Soc 126(51):16690–16691
37. Liu T (2002) Supramolecular structures of polyoxomolybdate-based giant molecules in aqueous solution. J Am Chem Soc 124(37):10942–10943
38. Liu G, Liu T (2005) Thermodynamic properties of the unique self-assembly of Mo72Fe30 inorganic macro-ions in salt-free and salt-containing aqueous solutions. Langmuir 21(7):2713–2720
39. Liu G, Liu T, Mal SS, Kortz U (2006) Wheel-shaped polyoxotungstate [Cu20Cl(OH)24(H2O)12(P8W48O184)]25- macroanions form supramolecular "Blackberry" structure in aqueous solution. J Am Chem Soc 128(31):10103–10110
40. Zhang J, Li D, Liu G, Glover KJ, Liu T (2009) Lag periods during the self-assembly of Mo72Fe30 macroions: connection to the virus capsid formation process. J Am Chem Soc 131(42):15152–15159
41. Casini GL, Graham D, Heine D, Garcea RL, Wu DT (2004) In vitro papillomavirus capsid assembly analyzed by light scattering. Virology 325(2):320–327
42. Mishra PP, Pigga J, Liu T (2008) Membranes based on "Keplerate"-type polyoxometalates: slow, passive cation transportation and creation of water microenvironment. J Am Chem Soc 130(5):1548–1549
43. Liu T, Langston MLK, Li D, Pigga JM, Pichon C, Todea AM, Müller A (2011) Self-recognition among different polyprotic macroions during assembly processes in dilute solution. Science 331(6024):1590–1592
44. Yin P, Zhang J, Li T, Zuo X, Hao J, Warner AM, Chattopadhyay S, Shibata T, Wei Y, Liu T (2013) Self-recognition of structurally identical, rod-shaped macroions with different central metal atoms during their assembly process. J Am Chem Soc 135(11):4529–4536
45. Yin P, Zhang Z-M, Lv H, Li T, Haso F, Hu L, Zhang B, Bacsa J, Wei Y, Gao Y, Hou Y, Li Y-G, Hill CL, Wang E-B, Liu T (2015) Chiral recognition and selection during the self-assembly process of protein-mimic macroanions. Nat Commun 6(1):6475
46. Raee E, Li H, Sun X, Ustriyana P, Luo J, Chen J, Sahai N, Liu T (2020) Strong enantiomeric preference on the macroion-counterion interaction induced by weakly associated chiral counterions. J Phys Chem B 124(44):9958–9966
47. Luo J, Ye S, Ustriyana P, Wei B, Chen J, Raee E, Hu Y, Yang Y, Zhou Y, Wesdemiotis C, Sahai N, Liu T (2020) Unraveling chiral selection in the self-assembly of chiral fullerene macroions: effects of small chiral components including counterions, co-ions, or neutral molecules. Langmuir 36(17):4702–4710
48. Li D, Yin P, Liu T (2012) Supramolecular architectures assembled from amphiphilic hybrid polyoxometalates. Dalton Trans 41(10):2853–2861

49. Li D, Song J, Yin P, Simotwo S, Bassler AJ, Aung Y, Roberts JE, Hardcastle KI, Hill CL, Liu T (2011) Inorganic-organic hybrid vesicles with counterion- and pH-controlled fluorescent properties. J Am Chem Soc 133(35):14010–14016
50. Yin P, Wu P, Xiao Z, Li D, Bitterlich E, Zhang J, Cheng P, Vezenov DV, Liu T, Wei Y (2011) A double-tailed fluorescent surfactant with a hexavanadate cluster as the head group. Angew Chem Int Ed 50(11):2521–2525
51. Misdrahi MF, Wang M, Pradeep CP, Li F-Y, Lydon C, Xu L, Cronin L, Liu T (2011) Amphiphilic properties of dumbbell-shaped inorganic–organic–inorganic molecular hybrid materials in solution and at an interface. Langmuir 27(15):9193–9202
52. Zhang J, Song Y-F, Cronin L, Liu T (2010) Reverse-vesicle formation of organic-inorganic polyoxometalate-containing hybrid surfactants with tunable sizes. Chem Eur J 16(37):11320–11324
53. Haso F, Yang P, Gao Y, Yin P, Li H, Li T, Kortz U, Liu T (2015) Exploring the Effect of Surface Functionality on the Self-Assembly of Polyoxopalladate Macroions. Chem Eur J 21(25):9048–9052
54. Izarova NV, Dickman MH, Biboum RN, Keita B, Nadjo L, Ramachandran V, Dalal NS, Kortz U (2009) Heteropoly-13-Palladates(II) [PdII13(AsVPh)8O32]6− and [PdII13SeIV8O32]6−. Inorg Chem 48(16):7504–7506
55. Yin P, Pradeep CP, Zhang B, Li FY, Lydon C, Rosnes MH, Li D, Bitterlich E, Xu L, Cronin L (2012) Controllable self-assembly of organic-inorganic amphiphiles containing dawson polyoxometalate clusters. Chem A Europ J 18(26):8157
56. Haso F, Luo J, Bassil BS, Artetxe B, Zhou J, Yin P, Reinoso S, Gutiérrez-Zorrilla JM, Kortz U, Liu T (2016) Effect of directional hydrogen bonding on the self-assembly of anisotropically-shaped macroions. ChemistrySelect 1(14):4345–4349
57. Ma JC, Dougherty DA (1997) The cation−π interaction. Chem Rev 97(5):1303–1324
58. Luo J, Chen K, Yin P, Li T, Wan G, Zhang J, Ye S, Bi X, Pang Y, Wei Y, Liu T (2018) Effect of cation–π interaction on macroionic self-assembly. Angew Chem Int Ed 57(15):4067–4072
59. Qi B, Guo X, Gao Y, Li D, Luo J, Li H, Eghtesadi SA, He C, Duan C, Liu T (2017) Strong co-ion effect via cation−π interaction on the self-assembly of metal-organic cationic macrocycles. J Am Chem Soc 139(34):12020–12026
60. Yin P, Bayaguud A, Cheng P, Haso F, Hu L, Wang J, Vezenov D, Winans RE, Hao J, Li T (2014) Spontaneous stepwise self-assembly of a polyoxometalate–organic hybrid into catalytically active one-dimensional anisotropic structures. Chem Europ J 20(31):9589–9595
61. Mariani G, Moldenhauer D, Schweins R, Gröhn F (2016) Elucidating electrostatic self-assembly: molecular parameters as key to thermodynamics and nanoparticle shape. J Am Chem Soc 138(4):1280–1293
62. Mariani G, Schweins R, Gröhn F (2016) Electrostatic self-assembly of dendrimer macroions and multivalent dye counterions: the role of solution ionic strength. Macromolecules 49(22):8661–8671
63. Gröhn F, Klein K, Koynov K (2010) A novel type of vesicles based on ionic and π–π interactions. Macromol Rapid Commun 31(1):75–80
64. Willerich I, Gröhn F (2010) Photoswitchable nanoassemblies by electrostatic self-assembly. Angew Chem Int Ed 49(44):8104–8108
65. Raee E, Yang Y, Liu T (2021) Supramolecular structures based on metal-organic cages. Giant 5:100050
66. Chen L-J, Ren Y-Y, Wu N-W, Sun B, Ma J-Q, Zhang L, Tan H, Liu M, Li X, Yang H-B (2015) Hierarchical self-assembly of discrete organoplatinum(II) metallacycles with polysaccharide via electrostatic interactions and their application for heparin detection. J Am Chem Soc 137(36):11725–11735

Chapter 4
Ionic Self-Assembly of Dendrimers

Alberto Concellón and Verónica Iguarbe

Abstract Dendrimers are highly branched macromolecules that possess a large number of functional groups at the periphery. Among the different types of dendrimers, those bearing charged sites in their structure, namely ionic dendrimers, attract increasing attention due to their exceptional self-assembling properties. These charged sites stimulate a cooperative binding mechanism that extends toward the formation of nanostructures both in bulk and in solution. Specifically, ionic dendrimers self-assemble in the solid state forming liquid crystal phases, even without being functionalized with liquid crystal units. Ionic dendrimers also self-assemble in solution leading to a wide variety of nanostructures, such as micelles or vesicles. The self-assembly of ionic dendrimers is a hot topic in materials science, and they have found several potential applications in ion conductive materials, opto-electronics, drug delivery, or gene transfection. The main objective of this chapter is to give a comprehensive overview of the functions, structures, and properties of these self-assembling ionic dendrimers.

4.1 Introduction to Dendritic Polymers

Dendritic polymers are macromolecules with a highly branched structure that display different properties compared with their linear analogues [1–3]. Dendritic polymers can be classified in five main categories depending on the control over the branching units: hyperbranched polymers, dendrons, dendrimers, dendrigraft polymers, and dendronized polymers (Fig. 4.1). The term dendrimer comes from the Greek words δένδρον (*dendron*) and μέρος (*meros*), which translate to *tree* and *part*, respectively. Although dendrimers were first called "cascade molecules" by Vögtle [4], and

A. Concellón (✉)
Department of Chemistry, Massachusetts Institute of Technology, Cambridge, MA 02139, USA
e-mail: aconcell@mit.edu

V. Iguarbe
Departamento de Química Orgánica, Universidad de Zaragoza, 50009 Zaragoza, Spain

M. A. Aboudzadeh and A. Frontera (eds.), *Supramolecular Assemblies Based on Electrostatic Interactions*,
https://doi.org/10.1007/978-3-031-00657-9_4

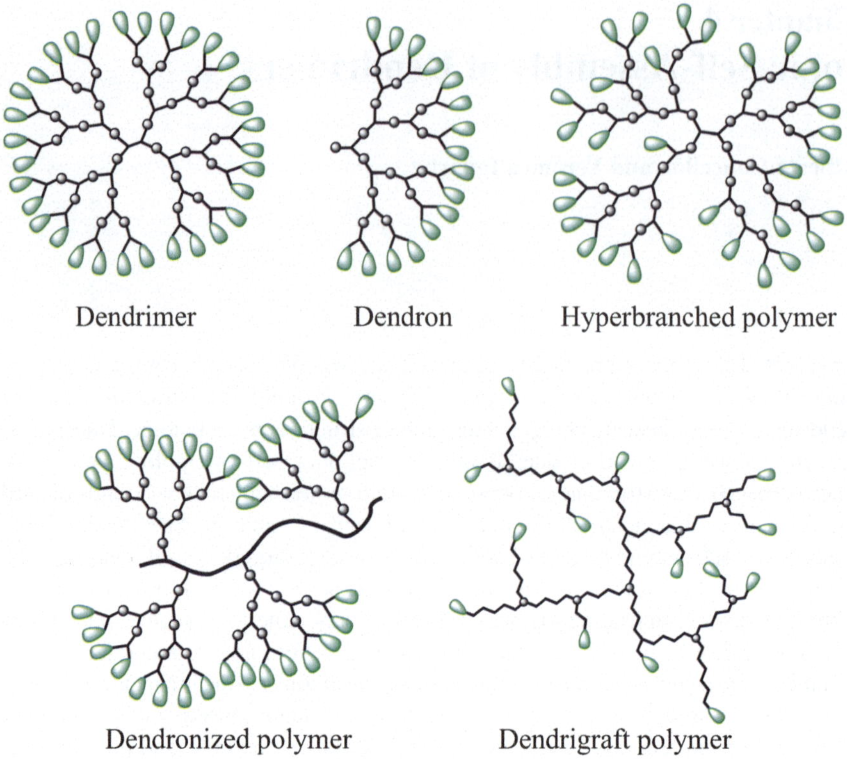

Fig. 4.1 Structure of common dendritic polymers

"arborols" by Newkome [5], the term "dendrimer", which was proposed by Tomalia [6], is widely used and preferred over the original ones.

Dendrimers are monodisperse macromolecules with a perfectly branched architecture that are prepared by multistep synthesis (Fig. 4.2). Dendrimers consist of layers of repeating units that are radially grown from a central multifunctional core. The branched structures attached to the multifunctional core are called dendrons, and each layer of repeating units is termed a generation (G1, G2, G3, etc.). The periphery of dendrimers is probably the most interesting part since it contains a large number of end-groups that are accessible for additional functionalization, allowing the modulation of the properties of the dendrimer. Moreover, upon increasing the generation of the dendrimer, the number of functional end-groups also increases, resulting in a magnification effect that is known as the dendritic effect [7]. Some of the most commonly recognized dendritic structures include poly(propylenimines) (PPI), carbosilanes, poly(amidoamines) (PAMAM), polyols, polyaliphatic esters based on 2,2-bis(hydroxymethyl)propionic acid (bis-MPA), or poly(benzyl ethers) (Fig. 4.3).

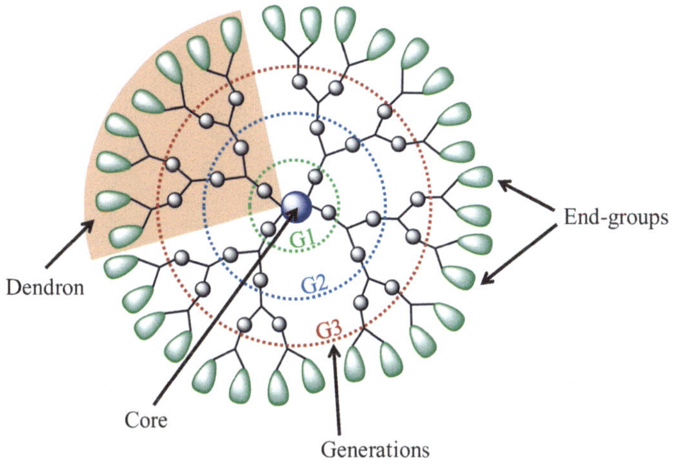

Fig. 4.2 Structure of a dendrimer and its parts

The latter three members of the dendritic polymer family (i.e. hyperbranched polymers, dendronized polymers, and dendrigraft polymers) are polydisperse and irregular branched macromolecular architectures. Hyperbranched polymers are almost identical to dendrimers but present defects in their branched structure. They are synthesized via one-step polymerization, resulting in less architectural control in comparison to dendrimers. Dendronized polymers, frequently called "rod-shaped polymers", are composed by a linear polymer backbone with dendrons as side-chains. Dendrigraft polymers (so-called "arborescent polymers") are a relatively new member of the dendritic family and consist of linear polymers with randomly distributed branching points.

Dendrimers are usually synthesized by using a divergent growth strategy [5, 8], or by a convergent growth strategy (Fig. 4.4) [9]. Both approaches involve stepwise synthesis and deprotection/activation reactions, requiring a careful purification when passing from one generation to the next one.

The divergent growth approach begins from a multifunctional core from which generations are grown. After each dendritic growth reaction, a careful purification of the dendrimer is needed to eliminate unreacted starting materials and partially functionalized dendrimers. The desired generation is achieved by repetitive growth steps. The most attractive characteristic of the divergent approach is the possibility of obtaining high-generation dendrimers. Nevertheless, with increasing the generation the risk of incomplete functionalization of the dendrimer periphery also increases, leading to non-perfect dendrimers with structural defects. In addition, the purification steps are very tedious due to the similar properties of structurally perfect and defective dendrimers. Nevertheless, the divergent growth is preferred for the preparation of some commercially available dendrimers because of the possibility of automation of the repetitive steps.

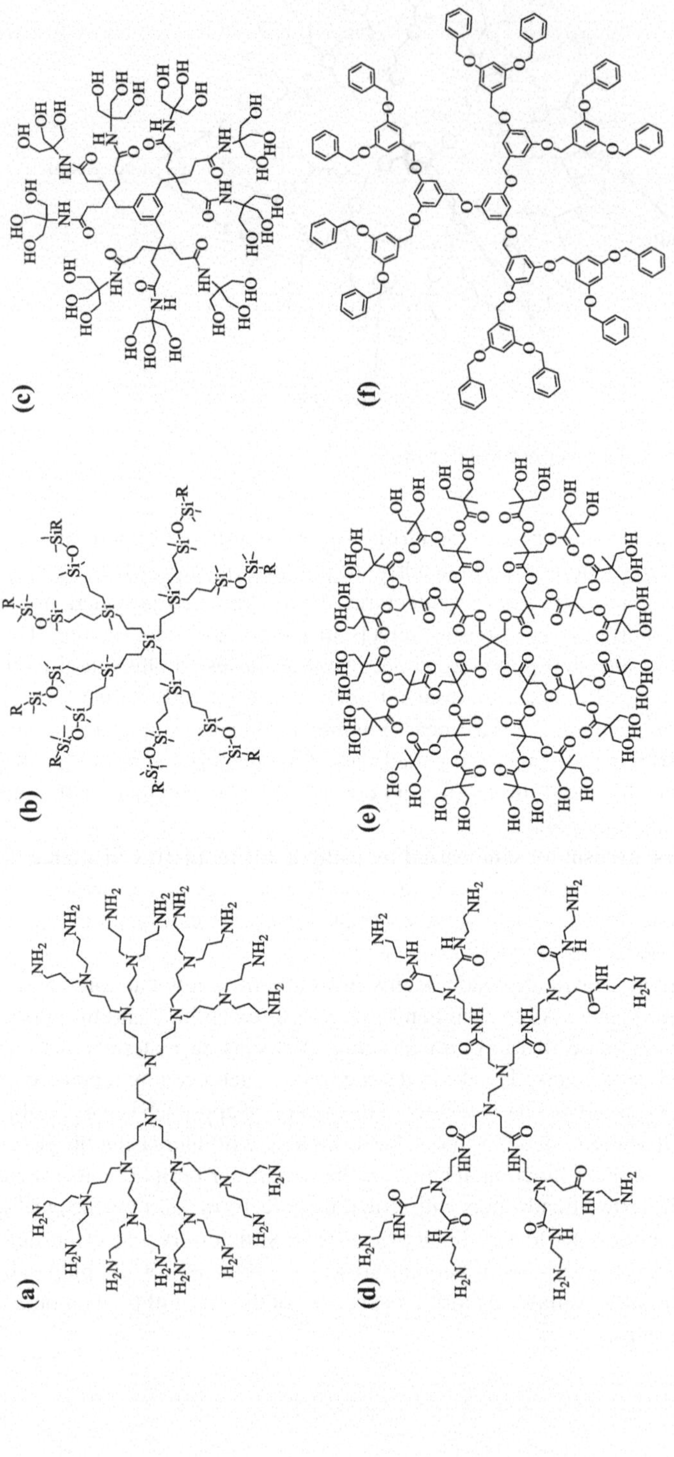

Fig. 4.3 Common dendritic structures: **a** PPI, **b** carbosilanes, **c** polyol, **d** PAMAM, **e** polyaliphatic ester, and **f** poly(benzyl ether)

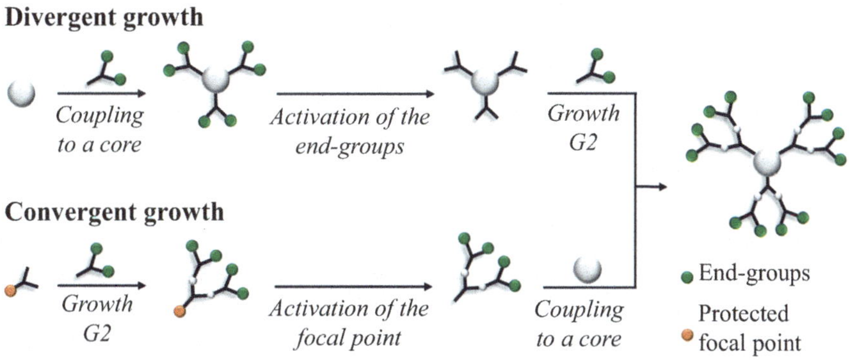

Fig. 4.4 Schematic representation of divergent and convergent growth

In the convergent growth strategy, single dendrons are first prepared through a divergent method and then coupled to a core. The main advantage is that the preparation of individual dendrons can be carefully controlled and thus, this strategy does not produce the structural defects that are often observed in the divergent method. Nevertheless, the last coupling step requires extremely efficient reactions due to steric hindrance. Therefore, this strategy is mostly employed for low-generation dendrimers since the preparation of dendrimers with high generations is complicated.

Dendritic polymers are ideal candidates for several high-value and nanoscale applications (e.g. target drug delivery, optoelectronics, catalysis, sensors, or biomarkers for imaging) due to their unique branched architecture with a high functional group density [10–12]. The commercial availability of dendritic materials was limited for some time due to their tedious and costly synthesis, but several improvements have recently been reported to simplify and fasten up their preparation [13, 14]. However, the increasing interest in dendrimers is due not only to their unusual branched structure, but also to the broad spectrum of possibilities of designing well-defined functional macromolecules, in which the functional groups can take part in physical or chemical processes. Although dendrimers can be categorized according to their function (e.g. catalytically active, photo-responsive, electrochemically active, or liquid crystalline dendrimers), the traditional way of classifying dendritic macromolecules is based on their chemical structure and physical characteristics (e.g. dendritic hydrocarbons, metallodendrimers, ionic dendrimers, glycodendrimers, silicon-based dendrimers, phosphorous-based dendrimers, etc.).

As a part of the present book, this chapter presents the fundamental principles of the self-assembly of dendrimers in bulk and in solution. Following this general introduction about dendritic polymers, Sect. 4.2 presents an overview of dendrimers bearing charges within the dendritic architecture. Section 4.3 describes some basic aspects of the ionic self-assembly in bulk, including theoretical and experimental studies, and phase morphologies achieved by the self-organization of charged dendrimers. Section 4.4 discusses the self-assembly of ionic dendrimers in solution, with special attention to the produced nanostructures and the main points that affect the morphologies, applications, among others. The final section gives a summary and outlook of the chapter.

4.2 Ionic Dendrimers

Dendrimers bearing charged sites in their structure, namely ionic dendrimers, have attracted interest in recent times due to their self-assembling properties. This kind of dendrimers are usually accompanied by a cooperative binding mechanism, wherein the charged sites promote additional assembly that extends toward the formation of self-assembled structures. Indeed, ionic self-assembly has been largely underestimated in supramolecular chemistry but this secondary interaction is very easy, reliable, flexible, and enables structural self-perfection [15].

The self-assembly of small molecules (such as charged surfactants, ionic liquids, charged dyes, etc.) driven by electrostatic interactions has been studied for decades [16]. The self-assembly of ionic dendrimers in bulk or in solution results in the formation of similar morphologies to those observed in low-molecular weight aggregates. Induction and stabilization of the self-assembled structures generally result from both the charge-charge coupling interactions overcoming the propensity of the dendrimer to display a globular conformation, and the microphase separation between incompatible apolar and polar parts of the dendritic structure. Aggregates of ionic dendrimers exhibit higher stability and durability than those of small molecules due to their mechanical and physical properties. In general, the electrostatic self-assembly of dendrimers has drawn extensive academic interest, but also has found application in many fields, including biomaterials, microelectronics, photoelectric materials, or catalysis [17, 18].

Ionic dendrimers can be classified depending on the nature and position of the charged species in the dendritic structure: those having ionic centers in the inner part of the molecule (branching units or core), or those holding ionic sites on the surface (terminal groups). Ionic dendrimers are usually prepared by introducing the charged sites during the synthesis, by transforming neutral groups into charged ones, or by incorporating transition metals into the dendrimer.

4.2.1 Dendrimers with Internal Charges

This kind of dendrimers has positive or negative charges throughout the covalent structure of the dendrimer. For instance, Stoddart and coworkers developed a convergent approach to gram quantities of ionic dendrimers based on mesitylene units (Fig. 4.5a) [19]. Dendrons with positive charges were prepared by using the high-yielding Menschutkin reaction. The first-, second-, and third-generation dendrons were subsequently attached to a 1,3,5-tris(diethylaminomethyl)benzene core. As new

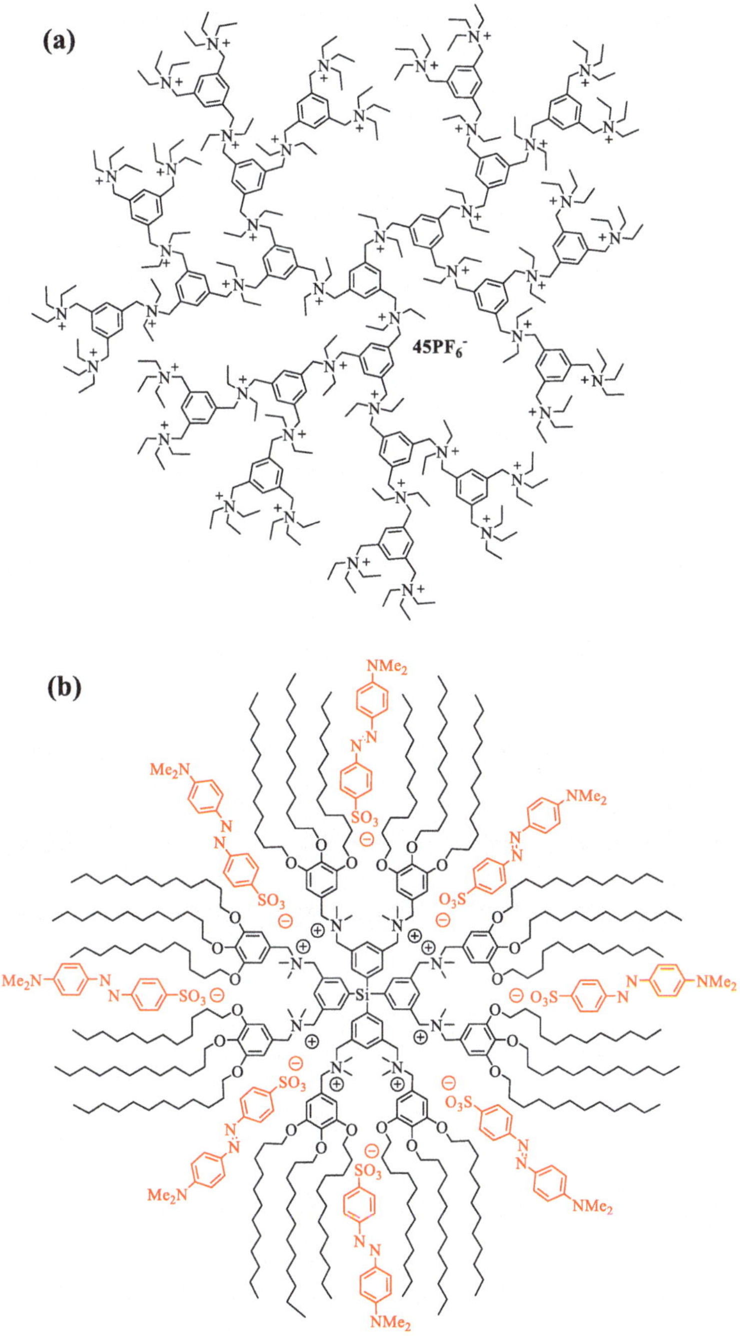

Fig. 4.5 a Polycationic dendrimers based on mesitylene units. **b** polycationic core–shell dendrimers and their host–guest complexes

cationic sites are created, the hexafluorophosphate anion was introduced to facilitate the solubility of the positively charged dendritic macromolecules in organic solvents.

In a different convergent approach, Koten and coworkers described a new class of ionic silicon-based dendrimers (Fig. 4.5b) [20–22]. They consist of a tetraarylsilane core surrounded by quaternary ammonium groups, which were decorated with poly(benzyl ether) dendrons. The authors performed a binding study with these polycationic dendrimers and observed that they bind a stoichiometric amount of methyl orange, an anionic guest molecule. Moreover, such electrostatic interactions allow quantitative and controlled release of the guest molecule by applying an external stimulus, such as acid or an excess of a salt with a competitive anion. These interesting dendritic structures were also employed as noncovalent supports for homogeneous catalysis. In this case, the ionic dendrimer was noncovalently decorated with a catalytically active arylpalladium complex bearing a tethered sulfate group [23, 24]. The catalytic performance of these metallodendritic assemblies was evaluated in the aldol condensation reaction between methyl isocyanoacetate and benzaldehyde using dichloromethane as solvent. A minor decrease in catalytic activity of the Pd(II) complex was observed upon encapsulation, whereas the product selectivity remained constant up to the third generation.

Dendritic structures with a redox-active 4,4′-bipyridinium (viologen) unit covalently attached to the apical position were prepared by Kaifer and coworkers [25]. Both polyol and poly(benzyl ether) dendrons from first to third generation were used in this study (Fig. 4.6). The electrochemical behavior of these dendrimers remains fast and reversible in all generations, unlike many other dendrimers with a redox-active center. Moreover, they also investigated the host–guest binding interactions between these viologen-containing dendrimers and a crown ether host (bis-*p*-phenylene-34-crown-10).

Metallodendrimers are another important group of charged dendrimers, in which transition-metal ions are regularly associated within the dendritic structure. Such species allow the incorporation of a large number of transition-metals into discrete

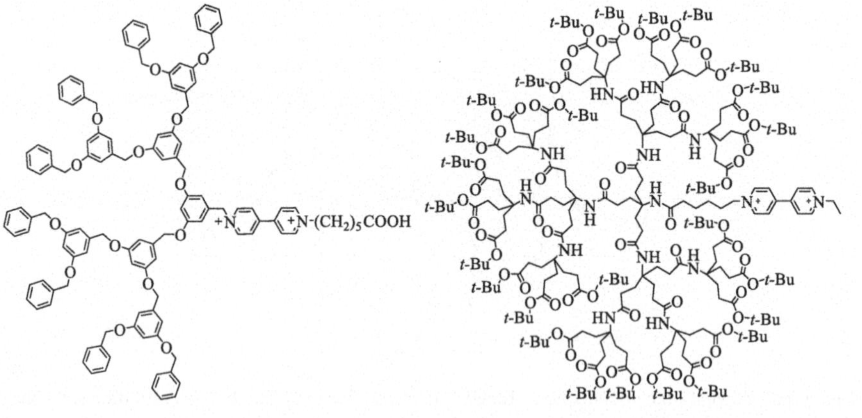

Fig. 4.6 Viologen-containing poly(benzyl ether) (left) and polyols (right) dendrimers

molecules, which have the potential to exhibit unique properties (redox, electronic, optical, magnetic, etc.). Therefore, it is not surprising that the investigation on metallodendrimers has attained increasing interest over the last years [26].

4.2.2 Surface-Charged Dendrimers

Surface-charged dendrimers consist of a classical dendritic core which bear cationic or anionic sites at the termini of the branches, i.e. at the surface of the dendrimer. These dendrimers are probably the most interesting class of ionic dendrimers since they possess polar groups on the dendrimer surface and an apolar core. The microphase separation at the nanoscale between chemically incompatible parts of the dendritic molecule leads to unique self-assembling properties both in bulk and in solution. The self-assembling properties of this type of ionic dendrimers are discussed in detail in Sect. 4.3 and Sect. 4.4.

4.3 Ionic Self-assembly of Dendrimers in Bulk

In the solid state, ionic dendrimers undergo microphase separation at the nanoscale resulting in liquid crystalline phases. Electrostatic interactions are essential in this self-assembly process since segregation between apolar and polar parts is the driving force for the formation of the liquid crystal ordering. Indeed, the phase formation is similar to that of block copolymers because incompatible constituent molecular regions segregate to minimize their interactions, occupying specific zones.

One of the most fascinating aspects of ionic dendrimers is their ability to show liquid crystal organizations even by being functionalized without any rigid, anisotropic liquid crystal unit. Moreover, electrostatic interactions are reversible interactions and thus, ionic dendrimers can adapt their conformation and their supramolecular organization in response to specific conditions. This interesting characteristic allows the formation of rare liquid crystal phases that cannot be formed by non-ionic dendrimers.

4.3.1 Liquid Crystals

Liquid crystals are dynamic soft materials that have a fascinating combination of crystalline order and liquid fluidity. Controlling the self-assembly of liquid crystals allows the association of discrete molecules into a diversity of nanostructures, which usually enhance the function of the constituent single molecules. Molecules with liquid crystal ordering, so-called *mesogens*, normally consist of flexible alkyl side chains and a wide variety of rigid cores. The liquid crystal self-assembly is facilitated by strong intermolecular π–π interactions, van der Waals interactions, hydrogen bonding, or charge transfer, while the disordered alkyl chains favor molecular mobility and prevent a "true long-range order". There are several ways of classifying liquid crystal organizations (so-called *mesophases*), being the most widely used the differentiation between thermotropic and lyotropic. Thermotropic mesophases are formed within a temperature range, while a solvent is required for the formation of lyotropic mesophases. Depending on the shape of the mesogenic units and supramolecular assemblies, liquid crystals are classified into calamitic (rod-shaped), discotic (disk-shaped) and bent-core (banana-shaped) liquid crystals (Fig. 4.7).

Calamitic liquid crystals typically form nematic or smectic mesophases (Fig. 4.7a). The nematic (N) phase is the least ordered liquid crystal organization,

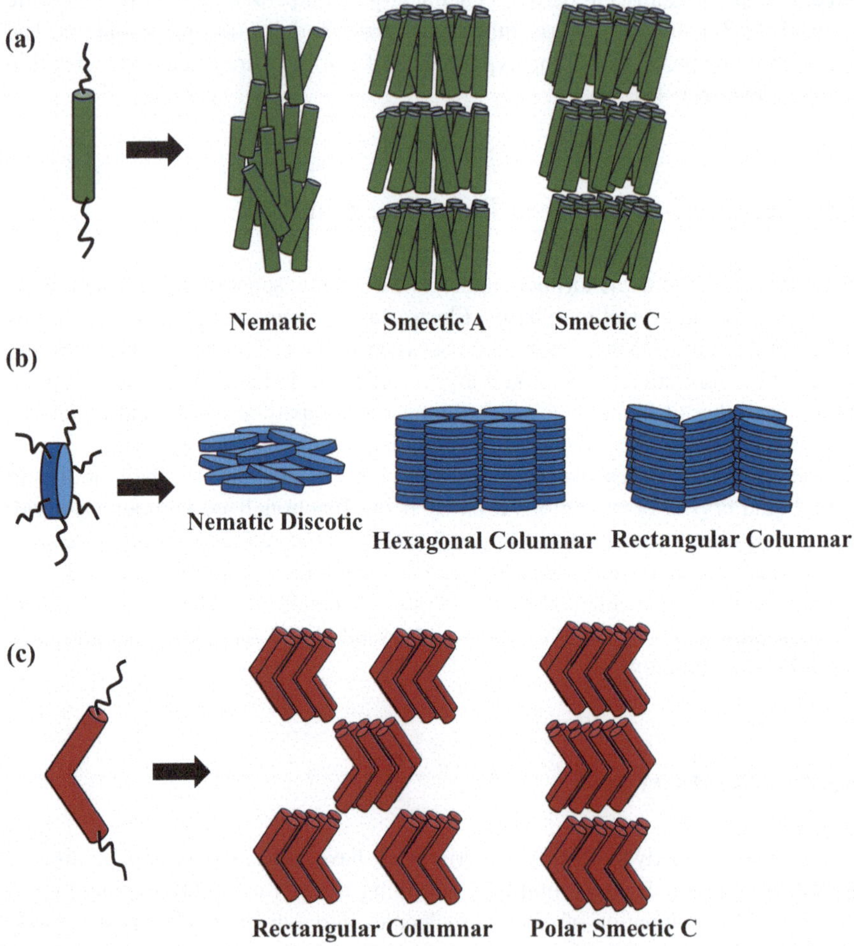

Fig. 4.7 Common liquid crystal phases created by the assembly of **a** calamitic, **b** discotic, and **c** bent-core molecules

wherein molecules are orientationally ordered and no positional order exists. Smectic phases are lamellar organizations, in which molecules are arranged in layers. There are different smectic phases depending on the molecular order within the layer. Smectic A (SmA, molecules are perpendicular to the layer plane) and smectic C (SmC, molecules are tilted and form an angle other than 90° to the layer plane) are the most common smectic organizations.

Discotic liquid crystals exhibit a range of mesophases that are distinct from those of rod-shaped molecules [27]. This class of materials arranges into nematic or columnar phases, in which a single molecule or self-assembled small molecule aggregates are disk-shaped (Fig. 4.7b). In the nematic discotic (N_D) phase, the disk-shaped units have orientational order but no long-range positional order. In columnar phases, the disks arrange primarily in columns by stacking on top of each other. In addition, the columns pack in several two-dimensional lattices, such as hexagonal (Col_h) or rectangular (Col_r) unit cells.

Bent-core liquid crystals self-organize in lamellar or columnar mesophases with their rotational freedom restricted due to the unique bent geometry that lead to a variety of polar mesophases with attractive properties (Fig. 4.7c) [28]. The most common phases are the polar smectic C (SmCP) and the rectangular columnar (Col_r). In the SmCP phase, molecules are tilted as in a calamitic SmC phase, but with polar order in each layer [29]. This polar order in adjacent layers can be opposite (antiferroelectric) or parallel (ferroelectric). The Col_r phase is a modulated smectic phase, which is different from the discotic Col_r phase as it consists of fragments of smectic layers [30].

Cubic phases are additional "exotic" liquid crystal organizations, which are optically isotropic while presenting long three-dimensional (3D) order (Fig. 4.8) [31]. The most common mesophases are: the bicontinuous cubic phase (Cub_V) that is composed of periodic, infinite 3D molecular networks and represents an intermediate state between lamellar and columnar phases, and the micellar cubic phase (Cub_I) which consists of ordered 3D arrays of spherical aggregates from small molecules. These liquid crystal phases, which have been less often observed, are usually found for some classical rod-like mesogens, polycatenars, or amphiphilic molecules.

Liquid crystal molecules display collective behavior and can self-assemble over length scales from molecular to macroscopic level [32, 33]. Their inherent anisotropy

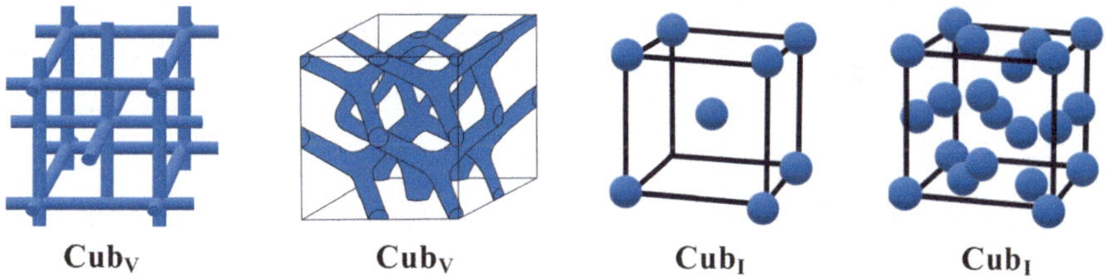

Cub_V *Im*3*m* | **Cub_V** *Pn*3*m* | **Cub_I** *Im*3*m* | **Cub_I** *Pm*3*n*

Fig. 4.8 Representation of the most common cubic phases

and the facility with which its alignment can be manipulated are the enabling features for several applications. The most representative example is the development of LCD (liquid crystal display) TVs, which have drastically revolutionized our daily life. Nonetheless, a plethora of "beyond display" applications have been discovered [34]. Specifically, liquid crystals have a prominent place in nanoscience and nanotechnology, and they can potentially be employed in soft robotics, optics, sensing, or electron/ion transport [35–39].

4.3.2 Liquid Crystal Dendrimers

The molecular-level engineering of liquid crystals can provide several opportunities for the development of functional materials. In particular, dendrimers are interesting building blocks for the creation of multifunctional liquid crystals due to their highly branched architecture. In addition, dendrimeric liquid crystals are quite unique because they can lead to the creation of rare supramolecular organizations that cannot be achieved with conventional liquid crystals [40]. Such liquid crystal behavior can be tuned and controlled by subtle modifications of the dendrimer chemical structure (i.e. generation number, branches multiplicity, focal core valency, or geometry of the interconnecting junctions). Moreover, liquid crystal dendrimers possess a highly branched architecture that provides high density of functional groups that can be used for the introduction of complex, function-bearing moieties. These active units self-assemble into a liquid crystal organization with a supramolecular order that usually enhances their activity [41, 42].

Liquid crystal dendrimers are commonly obtained by the incorporation of mesogenic moieties within a dendritic scaffold [43]. In the most widely used strategy, the mesogenic units are attached at the periphery of a preformed dendrimer; these dendrimers are called side-chain liquid crystal dendrimers (Fig. 4.9a). The incorpora-

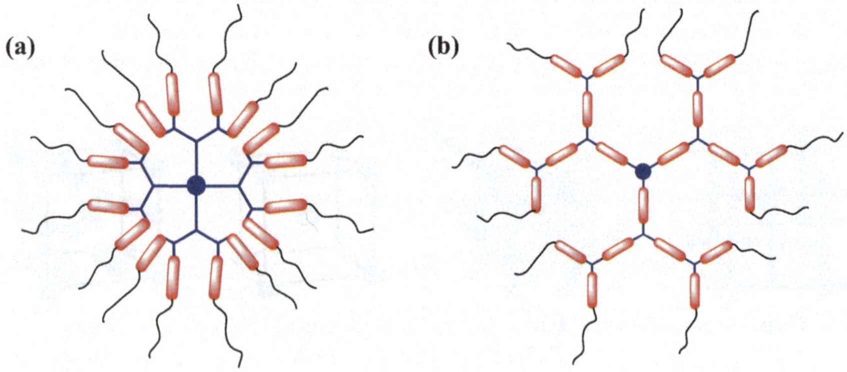

Fig. 4.9 Schematic representation of **a** side-chain and **b** main-chain LC dendrimers

tion of mesogenic units as repeating units in the dendritic scaffold is a less employed strategy for the preparation of liquid crystal dendrimers; these dendrimers are called main-chain liquid crystal dendrimers (Fig. 4.9b).

The incorporation of mesogenic units onto the periphery of a preformed dendrimer has been traditionally performed via covalent bonding. However, more complex dendritic architectures require major synthetic efforts, which often result in low yields and high polydispersity. An effective approach, which has been developed in recent years, takes advantage of supramolecular chemistry to functionalize a preformed dendritic core through noncovalent interactions. In particular, electrostatic interactions have been by far the most utilized interactions in this noncovalent strategy.

4.3.3 Ionic Liquid Crystal Dendrimers

The preparation of supramolecular dendrimers by using electrostatic interactions is a fascinating method, which allows a facile access to peripherally substituted dendrimers in a thermodynamically controlled self-assembly process. This strategy requires functional groups easy to charge electrostatically in the dendritic core, as well as in the mesogenic units. Therefore, PPI and PAMAM dendrimers are ideal dendritic scaffolds for electrostatic functionalization since they bear peripheral primary amines that can be ionically functionalized with carboxylic acid groups via a proton-transfer reaction (Fig. 4.10). One of the first examples was described by Tomalia and coworkers that showed lyotropic lamellar liquid crystal organizations with a G3 PPI dendrimer electrostatically functionalized with octanoic acid (Fig. 4.10a) [44]. Later, Tsiourvas and coworkers described several ionic dendrimers prepared by the protonation of PPI dendrimers of different generations with a carboxylic acid bearing a cholesterol unit (Fig. 4.10b) [45]. At low temperatures these dendrimers exhibited chiral smectic C organizations, whereas a smectic A phase was found at higher temperatures. Moreover, transition temperatures increase upon decreasing the generation of the dendrimer. Similarly, Ujiie and coworkers described ionic liquid crystal dendrimers based on G3 PAMAM dendrimer and myristic, palmitic, and stearic acids (Fig. 4.10c) [46]. All dendrimers showed smectic A phases, but mixtures of G3 PAMAM dendrimer with stearic acid in different molar ratios exhibited both smectic A and hexagonal columnar phases.

Serrano and coworkers reported pioneering research on the self-assembly of ionic dendrimers based on commercially available PAMAM or PPI dendrimers. The motivation of these studies was the rapid and simple synthesis of these dendrimers and the ability to create a large variety of self-organized structures. In their first work, the terminal amine groups of PAMAM and PPI dendrimers were reacted with fatty acids to form ammonium salts via a proton transfer reaction (Fig. 4.10d) [47]. The obtained ionic dendrimers were thermally stable at low temperatures, as progressive formation of amide bonds occurred over 100 °C. The formation of liquid crystalline phases in these derivatives can be explained by the strong segregation between the

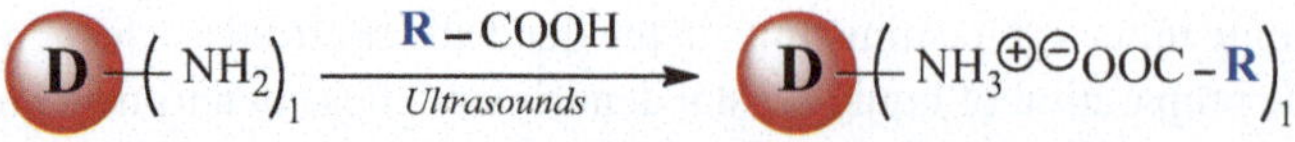

	Dendrimer (D)	Carboxylic acid
a	PPI (G3; n= 16)	$CH_3(CH_2)_6COOH$
b	PPI (G1-G5; n= 4, 8, 16, 32, 64)	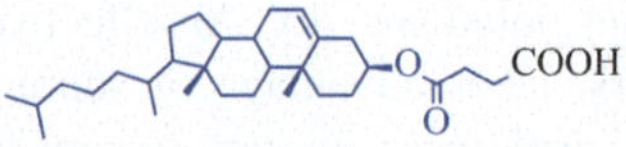
c	PAMAM (G3; n= 32)	$CH_3(CH_2)_{16}COOH$ $CH_3(CH_2)_{14}COOH$ $CH_3(CH_2)_{12}COOH$
d	PPI (G1-G5; n= 4, 8, 16, 32, 64) PAMAM (G0-G4; n= 4, 8, 16, 32, 64)	$CH_3(CH_2)_{16}COOH$ $CH_3(CH_2)_{12}COOH$ $CH_3(CH_2)_8COOH$
e	PPI (G1-G5; n= 4, 8, 16, 32, 64)	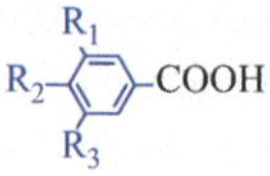 $R_1,R_3 = H, R_2 = OC_{10}H_{21}$ $R_1,R_2 = OC_{10}H_{21}, R_3 = H$ $R_1,R_2,R_3 = OC_{10}H_{21}$
f	PPI (G1-G5; n= 4, 8, 16, 32, 64) PAMAM (G0-G4; n= 4, 8, 16, 32, 64)	$CF_3(CF_2)_2(CH_2)_2COOH$
g	PPI (G3; n= 16)	$CF_3(CF_2)_7(CH_2)_2COOH$ $CH_3(CH_2)_9COOH$
h	PPI (G3; n= 16)	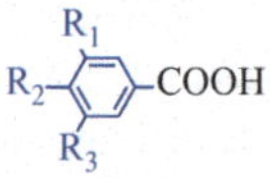 $R_1,R_3 = H, R_2 = O(CH_2)_2(CF_2)_7CF_3$ $R_1,R_3 = H, R_2 = OC_{10}H_{21}$ $R_1,R_2 = OC_{10}H_{21}, R_3 = H$ $R_1,R_2,R_3 = OC_{10}H_{21}$

Fig. 4.10 Ionic liquid crystal dendrimers prepared by proton transfer reaction between carboxylic acids and PPI and PAMAM dendrimers of different generations (G)

three constituent parts of the ionic dendrimers: the central dendrimeric matrix, the ionic continuum created by the ion pairs (ammonium carboxylates), and the aliphatic chains of the fatty acids. This segregation in well-defined layers resulted in smectic A liquid crystalline phases for most of these dendrimers, while columnar phases were found for the higher generation derivatives.

In the proposed self-assembly model, the ionic dendrimers adopt a cylindrical conformation, in which the central section is occupied by the dendrimer, while the ionic pairs extend up and down with the aliphatic chains distributed statistically (Fig. 4.11a). Theoretical calculations demonstrated that the diameter of the cylinder increases with the increase in the generation number of the central dendrimer because of the deformation of the dendritic branches in the direction parallel to the smectic

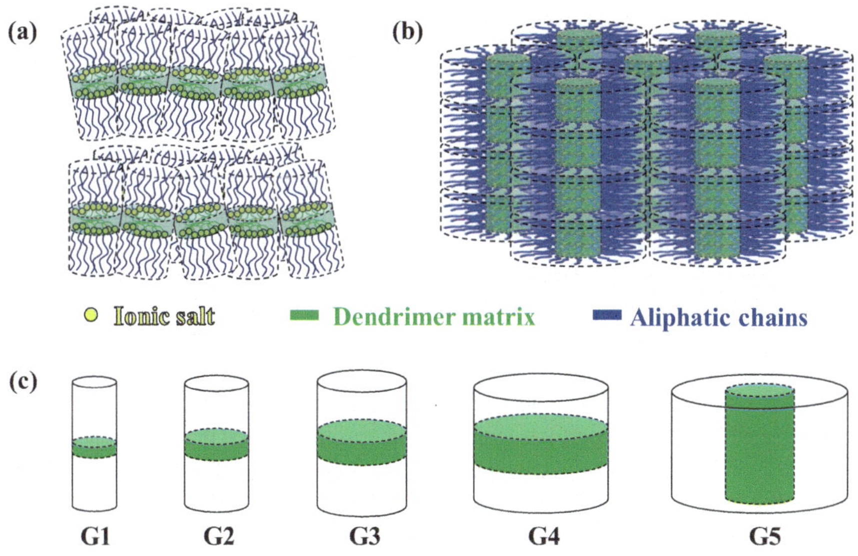

Fig. 4.11 Proposed self-assembly model for the ionic liquid crystal dendrimers in **a** the smectic A, and **b** the hexagonal columnar phase. **c** Variation of the elementary ionic dendrimer cylinder as a function of the generation number

layer to locate all the fatty acids (Fig. 4.11c). However, the height of these cylinders keeps constant for all generations. The diameter growth has a limit in which the dendrimeric matrix experiences a conformational change because the cylindrical dendrimer does not have enough space to locate the increasing number of fatty acids. Due to the flexibility of the dendrimer matrix, dendrimers adopt a disk-shaped conformation with the fatty acids extending radially from the central dendritic nucleus. Indeed, the shape of the dendrimer in the liquid crystal organization can be described more appropriately as a flattened cylinder. The arrangement of these disks (or flattened cylinders) into supramolecular organizations leads to the observed hexagonal columnar phases (Fig. 4.11b).

In a subsequent study, the same group investigated ionic dendrimers prepared by the introduction of mono-, di-, and tri-decyloxybenzoate counterions onto the periphery of PPI dendrimers (Fig. 4.10e) [48]. Dendrimers with mono- and di-substituted units displayed smectic A phases, while those with tri-substituted moieties displayed columnar mesomorphism, hexagonal or rectangular, depending on the dendrimer generation. The increasing number of terminal alkyl chains prevents their accommodation on the ideal cylinder model and thus, the flexible dendrimer matrix adopts a disk-shaped conformation, producing the observed columnar phases.

Fluorinated liquid crystals present enhanced mesophase stability since perfluorinated chains have more rigid, higher cross-sectional area, larger steric bulk, and lower van der Waals forces than the corresponding perhydrogenated analogues (fluorophobic effect). Serrano applied the concept of fluorination to ionic dendrimers by reacting PPI and PAMAM dendrimers with semifluorinated carboxylic acids (Fig. 4.10f) [49]. Smectic A liquid crystal phases were observed for all ionic dendrimers, with the exception of the fifth generation PPI dendrimer, which exhibited a rectangular columnar phase. Interestingly, columnar liquid crystal phases were not observed with PAMAM dendrimers due to the large volume of PAMAM dendritic structure. As expected, the thermal stability of the mesophases of the ionic dendrimers with fluorinated units increased in comparison to the analogues with hydrocarbon chains due to the fluorophobic effect.

Uncommon liquid crystal phases were found in ionic co-dendrimers derived from PPI dendrimer functionalized with mixtures of semifluorinated and perhydrogenated acids. (Fig. 4.10g) [50]. The corresponding homodendrimers exhibited smectic A phases, whereas intermediate proportions (from 30 to 70%) resulted in a frustrated smectic A phase (SmA^{+}). This less-organized smectic A phase originates from the nanosegregation between perhydrogenated and semifluorinated incompatible chains, resulting in a disruption of the lamellar order (Fig. 4.12a). However, such incompatibility effects are avoided and chains can mix regularly over the dendrimer surface, when the proportion of each unit is higher than 70% (Fig. 4.12b). Similar behavior was found for PPI co-dendrimers bearing semifluorinated or perhydrogenated benzoate counterions (Fig. 4.10h) [51]. In this case, some of the intermediate proportions showed modulated smectic A phases. The chemical incompatibility of the perhydrogenated and fluorinated chains precludes their arrangement together in the same layer, resulting into hydrocarbon and fluorocarbon sublayers (Fig. 4.12c).

Mezzenga and coworkers exploited the concept of electrostatic functionalization of dendrimers by reacting benzamide-based dendrons and dendrimers bearing terminal amino groups with alkylsulfate derivatives [52] (Fig. 4.13). Interestingly, they found inverted hexagonal columnar or lamellar phases, depending on the length of the alkyl chains. The second generation dendrimers adopted an uncommon inverted configuration, in which the column centers are formed by the confined alkyl units. Theoretical calculations predicted that this "frustrated" arrangement is opposed to the natural curvature of the dendron/dendrimer structure, but it is favored by an imbalance of conformational entropy between long dendron segments and shorter alkyl chains. Nevertheless, increasing the volume fraction of alkyl chains led to lamellar phases. In a subsequent study, they obtained lamellar and "normal-type" columnar phases using the same dendritic cores functionalized with a cholesterol derivative [53].

The electrostatic functionalization of the periphery of PAMAM/PPI dendrimers was also employed to decorate dendritic architectures with complex, function-bearing units (i.e. promesogenic, photoactive, fluorescent, or conductive units, among others). In general, the liquid crystalline behavior of these ionic dendrimers is very similar to those obtained with the covalent analogues. Nonetheless, ionic functionalization is an easy and effective strategy to prepare liquid crystal dendrimers, avoiding the time-consuming synthesis associated with the preparation of the covalent systems. For example, Tschierske and coworkers reported the preparation of ionic dendrimers with unusual rich liquid crystal properties [54]. These materials were prepared by electrostatic interactions between PPI dendrimers with facial amphiphilic

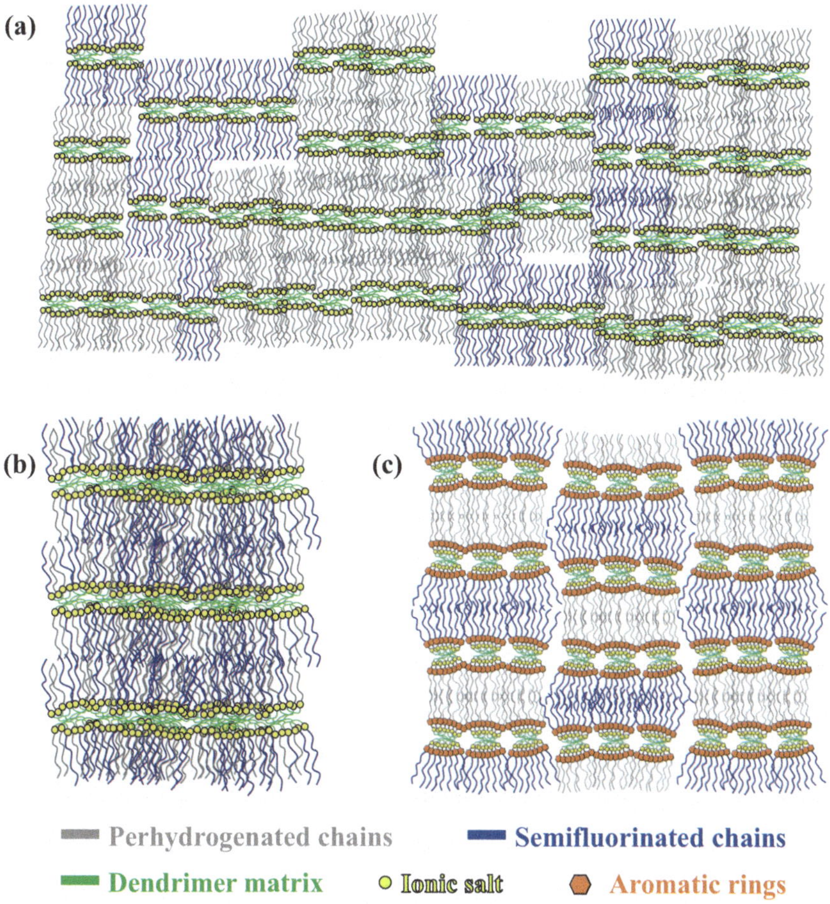

Fig. 4.12 Representation of the proposed arrangement of ionic co-dendrimers: **a** frustrated smectic A, **b** smectic A, and **c** modulated smectic A

carboxylic acids containing three different incompatible parts: a rigid terphenyl core, two terminal apolar alkyl chains and a polar lateral oligo(ethyleneglycol) chain (Fig. 4.14a). Six different liquid crystal organizations were found depending on the length of the oligo(ethyleneglycol) chain, the dendrimer generation, or the dendrimer/carboxylic acid ratio. In contrast, mesophases typical of bent-core liquid crystals were reported by Ros and coworkers by mixing carboxylic acids containing bent-shaped units with PPI dendrimers or random hyperbranched polyethyleneimine (PEI) polymer (Fig. 4.14b) [55]. The temperature range of the mesophases can be tuned with the number of aromatic rings (x) and the length of the inner spacer (m) of the bent-core acids. Moreover, the supramolecular arrangement of these ionic dendrimers is determined by the length of the terminal alkyl chains, obtaining rectangular columnar or polar smectic C phases. Sijbesma and coworkers also described

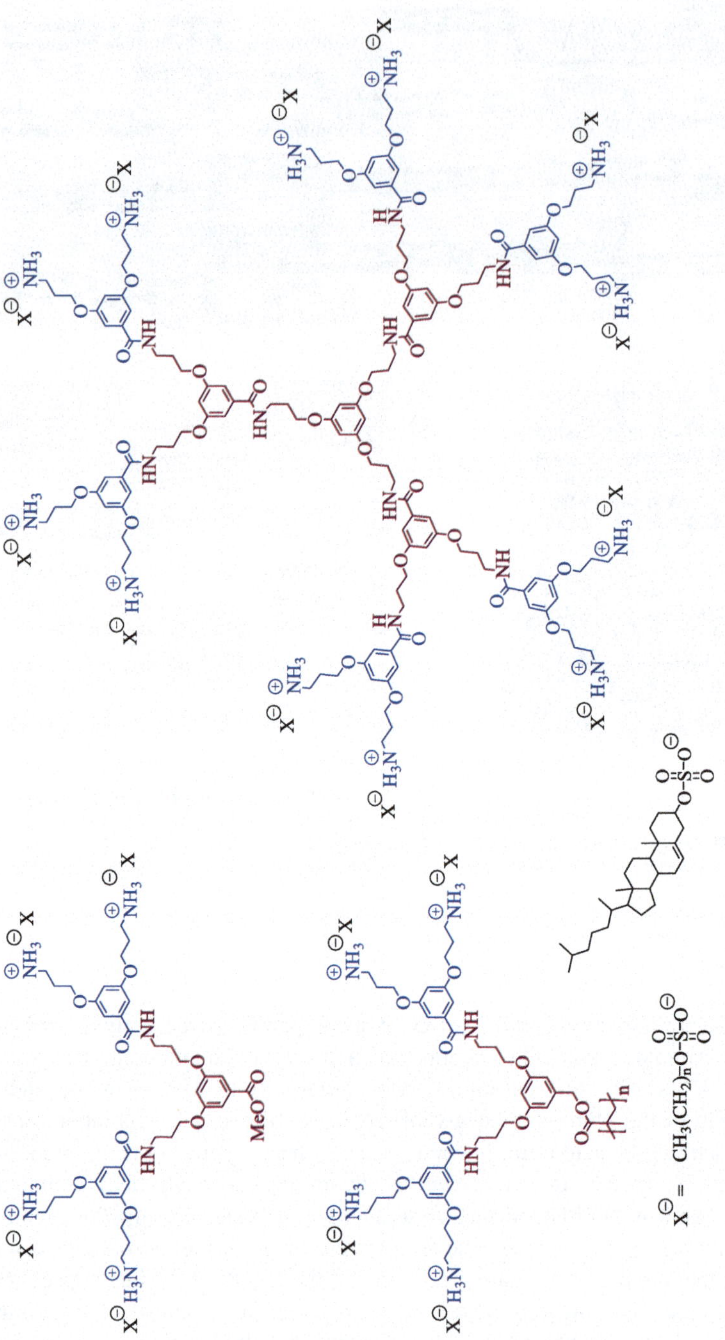

Fig. 4.13 Schematic representation of cationic benzamide-based dendrons, dendrimers and dendronized polymers with sulfate counterions

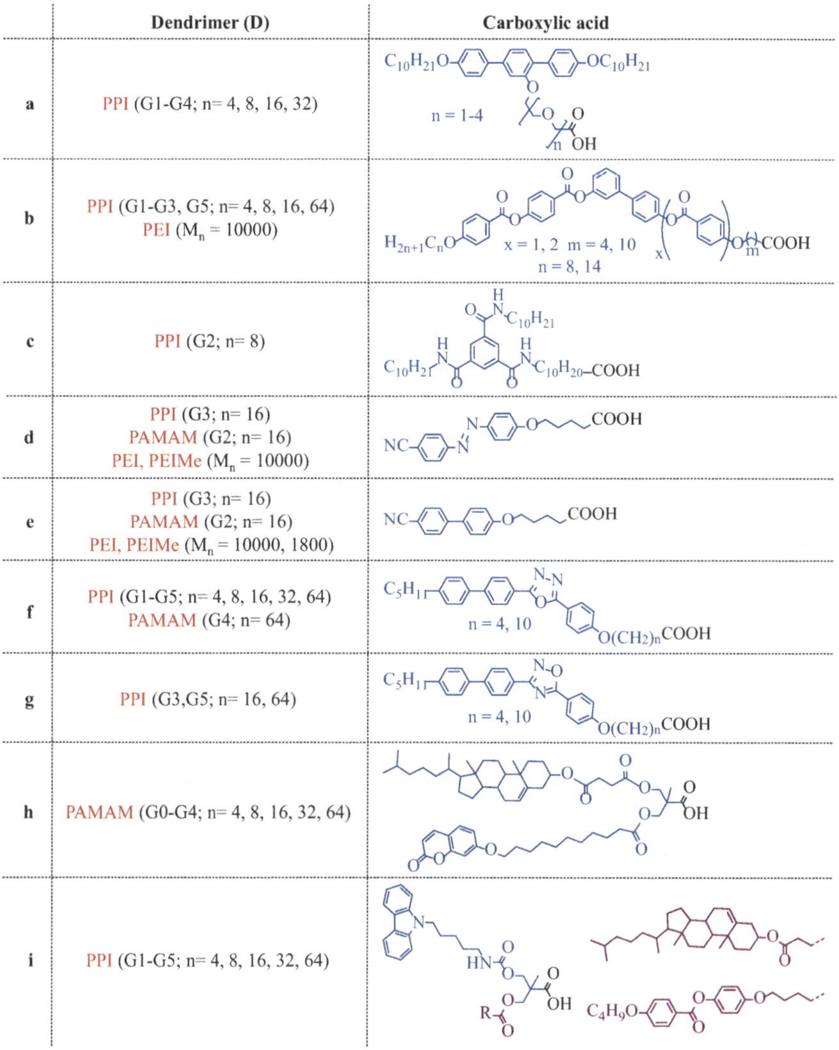

Fig. 4.14 Schematic representation of ionic dendrimers with function-bearing units

a very elegant approach by reacting a second generation PPI dendrimer with a carboxylic acid containing a disk-shaped tridecylbenzene-1,3,5-tricarboxamide unit (Fig. 4.14c) [56, 57]. The ionic dendrimers with dendrimer/carboxylic acid ratio between 1:4 and 1:8 exhibited a rare oblique columnar phase with a well-ordered superlattice formed by column-shaped dendrimers.

Marcos and coworkers reported photoresponsive liquid crystal dendrimers by using an azobenzene-functionalized carboxylic acid and PPI, PAMAM dendrimers, PEI hyperbranched polymer, or fully methylated PEI (PEIMe) (Fig. 4.14d) [58].

These dendrimers showed nematic organizations with large and stable photoinduced optical anisotropy. Interestingly, ionic dendrimers with smectic C phases were obtained by using carboxylic acids bearing cyanobiphenyl units instead of cyanoazobenzene moieties to ionically functionalize PPI or PAMAM dendrimers (Fig. 4.14e) [59, 60]. Moreover, several functional liquid crystal dendrimers have been described by electrostatic complexation of PPI dendrimers with carboxylic acids containing oxadiazole, carbazole, or coumarin fluorescent units (Fig. 4.14f–i) [61–64]. All these ionic dendrimers displayed good luminescent properties and liquid crystalline behavior with smectic and columnar organizations, depending on the dendrimer generation. Moreover, these ionic dendrimers showed promising proton conduction since the liquid crystal organization results in the formation of ionic nanosegregated areas that favor proton transport.

Although the ionic functionalization of the periphery of a preformed dendrimer is the most widely used approach, new dendritic architectures can be prepared through self-assembly of small dendritic building blocks containing charged sites at their focal point. For instance, Percec and coworkers reported the self-assembly of poly(benzyl ether) dendrons with triethylammonium, pyridinium, methylimidazolium, or tris(*n*-propyl)ammonium groups at their focal point (Fig. 4.15) [65, 66]. These dendrons self-organize into supramolecular spheres or columns, which further self-assemble into three-dimensional cubic, and two-dimensional columnar structures, respectively. The driving force for the formation of the observed liquid crystal phases are the electrostatic interactions between ionic moieties, as well as the nanosegregation of the aromatic, aliphatic, and ionic regions. The novelty of this strategy relies on the simplicity of the molecular geometry and the versatility of the structural modification. Similarly, Kato and coworkers made pioneering research on the self-assembly of imidazolium-, phosphonium-, ammonium-containing first-generation dendrons into columnar or cubic liquid crystal organizations that were used in the design of liquid crystals for ion transport [67–71].

4.4 Ionic Self-assembly of Dendrimers in Solution

Ionic dendrimers undergo phase separation in solution, resulting in various self-assembled aggregates. In this case, segregation between incompatible constituent molecular portions occurs when ionic dendrimers are dispersed in a solvent that is selective for one of the parts. Such ability is especially interesting in aqueous solution due to the amphiphilic character of ionic dendrimers, in which segregation gives rise to micelles or vesicles above a critical micellar concentration (similarly to surfactants) [72]. Alternatively, ionic dendrimers can also be decorated with oppositely charged ionic molecules or clusters, resulting in well-defined nano-assemblies with various shapes and functionalities [73–77]. These self-assembled systems were recently reported by Gröhn and coworkers, and will be discussed in detail in Chap. 5. Ionic

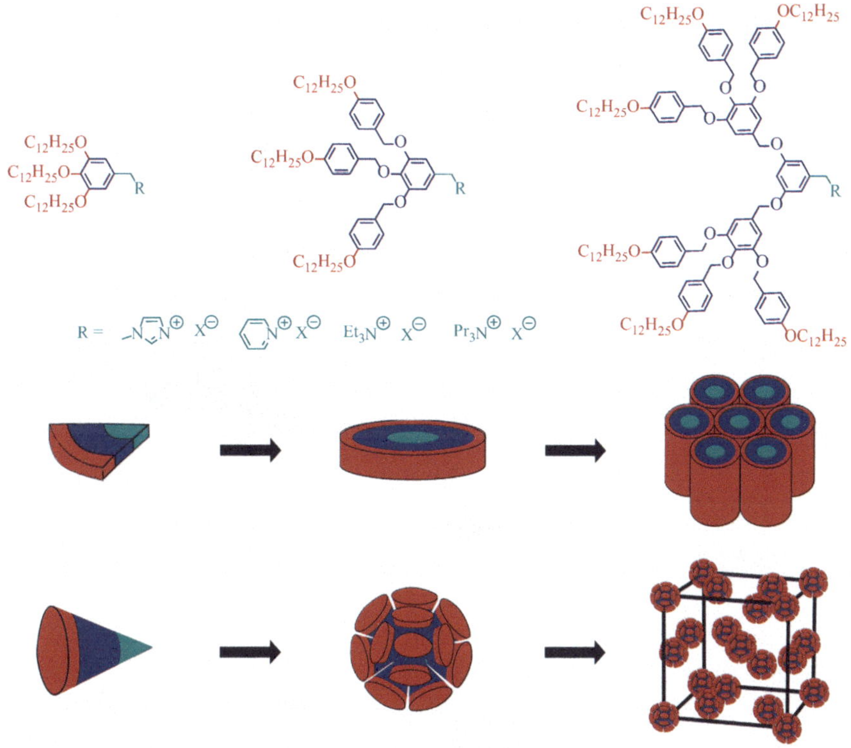

Fig. 4.15 Chemical structures of ionic poly(benzyl ether) dendrons and their self-assembly into columnar or cubic liquid crystal phases

dendrimers can also self-organize into unimolecular micelles, which possess excellent stability under diluted conditions. Unimolecular micelles are single-molecule micelles that have a core and a shell covalently bound together [78, 79].

4.4.1 Self-assembly of Dendritic Amphiphiles

4.4.1.1 Self-assembly of Amphiphilic Macromolecules

In aqueous solution, the self-assembly of amphiphilic macromolecules results in a wide range of morphologies, including vesicles or micelles (Fig. 4.16) [80–84]. Formation of these various morphologies is mostly driven by the attractive interactions between hydrophobic parts, and the repulsive or electrostatic interactions between the hydrophilic parts. Specifically, the hydrophobic segments of the amphiphile self-assemble and locate inside the supramolecular aggregate to minimize energetically unfavorable hydrophobic-water interactions, whereas the hydrophilic

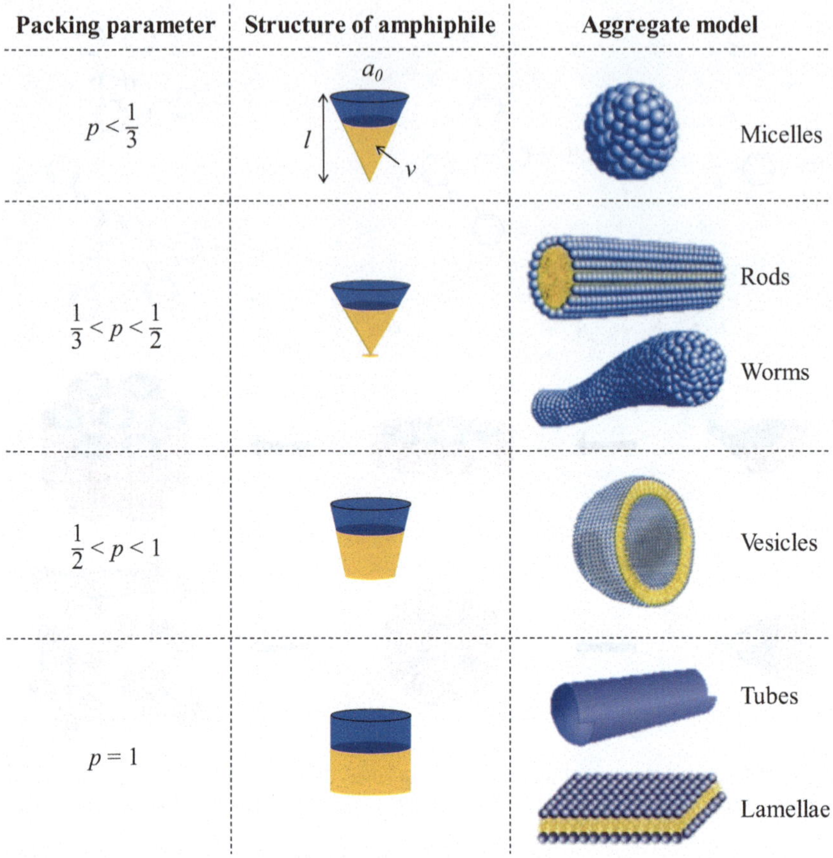

Fig. 4.16 Dependence of the nanostructure morphology on the packing parameter (p) of an amphiphile. Adapted with permission from Ref. [72]. Copyright 2016, American Chemical Society

parts locate at the periphery of the aggregate to maximize the interactions with water and stabilize the aggregate.

The morphology of these aggregates, as well as their size, mainly depends on the molecular structure, geometry, composition, and hydrophilic/hydrophobic ratio. Moreover, a prediction of the morphology of the aggregates can be made by using the packing parameter (p), which describes the geometry of an amphiphile and can be calculated by the equation [85]:

$$p = \frac{v}{a_0 \times l}$$

where v is the volume of the hydrophobic chain, a_0 is the area occupied by the hydrophilic headgroup, and l is the length of the molecule. For $p < 1/3$, amphiphilic molecules adopt a conic geometry and self-organize in spherical micelles. When $1/3 < p < 1/2$, the molecules adopt a truncated cone conformation and arrange in cylindrical micelles. For $1/2 < p < 1$, the amphiphiles also adopt a truncated cone conformation but with dimensions similar to a cylinder, and thus they self-assemble in a flexible bilayer that curves and generates a vesicle. When $p = 1$, molecules adopt a cylindrical shape that self-organizes in a rigid bilayer that might bend and form tubular structures [72].

4.4.1.2 Nanostructures from Ionic Dendrimers

Amphiphilic dendrimers are exceptional alternatives to polymers and block copolymers due to their perfect macromolecular architecture with a precise number of functional groups. Therefore, amphiphilic dendritic systems form in water well-defined nanostructures that have been thoroughly studied [86–88]. Most of these aqueous self-assembled nanostructures are prepared from covalent dendritic polymers, but their time-consuming synthesis might limit their utility and application. An alternative strategy to prepare aqueous self-assembled structures consists of using electrostatic interactions to decorate a preformed dendrimer. Moreover, introducing charged sites in a dendritic scaffold modifies the amphiphilic character of these macromolecules, thereby resulting in nanostructures that combine valuable properties from both polymeric and small molecular systems, i.e. they display membrane features as small-molecule assemblies, and are highly stable as polymeric assemblies.

Serrano and coworkers reported a library of PAMAM dendrimers ionically functionalized with aliphatic chains that self-assembled in water due to their amphiphilic character [89–91] (Fig. 4.17). Interestingly, the same supramolecular interactions which provide liquid crystal behavior in the condensed phase also appear in aqueous solution leading to a rich variety of nanostructures, such as nanospheres, micelles, or lamellae. The morphology of the nanostructures depends on the dendrimer generation, the length of the fatty acid, or the stoichiometric amount of the carboxylic acids attached to the PAMAM core. In the obtained nanostructures, the hydrophobic alkyl chains remain inside of the self-assemblies, while the hydrophilic PAMAM cores are located at the surface. Moreover, these nanostructures showed pH-responsive behavior and could encapsulate both hydrophobic and hydrophilic molecules. In a subsequent study, they prepared micellar nanocarriers with traceable fluorescent units [92]. This was accomplished with a family of amphiphilic dendrimers prepared from a PAMAM dendritic core which was functionalized with bis-MPA dendrons containing coumarin and cholesterol moieties. These ionic dendrimers self-assembled in water forming stable spherical micelles that could be used as versatile capsules for Nile Red, a model hydrophobic molecule. Moreover, the fluorescent traceability of the micelles was demonstrated in vitro, concluding that these ionic dendrimers may be useful as fluorescent polymeric nanocarriers for drug delivery.

Ros and coworkers described unprecedented results on the self-assembly of ionic dendrimers based on a PPI dendritic core functionalized with bent-core molecules through electrostatic interactions (Fig. 4.18) [93]. The compact packing promoted by

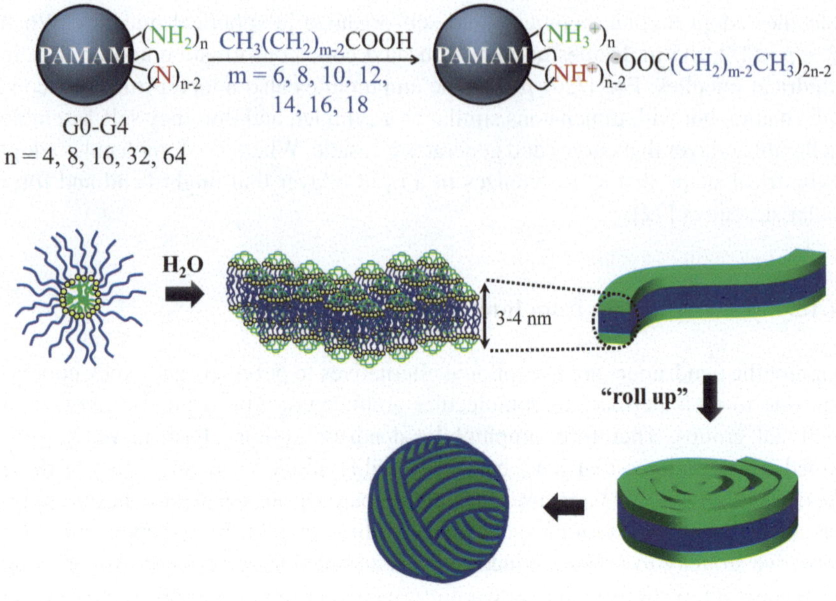

Fig. 4.17 Synthetic route of ionic dendrimers and schematic representation of the self-assembly process in nanospheres

the bent-core structures resulted in the formation of bent-core liquid crystal phases in bulk, and also allowed their self-assembly in water forming a rich variety of nanostructures, including spheres, rods, fibers, helical ribbons, or tubules. The morphology of the obtained nanostructures can be controlled by the appropriate selection of the PPI dendritic core, the length of the inner spacer, or the flexible aliphatic chains. Interestingly, it was possible to create chiral nano-objects formed by achiral molecules as well as to obtain helical structures by a local saddle-splay of the lamellae [94]. In a subsequent study, the same authors demonstrated that these nanostructures can also be produced by microfluidics [95].

Another type of self-assembled nanostructures are dendrimer-based polyion complex micelles, which consist of the electrostatic assembly of ionic polymers (or diblock copolymers) with oppositely charged dendritic cores (Fig. 4.19) [96]. The globular and rigid dendritic structure improves both aqueous solubility and the stability of the micellar assembly in comparison to polyion complex micelles containing linear polymers. Dendrimer-based polyion complex micelles are excellent platforms for anticancer drug delivery [97], photodynamic therapy [98], and nasal administration of drugs [99]. For instance, Qiu and coworkers recently reported the variation of the micellar size with pH in polycationic G2-G7 PAMAM dendrimers with diblock polymers [100]. At pH 7, all dendrimers formed well-defined and stable spherical micelles, which could encapsulate doxorubicin. Nevertheless, at pH 3, the seventh generation PAMAM dendrimer showed large and more polydisperse assemblies. Fernandez-Megia and coworkers studied the influence of the length of the

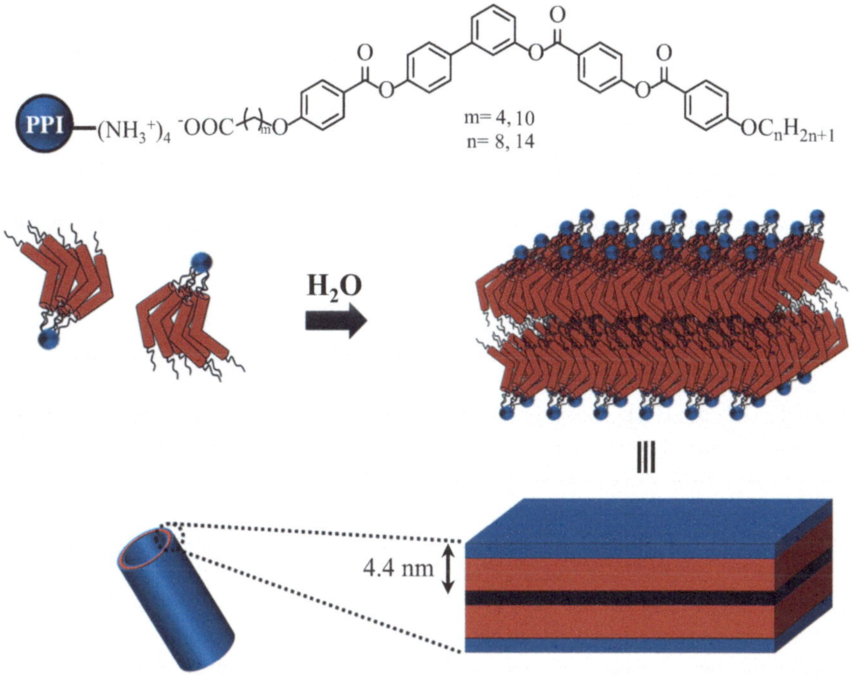

Fig. 4.18 Chemical structure of ionic dendrimers containing bent-core molecules, and the proposed self-assembly model

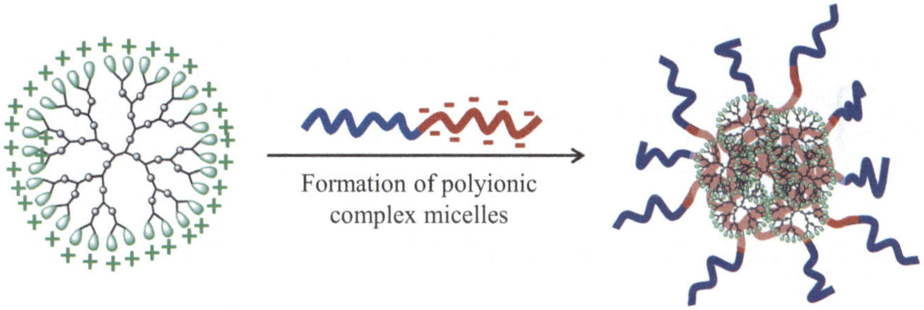

Fig. 4.19 Schematic representation of the formation of polyionic complex micelles

polyethylene glycol (PEG) block and dendrimer generation in the stability of micellar assemblies based on anionic PEG-dendritic copolymers and cationic poly-l-lysine [101]. G3 dendrimers and PEG (Mn ≈ 5 and 10 kDa) gave the best spherical micelles, which were decorated with different functional groups (e.g. carboxylates, sulfates, and sulfonates) at the periphery.

4.4.2 Unimolecular Micelles

As previously explained, spherical micelles typically exhibit core–shell architectures that consist of a hydrophobic core surrounded by a water-soluble shell. Dendrimers are three-dimensional globular macromolecules that have a large number of peripheral units that enable the introduction of hydrophilic charged sites. When this surface modification is performed into a hydrophobic dendrimer, the chemical structure of the dendrimer resembles to a micelle with a hydrophobic core surrounded by a hydrophilic charged shell.

Tomalia reported one of the first examples of unimolecular micelles which consisted of a hydrophobic dendrimer surrounded by 36 carboxylic acids that were converted into their corresponding ammonioum and tetramethylammonium carboxylates (Fig. 4.20a) [102]. These dendrimers generated unimolecular micelles in water that could encapsulate hydrophobic molecules within the dendritic core. Similarly, Hawker and coworkers described the synthesis of water-soluble poly(benzyl ether)-based dendrimers having carboxylate groups as end-groups (Fig. 4.20b) [103]. These dendrimers acted as unimolecular micelles which were able to solvate hydrophobic molecules. Interestingly, a linear relationship between the solubilizing ability and the dendrimer concentration, even at very low concentration of dendrimers, demonstrated the strong encapsulation capabilities of these dendrimers, which did not have a critical micellar concentration. Since then, a plethora of amphiphilic dendrimers have been reported for the encapsulation or extraction of different small guest molecules [79, 104].

Dendrimers with protonable amine groups can interact with nucleic acids by electrostatic interactions to form complexes, which are commonly named "dendriplexes" (Fig. 4.21). Pioneering work on gene delivery with PAMAM dendrimers by Szoka and coworkers has been followed by a growing interest in these systems [105]. Despite the fact that PAMAM dendrimers are cytotoxic and not fully biocompatible, they are probably the most widely used dendrimers for gene delivery because of their commercial availability and their exceptional gene transfection properties with several cell lines [106, 107]. However, alternative strategies were developed to overcome the cytotoxicity issue. For instance, PAMAM dendrimers were decorated with amino acids [108, 109], cyclodextrins [110, 111], or poly(ethylene glycol) units [112].

Moreover, a wide number of dendrimers were studied as vectors for gene transfection, including PPI/PEI, polyesters, or phosphorous dendrimers. PPI dendrimers were selected due to their commercial availability and their structural similarities with PAMAM. Nonetheless, the intrinsic toxicity of PPI dendrimers and the low transfection levels have limited their application [113]. Alternatively, Majoral and coworkers described a new class of water-soluble phosphorous-containing dendrimers with protonated or methylated terminal tertiary amines [114]. These dendrimers were explored as transfection vectors for a luciferase gene. This fireflies' gene makes bacteria produce luciferase, an enzyme that oxidizes luciferin to strongly luminescent oxyluciferin. The transfection ability of these dendrimers was highly dependent

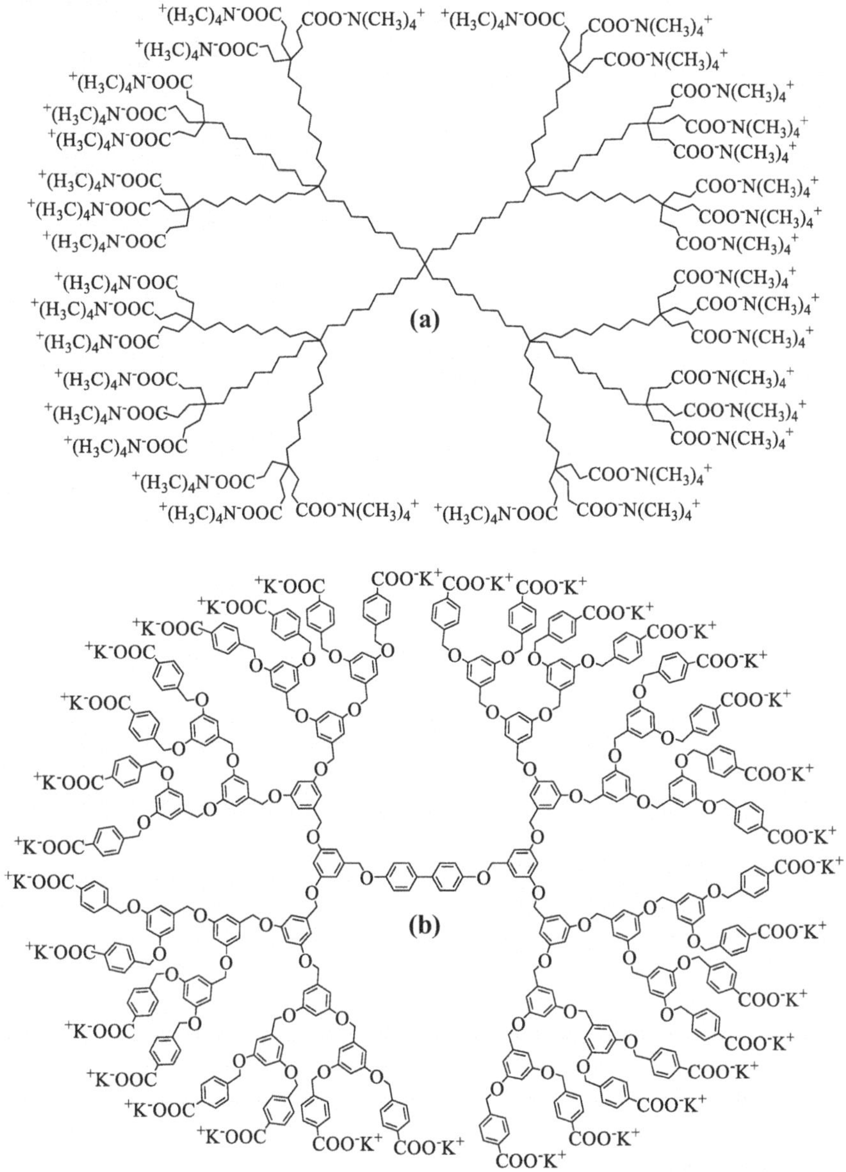

Fig. 4.20 Chemical structures of dendrimers that form unimolecular micelles in water

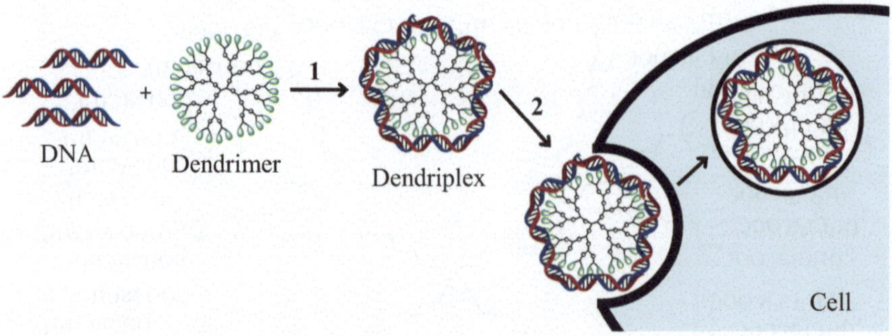

Fig. 4.21 Schematic representation of the proposed mechanism for dendriplexes formation and action: (1) dendrimer forms an ionic complex with the nucleic acid, (2) cell internalization

on the generation of the dendrimer (the first and second generation were less efficient than higher generations) and the chemical nature of the terminal amines (protonated derivatives showed higher efficiency than the methylated analogues). This kind of phosphorous polymers has demonstrated promising properties as gene transfection vectors [115]. Efficient transfection was also found with polyester dendrimers by Serrano and coworkers [116]. They prepared amphiphilic hybrid dendrimers using commercial Boltorn® (bis-MPA hyperbranched polymer) functionalized with amino-terminated bis-MPA dendrons. These dendrimers were non-cytotoxic and biocompatible. Due to the high number of terminal amino groups, they formed complexes with DNA that could be internalized in mesenchymal stem cells.

4.5 Conclusions and Outlook

Dendritic polymers are highly branched macromolecular architectures that present different properties when compared with their linear analogues. Their unique flawless structure with a high functional group density makes dendrimers ideal candidates for high-value applications, including drug delivery, sensors, or catalysis. Moreover, the decoration of dendrimers by means of electrostatic interactions has turned into a powerful and versatile alternative to create new nanostructured materials, since such ionic coupling is usually accompanied by a cooperative binding process, in which the charged units promote further assembly.

In the solid state, ionic dendrimers undergo microphase separation resulting in liquid crystal organizations. In most cases, they form phases that are not observed with conventional non-ionic dendrimers. The electrostatic interactions are essential since the nanosegregation between polar and apolar regions is the driving force for this self-assembly process. Current research is not focused on obtaining liquid crystal properties in dendrimers and thus, the search of applications in these materials is the main objective. The most promising application of ionic liquid crystalline dendrimers is the preparation of ion conductors. Such systems display liquid crystal organizations such that the ionic segregated areas form the continuous channels required for ion transport.

The same nanosegregation phenomenon that occurs in bulk also appears in aqueous solution, leading to a wide range of self-assembled nanostructures, such as micelles (spherical or cylindrical), vesicles, nanotubes, etc. The morphology of the self-assemblies, as well as their size, mostly depends on the chemical structure, composition, and hydrophilic/hydrophobic balance. These nanostructures have been thoroughly investigated for drug delivery. In addition, due to the large molecular weight of ionic dendrimers, they can also self-organize into unimolecular micelles, i.e. micelles composed of a single molecule, which have demonstrated promising properties as vectors for gene transfection.

In conclusion, ionic dendrimers present attractive properties in the design of a wide range of functional materials as they combine the intrinsic properties of the dendritic architecture with the self-assembling properties provided by the electrostatic interactions. Although their application in real world is still a challenge, we firmly believe that this research field has a bright future, and the increasing number of research groups working on ionic dendrimers will serve to demonstrate their true potential in materials science and biomedicine.

References

1. Tomalia DA, Fréchet JMJ (2002) Dendrimers and other dendritic polymers. Wiley, Chichester
2. Caminade AM, Yan D, Smith DK (2015) Dendrimers and hyperbranched polymers. Chem Soc Rev 44(12):3870–3873
3. Yan D, Gao C, Frey H (2011) Hyperbranched polymers: synthesis, properties, and applications. Wiley, Hoboken
4. Buhleier E, Wehner W, Vögtle F (1978) "Cascade"- and "Nonskid-Chain-like" syntheses of molecular cavity topologies. Synthesis 2:155–158
5. Newkome GR, Yao Z, Baker GR, Gupta VK (1985) Micelles. Part 1. Cascade molecules: a new approach to micelles. A [27]-arborol. J Org Chem 50(11):2003–2004
6. Tomalia DA, Naylor AM, Goddard WA (1990) Starburst dendrimers: molecular-level control of size, shape, surface chemistry, topology, and flexibility from atoms to macroscopic matter. Angew Chem Int Ed 29(2):138–175
7. Caminade AM, Ouali A, Laurent R, Turrin CO, Majoral JP (2015) The dendritic effect illustrated with phosphorus dendrimers. Chem Soc Rev 44(12):3890–3899
8. Tomalia DA, Baker H, Dewald J, Hall M, Kallos G, Martin S, Roeck J, Ryder J, Smith P (1986) Dendritic macromolecules: synthesis of starburst dendrimers. Macromolecules 19(9):2466–2468
9. Hawker C, Frechet JMJ (1990) A new convergent approach to monodisperse dendritic macromolecules. J Chem Soc Chem Comm (15), 1010–1013
10. Newkome GR, Moorefield CN, Vögtle F (2001) Dendrimers and dendrons: concepts, syntheses. Applications. Wiley-VCH Verlag GmbH & Co. KGaA, Weinheim
11. Vögtle F, Richardt G, Werner N (2009) Dendrimer chemistry. Wiley-VCH Verlag GmbH & Co. KGaA, Weinheim
12. Astruc D, Boisselier E, Ornelas C (2010) Dendrimers designed for functions: from physical, photophysical, and supramolecular properties to applications in sensing, catalysis, molecular electronics, photonics, and nanomedicine. Chem Rev 110(4):1857–1959

13. Walter MV, Malkoch M (2006) Accelerated approaches to dendrimers. In: Materials science and technology, pp 1027–1056
14. Lundberg P, Hawker CJ, Hult A, Malkoch M (2008) Click assisted one-pot multi-step reactions in polymer science: accelerated synthetic protocols. Macromol Rapid Commun 29(12–13):998–1015
15. Faul CFJ (2014) Ionic self-assembly for functional hierarchical nanostructured materials. Acc Chem Res 47(12):3428–3438
16. Faul CFJ, Antonietti M (2003) Ionic self-assembly: facile synthesis of supramolecular materials. Adv Mater 15(9):673–683
17. Salikolimi K, Sudhakar AA, Ishida Y (2020) Functional ionic liquid crystals. Langmuir 36(40):11702–11731
18. Goossens K, Lava K, Bielawski CW, Binnemans K (2016) Ionic liquid crystals: versatile materials. Chem Rev 116(8):4643–4807
19. Stoddart JF, Ashton PR, Shibata K, Shipway AN (1997) Polycationic dendrimers. Angew Chem Int Ed 36(24):2781–2783
20. Kleij AW, van de Coevering R, Klein Gebbink RJM, Noordman A-M, Spek AL, van Koten G (2001) Polycationic (Mixed) core-shell dendrimers for binding and delivery of inorganic/organic substrates. Chem Eur J 7(1):181–192
21. van de Coevering R, Bruijnincx PCA, Lutz M, Spek AL, van Koten G, Klein Gebbink RJM (2007) Ionic core–shell dendrimers with a polycationic core: structural aspects and host–guest binding properties. New J Chem 31(7):1337–1348
22. van de Coevering R, Bruijnincx PCA, van Walree CA, Klein Gebbink RJM, van Koten G (2007) Dendritic host molecules with a polycationic core and an outer shell of dodecyl groups. Eur J Org Chem 2007(18):2931–2939
23. van de Coevering R, Alfers AP, Meeldijk JD, Martínez-Viviente E, Pregosin PS, Klein Gebbink RJM, van Koten G (2006) Ionic core−shell dendrimers with an octacationic core as noncovalent supports for homogeneous catalysts. J Am Chem Soc 128(39):12700–12713
24. van de Coevering R, Kuil M, Gebbink RJMK, van Koten G (2002) A polycationic dendrimer as noncovalent support for anionic organometallic complexes. Chem Commun 15:1636–1637
25. Ong W, Grindstaff J, Sobransingh D, Toba R, Quintela JM, Peinador C, Kaifer AE (2005) Electrochemical and guest binding properties of fréchet- and newkome-type dendrimers with a single viologen unit located at their apical positions. J Am Chem Soc 127(10):3353–3361
26. Caminade AM (2014) Metallodendrimers. Inorg Chim Acta 409:1–184
27. Wöhrle T, Wurzbach I, Kirres J, Kostidou A, Kapernaum N, Litterscheidt J, Haenle JC, Staffeld P, Baro A, Giesselmann F, Laschat S (2015) Discotic liquid crystals. Chem Rev 116(3):1139–1241
28. Gimeno, N.; Ros, M. B., (2014) Chemical structures, mesogenic properties, and synthesis of liquid crystals with bent-core structures. In: Goodby JW, Collings PJ, Kato T, Tschierske C, Gleeson H, Raynes P (eds) Handbook of liquid crystals, vol 4. Wiley-VCH Verlag GmbH & Co. KGaA, pp 1–75
29. Gorecka E, Vaupotič N, Pociecha D (2014) Smectic phases of bent-core liquid crystals. In: Goodby JW, Collings PJ, Kato T, Tschierske C, Gleeson HF, Raynes P (eds) Handbook of liquid crystals, vol 4. Wiley-VCH, Weinheim
30. Gorecka E, Pociecha D, Vaupotič N (2014) Columnar liquid crystalline phases made of bent-core mesogens. In: Goodby JW, Collings PJ, Kato T, Tschierske C, Gleeson HF, Raynes P (eds) Handbook of liquid crystals, vol 4. Wiley-VCH, Weinheim
31. Ungar G, Liu F, Zeng X (2014) Cubic and Other 3D thermotropic liquid crystal phases and quasicrystals. In: Goodby JW, Collings PJ, Kato T, Tschierske C, Gleeson HF, Raynes P (eds) Handbook of liquid crystals, vol 5. Wiley-VCH Verlag GmbH & Co. KGaA, pp 363–436
32. van der Asdonk P, Kouwer PHJ (2017) Liquid crystal templating as an approach to spatially and temporally organise soft matter. Chem Soc Rev 46(19):5935–5949
33. Tschierske C (2013) Development of structural complexity by liquid-crystal self-assembly. Angew Chem Int Ed 52(34):8828–8878
34. Li Q (2012) Liquid crystals beyond displays. Wiley, Hoboken

35. Kato T, Uchida J, Ichikawa T, Sakamoto T (2018) Functional liquid crystals towards the next generation of materials. Angew Chem Int Ed 57(16):4355–4371
36. Kato T, Yoshio M, Ichikawa T, Soberats B, Ohno H, Funahashi M (2017) Transport of ions and electrons in nanostructured liquid crystals. Nat Rev Mater 2:17001
37. Concellón A, Zentner CA, Swager TM (2019) Dynamic complex liquid crystal emulsions. J Am Chem Soc 141(45):18246–18255
38. Nayani K, Yang Y, Yu H, Jani P, Mavrikakis M, Abbott N (2020) Areas of opportunity related to design of chemical and biological sensors based on liquid crystals. Liq Cryst Today 29(2):24–35
39. Concellón A, Fong D, Swager TM (2021) Complex liquid crystal emulsions for biosensing. J Am Chem Soc 143(24):9177–9182
40. Hernández-Ainsa S, Marcos M, Serrano JL (2014) Dendrimeric and hyperbranched liquid crystal structures. In: Goodby JW, Collings PJ, Kato T, Tschierske C, Gleeson H, Raynes P (eds) Handbook of liquid crystals, vol 7. Wiley-VCH Verlag GmbH & Co. KGaA, pp 259–300
41. Donnio B, Buathong S, Bury I, Guillon D (2007) Liquid crystalline dendrimers. Chem Soc Rev 36(9):1495–1513
42. Barberá J, Donnio B, Gehringer L, Guillon D, Marcos M, Omenat A, Serrano JL (2005) Self-organization of nanostructured functional dendrimers. J Mater Chem 15(38):4093–4105
43. Marcos M, Martín-Rapún R, Omenat A, Serrano JL (2007) Highly congested liquid crystal structures: dendrimers, dendrons, dendronized and hyperbranched polymers. Chem Soc Rev 36(12):1889–1901
44. Friberg SE, Podzimek M, Tomalia DA, Hedstrand DM (1988) A non-aqueous lyotropic liquid crystal with a starburst dendrimer as a solvent. Mol Cryst Liq Cryst 164(1):157–165
45. Tsiourvas D, Felekis T, Sideratou Z, Paleos CM (2004) Ionic liquid crystals derived from the protonation of poly(propylene imine) dendrimers with a cholesterol-based carboxylic acid. Liq Cryst 31(5):739–744
46. Ujiie S, Yano Y, Mori A (2004) Liquid-crystalline branched polymers having ionic moieties. Mol Cryst Liq Cryst 411(1):483–489
47. Martín-Rapún R, Marcos M, Omenat A, Barberá J, Romero P, Serrano JL (2005) Ionic thermotropic liquid crystal dendrimers. J Am Chem Soc 127(20):7397–7403
48. Marcos M, Martín-Rapún R, Omenat A, Barberá J, Serrano JL (2006) Ionic liquid crystal dendrimers with mono-, di- and trisubstituted benzoic acids. Chem Mater 18(5):1206–1212
49. Martín-Rapún R, Marcos M, Omenat A, Serrano JL, de Givenchy ET, Guittard F (2007) Liquid crystalline semifluorinated ionic dendrimers. Liq Cryst 34(3):395–400
50. Hernández-Ainsa S, Marcos M, Barberá J, Serrano JL (2010) Philic and phobic segregation in liquid-crystal ionic dendrimers: an enthalpy-entropy competition. Angew Chem Int Ed 49(11):1990–1994
51. Hernández-Ainsa S, Barberá J, Marcos M, Serrano JL (2010) Effect of the phobic segregation between fluorinated and perhydrogenated chains on the supramolecular organization in ionic aromatic dendrimers. Chem Mater 22(16):4762–4768
52. Mezzenga R, Ruokolainen J, Canilho N, Kasëmi E, Schlüter DA, Lee WB, Fredrickson GH (2009) Frustrated self-assembly of dendron and dendrimer-based supramolecular liquid crystals. Soft Matter 5(1):92–97
53. Soininen AJ, Kasëmi E, Schlüter AD, Ikkala O, Ruokolainen J, Mezzenga R (2010) Self-assembly and induced circular dichroism in dendritic supramolecules with cholesteric pendant groups. J Am Chem Soc 132(31):10882–10890
54. Cook AG, Baumeister U, Tschierske C (2005) Supramolecular dendrimers: unusual mesophases of ionic liquid crystals derived from protonation of DAB dendrimers with facial amphiphilic carboxylic acids. J Mater Chem 15(17):1708–1721
55. Vergara J, Gimeno N, Cano M, Barberá J, Romero P, Serrano JL, Ros MB (2011) Mesomorphism from bent-core based ionic dendritic macromolecules. Chem Mater 23(22):4931–4940
56. Fitié CFC, Mendes E, Hempenius MA, Sijbesma RP (2011) Self-assembled superlattices of polyamines in a columnar liquid crystal. Macromolecules 44(4):757–766

57. Fitié CFC, Tomatsu I, Byelov D, de Jeu WH, Sijbesma RP (2008) Nanostructured materials through orthogonal self-assembly in a columnar liquid crystal. Chem Mater 20(6):2394–2404
58. Marcos M, Alcalá R, Barberá J, Romero P, Sánchez C, Serrano JL (2008) Photosensitive ionic nematic liquid crystalline complexes based on dendrimers and hyperbranched polymers and a cyanoazobenzene carboxylic acid. Chem Mater 20(16):5209–5217
59. Hernández-Ainsa S, Barberá J, Marcos M, Romero P, Serrano JL (2012) Thermotropic mesomorphism via self-assembly of cationic dendritic polymers with an anionic polar carboxylic acid. Macromol Chem Phys 213(3):270–277
60. Chen Y, Shen Z, Gehringer L, Frey H, Stiriba S-E (2006) Supramolecular thermotropic liquid crystalline materials with nematic mesophase based on methylated hyperbranched polyethylenimine and mesogenic carboxylic acid. Macromol Rapid Commun 27(1):69–75
61. Concellón A, Hernández-Ainsa S, Barberá J, Romero P, Serrano JL, Marcos M (2018) Proton conductive ionic liquid crystalline poly(ethyleneimine) polymers functionalized with oxadiazole. RSC Adv 8(66):37700–37706
62. Hernández-Ainsa S, Barberá J, Marcos M, Serrano JL (2012) Liquid crystalline ionic dendrimers containing luminescent oxadiazole moieties. Macromolecules 45(2):1006–1015
63. Castelar S, Romero P, Serrano JL, Barberá J, Marcos M (2015) Multifunctional ionic hybrid poly(propyleneimine) dendrimers surrounded by carbazole dendrons: liquid crystals, optical and electrochemical properties. RSC Adv 5(81):65932–65941
64. Concellón A, Liang T, Schenning APHJ, Serrano JL, Romero P, Marcos M (2018) Proton-conductive materials formed by coumarin photocrosslinked ionic liquid crystal dendrimers. J Mater Chem C 6(5):1000–1007
65. Imam MR, Peterca M, Edlund U, Balagurusamy VSK, Percec V (2009) Dendronized supramolecular polymers self-assembled from dendritic ionic liquids. J Polym Sci Part A: Polym Chem 47(16):4165–4193
66. Ungar G, Liu Y, Zeng X, Percec V, Cho W-D (2003) Giant supramolecular liquid crystal lattice. Science 299(5610):1208–1211
67. Yoshio M, Mukai T, Ohno H, Kato T (2004) One-dimensional ion transport in self-organized columnar ionic liquids. J Am Chem Soc 126(4):994–995
68. Masafumi Y, Takahiro I, Harutoki S, Takayoshi K, Atsushi H, Tomohiro M, Hiroyuki O, Takashi K (2007) Columnar liquid-crystalline imidazolium salts. Effects of anions and cations on mesomorphic properties and ionic conductivities. Bull Chem Soc Jpn 80(9):1836–1841
69. Soberats B, Uchida E, Yoshio M, Kagimoto J, Ohno H, Kato T (2014) Macroscopic photo-control of ion-transporting pathways of a nanostructured imidazolium-based photoresponsive liquid crystal. J Am Chem Soc 136(27):9552–9555
70. Ichikawa T, Yoshio M, Hamasaki A, Mukai T, Ohno H, Kato T (2007) Self-organization of room-temperature ionic liquids exhibiting liquid-crystalline bicontinuous cubic phases: formation of nano-ion channel networks. J Am Chem Soc 129(35):10662–10663
71. Ichikawa T, Yoshio M, Hamasaki A, Taguchi S, Liu F, Zeng XB, Ungar G, Ohno H, Kato T (2012) Induction of thermotropic bicontinuous cubic phases in liquid-crystalline ammonium and phosphonium salts. J Am Chem Soc 134(5):2634–2643
72. Thota BNS, Urner LH, Haag R (2016) Supramolecular architectures of dendritic amphiphiles in water. Chem Rev 116(4):2079–2102
73. Krieger A, Fuenzalida Werner JP, Mariani G, Gröhn F (2017) Functional supramolecular porphyrin-dendrimer assemblies for light harvesting and photocatalysis. Macromolecules 50(9):3464–3475
74. Mariani G, Moldenhauer D, Schweins R, Gröhn F (2016) Elucidating electrostatic self-assembly: molecular parameters as key to thermodynamics and nanoparticle shape. J Am Chem Soc 138(4):1280–1293
75. Düring J, Hölzer A, Kolb U, Branscheid R, Gröhn F (2013) Supramolecular organic-inorganic hybrid assemblies with tunable particle size: interplay of three noncovalent interactions. Angew Chem Int Ed 52(33):8742–8745
76. Willerich I, Gröhn F (2010) Photoswitchable nanoassemblies by electrostatic self-assembly. Angew Chem Int Ed 49(44):8104–8108

77. Willerich I, Gröhn F (2008) Switchable nanoassemblies from macroions and multivalent dye counterions. Chem Eur J 14(30):9112–9116
78. Lukowiak MC, Thota BNS, Haag R (2015) Dendritic core–shell systems as soft drug delivery nanocarriers. Biotechnol Adv 33(6, Part 3):1327–1341
79. Fan X, Li Z, Loh XJ (2016) Recent development of unimolecular micelles as functional materials and applications. Polym Chem 7(38):5898–5919
80. Riess G (2003) Micellization of block copolymers. Prog Polym Sci 28(7):1107–1170
81. Letchford K, Burt H (2007) A review of the formation and classification of amphiphilic block copolymer nanoparticulate structures: micelles, nanospheres, nanocapsules and polymersomes. Eur J Pharm Biopharm 65(3):259–269
82. Smart T, Lomas H, Massignani M, Flores-Merino MV, Perez LR, Battaglia G (2008) Block copolymer nanostructures. Nano Today 3(3–4):38–46
83. Blanazs A, Armes SP, Ryan AJ (2009) Self-assembled block copolymer aggregates: from micelles to vesicles and their biological applications. Macromol Rapid Commun 30(4–5):267–277
84. Mai Y, Eisenberg A (2012) Self-assembly of block copolymers. Chem Soc Rev 41(18):5969–5985
85. Israelachvili JN, Mitchell DJ, Ninham BW (1976) Theory of self-assembly of hydrocarbon amphiphiles into micelles and bilayers. J Chem Soc Faraday Trans 72:1525–1568
86. Wei T, Chen C, Liu J, Liu C, Posocco P, Liu X, Cheng Q, Huo S, Liang Z, Fermeglia M, Pricl S, Liang XJ, Rocchi P, Peng L (2015) Anticancer drug nanomicelles formed by self-assembling amphiphilic dendrimer to combat cancer drug resistance. PNAS 112(10):2978–2983
87. Sherman SE, Xiao Q, Percec V (2017) Mimicking complex biological membranes and their programmable glycan ligands with dendrimersomes and glycodendrimersomes. Chem Rev 117(9):6538–6631
88. Liko F, Hindré F, Fernández-Megía E (2016) Dendrimers as innovative radiopharmaceuticals in cancer radionanotherapy. Biomacromol 17(10):3103–3114
89. Fedeli E, Hernández-Ainsa S, Lancelot A, González-Pastor R, Calvo P, Sierra T, Serrano JL (2015) Nanoobjects formed by ionic PAMAM dendrimers: hydrophilic/lipophilic modulation and encapsulation properties. Soft Matter 11(30):6009–6017
90. Hernández-Ainsa S, Fedeli E, Barberá J, Marcos M, Sierra T, Serrano JL (2014) Self-assembly modulation in ionic PAMAM derivatives. Soft Matter 10(2):281–289
91. Hernández-Ainsa S, Barberá J, Marcos M, Serrano JL (2011) Nanoobjects coming from mesomorphic ionic PAMAM dendrimers. Soft Matter 7(6):2560–2568
92. Concellón A, San Anselmo M, Hernández-Ainsa S, Romero P, Marcos M, Serrano JL (2020) Micellar nanocarriers from dendritic macromolecules containing fluorescent coumarin moieties. Polymers 12(12):2872
93. Cano M, Sánchez-Ferrer A, Serrano JL, Gimeno N, Ros MB (2014) Supramolecular architectures from bent-core dendritic molecules. Angew Chem Int Ed 53(49):13449–13453
94. Castillo-Vallés M, Cano M, Bermejo-Sanz A, Gimeno N, Ros MB (2020) Towards supramolecular nanostructured materials: control of the self-assembly of ionic bent-core amphiphiles. J Mater Chem C 8(6):1998–2007
95. Castillo-Vallés M, Romero P, Sebastián V, Ros MB (2021) Microfluidics for the rapid and controlled preparation of organic nanotubes of bent-core based dendrimers. Nanoscale Adv 3(6):1682–1689
96. Mignani S, Shi X, Zablocka M, Majoral J-P (2021) Dendritic macromolecular architectures: dendrimer-based polyion complex micelles. Biomacromol 22(2):262–274
97. Lopez-Blanco R, Fernandez-Villamarin M, Jatunov S, Novoa-Carballal R, Fernandez-Megia E (2019) Polysaccharides meet dendrimers to fine-tune the stability and release properties of polyion complex micelles. Polym Chem 10(34):4709–4717
98. Nishiyama N, Jang W-D, Kataoka K (2007) Supramolecular nanocarriers integrated with dendrimers encapsulating photosensitizers for effective photodynamic therapy and photochemical gene delivery. New J Chem 31(7):1074–1082

99. Mignani S, Shi X, Karpus A, Majoral JP (2021) Non-invasive intranasal administration route directly to the brain using dendrimer nanoplatforms: an opportunity to develop new CNS drugs. Eur J Med Chem 209, 112905
100. Qiu Z, Huang J, Liu L, Li C, Cohen Stuart MA, Wang J (2020) Effects of pH on the formation of PIC micelles from PAMAM dendrimers. Langmuir 36(29):8367–8374
101. Fernandez-Villamarin M, Sousa-Herves A, Porto S, Guldris N, Martínez-Costas J, Riguera R, Fernandez-Megia E (2017) A dendrimer–hydrophobic interaction synergy improves the stability of polyion complex micelles. Polym Chem 8(16):2528–2537
102. Newkome GR, Moorefield CN, Baker GR, Saunders MJ, Grossman SH (1991) Unimolecular micelles. Angew Chem Int Ed 30(9):1178–1180
103. Hawker CJ, Wooley KL, Fréchet, JMJ (1993) Unimolecular micelles and globular amphiphiles: dendritic macromolecules as novel recyclable solubilization agents. J Chem Soc Perkin Trans 1(12):1287–1297
104. Hayouni S, Robert A, Maes C, Conreux A, Marin B, Mohamadou A, Bouquillon S (2018) New dendritic ionic liquids (DILs) for the extraction of metallic species from water. New J Chem 42(22):18010–18020
105. Haensler J, Szoka FC (1993) Polyamidoamine cascade polymers mediate efficient transfection of cells in culture. Bioconjugate Chem 4(5):372–379
106. Kesharwani P, Banerjee S, Gupta U, Mohd Amin MCI, Padhye S, Sarkar FH, Iyer AK (2015) PAMAM dendrimers as promising nanocarriers for RNAi therapeutics. Mater Today 18(10):565–572
107. Labieniec-Watala M, Watala C (2015) PAMAM dendrimers: destined for success or doomed to fail? Plain and modified PAMAM dendrimers in the context of biomedical applications. J Pharm Sci 104(1):2–14
108. Choi JS, Nam K, Park J-Y, Kim J-B, Lee J-K, Park J-S (2004) Enhanced transfection efficiency of PAMAM dendrimer by surface modification with l-arginine. J Control Release 99(3):445–456
109. Kono K, Akiyama H, Takahashi T, Takagishi T, Harada A (2005) Transfection activity of polyamidoamine dendrimers having hydrophobic amino acid residues in the periphery. Bioconjugate Chem 16(1):208–214
110. Kihara F, Arima H, Tsutsumi T, Hirayama F, Uekama K (2003) In vitro and in vivo gene transfer by an optimized α-cyclodextrin conjugate with polyamidoamine dendrimer. Bioconjugate Chem 14(2):342–350
111. Arima H, Kihara F, Hirayama F, Uekama K (2001) Enhancement of gene expression by polyamidoamine dendrimer conjugates with α-, β-, and γ-cyclodextrins. Bioconjugate Chem 12(4):476–484
112. Luong D, Kesharwani P, Deshmukh R, Mohd Amin MCI, Gupta U, Greish K, Iyer AK (2016) PEGylated PAMAM dendrimers: enhancing efficacy and mitigating toxicity for effective anticancer drug and gene delivery. Acta Biomater 43:14–29
113. Nam K, Jung S, Nam J-P, Kim SW (2015) Poly(ethylenimine) conjugated bioreducible dendrimer for efficient gene delivery. J Control Release 220:447–455
114. Loup C, Zanta M-A, Caminade A-M, Majoral J-P, Meunier B (1999) Preparation of water-soluble cationic phosphorus-containing dendrimers as DNA transfecting agents. Chem Eur J 5(12):3644–3650
115. Shcharbin D, Dzmitruk V, Shakhbazau A, Goncharova N, Seviaryn I, Kosmacheva S, Potapnev M, Pedziwiatr-Werbicka E, Bryszewska M, Talabaev M, Chernov A, Kulchitsky V, Caminade A-M, Majoral J-P (2011) Fourth generation phosphorus-containing dendrimers: prospective drug and gene delivery carrier. Pharmaceutics 3(3):458–473
116. Lancelot A, González-Pastor R, Concellón A, Sierra T, Martín-Duque P, Serrano JL (2017) DNA transfection to mesenchymal stem cells using a novel type of pseudodendrimer based on 2,2-bis(hydroxymethyl)propionic acid. Bioconjugate Chem 28(4):1135–1150

Chapter 5
Nano-Objects by Spontaneous Electrostatic Self-Assembly in Aqueous Solution

Alexander Zika, Anja Krieger, and Franziska Gröhn

Abstract Electrostatic self-assembly for the formation of well-defined nano-objects in solution represents an emerging field. The key is to use building blocks with a certain geometry and/or a combination of noncovalent interaction types. Polyelectrolytes providing both stability and designability are especially valuable. Central for a targeted design is the fundamental understanding of structure-directing effects. This chapter addresses both an introduction to the field of spontaneous electrostatic self-assembly as well as the state of the art of its understanding. It commences from the special effects of polyelectrolytes, shortly pictures the established areas of polyelectrolyte complexes and block polyelectrolyte micelles, before focusing on novel approaches: Besides discussing the interplay of ionic and π-π interaction in a dendrimer-dye model system, we elucidate how thermodynamics encodes the nanoscale structure: the free energy determines the aggregation number but the entropy/enthalpy ratio the nano-object's shape. The approach's versatility applicable for building blocks from linear or cylindrical brush polyelectrolytes, proteins, DNA, polyoxometalate clusters to micelles is demonstrated. We provide examples of photocatalytic activity and triggering of the nano-objects' size, shape, or function (e.g., enzyme activity) by addressing multiple stimuli such as pH and light. This includes the novel use of photoacids as interconnecting counterions with light-switchable valency.

Alexander Zika and Anja Krieger: These authors have contributed equally to this work and share first authorship.

A. Zika · A. Krieger · F. Gröhn (✉)
Department of Chemistry and Pharmacy and Interdisciplinary Center for Molecular Materials (ICMM), Bavarian Polymer Institute (BPI), Friedrich-Alexander University (FAU) Erlangen-Nürnberg, Egerlandstraße 3, 91058 Erlangen, Germany
e-mail: franziska.groehn@fau.de

A. Zika
e-mail: alexander.az.zika@fau.de

A. Krieger
e-mail: anja.krieger@fau.de

M. A. Aboudzadeh and A. Frontera (eds.), *Supramolecular Assemblies Based on Electrostatic Interactions*,
https://doi.org/10.1007/978-3-031-00657-9_5

5.1 Introduction

Self-assembly, the structure formation by intermolecular interaction, is responsible for forming almost all architectures and functions in natural systems and self-assembled structures are ubiquitous in nature. Often, polyelectrolytes, polymers that carry many charges, play a key role — DNA, polysaccharides, and proteins. For example, the electrostatic interaction of DNA and small histones, small peptides, causes the DNA to fit into the cell nucleus and enables the DNA readout. Inspired by the principles of Mother Nature, also synthetic self-assembly has elicited a fair bit of importance. Potential exists in a broad range of areas such as drug delivery, sensors, and catalysis.

Presently, there is plenty of scope for designing both switchable as well as functional structures by electrostatic self-assembly in solution. “Electrostatic interaction” or “ionic interaction”, that is, the Coulomb attraction or repulsion between charged particles, is an important intermolecular force. It denotes the interaction of oppositely charged small ions that hold together salt crystals such as sodium chloride. Pauling distinguished three types of chemical bonds: the electrostatic, the covalent, and the metallic bond, while he already emphasized that the transition from one to the other type is gradual [1]. According to Pauling, the “ionic bond” results from the excess electric charge of oppositely charged ions, while the expression “electrostatic bond” typically refers to a situation where “each of two atoms or groups of atoms have a definite electronic structure such that electrostatic interactions are set up”. In other words, the electrostatic bond of Pauling includes ion-dipole and dipole-dipole interactions with induced and permanent dipoles.

The premise of electrostatic self-assembly is the attraction of ions that are oppositely charged. Despite the robust and long-ranged interaction and the widespread availability of ionic species of various kinds, defined nanostructures in solution that can be formed by electrostatic self-assembly is only a more recent phenomenon than say, hydrogen bonding or π-π interaction. This is attributed to the electrostatic potential’s missing directionality since a single charge inevitably shows the potential of spherical interaction across all directions instead of hydrogen bonding and π-π interaction that are associated with directed binding motives. However, developments in the recent past suggest that electrostatic self-assembly is capable of yielding nano-objects that are well-defined in solution. Here, it is imperative to implement electrostatics along with other structure-directing impacts like steric effects or other types of noncovalent interactions. Existing research focuses on developing a foundational comprehension of structural control and fabricating an increasing number of architectures that are capable of performing a specific function. A great versatility of electrostatic self-assembly originates in a large variety of ionic building blocks that can facilitate effective function design and targeted structure.

5.2 Polyelectrolytes as Building Bocks

Building blocks of especial value are polyelectrolytes, polymers that carry many charges, as they lend both stability and "designability". The many charges can partly serve as interconnection and partly remain as charges to electrostatically stabilize assemblies in solution. In addition, polyelectrolytes exist as polymers in various architectures and with various chemistries. Figure 5.1 sketches different polyelectrolyte architectures. Figure 5.2 shows the chemistry of typical polyelectrolytes and polymers which become polyelectrolytes in certain pH ranges.

Polyelectrolytes (or macroions) in solution are always accompanied by many small oppositely charged ions, so-called counterions. This leads to a complex system of overlaying attractive and repulsive, enthalpic, and entropic effects in a multicomponent system, even for a "simple" polyelectrolyte solution. The conformation of a polyelectrolyte in solution is the result of a complex interplay of electrostatic interaction, intrinsic excluded volume, hydrophobic effects, and specific counterion properties, and thus a complete theoretical understanding is lacking despite many prior studies. In solution, an equilibrium distribution of counterions around a polyelectrolyte is realized where a certain number of counterions usually is closely associated with the polymer backbone, sometimes referred to as "counterion condensation" yielding a polyelectrolyte with an effective charge that is lower than the number of chemically

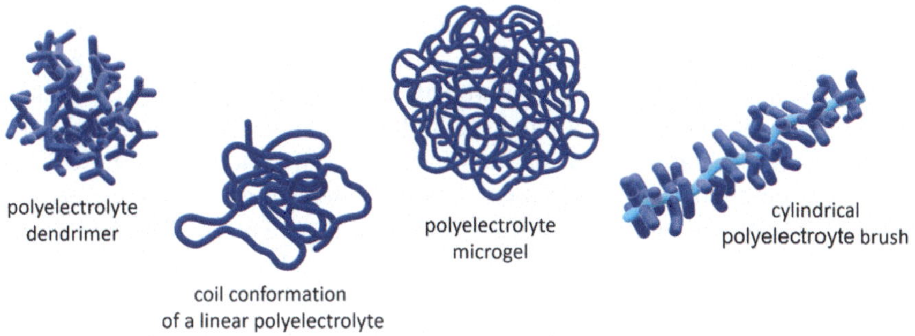

Fig. 5.1 Schematic representation of different polyelectrolyte architectures

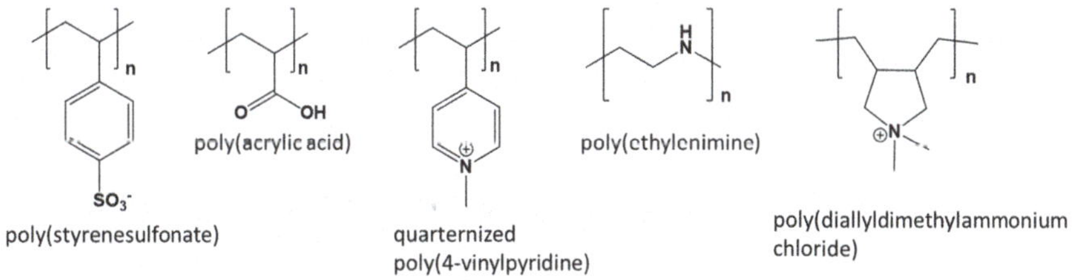

Fig. 5.2 Selection of typical polyelectrolytes and polymers that can be deprotonated or protonated into polyelectrolytes

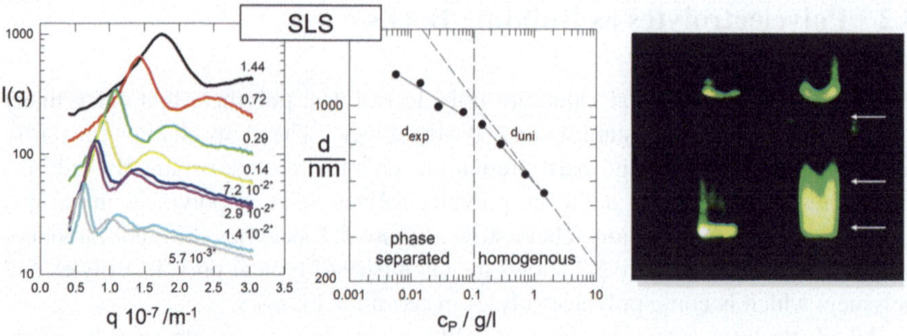

Fig. 5.3 Structure formation of polyelectrolyte microgels in salt-free solution: Static light scattering (SLS) shows expressed structure factor peaks *(left)* due to an ordering of the macroions in solution. Quantitative analysis reveals that the interparticle distance is smaller than that of uniform distribution (*middle*); i.e., domains with higher particle density. At certain concentrations, this is also visible macroscopically under irradiation (*right*): only the lower part of the solution exhibits expressed scattering. In this case, 50 nm microgels show interparticle distances from 400 to 1500 nm. Reprinted with permission from ref. [5]. Copyright © 2000 American Chemical Society

charged repeat units, along with a distribution of further counterions ensconcing the polyelectrolyte as "counterion cloud" [2, 3]. Yet, beyond the individual and symmetric counterion clouds, this complex interplay causes like-charged polyelectrolytes to form domains in solution, as shared counterions and/or temporarily unsymmetrical counterion clouds mediate attraction [4, 5]. These phenomena are far beyond what a mean-field Debye-Hückel theory can access. This gives rise to polyelectrolyte effects, that is, the effects that are only observed for macroions but widely found for any kind of organic or inorganic, or biologic macroion. This includes the occurrence of a fast and slow mode in diffusion as well as an increase in reduced viscosity upon dilution [5–8]. However, even in these denser solution regions, the distances in-between macroions still are about ten times greater than their diameter, and these effects are observed with very low to no added low molecular salt in solutions. Figure 5.3 illustrates these effects based on results for a model system of polyelectrolyte microgels, i.e., small highly swollen hydrophilic spherical polyelectrolytes with sizes ranging from 5 to 500 nm — without hydrophobic interiors in the absence of any Hamaker interactions or hydrophobic effects, and in the absence of sedimentation issues, as a pure electrostatic model system [5].

Nevertheless, regardless of the salt concentration, a polyelectrolyte will be ensconced by counterions, as indicated on the left in Fig. 5.4. This sketches a polyelectrolyte with its monovalent counterions in an aqueous solution. In contrast to the counterion cloud with monovalent counterions, it is well known that upon the addition of multivalent counterions, polyelectrolytes collapse, aggregate and precipitate [9–12]. The effect of the valency of added salt is immense. The electrostatic potential scales with the valency to the 6th power, known as the Schulze Hardy rule [13]. Simultaneously, it is also a question of the concentration of macroions and counterions and a question of the interparticle distance. In addition, this association was shown to be entropy-driven, due to the release of a larger number of initial

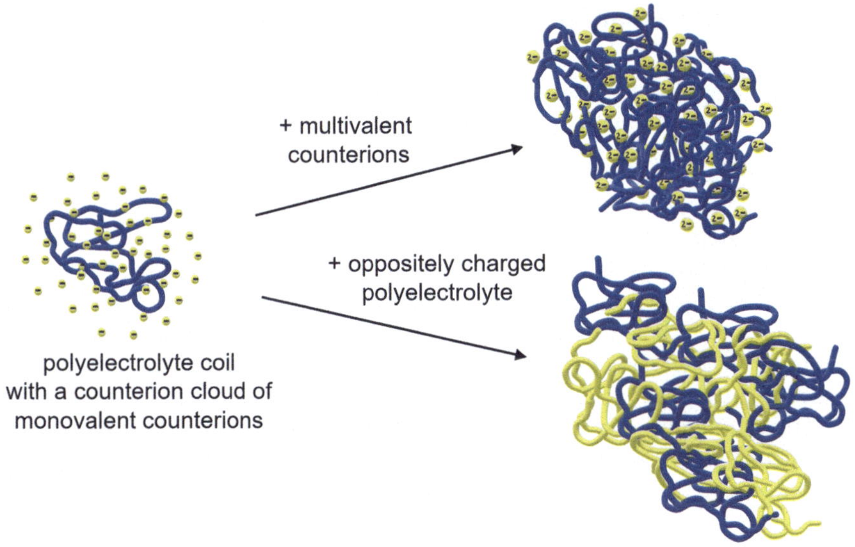

Fig. 5.4 Polyelectrolyte coil with its counterion cloud (*left*) and aggregation upon addition of divalent counterions (*top right*) and with an oppositely charged polyelectrolyte (*bottom right*)

monovalent counterions upon addition of the multivalent ions, and occurs although the process is endothermic [14]. However, from the standpoint of structure formation, the linear flexible polyelectrolyte with spherical symmetric counterions does not represent a system that entails any structure-directing effects which could be exploited significantly.

Similarly, the association of two flexible polyelectrolytes occurs. Here, depending on the molecular masses, the association often is kinetically controlled. With some exceptions that will be discussed below, this association, while potentially being of fundamental interest, also offers limited structural design principles.

These are the reasons why the formation of stable defined assemblies in solution by electrostatic interaction has been considered almost impossible by many. Yet, due to the immense potential for self-assembled structures in aqueous solution, it was imperative to find new concepts for self-assembly exploiting electrostatic interactions for the formation of defined nano-objects. When solid structured polyelectrolyte-surfactant materials, as well as multilayered polyelectrolyte structures, were already well known, the structuring in solution by the spontaneous assembly was still a challenge. This topic has elicited considerable attention over the past few years. The solution here is to combine electrostatics with other structure-directing effects, such as geometric constraints or secondary additional noncovalent interactions.

5.3 Intermolecular Interactions for Self-Assembly

As the combination of electrostatic and other interactions plays a key role in these newer concepts, we first revisit different noncovalent interactions separately.

The hydrophobic effect involves amphiphilic components such as surfactants possessing hydrophilic as well as hydrophobic attributes to create micelles in water [15]. The system's free energy reduces as the water molecules' entropy rises when they no longer have to order around hydrophobic tails upon the formation of micelles [16, 17]. The hydrophobic effect assumes great importance in protein folding [18, 19]. Synthetically, both vesicles [20] and intricate structures like trefoil knots could be formed [21], whereas the supramolecular materials' development extends from surfactants to amphiphilic block copolymers [22, 23]. In addition, the hydrophobic effect plays a significant role in reactions involving ligand-metal catalysts such as the Suzuki-coupling where hydrophobic oxygen-free cores are generated by the ligands for the metal. [24]

Hydrogen bonds refer to the interplay between positively polarized hydrogen atoms and highly electronegative atoms like nitrogen or oxygen atoms. Here, the nitrogen or oxygen atom serves as the electron donor whereas the hydrogen atom acts as the electron acceptor. [25] The strength of hydrogen bonds ranges from 5 kJ mol^{-1} up to >100 kJ mol^{-1} [26, 27]. These bonds are responsible for folding proteins, for DNA base pair association [28], and also causing viruses such as SARS-CoV-2 for binding to the receptors [29, 30]. Often, these bonds' directing properties are utilized for self-assembly to form polymeric structures or dimers [31, 32]. Furthermore, hydrogen bonds can also form even more complex structures such as helix-turn-helix comprising helices of varied orientations and handedness [33]. Polymers with high tensile strength and self-healing properties are synthesized by implementing hierarchical single, double, and quadruple hydrogen-bonding moieties in a polymer backbone mimicking the folding of proteins [34–36].

π-π interactions are premised on aromatic systems that are electron-intensive. The interaction's strength, which can be up to 50 kJ mol^{-1} is predicated on substituents attached to the π-system [37]. Geometrically, the possibilities include an edge-to-face/face-to-face/side-by-side staircase. For example, porphyrins that interact by π-π stacking can form stacks that can be grouped into face-to-face H-aggregates, side-by-side J-aggregates, and couple electronically, depending on the substituents, the concentration, as well as the added molecules [38–40]. Moreover, π-π stacking is also known to provide benefits for supramolecular electronics. Fiber structures are formed by oligo(thiophene) and copolymers of fluorene or indenofluorene with oligo(thiophene) [41, 42]. Nanowires from self-assembled polythiophenes reveal high conductivities (40 S/cm) [43]. Efficient charge transport is also achieved with small molecules such as planar perylene-bis(dicarboximide) (PDI) derivatives, which, in turn, form millimeter-long fibers with a constant cross-section of a few hundred nanometers by means of solvent-vapor annealing [44]. π-π interactions also play a key role for aromatic-rich dipeptides, which form

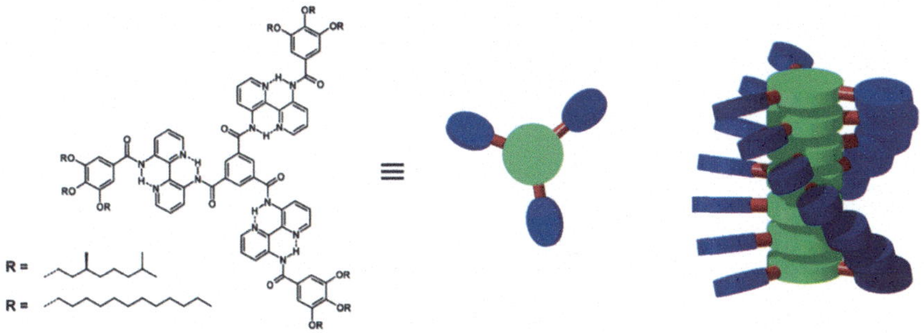

Fig. 5.5 Bipyridine-based C3-symmetrical disks and a cartoon representing their helical supramolecular stacking. Reprinted with permission from ref. [50]. Copyright © 2002 American Chemical Society

organogels and reveal the potential for wastewater treatment and oil spill recovery [45]. Meijer and colleagues introduced polymer-linked self-assembled helical structures of bifunctional triazines [46–48]. π-π stacking helps in forming the backbone of the helical polymer. For example, bipyridine-based disks and functionalized oligo(p-phenylenevinylene)s, both of which have a C3-symmetry are capable of acting as the backbone of supramolecular fibers [49, 50]. Perylene diimides and polycyclic aromatic hydrocarbons are other such backbones (Fig. 5.5) [51].

5.4 Polyelectrolyte-Polyelectrolyte Aggregates

There is a long history of studying the aggregation of two polyelectrolytes that are charged oppositely [52–62]. When such (that is, oppositely charged polymers) engage with each other, this leads to the formation of polyelectrolyte complexes. The polyelectrolyte complexes' stoichiometry, composition, and structure depend on various factors like concentration and molecular mass, polymer structure, the mixing ratio, pH, salt concentration, ionic strength, as well as hydrophobicity [53–57] Figure 5.6 considers complex particles consisting of poly(styrene) sulfonate (PSS) and poly(diallyldimethylammonium chloride) (PDADMAC), displaying the dependence of the complexes' radius as the function of salt concentration (Debye screening length) and PSS molar mass [55]. Depending on the molar mass and charge ratio, so-called "ladder-like" or "scrambled egg-like" aggregates are formed from oppositely charged polyelectrolytes. Furthermore, it has been discussed that for smaller molar masses, i.e., less highly charged components, more uniform structures can be formed, while kinetic effects usually lead to a broader size distribution for higher molar masses. The formation of the primary complex is attributed to the electrostatic interactions and is known to occur at a rapid pace. It was suggested that the polymers rearrange within the complex under the second step. Similarly, under the third step, the polyelectrolyte complexes would be interacting with each other [53,

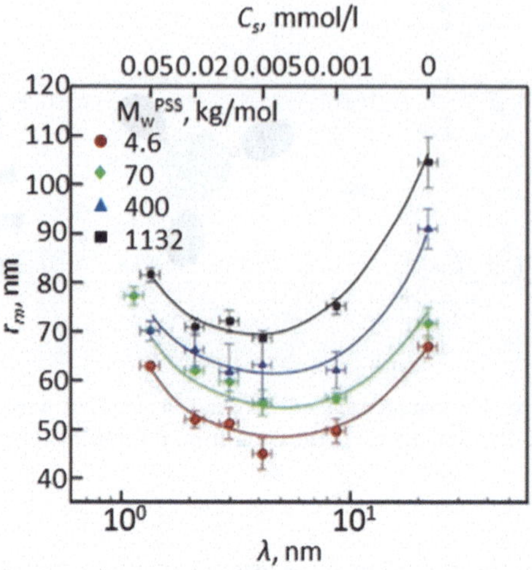

Fig. 5.6 Dependence of the size of PDADMAC-PSS-PEC particles r_m on the Debye length λ (lower x-axis) and salt concentration Cs (upper x-axis) for different molecular masses of PSS, M_{wPSS} (C_{PEL} = 0.001 $n-/n+$ = 1.5). Reprinted with permission from ref. [55]. Copyright © 2012 American Chemical Society

58–61]. Recently, different uptake properties were found for polyelectrolyte complex nanoparticles composed of cationic poly(l-lysine) (PLL) and various anionic polysaccharides with human vascular endothelial cells [57]. Therefore, it was postulated that the different complexes may either serve to deliver growth factors to endothelial cells of bone reconstitution material, whereas others have an advantage for substituting biomimetic bone scaffold materials.

In summary, in addition to usually having an undefined size and form, these polyelectrolyte assemblies tend to agglomerate and precipitate with time, owing to the high charge numbers and kinetic processes associated with the assembly. To introduce structural control to a greater extent, cylindrical polyelectrolyte brushes have also been used in conjunction with linear polyelectrolytes, thus resulting in structures ranging from wormlike to spheres, based on the composition [62]. Furthermore, it is possible for polyelectrolyte-polyelectrolyte materials to originate from the coating of spherical thermoresponsive microgels using linear polyelectrolytes [63].

5.5 Polyelectrolyte Assemblies with Block-Polyelectrolytes

A specific kind of polyelectrolyte-polyelectrolyte materials is those involving double-hydrophilic block copolymers. In this case, polyelectrolyte blocks that are charged oppositely get attached and create an unsoluble interior of a micelle, with an uncharged hydrophilic block forming the micelle corona, thus facilitating the formation of stable/well-defined micelle-such as complexes in solution. Building on the

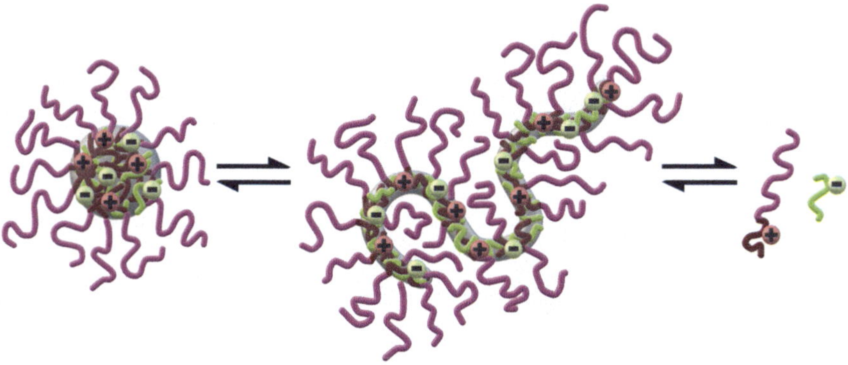

Fig. 5.7 Electrostatic self-assembly with double-hydrophilic block-polyelectrolytes: Formation of spherical complex coacervate core micelles from anionic poly(acrylic acid) (red) and block copolymer comprising a cationic poly(N-methyl-2-vinylpyridinium) (blue) and a neutral poly(ethylene oxide) block (green); an increase in the salt concentration leads to the formation of an elongated wormlike structure; an even further increase in the salt concentration above the critical salt concentration leads to the separation of the polymers. Figure based on results reported in reference 66

pioneering work of Štěpánek and Procházka [64, 65], Cohen Stuart et al. examined numerous systems forming myriad defined structures [66–74]. It is possible for these shapes to range from spherical micelles to more intricate disks such as micelles with an asymmetric corona and a coacervate core [66–68]. As far as complex coacervate core micelles created from poly(acrylic acid) and poly(N-methyl-2-vinylpyridinium)-b-poly(ethylene oxide) are concerned, it is possible to change the shape via the added salt concentration from spherical micelles to elongated structures below the critical salt concentration in case the poly(acrylic acid) homopolymer is not overtly long, as shown in Fig. 5.7. Meanwhile, due to electrostatic interactions, other copolymers form raspberry-like precipitates [75]. Moreover, semipermeable polymer vesicles can also be formed from two block copolymers (oppositely charged), which can serve as oxygen carriers, as demonstrated by Kataoka et al. in Fig. 5.8 [76–79].

In addition to the utilization of polyelectrolyte-polyelectrolyte complexes in the form of carrier systems, block copolymers' electrostatically self-assembled materials reveal an efficient ion transport and impart mechanical stability to render them potentially suitable materials for lithium-ion batteries [80]. Furthermore, electrostatically self-assembled conjugated polymers can serve as excitonic donor/acceptor pairs, thereby acting as light-harvesting antennae because the energy transfer from the donor to the acceptor conducting polyelectrolyte (CPE) is less than 250 fs [81].

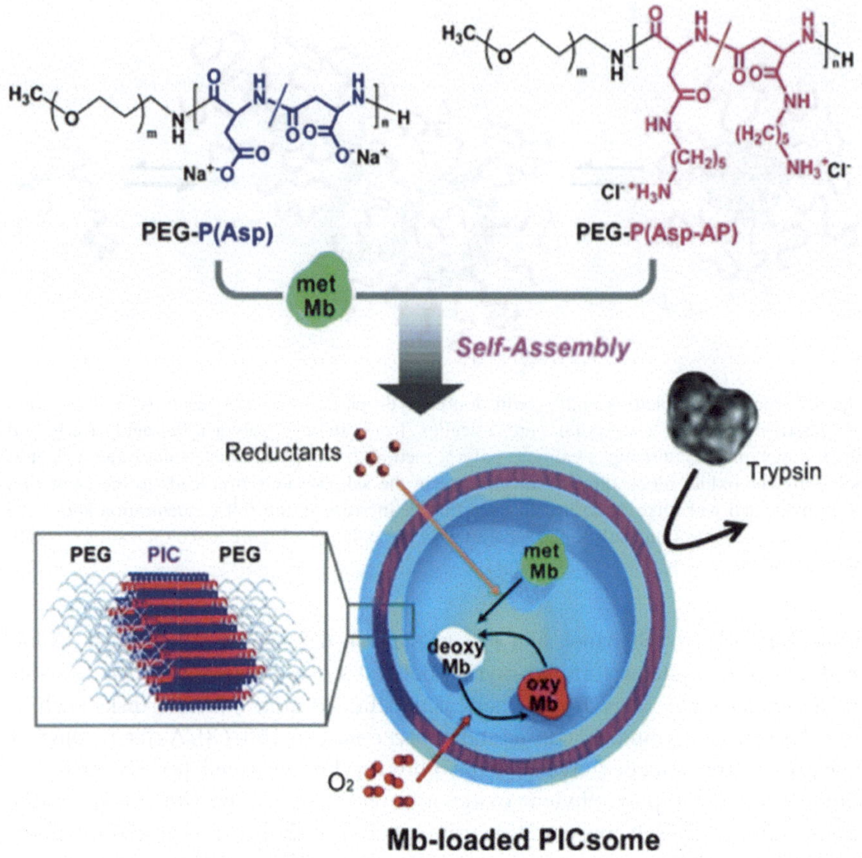

Fig. 5.8 Electrostatic self-assembly with double-hydrophilic block-polyelectrolytes: Reversible myoglobin (Mb) oxygenation inside the polyion complex membrane self-assembled from a pair of oppositely charged block-polyelectrolytes. Reprinted with permission from ref. [78]. Copyright © 2007 WILEY–VCH Verlag GmbH & Co. KGaA, Weinheim

5.6 Nano-objects by Electrostatic Self-assembly in Solution: The Principle

Figure 5.9 shows the novel concept introduced by the Gröhn group for forming electrostatically self-assembled nano-objects in solution via the usage of "structural counterions." The authors presented a plethora of supramolecular structures in solution that were created by electrostatic self-assembly [82–97]. This concept is premised on self-assembly by electrostatic interactions. However, it makes use of structural counterions that can reveal secondary impacts like mutual π-π interactions or geometric factors for directing the structure formation. This results in stable nanoscale assemblies in solution that have a specific shape and size. Concerning the objective to form well-defined nanoscale objects, the approach of using two polyelectrolytes

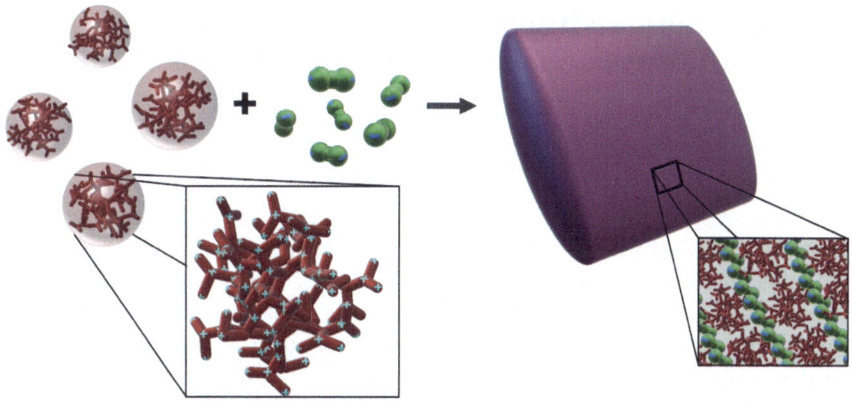

Fig. 5.9 Electrostatic self-assembly in solution: polyelectrolytes interact with oppositely charged stiff, multivalent counterions to form nano-objects with a defined size and shape; for example, cationic 5 nm-sized G4 PAMAM dendrimers interact with dianionic dye molecules to form a 200 nm elongated structures with elliptical cross-section and a layered internal structure

or block polyelectrolyte complexes is extended so that it ensures a more generic use of several building blocks. Directional effects could accommodate geometric attributes such as the building blocks' rigidity or symmetry in solution or secondary noncovalent forces, including hydrogen bonds, or π-π interaction that help safeguard structure control in solution. As a case in point, the macroion in a model system is a highly branched polyelectrolyte poly(amidoamine) (PAMAM) dendrimer that has a recognizable geometry with several primary and ternary charged groups. There is an interaction of this dendrimeric macroion with small organic dye molecules that carry two negative charges. After the association of one dye molecule caused by ionic interaction, further dye molecules are attached to the macroion adjacently to each other undergoing mutual π-π interactions in addition to the electrostatic interaction [83, 86]. The overall shape and internal structure of the resulting supramolecular assembly are defined, thus ensuring its stability in solution. Figure 5.9 illustrates this principle. After studying various linear and branched polymers as well as their interaction with numerous rigid ionic molecules, stable supramolecular particles over wide parameter ranges with narrow size distribution were created in an aqueous solution

The supramolecular structures are ready to use. Advantages of well-defined supramolecular structures include straightforward "synthesis" and handling, which obviates the need to involve centrifugation or separation of the precipitate and purification processes. This is because the assemblies are stable in solution. Owing to the large variety of possible building blocks, the assemblies' applications range from catalysts to light-harvesting materials for converting solar energy to switchable structures that can be deployed as sensors [88, 91, 94, 98–100].

5.7 The Dendrimer-Dye Model System: Understanding the Structure Formation in Electrostatic Self-assembly

5.7.1 Structural Analysis

The first proof of concept for this type of structure formation was based on assemblies of cationic PAMAM dendrimers of the fifth generation (G5) forming nano-objects with defined size and spherical or cylindrical shape in methanolic solution with 1,4- or 2,3-naphthalene dicarboxylic acid, respectively [82]. With a view to deriving further insights, a systematic study of cationic PAMAM dendrimer-anionic dye assemblies was done using a set of aromatic and organic azo dyes involving 2–3 sulfonate groups at several positions (as well as corresponding aliphatic organic molecules) as counterion in an aqueous solution. According to the findings, the dyes' molecular structure and the dye to dendrimer charges' molar charge ratio play a key role in ascertaining the assemblies' size and shapes [86, 101–104]. Figure. 5.10 shows atomic force microscopy (AFM) images of two samples formed from ionic dendrimers of different generations with the same ionic azo dye, thus demonstrating the crucial influence of the building blocks on the supramolecular aggregate's shape. Despite the varying sizes and shapes, dynamic light scattering (DLS) revealed well-defined sizes in the range of 20–440 nm (Fig. 5.11).

Thereafter, a detailed nanostructure characterization was carried out on this dendrimer-dye model system by altering the dendrimer generation (as well as the polyelectrolyte architecture, but this paragraph focuses on the dendrimer-dye model system). Figure 5.12 illustrates the molecular structures. DLS and static light scattering (SLS), small-angle neutron scattering (SANS), and atomic force microscopy (AFM) were carried out to characterize the nanoscale structure, and UV-Vis spectroscopy was performed to reveal π-π stacking characteristics. Meanwhile ζ-potential measurements revealed the corresponding charge properties. Isothermal

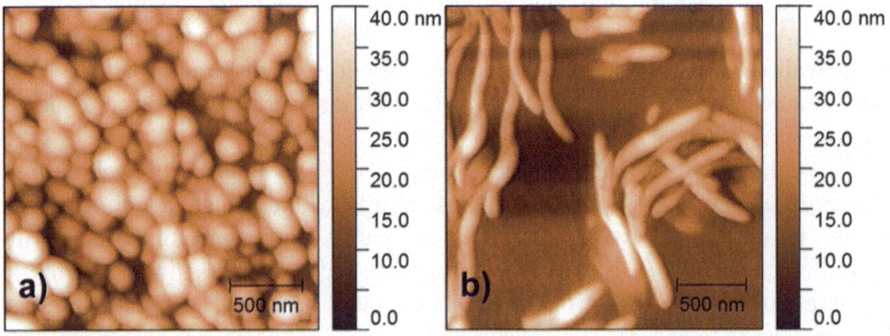

Fig. 5.10 AFM of electrostatically assembled dendrimer-dye structures: **a** Ar26 and G4-dendrimer; **b** Ar26 and G8-dendrimer, both loading ratio (c(primary dendrimer amine groups)/c(dye sulfonate groups) $l = 1.8$). Reprinted with permission from ref. [104]. Copyright © 2016 American Chemical Society

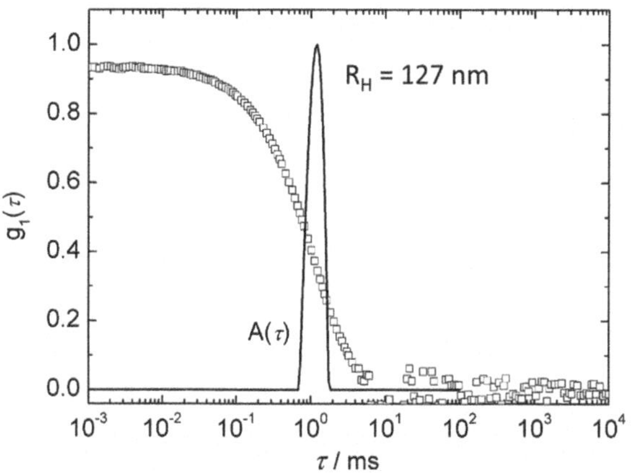

Fig. 5.11 DLS: electric field autocorrelation function and decay time distribution at a scattering angle $\theta = 90°$ for acid red 44 (Ar44)-G4 dendrimer with a ratio $l = 2.1$: nano-objects with a defined size and a narrow size distribution are detected. Reprinted with permission from ref. [86]. Copyright © 2016 Springer-Verlag GmbH Germany

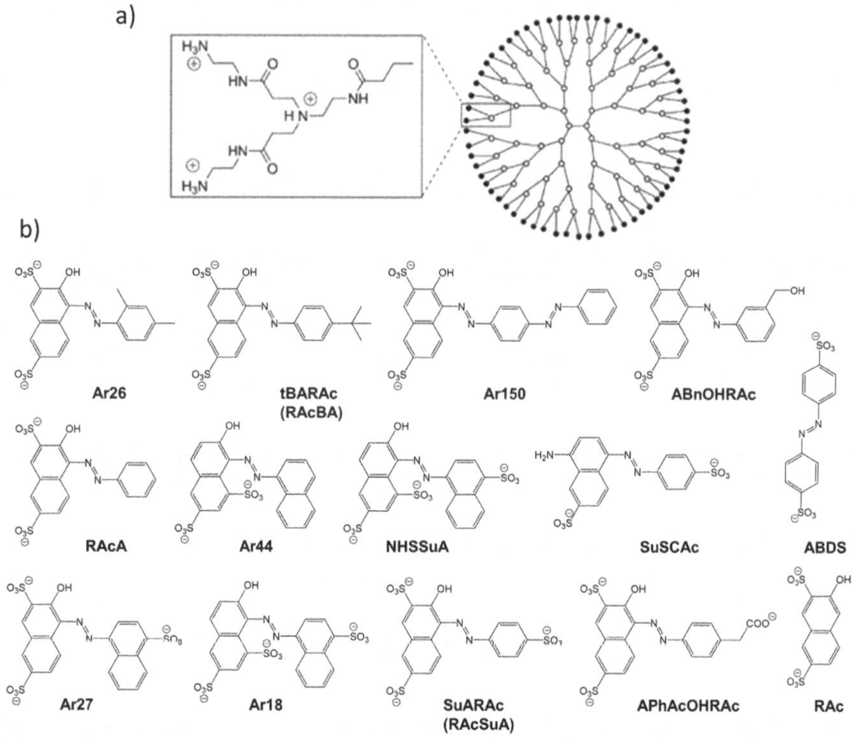

Fig. 5.12 Building blocks in the dendrimer-dye model system for electrostatic self-assembly: **a** PAMAM dendrimer structure; **b** structural formulas of structurally related divalent and trivalent azo dyes

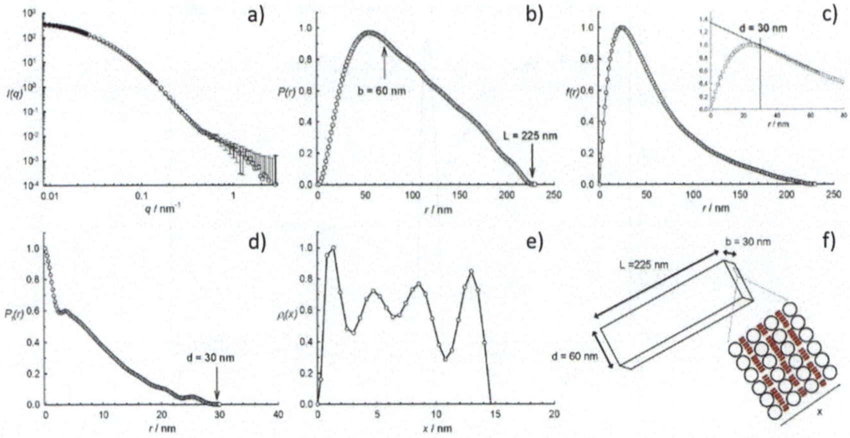

Fig. 5.13 LS-SANS characterization of supramolecular dendrimer-dye assemblies (G4-Ar26): **a** scattering curve I(q) with fit to the data, **b** pair distance distribution functions P(r), **c** f(r) = P(r)/r, **d** thickness pair distance distribution function $P_t(r)$, **e** thickness-density profile $\rho_t(x)$, x = r/2. P(r) indicates anisotropic aggregates of 225 nm in length. In this case, the cylindrical cross-section is also anisotropic with a thickness of 30 nm and a diameter of 60 nm. The density profile across the flat structure $\rho_t(x)$ is consistent with an internal structure of alternating dendrimers and dye columns, as indicated in **f**. Reprinted with permission from ref. [83]. Copyright © 2008 WILEY–VCH Verlag GmbH & Co. KGaA, Weinheim and from ref. [89]. Copyright © 2010 The Royal Society of Chemistry

titration calorimetry (ITC) tracking was undertaken of the heat evolution that occurred during the assembly formation process.

SLS, in conjunction with SANS, reveals the nanoscale particle shape. The form factors measured can be analyzed either via Fourier transformation into the pair distance distribution function and further deconvolution into the density profile and/or by fitting the experimental form factors with structural models. Figure 5.13 illustrates an example of the structural analysis according to the first method [83, 99]. From the P(r) and cross-section density profile analysis in Fig. 5.13, it is evident that entities with different neutron scattering contrast alternate within the particle, confirming the internal structure sketched in Fig. 5.9. The structural analysis of many samples is given in Fig. 5.14, for varying dendrimer-dye systems and in dependence on the loading ratio (scattering curves within one plot). It reveals that the general nano-object symmetry is encoded in the choice of building block, while the component ratio determines the size.

Considering the sheer diversity of possible shapes, it becomes necessary to determine what provides control over the shape, size, and stability of self-assembled aggregates in solution. In particular, we focus on equilibrium structures here. With the divalent counterions used here, equilibrium structures form independent of the pathway of formation. This becomes more complicated when moving to tetravalent porphyrins below, where equilibrium structures or kinetically controlled structures

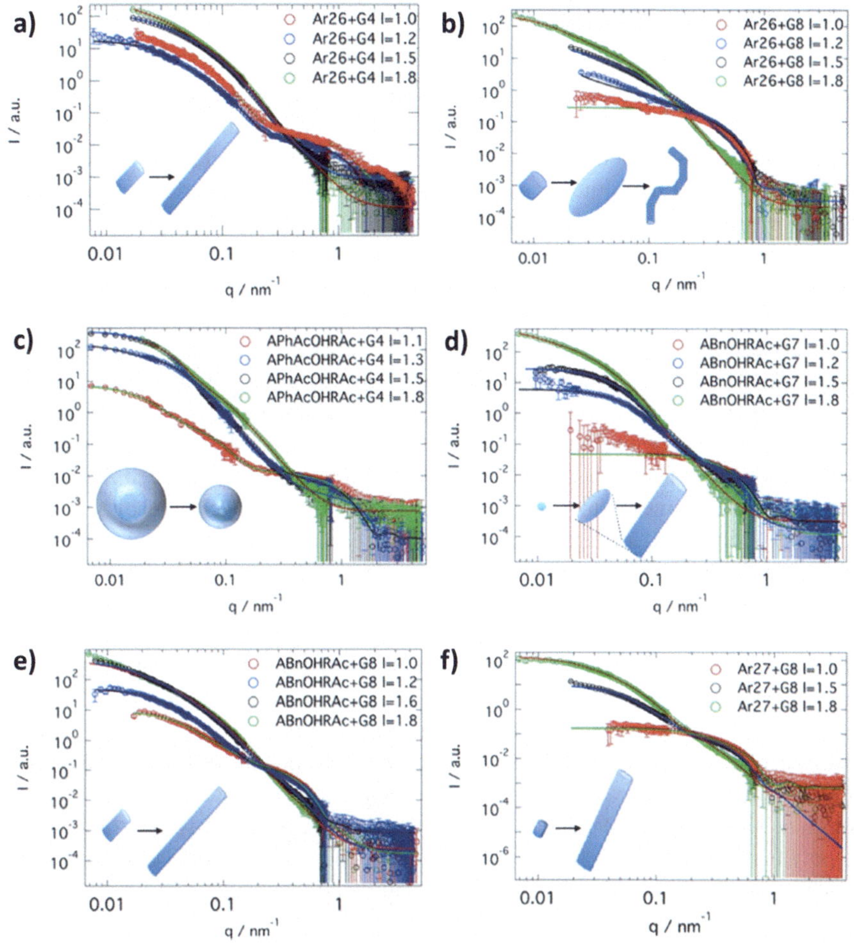

Fig. 5.14 SANS analysis of a dendrimer-dye model system for electrostatic self-assembly in solution: Scattering curves and resulting nanoparticle shapes as a function of dye type, dendrimer generation, and loading ratio: SANS results for **a** Ar26 + G4, **b** Ar26 + G8, **c** APhAcOHRAc + G4, **d** ABnOHRAc + G7, **e** ABnOHRAc + G8, and **f** SuACAc + G8, each at varying loading ratio. (The loading ratio l given in the legend is twice the charge ratio, i.e., a loading ratio $l = 2.0$ corresponds to charge stochiometry). Continuous lines represent the best structural fit. Reprinted with permission from ref. [104]. Copyright © 2016 American Chemical Society

form depending on the pathway. In particular, it is interesting to determine what stabilizes the assemblies with a finite size in an aqueous solution.

5.7.2 *Particle Stabilization*

The charge ratio of the cationic groups to anionic groups influences the stability of the assemblies, as indicated in Fig. 5.15a. The charge status of the nano-objects can be analyzed by ζ-potential measurements, given in Fig. 5.15b for the example of a G4 dendrimer with the divalent sulfonate-group carrying azo dyes Ar26 and Ar27 (for structures see Fig. 5.12b. Note that in some publications a pH independent loading ratio l was defined where $l = 2 \cdot$charge ratio for pH = 3.5 where both, primary and tertiary dendrimer amine groups protonated and thus charge stoichiometry lying at $l = 2.0$). In this case, it is the charge ratio that also determines the presence of co-existing free molecules apart from the dendrimer-dye assemblies [105]. Individual dendrimers loaded with dye molecules are found only at extreme excess of dendrimers (loading of just 25% of the dendrimer charges in case of Ar26, Ar27). At more moderate dendrimer excess, dendrimer-dye assemblies with dendrimer in excess show a positive ζ-potential exists. The charge density is reduced in comparison to the dendrimers only because of the linkage with dye molecules that are oppositely charged. In difference, excessive dye results in negatively charged assemblies, which again find stability in solution. Assemblies with other dyes such as acid yellow 38 (Ay38) and direct yellow 12 are known to exhibit similar behaviors [101]. The assemblies have a positive surface charge at a charge ratio of 1.0, because of excessive cationic dendrimer alterations in the actual assembly, as shown by ζ-potential measurements. Accordingly, ITC (more results of this method below) also makes the slightly under-stoichiometric nature of the assembly evident. Therefore, not all dendrimer charges become neutralized at a stoichiometric mixing ratio in equilibrium, but some excess charge remains, which gets the assemblies stabilized in solution. This is also applicable to other systems with different polyelectrolytes such as polymer brushes or DNA [84, 85]. The dendrimer-dye assemblies are stabilized in a manner that is similar to that of classical electrostatically stabilized colloids.

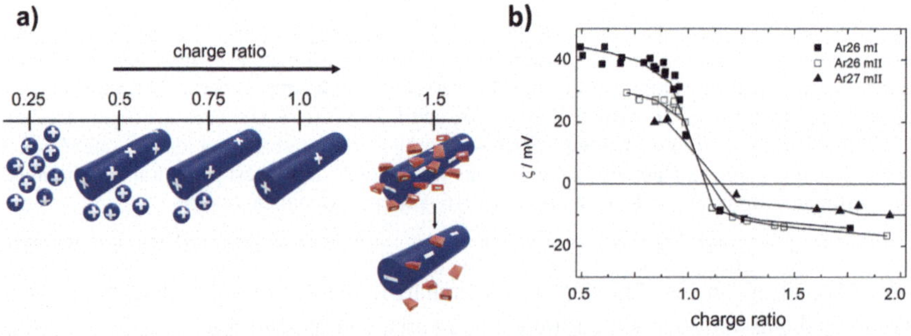

Fig. 5.15 Charge ratio-dependent assembly formation in a cationic dendrimer-anionic dye system: **a** Scheme: small spheres represent individual dendrimers, cylinders larger assemblies, and red rectangles dye molecules. Bound dye molecules are not sketched for simplicity; " + " and"- " represent sign and relative magnitude of the net charge; **b** ζ-potential on the dependence on charge ratio. Reprinted with permission from ref. [105]. Copyright © 2010 American Chemical Society

However, in contrast, they denote internally swollen hydrophilic species (as illustrated by SANS, light scattering, and AFM in conjunction) and lack a compact hydrophobic interior. When samples are prepared at a slight excess of dendrimer charges (charge ratio 0.7 to 0.95), this leads to the formation of the most stable assemblies.

5.7.3 Cooperative Binding of Dye Molecules

Secondary dye-dye π-π interactions play an important role in addition to electrostatic forces, as is evident already by eye when observing the dye solution upon dendrimer addition (Fig. 5.16a), as well as from the possibility of building stable nano-objects with excess dye. UV-Vis spectra exhibit dyes' interactions with each other owing to stacking observed as bathochromic or hypochromic shifts. Figure 5.16 displays spectra where a gradual alteration in absorption upon dendrimer addition is indicative of the "stacked dye" appearance as H-aggregates, which is induced by the dendrimer. The anionic dyes electrostatically associate with the dendrimer after

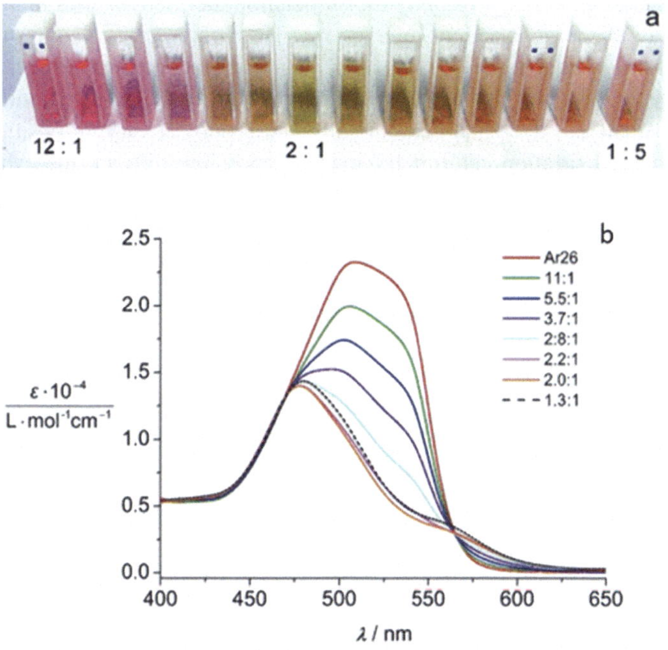

Fig. 5.16 **a** Photograph of Ar26-G4 samples for different dye: dendrimer ratios. **b** UV/Vis spectra of dye–dendrimer samples with c(Ar26) = 2.2×10^{-6} mol L^{-1} at varying loading ratios (pH 3.5). Reprinted with permission from ref. [83]. Copyright © 2008 WILEY–VCH Verlag GmbH & Co. KGaA, Weinheim

adding positively charged dendrimer, thus facilitating their interaction by means of spatial proximity. Until a charge ratio of 1 is reached, an isosbestic point is observed. The absorption increases again slightly upon further addition of dendrimer, which denotes an alteration in aggregate type. Quantitative analysis reveals that the ionic dye molecule initially binds electrostatically to the polyelectrolyte; thereafter, additional dye molecules preferably bind adjacently in a manner that allows them to undergo electrostatic and π-π interaction. In this instance, the π-π interaction is induced as a "secondary" interaction through the electrostatic interaction at dye concentrations where such molecules do not stack otherwise (and for dye molecules that do not stack at all in the absence of polyelectrolyte). Simultaneously, the multivalent dye molecules get the dendrimer connected to larger assemblies, instead of forming host–guest complexes, which is sometimes observed for mono as well as uncharged molecules.

5.7.4 Assembly Thermodynamics: Understanding Particle Size

It is essential to quantitatively study the association thermodynamics to understand the interplay of the noncovalent interactions. The Gröhn group demonstrated that it is possible to understand the assembly from thermodynamic considerations by a systematic model study via the above-mentioned family of dyes with various di- and trivalent azo dyes that contain sulfonate groups at varying positions [102, 104]. Isothermal titration calorimetry makes it possible to track the evolution of heat during the process of forming assembly for dendrimer-dye interactions. The enthalpies $\Delta H_{\text{dendrimer-dye}}$ and stoichiometries directly result from the titration experiments, whereas fitting the titration curve yields the binding equilibrium constants $K_{\text{dendrimer-dye}}$.

$$\Delta G = -RT \cdot \ln K \tag{5.1}$$

and from

$$\Delta G = \Delta H - T\Delta S \tag{5.2}$$

with ΔH representing the enthalpy change, T signifies the temperature and ΔS denotes the entropy change — from ΔH and ΔG; ΔS is accessible as well. Notably, the process of binding the dye to polyelectrolyte is highly exothermic, for instance, with around $\Delta H = -3000$ kJ/mol per G4 dendrimer — in other words, for binding 62 Ar26 molecules (slightly under-stoichiometric) that corresponds to $\Delta H = -45$ kJ/mol per dye molecule being bound to the dendrimer. This is in difference to polyelectrolyte aggregation with multivalent counterions which is often endothermic and entropically driven [14]. After taking various dyes into consideration, it is necessary to have a free energy gain of at least $\Delta G \approx -32$ kJ mol for the dendrimer and dye to be interconnected.

The free energy of the dye-dye interaction can be compared to the free energy of the assembly formation. For the dyes that undertake self-association — at much higher concentrations than what is used for assembly — it is possible to extricate the dye-dye interaction's thermodynamic parameters from dye dilution experiments. Concordantly, information on the electrostatic interaction is provided by comparison ITC experiment on the linkage of the dendrimer with a non-aromatic disulfonate — which leads to undefined aggregation with broad size distribution as opposed to the case involving aromatic counterion. It turns out that

$$\Delta G_{\text{dendrimer-dye}} = \Delta G_{\text{dye-dye}} + \Delta G_{\text{electrostatic}} \tag{5.3}$$

This also provides quantitative confirmation of the interplay of both interaction forces, π-π and electrostatic interaction.

Figure 5.17 shows that the aggregation number N_{Den} derived from scattering experiments can be related to the thermodynamic results. One master curve of N_{Den} versus $\Delta G_{\text{dendrimer-dye}}$ (and likewise versus $\Delta H_{\text{dendrimer-dye}}$ and $\Delta G_{\text{dye-dye}}$) results for all the dyes in unison. Thus, the aggregation number of the corresponding dendrimer-dye assembly can be predicted if the $\Delta G_{\text{electrostatic}}$ is approached to only depend on valence for a new divalent dye if the information is determined only based on $\Delta G_{\text{dye-dye}}$, e.g., from spectroscopy on the basis of master curve.

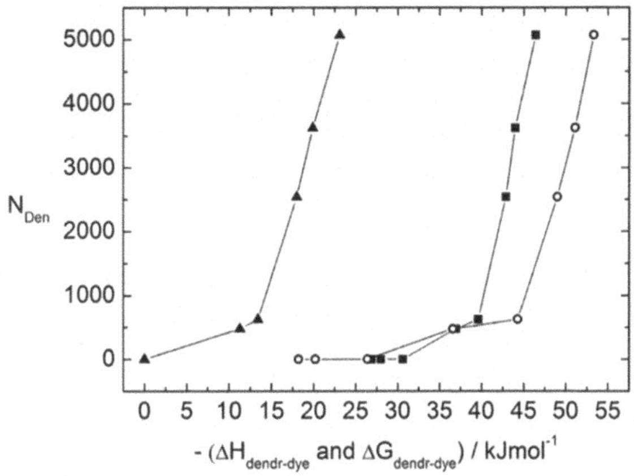

Fig. 5.17 Thermodynamic study of a dendrimer-dye model system for electrostatic self-assembly in solution explicating the size control of nano-objects: Dendrimer aggregation number N_{Den} in dependence on the interaction (of dendrimer and dye) free energy $\Delta G_{\text{dendrimer-dye}}$ (squares), dendrimer-dye interaction enthalpy $\Delta H_{\text{dendrimer-dye}}$ (open circles), as well as dye-dye interaction free energy $\Delta G_{\text{dye-dye}}$ (triangles) for divalent dyes. Reprinted with permission from ref. [102]. Copyright © 2011 American Chemical Society

Furthermore, it is possible to calculate the free energy for the self-assembly as follows:

$$\begin{aligned}\Delta G_{\text{dendrimer-dye}} &= \Delta G_{\text{electrostatic interactions}} + \Delta G_{\text{secondary interactions}} \\ &= \Delta G_{\text{dendrimer-sulfonate groups}} + \Delta G_{\text{dye-dye}}\end{aligned} \tag{5.4}$$

In equilibrium, it is as follows:

$$\Delta G_{\text{attraction}} = \Delta G_{\text{repulsion}} \tag{5.5}$$

In other words, when all repulsive and attractive contributions of the association process are considered:

$$\begin{aligned}N_{\text{Den}} N_{\text{dye}} (\Delta G_{\text{electrostatic}} + \Delta G_{\text{dye-dye}}) = &\frac{q^2_{\text{dendrimer}}}{4\pi \varepsilon_0 \varepsilon_R R_{\text{H-den}}} \\ &\cdot \sum_{i=1}^{N_{\text{Den}}-1} \left(\frac{i}{\sqrt[3]{i}+1} \exp(-\kappa(\sqrt[3]{i}+1)R_{\text{H-den}})\right)\end{aligned} \tag{5.6}$$

or

$$\begin{aligned}N_{\text{Den}} N_{\text{dye}} \Delta G_{\text{dendrimer-dye}} = &\frac{q^2_{\text{dendrimer}}}{4\pi \varepsilon_0 \varepsilon_R R_{\text{H-den}}} \\ &\cdot \sum_{i=1}^{N_{\text{Den}}-1} \left(\frac{i}{\sqrt[3]{i}+1} \exp(-\kappa(\sqrt[3]{i}+1)R_{\text{H-den}})\right)\end{aligned} \tag{5.7}$$

Equation 5.6, as solution, yields one finite aggregation number $N_{Den.}$ for each $\Delta G_{dye\text{-}dye}$ (that is, for each dye), thereby confirming the formation of finite-size assemblies with a specific aggregation number (size). Equation 5.7 makes it possible to calculate the aggregation number N_{Den} as a function of the free energy associated with the assembly formation $\Delta G_{dendrimer\text{-}dye}$, thus exhibiting an increased aggregation number with the rising $\Delta G_{dendrimer\text{-}dye}$ in consonance with the result of the experiment (Fig. 5.17). This implies that the aggregation number N_{Den} is a value that is thermodynamically favored and does not rise infinitely, which would then lead to precipitation. However, this calculation is excessively simple, as it is not possible to elucidate spherical polyelectrolytes in solution merely by adopting a simple Debye-Hückel approach; a sound approximation can also be provided for the general consideration. Moreover, the principle outcome will not undergo a change even with a more refined expression for the aforementioned electrostatic interaction.

Systematic research studies have been carried out on changing pH and salt concentration as well as on altering the dendrimer generation [86, 102, 108]. The pH alters the dendrimer's charge status. Thereby the loading ratio changes with the same amount of dye present. In addition, the dendrimers of other generations possess various charge numbers. An additional role is played by the altered size and charge distribution geometry. The pH was explained to be a salient trigger for tuning the stability/shape/size of these building blocks' nano-assemblies. By coalescing the findings of DLS and ζ-potential, the surface charge density was found to be the crucial parameter in determining the nanoparticles' style and stability. Stable and well-defined nanoparticles are found at low pH values when dye molecules also bind to the ternary amines "inside" the dendrimer. At higher pH, the dye molecules are only capable of binding to the primary amines, thus causing a reduced stabilization of particles less than at lower pH values. In this system, the pH value triggers the structure of the Ar26-dendrimer assemblies from nanoscale cylinders with elliptical cross-sections with $R_H = 600$ nm to spherical nanoparticles with $R_H = 57$ nm. [106, 107] For another structurally different but related dye (disodium 4-((3-(hydroxymethyl)phenyl)diazenyl)-3-hydroxynaphthalene-2,7-disulfonate, ABnOHRA), it can be possible to use the pH as a trigger to switch between 2 μm long flexible cylinders and shorter (100–500 nm) stiff cylinders each with well-defined shape and size. Here, the control of the size and shape of the nanostructure again follows the above-discussed thermodynamic structure encoding, where the pH alters the dendrimer's charge status, thus causing a change in counterion/macroion charge ratio. Simultaneously, there is a change in the charge distribution, which leads to a change in the macroion's structure.

It is notable that assemblies are stable up to a relatively high concentration of added salt. Moreover, in all the sample solutions expounded on here, a certain amount of salt originates from the pH adjustment and initial counterions. Therefore, we discuss cases in the presence of additional salt, which might assume significance for some possible applications in environments containing salts. This salt also helps avoid the extraordinary polyelectrolyte impacts seen in salt-free solutions. This is necessary as an additional structure factor would complicate or make it impossible to properly analyze the size distribution and shape of nano-objects in complex systems. Furthermore, a drastic rise in salt concentration causes an interplay of screening effects by screening both the attractive (assembly forming) electrostatic force between building blocks as well as the repulsive (assembly stabilizing) electrostatic force between assemblies. This leads to an intersection of a nano-object "coagulation" and a nano-object dissolution proclivity, thus resulting in assemblies with ionic strength-dependent size and surface charge density [108].

Meanwhile, corresponding results are yielded by systems with other polyelectrolytes and other dyes: predominantly, the assembly formation is enthalpically driven by a building block and loading ratio controlled assembly size reflected in a ΔG-determined aggregation number [92, 101, 109]. Here, the attractive forces' additive nature also becomes evident in other cases, thereby exemplifying its general attribute. Interestingly, when attractive Hamaker interaction occurs in-between gold nanoparticles included inside of the dendrimers, it denotes the third contribution besides electrostatics and π-π interactions. It is as follows:

$$\Delta G_{\text{dendrimer-dye}} = \Delta G_{\text{electrostatic interactions}} + \Delta G_{\text{dye-dye}} + \Delta G_{\text{Hamaker}} \tag{5.8}$$

in such a manner that owing to the heightened $\Delta G_{dendrimer\text{-}dye}$, still, well-defined-assemblies form which is seen to be larger when compared to the ones with only dendrimers [92].

5.7.5 *Assembly Thermodynamics: Understanding Particle Shape*

As the thermodynamic consideration demonstrated that the aggregation number and, as a consequence, the size of nano-objects is determined by free energy ΔG, it becomes necessary to comprehend how the encoding of various shapes of the assemblies takes place. Is it also possible for thermodynamics to play a key role in understanding particle shape? For this model system, it can be inferred that the entropy-enthalpy balance determines the nanoparticle form, as illustrated in Fig. 5.18, which shows the so-called effect of entropy-enthalpy compensation as far as the dendrimer-dye assembly is concerned. The effect postulates that a significant portion of the enthalpic gain is offset by entropic effects $T\Delta S$ and does not get reflected in the free energy $\Delta G_{dendrimer\text{-}dye}$. The enthalpy $\Delta H_{dendrimer\text{-}dye}$ rises with increasing entropy by a factor of 1.3. Meanwhile, an entropy exchange of $T\Delta S = -10$ kJ/mol also denotes the threshold for shape change. Here, it is possible to separate the contributions among divalent dyes with all having almost the same electrostatic interaction. It may also be pertinent to point out that the shape of the nanoscale particle is contingent on the dye-dye interaction parameters. A $\Delta H_{dye\text{-}dye} > -21$ kJ/mol (less negative $\Delta H_{dye\text{-}dye}$) leads to spherical particles, whereas more negative $\Delta H_{dye\text{-}dye}$ values cause the creation of anisotropic and elongated particles as shown in Fig. 5.18b [104]. This implies that a robust π-π contribution increases the size as well as the anisotropic character of dendrimer-dye particles, both of which have been quantified here.

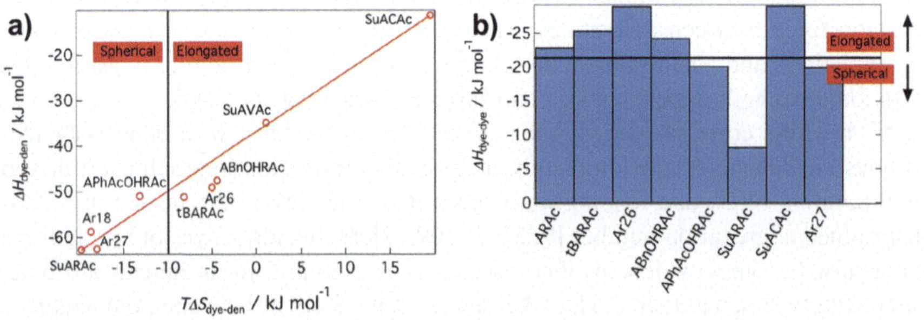

Fig. 5.18 Elucidating the nanoscale shape control in electrostatic self-assembly by an experimental thermodynamics study: **a** Enthalpy-entropy relation for dye-dendrimer interaction; **b** assembly symmetry depending on dye-dye interaction enthalpy. Reprinted with permission from ref. [104]. Copyright © 2016 American Chemical Society

After showcasing this extraordinary relationship between thermodynamics and the shape of nanoparticles, it would be interesting to determine thermodynamics and understand how it can be related to molecular properties. Notably, the dye molecules' molecular modeling by DFT calculations makes it possible to stimulate not only the electrostatic potential at the molecular surface but also numerous related parameters like the dipole moment. Figure 5.19 shows a gamut of dye molecules. The dye molecules' polar surface area (PSA) is found to relate to the $\Delta H_{dye\text{-}dye}$ in a master curve (Fig. 5.19d).

A higher polar surface area reduces the π-π driven dye-dye interaction enthalpy for the same aromatic backbone with differing substituents. It is interesting to note

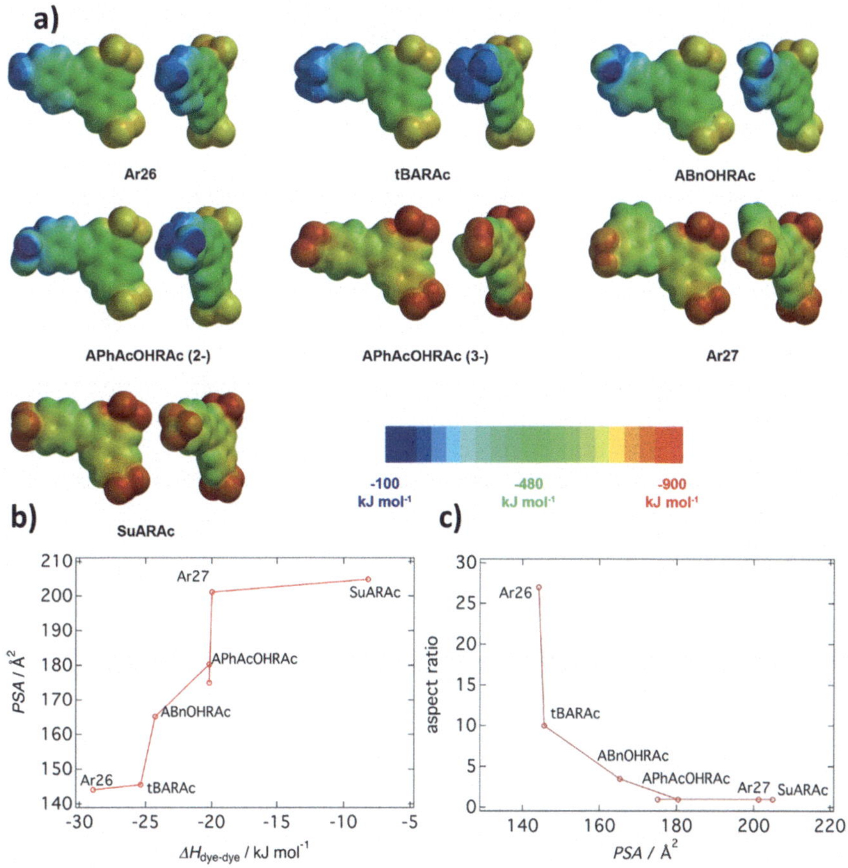

Fig. 5.19 Elucidating the nanoscale shape control in electrostatic self-assembly by an experimental thermodynamics study: **a** electrostatic potential at the molecular surface for the dye molecules from DFT calculation: each top view (*left*) and front view (*right*); **b** master plot of the polar surface area of the molecules as a function of $\Delta H_{dye-dye}$; **c** master plot of the nano-object aspect ratio as a function of the polar surface area of the molecules. Reprinted with permission from ref. [104]. Copyright © 2016 American Chemical Society

that this very elemental relationship of the substituents' influence on the π-π interaction had not been observed before. It was additionally related to the inter-dye stacking distance and tilt angle in the study of the Gröhn group [104]. Here, it is necessary to emphasize the observation that thermodynamics plays a key role in understanding the formation of nanoscale structures. An explanation has been given on how thermodynamics relates to the ensuing nanoscale architecture: ΔG gets the size determined, $\Delta H/\Delta S$ gets the shape determined—as well as the manner in which molecular properties determine the thermodynamics. Collectively, the latter results in a master plot of particle anisotropy versus PSA.

Thus, a well-rounded understanding of both structure formation and structure-directing principles has been developed for this model system. Douglas et al. carried out various studies wherein they considered self-assembly beyond classical micelles in an in-depth manner. They recently stated that a diverse range of supramolecular assembly processes emerges from the competition between directional and isotropic intermolecular interactions, concentrating on the essence of the spherical charged particles' self-assembly with competing van der Waals interactions bound to cylindrical surfaces, as a model system [110]. On the other hand, Glotzer et al. adopted a different theoretical approach by making use of blocks using a highly intricate shapes. They forecasted different shaped structures formed of anisotropic building-block structures via the calculation of the symmetry of ordered arrays formed via these building blocks [111, 112]. However, a comprehensive understanding of the structure-directing effects is still found lacking in several other complex architectures, which is unsurprising since the entire field of building defined structures by means of electrostatic self-assembly is far younger than the examination of amphiphilic or amphiphilic-analogous concepts. Furthermore, it is imperative to gain an incisive understanding from detailed model studies to exploit the latent capacity of the versatile electrostatic self-assembly for functionality purposes, for example, as photocatalysis and switching for conversion of solar energy and delivery of drugs, as shall be expounded on below.

5.8 Versaility of the Approach: Structural Variety Using Polyelectrolytes

For anionic cylindrical polystyrene sulfonate brushes, that is, polymers with a wormlike shape, the tetravalent cationic porphyrin meso-tetrakis(4-N-methylpyridinium)porphyrin (TMPyP) or meso-tetrakis-(4-(trimethylammonium)phenyl)-porphyrin (TAPP) has been used as a linker [84, 91, 113]. Electrostatic interactions between the porphyrin and the polymer brush, as well as π-π interactions among the porphyrin molecules end up forming interconnected networks of the polymer brushes with further stacked porphyrin molecules within the brush, as shown in Fig. 5.20.

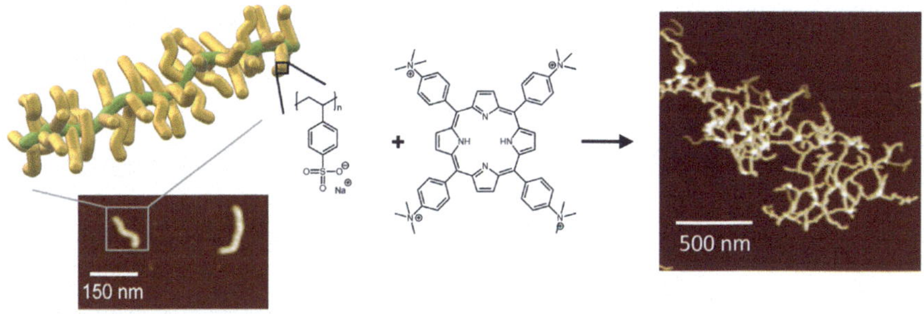

Fig. 5.20 Wormlike bottle-brush polyelectrolytes and tetravalent porphyrin counterions form finite-size networks; AFM of brush and brush-porphyrin aggregates spin-coated on mica. Reprinted with permission from ref. [84]. Copyright © 2009 American Chemical Society

The addition of salt disconnects the brush networks to form single elongated worms of porphyrin-loaded brushes due to the weakening of electrostatic interactions [84]. While assemblies with the metalated porphyrins exhibit similar structures, for Cu^{2+}, these interactions are a bit more robust owing to a side-by-side stacking, which, in turn, results in H- and J-aggregates [110]. Such poylectrolyte-porphyin structures are known to demonstrate particular photocatalytic attributes [91].

In another study, the Keggin-type polyoxotungstate $SiW_{12}O_4{}^{4-}$ with four negative charges is utilized as interconnecting structural counterion with G4 dendrimer, thus yielding assemblies with expressed photocatalytic activity [94]. These particles comprise a single dendrimer strain ribbon interconnected and embellished by polyoxometalate clusters, and the total hydrodynamic radius of ranges from 120 nm < R_H <180 nm (depending on loading ratio), with a defined size. SANS shows assemblies that have the shape of a flexible cylinder with a diameter of 6.2 nm, that is, a chain of dendrimers interconnected by polyoxometalate (POM) clusters (Fig. 5.21). This is extraordinary from a structural standpoint, since, in difference to the electrostatically self-assembled dendrimer-dye nano-objects introduced above, no secondary π-π interaction takes place in this system. Instead, only electrostatics in combination with geometric factors and the probable dipole effects direct the structure formation.

The combination of ionic interaction and geometric effects can also yield defined aggregates. DNA was combined with different organic counterions [85]. With a tetravalent perylene-based molecule carrying side groups that prevent stacking, a transition from flower-like structures at low charge ratios to rods and toroids at higher charge ratios was found (Fig. 5.22). The stiff structure of this counterion with one of its charge distances matching the distance of phosphate groups on the DNA was crucial. Thus, with this system, as for the POM system, it is also demonstrated that the counterplay between electrostatics and geometric factors can also direct the structure formation if no secondary π-π-interactions between counterions occur. This may be seen in analogy to theoretical considerations on the formation of biomacromolecule bundles with a certain diameter if counterions exhibit a certain size [114].

Other instances of polyelectrolyte-porphyrin assemblies that are photocatalytically active include aggregates of cationic TAPP and anionic PAMAM dendrimer

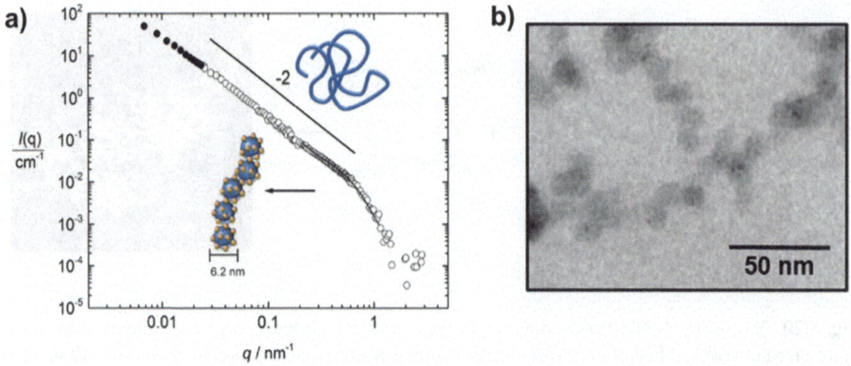

Fig. 5.21 POM–dendrimer assemblies: **a** static light scattering and SANS of POM–dendrimer assemblies with $l = 0.7$; filled symbols: SLS data points, open symbols: SANS data points, a black line at high q: flexible cylinder fit; **b** TEM image. Reprinted with permission from ref. [94]. Copyright © 2018 Royal Society of Chemistry

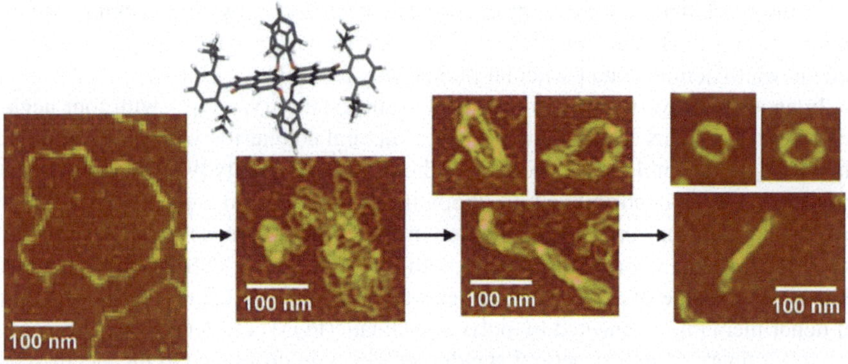

Fig. 5.22 Assemblies of pure 19 DNA with a tetravalent perylene-based non-stacking counterion: from *left* to *right*: supercoiled DNA on mica, PSPDI/DNA samples with charge ratios $l = 0.5$, $l = 1$, and $l = 1.4$ on modified mica. AFM images. Reprinted with permission from ref. [85, 89]. Copyright © 2009 American Chemical Society. Copyright © 2009 American Chemical Society. Copyright © 2010 The Royal Society of Chemistry

of the generation 7.5 at basic pH (Fig. 5.23) [98]. Here too, the size of cylindrical aggregates and the porphyrins' internal aggregation into stacks within the assemblies is predicated on the charge ratio, which is illustrated in Fig. 5.23. The change in UV-Vis spectra, as well as fluorescence spectra, demonstrates the porphyrins' internal stacking in the assemblies. In turn, this directly affects photocatalytic activity (Fig. 5.23d). In excess of dendrimer ($l = 0.05$), the porphyrins develop J-aggregates that reveal a higher quantum yield owing to a smaller HOMO–LUMO gap and their length is 180 nm, whereas H-aggregates predominate at equal charge stoichiometry ($l = 1.0$) and at an excess of porphyrin ($l = 1.6$) and have a length of 100 nm and

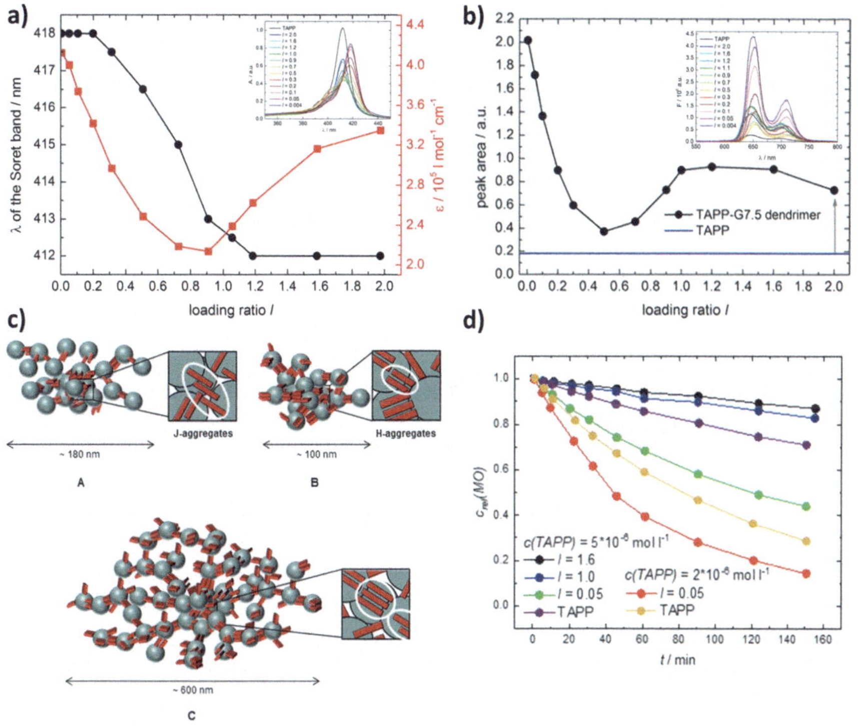

Fig. 5.23 TAPP-G4 dendrimer aggregates and the dependence of the catalytic activity on the internal stacking of the porphyrins: **a** UV/Vis analysis in dependence on the loading ratio l: shift of the Soret band and changes in the extinction coefficient ε of the Soret band at pH 11; inset: Soret band; **b** fluorescence analysis in dependence on the loading ratio l: area of the fluorescence peak at 574 nm $\leq \lambda \leq$ 800 nm; inset: fluorescence; **c** Schematic illustration of TAPP–G7.5 dendrimer assemblies with $l = 0.05$, $l = 1.0$, and $l = 1.6$ (red: TAPP porphyrin molecules; blue: G7.5 dendrimers); **d** photocatalytic degradation of methyl orange upon irradiation with visible light similar to the sun spectrum as photocatalytic model reaction: Decrease of the methyl orange concentration measured at $\lambda = 464$ nm for TAPP–G7.5 dendrimer assemblies with different loading ratios and TAPP only. Reprinted with permission from ref. [98]. Copyright © 2017 American Chemical Society

500 nm, respectively. For dendrimer-porphyrin assemblies, the photocatalytic degradation of methyl orange with visible light increases with $l = 0.05$ when compared with pure TAPP, whereas aggregates with a charge ratio of $l = 1.0$ and $l = 1.6$ reveal a lower activity. The heightened photocatalytic activity for the assemblies with an excess of dendrimers is attributed to two reasons. The first reason is the large surface area caused by the small particle size. The second reason is the higher number of porphyrin J-aggregates. Thus the internal assembly structure, as well as the size of assemblies, are responsible for methyl orange's degradation rate.

The so-called coiled-coil structures take place in a designed peptide with porphyrins depending on the pH, thus impacting the protonation and by extension, the porphyrins' charges. The peptides reveal secondary helical folding in addition to the electrostatic interaction, the π-π interaction between porphyrin-peptide molecules as well as porphyrin-porphyrin [115, 116]. Phycocyanin combined with a four-armed porphyrin star macroion assembling in a pomegranate-like structure, which can act as a light-harvesting system, has created an interesting and functional architecture [117].

Further, surfactants can also be combined with oppositely charged polyelectrolytes to provide structures in solution. In this regard, Ayzner et al. propounded polyelectrolyte-surfactant complexes created by interconjugated polyelectrolyte complex capable of being separated by cationic as well as anionic surfactants [118].

5.9 Nano-Objects by Electrostatic Self-assembly: Beyond Polyelectrolytes

The group of Tianbu Liu adopted a different approach and explored the interaction between cationic surfactants and a type of "Keplerate" polyoxometalate macroanionic molecular clusters. Here, polyoxometalate acts more like a polyelectrolyte, where a large number of surfactants can interact with the macroion (Fig. 5.24). Subsequently, long-chain surfactants (cetyl trimethylammonium bromide CTAB, and cetyl trimethylammonium tosylate CTAT) can lower their charge density. Consequently, the macroions aggregate into bigger "blackberries", which even start to precipitate at a specific amount of surfactant [118].

Although all these assemblies discussed above do consist of larger macroions, the combination of several secondary interaction forces also allows for small molecules that are oppositely charged to form assemblies in solution. Figure 5.25 shows that alkyltrimethylammonium surfactants and Ar 26 build cylindrical particles of

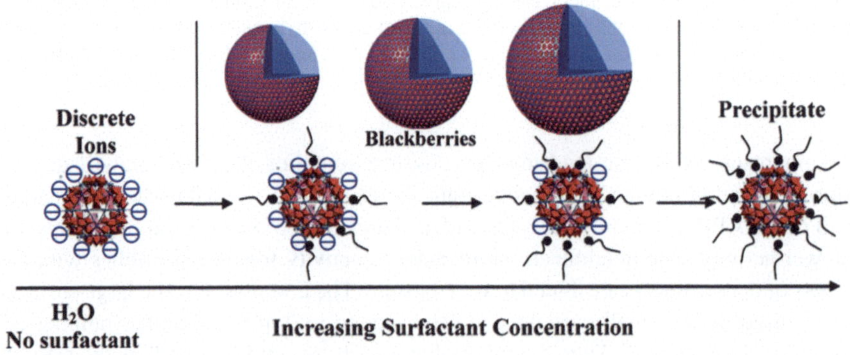

Fig. 5.24 Scheme of the introduction of alkyltrimethylammonium halide cationic surfactants into the dilute aqueous solution of $\{Mo_{72}V_{30}\}$ clusters. Transition from discrete $\{Mo_{72}V_{30}\}$ macroanions to blackberry structures to finally insoluble, surfactant-encapsulated (almost neutrally charged) clusters can be achieved during the process. Reprinted with permission from ref. [119]. Copyright © 2009 American Chemical Society

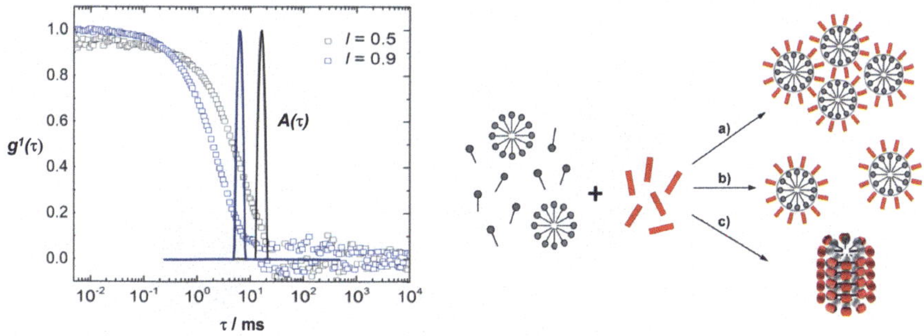

Fig. 5.25 Assemblies of ionic surfactants with ionic dyes as counterions, left: DLS: electric field autocorrelation function $g^1(\tau)$ and distribution of relaxation times $A(\tau)$ for C_{12}TAB-Ar26 assemblies; right: overview on the structures formed: associated and individual spherical surfactant micelles with Ar26 acting connector and counterions near the micelle surface, respectively cylindrical surfactant–dye aggregates from cylindrical surfactant micellization with condensed mutually π-π interacting Ar26 counterions. Reprinted with permission from ref. [120]. Copyright © 2016 Springer-Verlag GmbH Germany

stacked micelles. The dye molecules and charged surfactant interact by electrostatic interaction; stacking between the dyes occurs owing to π-π interactions and it is due to the hydrophobic effect that the surfactants form micelles [120]. Thus, this system denotes an analogy to Faul's dye-surfactant ionic self-assembly for material design while simultaneously signifying polyelectroylte-dye nano-objects. As demonstrated by Sun et al., Keggin-type polyoxotungstates $SiW_{12}O_4{}^{4-}$ with four negative charges develops onion-like layered spherical particles with cationic dimethyldioctadecylammonium surfactants [121].

Figure 5.26 provides a summary of the numerous shapes and the structural variety of electrostatically self-assembled structures. These range from spherical particles to elongated wires and rods to even more anisotropic raspberry structures of multilayer capsules.

5.10 Switching Nanostructure and Properties

The possibility of triggering the structures via external forces is a great benefit associated with self-assembly by non-covalent interactions. Our body exemplifies the importance of this process. Our ability to see is made possible by the light-induced isomerization of 11-*cis*-retinal to all-*trans*-retinal [122]. For electrostatic self-assembly, switching the structures has elicited attention only in the more recent past. In this context, it is possible to take advantage of myriad extraneous triggers like pH, light irradiation, temperature, and electrochemical or magnetic stimuli to change the photo-physical attributes of the assemblies.

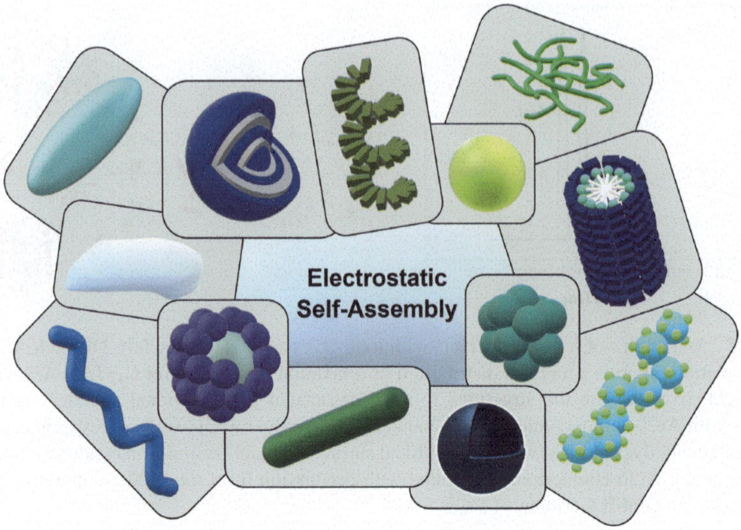

Fig. 5.26 Shape variety achieved by electrostatic self-assembly: networks, decorated wires, stacked micelles, vesicles, tubes, ellipsoids, helices, pomegranate-like structures, disks, onion-like structures, raspberry-like structures, spheres, and rods

5.10.1 pH-Switchability

As one of the first researchers using electrostatic interaction for forming pH-dependent nanoparticles Andrade et al. combined meso-tetrakis(p-sulfonatophenyl)porphyrin (TPPS) sodium salt with the human serum albumin and β-lactoglobulin. Here, the binding affinity was strongly reliant on the solution's pH value [123]. To that end, the Gröhn group showcased a pH-switchable creation of electrostatically self-assembled nanoparticles in solution. Figure 5.27 illustrates switching between building blocks and assemblies [83]. Fourth-generation (G4) PAMAM dendrimers were used as the cationic polyelectrolyte and Ar26 as the counterion. At a pH value of pH = 3.0, stable nano-objects form, in this case, of a cylindrical shape with a length of about 225 nm and a cross-section dimension of about 60 nm. By altering the pH values from pH = 3.0 to pH = 10, these assemblies dissolve because of the polyelectrolyte's deprotonation as well as the missing electrostatic interaction. After adding an acid, the nanoparticles are formed again and it is possible to repeat the switching back and forth a minimum of 10 times. Figure 5.27 exhibits this scenario.

It is also possible to use this switch (on and off) to form nano-assemblies for drug release. Cationic G8 dendrimer with the trianionic azo dye Ar27 is known to form hollow capsules that are well-defined with a size of $R_H = 130$ nm (Fig. 5.28) [90]. The attribute that these hollow spheres can be dissolved by increasing the pH value is particularly interesting. Moreover, it is also possible to include guest molecules inside the carriers, as demonstrated in fluorescently labeled peptides.

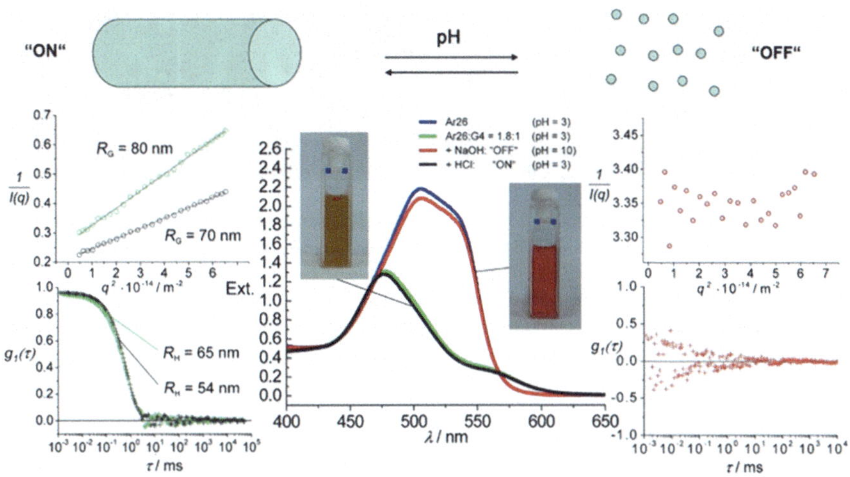

Fig. 5.27 Switching "on" and "off" of electrostatically self-assembled structures by pH: G4-Ar26 assemblies: "on" and "off" means aggregates and single dendrimers, respectively; *center*: UV/Vis spectra; *left and right*, static and dynamic light scattering. Reprinted with permission from ref. [83]. Copyright © 2008 WILEY–VCH Verlag GmbH & Co. KGaA, Weinheim

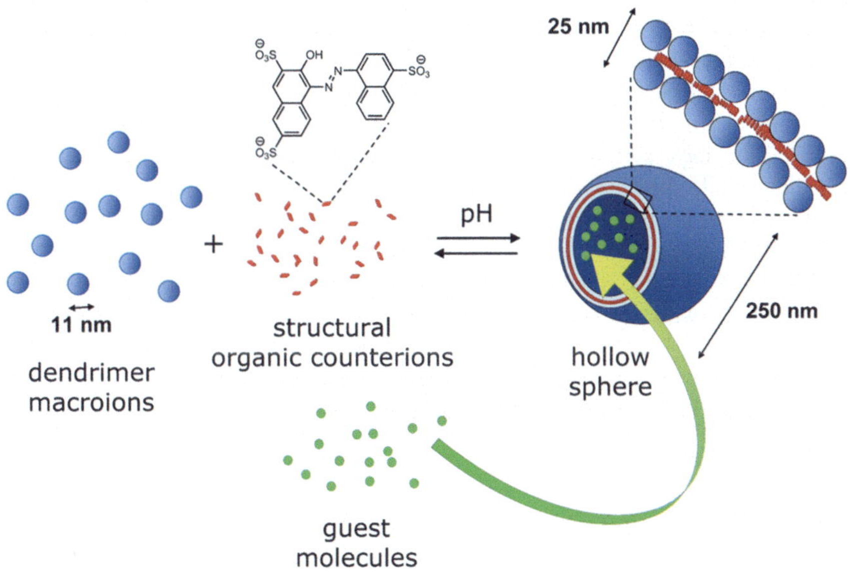

Fig. 5.28 Switching "on" and "off" of electrostatically self-assembled structures: capsule formation and pH on-off switching of G8 dendrimer with Ar27. Adapted with permission from ref. [89, 90]. Copyright © 2010 WILEY–VCH Verlag GmbH & Co. KGaA, Weinheim. Copyright © 2009 American Chemical Society. Copyright © 2010 The Royal Society of Chemistry

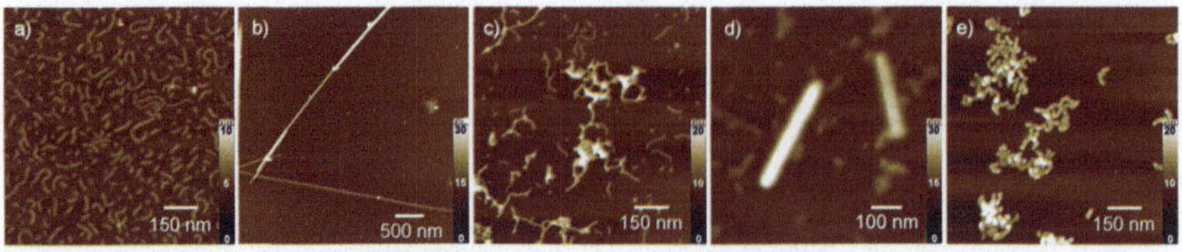

Fig. 5.29 pH-responsive structuring in electrostatic self-assembly: Cylindrical brush-nanorod-network switching in an electrostatically self-assembled polymer-brush porphyrin system; AFM: **a** wormlike PVP brushes, pH = 7; **b** porphyrin nanorods, pH = 2; **c** porphyrin-brush networks, pH = 7; **d** nanorod-brush system, set from pH = 7 to pH = 2; **e** switched networks, pH = 7; **f** nanorod-brush system, prepared at pH = 2. Reprinted with permission from ref. [124]. Copyright © 2011 WILEY-VCH Verlag GmbH & Co. KGaA, Weinheim

Notably, these capsules are very different from amphiphilic vesicles. In this instance, two hydrophilic water-soluble molecules coalesce into hollow capsules because of the interplay of the interaction effects mentioned above.

A system similar to the above-mentioned polyelectrolyte brushes and ionic porphyrins consists of oppositely charged components, anionic meso-tetrakis(4-sulfonatophenyl)porphyrin (TPPS), as well as cationic poly(2-vinylpyridine) cylindrical brushes, also creating network-like structures [124]. By changing the pH from pH = 7.0 to pH = 2.0, it is found that the network structure disassembles into porphyrin nanorods as well as individual polymer brushes hosting TPPS J-aggregates and dianionic TPPS, as shown in Fig. 5.29. The network structure reforms after readjusting the pH to pH = 7.0. Switching from pH = 7.0 to pH = 3.5 was of further interest. The shape and size of these assemblies remain unchanged directly after switching, while the network structure disassembles to individual brushes after a few days. Here, the porphyrin remains bound to the polymer, albeit with an altered ratio of H/J aggregates.

Helseth et al. also combined the cationic polyelectrolyte poly(allylamine-hydrochloride) with the anionic dye pyranine as counterion to yield a sheet-like colloidal structure [125]. The addition of NaOH leads to disaggregation of the initial complex and the formation of small spherical assemblies, which was accompanied by a change in the fluorescence of the particles.

5.10.2 Light-Switchable Particle Size

When compared with pH-responsive nano-assemblies, light-switchable structures have recently garnered more attention as light is an elegant and non-invasive external trigger that makes it possible to impact the nano-assemblies' properties which could be interesting for switches or optical materials or in biological systems. The light-responsive reaction, which is most commonly used in nano-assemblies, is the *cis-trans* isomerization of azo benzenes. Altering the secondary structure of a polymer was the first time that azo benzenes were utilized for changing the form and size of a structure [126]. Higashi et al. are credited with the forming of the first light-responsive electrostatically self-assembled nanostructures based on azobenzene [127]. The researchers assembled an anionic azobenzenes derivate using a cationic β-sheet peptide into nanofibers. After irradiation and subsequent isomerization of the azo dye to its *cis*-state, the assemblies dissolved, and the particles reassembled to their *trans*-state upon isomerization of the azobenzenes.

Although it was possible to achieve assembly/disassembly and gelation through irradiation by light via the use of various assembly triggers/concepts (for the various triggers see below), it had long been challenging to access a light-triggered nano-object size in solution. The Gröhn group first achieved this in 2010 [88]. The model system comprised azo-switchable anionic dyes in conjunction with a cationic dendrimer macroion in an aqueous solution (Fig. 5.30). Unlike the electrostatically self-assembled structures of this group focused on above, the supramolecular particles with an excess of dye were found to be of key interest for this size switchability in solution. The size of the spherical assemblies ranged between 31 nm $\leq R_H \leq$ 52 nm, depending on the loading ratio between the sulfonate groups of the dye and the primary dendrimer amine groups. At excessive dye molecules, irradiation resulted in an increase in size, for instance, from $R_H = 32$ nm to $R_H = 165$ nm with a narrow size distribution. Figure 5.30 displays the DLS and AFM findings.

The particles' stability of the non-irradiated nano-assemblies is found to stem from the excess and overbinding of anionic *trans* azo dyes to the cationic polyelectrolyte. This causes a highly negative surface charge density, which, in turn, electrostatically stabilizes the particles. The azo dye's isomerization upon irradiation leads to an increase in the size of the nano-assemblies. As derived from ζ-potential measurements, the effective surface charge density is less expressed after irradiation as compared to the assemblies with the *trans*-dye before irradiation. The *cis* dye after irradiation is bound to the dendrimer like the *trans* isomer before irradiation, albeit to a lesser extent. Figure 5.30 shows that ITC quantifies this effect and provides further insights. While binding isotherm shows one step before irradiation, after irradiation, the dye-binding showed two steps and was elucidated by two sets of site models since *tans* and *cis*-isomer coexist after irradiation. ITC measurements also reveal that the binding enthalpy of *trans* azo dye is higher. Similarly, its binding equilibrium constant is higher than the *cis* dye, as per the ITC and ζ-potential results. This is because *cis*-isomer is more hydrophilic and polar as compared to the trans-isomer. Thus, for the *cis*-isomer, the *cis*-state's affinity to the aqueous phase relative to the development of mutual π-π interaction is higher. After irradiation, dye molecules get detached from the nano-assemblies, lowering their negative surface charge density, which leads to the nanoparticles' coagulation into larger assemblies with a lower total surface that again becomes stable in an aqueous solution. The difference between the binding enthalpy of the *trans*-isomer and *cis*-isomer is imperative to determine the extent of the changes in size irradiation and is capable of being exploited to adjust the cis/trans ratio by choice of the irradiation wavelength [101].

Moreover, the Ay38-dendrimer system can be extended into a multi-responsive scenario (Fig. 5.31) [88]: As discussed above, nano-assemblies with a size of $R_H =$ 32 nm and light irradiation of the azo dye cause an increase in size to $R_H = 165$ nm.

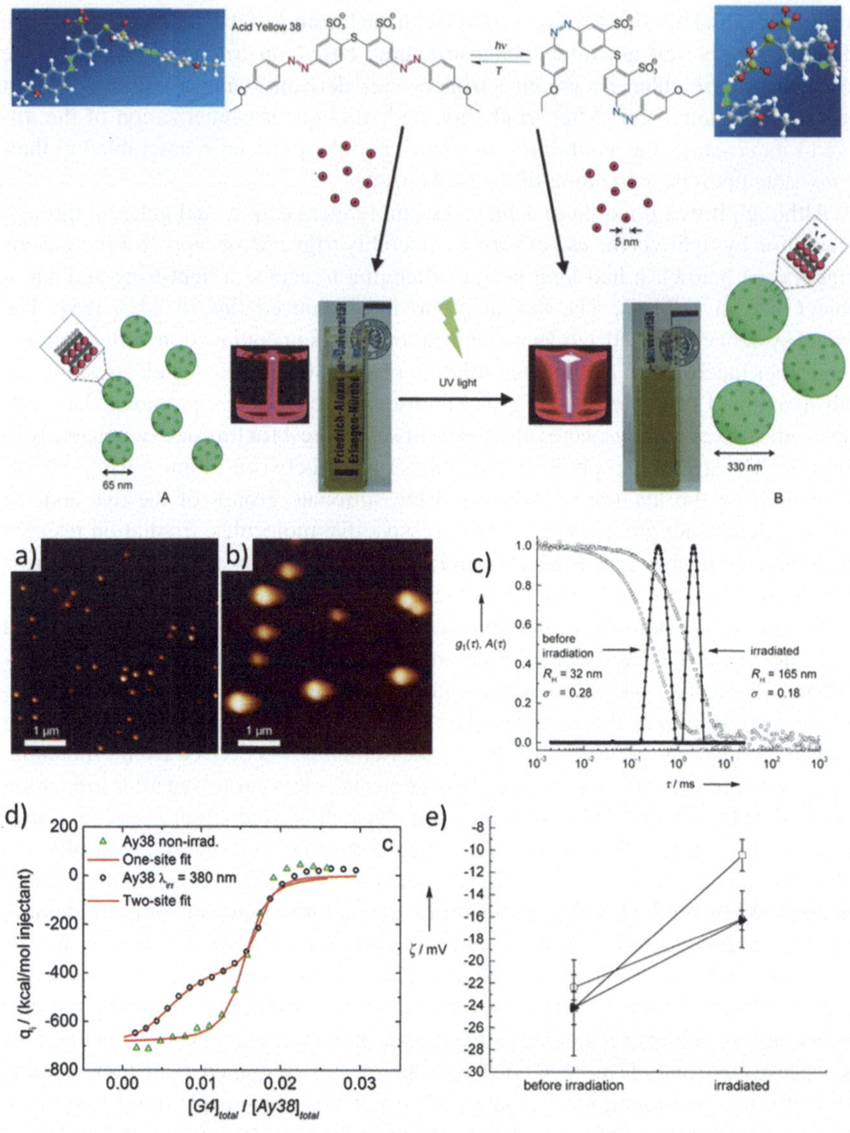

Fig. 5.30 A Light-switchable electrostatically self-assembled system with photo addressable particle size; top: schematic representation. Starting from a divalent anionic azobenzene dye and cationic dendrimer macroions, self-assembled nanoparticles result (A), which grow in size upon irradiation with UV-light (B); middle: AFM **a** before and **b** after UV irradiation and **c** corresponding DLS for a loading ratio $l = 4.5$; **d** Isothermal titration calorimetry revealing the different binding enthalpies of the cis and the trans dye isomer; **e** ζ-potential for dendrimer-dye assemblies before and after irradiation for different loading ratios confirming the charge density control of the particle size. Reprinted with permission from ref. [88, 101]. Copyright © 2010 WILEY–VCH Verlag GmbH & Co. KGaA, Weinheim. Copyright © 2011 American Chemical Society

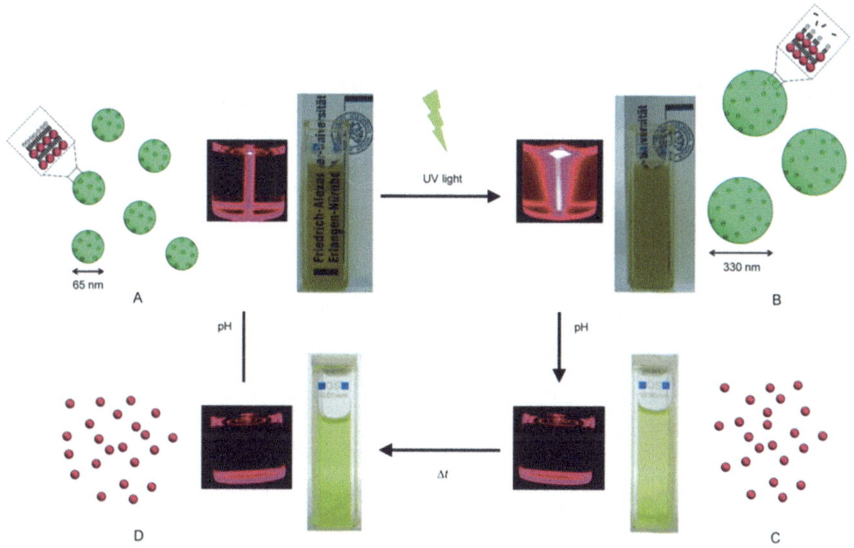

Fig. 5.31 Representation and photographs of a light- and pH-switchable electrostatically self-assembled system with photoaddressable particle size. Starting from a divalent anionic azobenzene dye and cationic dendrimeric macroions, self-assembled nanoparticles result (**a**), which grow in size upon irradiation with UV light (**b**) and further disassemble, triggered by the pH (**c**); after re-isomerization (**d**), building blocks can be reassembled into the original structure (**a**) by another pH change. Photographs indicate scattering power with a red laser and turbidity of the solution. Results are shown for a loading ratio (dye sulfonate groups/dendrimer amine groups) $l = 4.5{:}1$. Reprinted with permission from ref. [88]. Copyright © 2010 WILEY–VCH Verlag GmbH & Co. KGaA, Weinheim

It is possible to dissolve these larger assemblies by changing the pH value from pH = 3.5 to pH = 10.5. This rise in pH causes the deprotonation of the polyelectrolyte's amine groups; the assemblies dissolve as the macroion does not entail any charge. Afterward, the dye can more easily isomerize to its *trans*-state by keeping the assemblies in the dark. Altering the pH back to pH = 3.5 reforms the assemblies with the same size as in the beginning, that is, a dually responsive system that could be triggered by pH and light has been developed.

Variation of the building blocks further demonstrated the generality and versatility of this type of light-switchable nano-objects created by electrostatic self-assembly [128]. Various linear flexible cationic polyelectrolytes got combined with the same azo dye Ay38. Stable nanoparticles are formed except for the loading ratio of $l = 1.0$ where the assemblies precipitate directly owing to the assemblies' missing electrostatic stabilization in solution. As shown in Fig. 5.32, irradiation for the samples with $l < 1.0$ leads to a decrease in size, which is different from the dendrimer system. At $l > 2.0$ irradiation causes an increase in size due to a similar reason as pinpointed for the dendrimer-Ay38 system. The density of surface charge also plays an important role for $l < 1.0$ and the majority of stable particles are at an excess of positive charges, which can be compared with the (non-switchable) model system explained

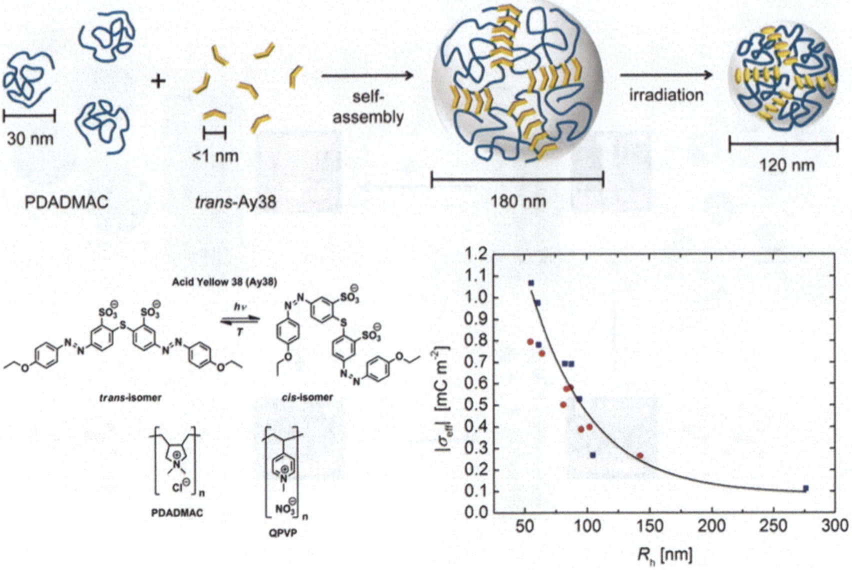

Fig. 5.32 Schematic representation of light-responsive dye-polyelectrolyte assemblies with under-stoichiometric dye-load. Structure of Ay38 and the different linear flexible polyelectrolytes. Right bottom: Effective surface charge density $|\sigma_{eff}|$ as derived from ζ-potential measurements in dependence on hydrodynamic radius R_H for Ay38-PDADMAC (blue squares) and Ay38-QPVP (red circles) for non-irradiated and irradiated samples. Reprinted with permission from ref. [128]. Copyright © 2012 Wiley Periodicals, Inc

previously. However, AFM measurement and light scattering further demonstrated that the alterations caused by irradiation here are attributed to an alteration in density as opposed to a change in aggregation number. It is possible that the assemblies with the linear polyelectrolytes are more flexible and looser as compared to the dendrimer assemblies, which explains why shrinking denotes an additional response possibility after light irradiation. Two reasons are co-playing here: the *cis*-form present following irradiation being more compact as compared with the *trans*-form of the azo dye. In addition, the interaction before the irradiation is more enthalpy-driven than entropically. Upon irradiation, the interaction is controlled more entropically, i.e., a bigger gain in water's entropy can account for entropy loss through polymer chain compaction.

Yu et al. utilized the electrostatic self-assembly for the purpose of combining 4-ethyl-4′-(trimethylaminohexyloxy) azobenzene bromide with an original Keggin-type POM [129]. The structure alters from a coral-like shape to smaller-sized nanospheres upon irradiation of the azo dye. This behavior is caused by the changes in hydrophobic effect and stacking matter of the surfactant molecules after irradiation. The group of Li also carried out a similar study [130]. In another instance, for light-triggered assembly/disassembly explored by the group of Samori, a cationic azo benzene derivate was directly anchored to a spherical Keplerate-type POM through electrostatic self-assembly. The assemblies here form a spherical structure which results in larger aggregates of nanostructures via π-π interaction called supramolecules [131]. A hydrodynamic radius of $R_H = 76.5$ nm was found by dynamic light scattering. The self-assembled structures start to disassemble and aggregates with a size of $R_H = 9.5$ nm build after irradiation.

5.10.3 Light-Triggerable Particle Shape

Another key area of interest is light-triggerable particle shape through modern self-assembly approaches, beyond the classical micelle shapes. Here again, electrostatic self-assembly in solution played a significant role. The Gröhn group studied linear cationic polyelectrolytes with different azo dyes in the first and impressive model system (Fig. 5.33) [100]. Poly(diallyldimethyl-ammonium chloride) combines with Ay38 to form flexible cylinders at a micrometer length, whereas the cross-section has a major axis $R_{maj} = 14$ nm, whereas a minor axis $R_{min} = 5$ nm. The particles' shape changes drastically into a core–shell ellipsoid with a major axis $R_{maj} = 200$ nm and a minor axis $R_{min} = 20$ nm upon irradiation. Here the inner core has $r_{maj} = 60$ nm and $r_{min} = 10$ nm. Via UV/Vis spectroscopy, the stacking angle was found to change after irradiation from $\beta = 60°$ to $\beta = 90°$. This implies that the observed change

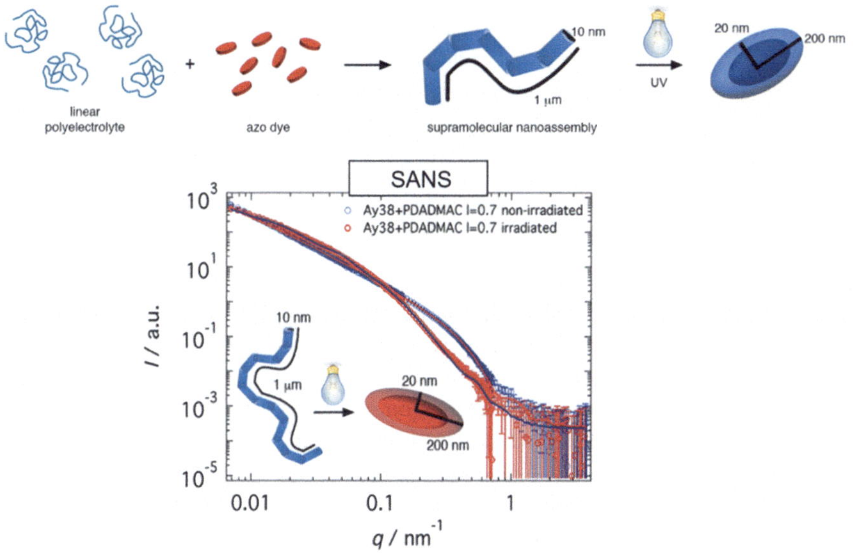

Fig. 5.33 Light-switchable nano-object shape by electrostatic self-assembly: An anionic azo dye (Ay38) and a linear flexible polyelectrolyte (PDADMAC) for micrometer long thin fibers which compact into ellipsoids upon irradiation. Reprinted with permission from ref. [100]. Copyright © 2018 WILEY–VCH Verlag GmbH & Co. KGaA, Weinheim

upon irradiation is associated with the dye configuration's alteration. It is possible to understand this behavior with the varying polarities of both isomer states of the azo dye, which is also known to impact the dye-dye stacking. The same experiment was carried out with a different polymer in the form of poly(N-methyl-4-vinylpyridinium nitrate). In this case, the irradiation was not found to cause any change in shape. The additional π-interaction, which could take place due to the polymer's aromatic ring (preventing a shape change) is one plausible explanation for this.

Huber and colleagues recently adopted a similar approach with the anionic polyelectrolyte polyacrylate (PA) as well as the *trans* isomer of a divalent diazophenol cation with an emphasis on the rearrangement mechanism. In an aqueous solution well-defined spherical structures in the order of $R_H = 75$ nm are formed [132]. These assemblies can be assembled and disassembled by utilizing the azo dye's isomerization. In addition, by varying the temperature at the reassemble process, the size and structure can also be controlled. Via light/neutron scattering, they stated that the first step, namely the *cis -trans* isomerization, is the rate-limiting measure. This is inclusive of the binding of the *trans*-azo dye to the anionic polyelectrolyte. The chain becomes too hydrophobic when a specific number of the azo dye molecules interacts with the polyelectrolyte, which causes nucleation. This results in the occurrence of the second step by means of monomer-addition-mechanism. In this case, the aggregate's morphology alters from a loosely packed structure to a dense/homogeneous spherical aggregate. These assemblies grow via a monomer-addition-mechanism in the third step. However, under the system explored by Gröhn et al., the irradiation to the *cis*-isomer only causes an increase or decrease in size, depending on the loading ratio. Moreover, the samples begin precipitating at a charge ratio greater than $l > 1.0$. The assemblies from the group of Gröhn were stable in solution. This was possibly caused by a difference in both systems' hydrophilicity.

5.10.4 Switching Enzyme Activity

The light-responsive property of spiropyrans was used to build supramolecular enzyme-spiropyran structures by electrostatic self-assembly [133]. The Gröhn group combined anionic spiropyran sulfonate with the enzyme lysozyme. In the dark, the spiropyran is available in the thermodynamic stable open-ring merocyanine isomer. After irradiation, the merocyanine alters from the open-ring isomer to the closed-ring isomer, which thermally converts back in the dark or by UV irradiation. The ring closure changes the molecular structure, the number of charges, and the dipole moment, thus also changing the nanoscale assembly structure. While it decreases the basicity of the oxygen in the closed-ring isomer, the open-ring isomer has a much larger molecular volume. After mixing the building blocks in an aqueous solution, aggregates with a size of $R_H = 103$ nm were formed. Upon irradiation of the merocyanine, the size nearly doubles to $R_H = 198$ nm, which changes back by storing the samples in the dark. This size-changing cycle can be repeated many times. The differences can be understood with the different binding of the two spiropyrans isomers to the protein. The merocyanine is planar and thus binds more internally to the protein, whereas the closed-ring isomer binds more externally due to its bulky form and therefore, can interconnect more lysozymes. Yet, irradiation not only influences the size of the aggregates but also the functionality of the enzyme. The formation of the aggregates and the interaction with the merocyanine nearly does not change the enzymatic activity of lysozyme. After irradiation, this activity decreases significantly, which is a proof of concept of supramolecular structures where the enzymatic activity can be switched [133]. The scenario is outlined in Fig. 5.34.

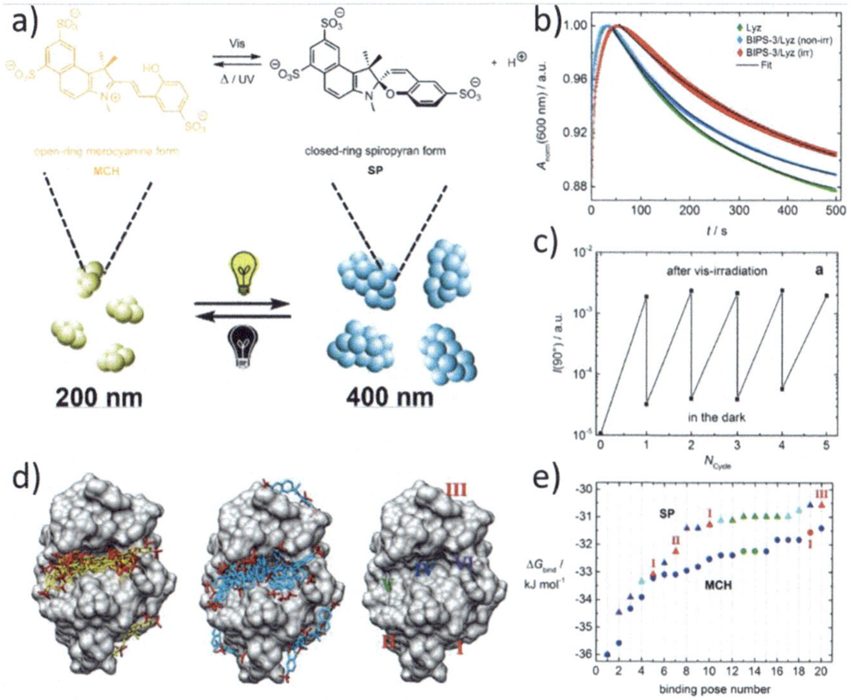

Fig. 5.34 Light-triggerable enzyme aggregation and activity; top: **a** interconversion of the two spiropyran isomers in inverse photochromism and schematic representation of the assemblies of an ionic spiropyran and lysozyme, interconnected with spiropyran, **b** Enzyme activity in the degradation of lyophilized cells, as detected via the scattering intensity, here detected as "absorption" A_{norm} (600 nm): the enzyme activity remains almost the same in the aggregate as compared to the individual lysozyme and it decreases upon irradiation; **c** repetitive switching of the assembly size as detected by the scattering intensity; **d** different binding sites of the two spiropyran states calculated by a docking model; **e** thermodynamics of the different binding sites. Reprinted with permission from ref. [133]. Copyright © 2018 American Chemical Society

5.10.5 Photoacid Counterions as Switch in Electrostatic Self-assembly

In 2015, the Gröhn group put forward a completely different concept for building light-responsive nano-assemblies by getting the unique properties of photoacids utilized. [134] Photoacids are only weak Brønsted acids in the ground state, and exhibit an increased acidity of the hydroxyl proton upon irradiation and go through an excited-state intermolecular proton transfer reaction [135]. This causes the formation of a molecule that is charged more highly. A linear oligomeric ethylene imine and sodium 1-naphthol-4-sulfonate were combined in the first project. Since the counterion has just one charge, no assemblies are created directly after mixing the building blocks. After irradiation, DLS and AFM displayed nano-assemblies with a size of $R_H = 170$ nm. A mechanism for forming assemblies was proposed in combination with UV/Vis spectroscopy, as shown in Fig. 5.35: Prior to irradiation, adding counterion causes an electrostatic interaction between the cationic polyelectrolyte and the anionic naphthol derivate. Photoacid dissociation takes place upon irradiation, and the polyelectrolyte gets protonated, thus triggering a strong interconnection caused by electrostatic interaction. This interaction changes from the strong ionic interaction to hydrogen bonds after decaying to the ground state. A stronger hydrophobization of the polymer chains is to be expected due to the desolvation of the amine groups. Subsequently, the closer approximation of building blocks gives rise to rearrangements caused by hydrophobic, π-π, and electrostatic interactions.

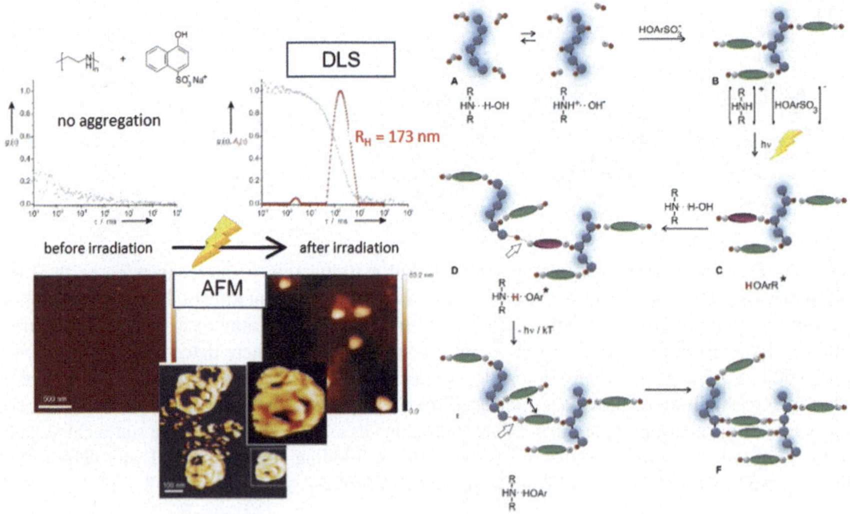

Fig. 5.35 Photoacid with changes pKa in the excited state acting as building block in electrostatic self-assembly with linear oligo(ethylene imine), left: DLS and AFM before (left) and after (right) UV irradiation. right: light-induced assembly mechanism. Reprinted with permission from ref. [134]. Copyright © 2015 American Chemical Society

In the aftermath of this study, Zika and Gröhn expanded the scope of the systems to facilitate a stronger understanding. In this process, different photoacids were combined with G4 dendrimer [136]. Here, the photoacids already have two negative charges in the ground state, which is why nano-assemblies are created before the irradiation. In the case of the strongest photoacid, 1-naphthol-3,6-disulfonate (1N36S) nanostructures are created with the size ranging from 89 nm $\leq R_H$ to $\leq$ 190 nm, depending on the loading ratio. The size increases ranging from 130 nm $\leq R_H$ $\leq$ 244 nm after irradiation. Thus, it is possible to use a photoacid switch not only to trigger assembly from building blocks but also to change the size of the assembly. Additionally, Zika and Gröhn recently developed a three-component (ternary) pH- and light-responsive system through the combination of a photoacid — an anionic disulfonated naphthol derivative — neutral hydroxyflavylium ions as well as a linear cationic poly(allylamine) in an aqueous solution [137]. Depending on pH as well as light irradiation, hydroxyflavylium can showcase a gamut of different reactions. This system offers multiple ways to trigger the nanostructures owing to the unique light-responsive property of the photoacid and the pH and light-dependent hydroxyflavylium. Based on the switching step, the size of the assembly is found to range between R_H = 142 nm and R_H = 332 nm. Furthermore, both the size and shape of the nano-objects could be switched.

5.10.6 Further Triggers

Additional triggers for electrostatically self-assembled polyelectrolyte systems include temperature [138], magnetic field [139], and redox responsiveness [140]. An electrostatic self-assembly-constructed light-pH-copper ion tri-stimuli responsive system, as illustrated in Fig. 5.35, is also promising [141]. Here, Yu et al. combined the cationic azobenzene-containing surfactant (4ethyl-4′-(trimethylaminohexyloxy) azobenzene bromide) with an anionic Eu-containing polyoxometalate, which led to a spherical aggregate structure sized d = 200 nm. The size of the particle somewhat decreases after light irradiation, and the POM's fluorescence is quenched strongly. This is probably ascribed to the more hydrophilic nature around luminescence center Eu^{3+}. Notably, the *cis*-isomer is more hydrophilic, thereby decreasing the hydrophobic interaction. In addition, it is hindered more sterically, which results in a looser alignment. By adding H^+, the morphology alters from a spherical structure to a necklace-like geometry, also quenching the fluorescence due to a freshly-created hydrogen bond W–O···H···O–H. Adding Cu^{2+} with an O $\rightarrow$ W energy transfer, a ligand–metal charge transfer to the Cu^{2+} ion in the excited state is another way of quenching the assemblies' fluorescence Fig. 5.36.

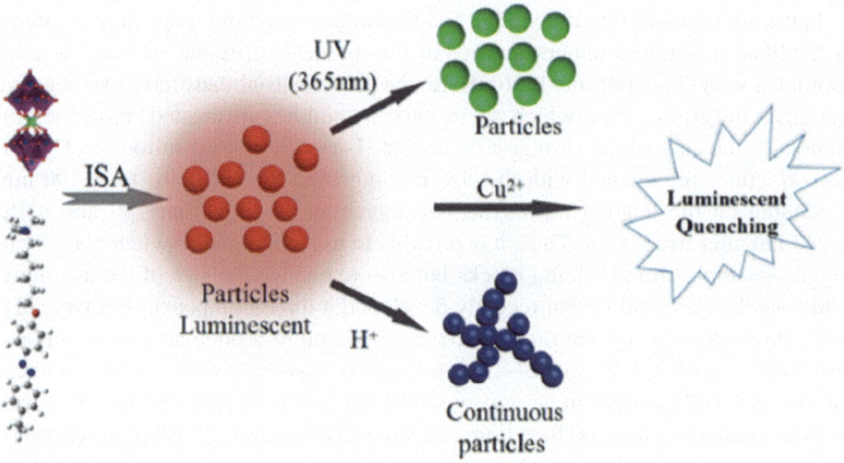

Fig. 5.36 Formation of ETAB/ Eu-POM Supramolecular materials and their responsiveness to UV-light, Cu^{2+}, and H^{+}. Reprinted with permission from ref. [141]. Copyright © 2016 American Chemical Society

5.11 Conclusion

To conclude, using electrostatic self-assembly in solution to design functional nano-objects signifies an emerging field that has a great potential to be further explored. Polyelectrolytes play a special role here, as they allow for the formation of larger assemblies and for the electrostatic stabilization of the nano-objects in an aqueous solution across a wide variety of designs. More specifically, the targeted combination of electrostatic interaction with other interactions and effects, like the positioning of charges on stiff building blocks, using further amphiphilic, polyelectrolytes with certain architectures, or π-π stacking building blocks have promulgated electrostatic self-assembly to a principle for structure formation that is versatile and well-defined. A variety of structures from spheres over rods and hollow spheres to networks in the size range of a few tenths to a few hundred nanometers can be formed. Even though the electrostatic self-assembly for the formation of finite-size nano-objects in solution represents a new and emerging field, in contrast to, for instance, the classical micelle formation that has been investigated for 60 years, an understanding of structure-directing effects is increasingly developed. Here, interaction thermodynamics plays a key role in linking molecular building-block features with the properties of self-assembled objects on the nanoscale. Triggering assembling/disassembling, assembly size, and shape by pH and light has also been addressed, whereas the capability of electrostatic self-assembly for photocatalysis has also been indicated. However, this contribution has not explored the ability of electrostatic self-assembly for the purpose of photocatalysis in detail. Deriving a foundational understanding of interplaying interactions and property/structure-directing effects is imperative for establishing electrostatic self-assembly, or novel self-assembly approaches in new areas — as evidenced in this case for forming functional and switchable nano-assemblies. A targeted design of functional structures will be possible with this approach, and only then will it be possible to take advantage of the great capability of electrostatic self-assembly representing a concept that renders it possible to form a gamut of functions and structures encompassing various applications.

References

1. Pauling L (2000) The nature of the chemical bond and the structure of molecules and crystals: an introduction to modern structural chemistry. Ithaca, Cornell Univ. Press, New York
2. Förster S, Schmidt M (1995) Polyelectrolytes in solution. In: Abe A, Benoit H, Cantow H-J, Corradini P, Dušek K, Edwards S et al (eds) Physical Properties of Polymers: Berlin, Heidelberg: Springer Berlin Heidelberg, pp 51–133
3. Schmitz KS (1993) Macroions in solution and colloidal suspension. VCH, New York
4. Ito K, Yoshida H, Ise N (1994) Void structure in colloidal dispersions. Science 263(5143):66–68
5. Gröhn F, Antonietti M (2000) Intermolecular structure of spherical polyelectrolyte microgels in salt-free solution. 1. Quantification of the attraction between equally charged polyelectrolytes. Macromolecules 33(16):5938–5949
6. Förster S, Schmidt M, Antonietti M (1990) Static and dynamic light scattering by aqueous polyelectrolyte solutions: effect of molecular weight, charge density and added salt. Polymer 31(5):781–792
7. Sedlák M, Amis EJ (1992) Concentration and molecular weight regime diagram of salt-free polyelectrolyte solutions as studied by light scattering. J Chem Phys 96(1):826–834
8. Antonietti M, Briel A, Gröhn F (2000) Spherical polyelectrolyte microgels in salt free solution. 2. combined analysis of static structure and viscosity. Macromolecules 33(16): 5950– 5953
9. Ikeda Y, Beer M, Schmidt M et al (1998) Ca 2+ and Cu 2+ induced conformational changes of sodium polymethacrylate in dilute aqueous solution. Macromolecules 31(3):728–733
10. Schweins R, Huber K (2001) Collapse of sodium polyacrylate chains in calcium salt solutions. Eur Physi J E 5(1):117–126
11. Goerigk G, Huber K, Schweins R (2007) Probing the extent of the Sr2+ ion condensation to anionic polyacrylate coils: a quantitative anomalous small-angle x-ray scattering study. J Chem Phys 127(15):154908
12. Lages S, Michels R, Huber K (2010) Coil-Collapse and coil-aggregation due to the interaction of Cu^{2+} and Ca^{2+} ions with anionic polyacrylate chains in dilute solution. Macromolecules 43(6):3027–3035
13. Evans DF, Wennerström H (1999) The colloidal domain: where physics, chemistry, biology, and technology meet. Wiley-VCH, New York, NY
14. Sinn CG, Dimova R, Antonietti M (2004) Isothermal titration calorimetry of the polyelectrolyte/water interaction and binding of Ca^{2+} effects determining the quality of polymeric scale inhibitors. Macromolecules 37(9):3444–3450
15. Tadros TF (2005) Applied Surfactants. Wiley
16. Chandler D (2005) Interfaces and the driving force of hydrophobic assembly. Nature 437(7059):640–647
17. Southall NT, Dill KA, Haymet ADJ (2002) A view of the hydrophobic effect. J Phys Chem B 106(3):521–533
18. Tanford C (1962) Contribution of hydrophobic interactions to the stability of the globular conformation of proteins. J Am Chem Soc 84(22):4240–4247
19. Tanford C (1978) The hydrophobic effect and the organization of living matter. Science 200(4345):1012–1018

20. Kukula H, Schlaad H, Antonietti M et al (2002) The formation of polymer vesicles or "peptosomes" by polybutadiene-block-poly(L-glutamate)s in dilute aqueous solution. J Am Chem Soc 124(8):1658–1663
21. Cougnon FBL, Caprice K, Pupier M et al (2018) A strategy to synthesize molecular knots and links using the hydrophobic effect. J Am Chem Soc 140(39):12442–12450
22. Antonietti M, Förster S, Hartmann J et al (1996) Novel amphiphilic block copolymers by polymer reactions and their use for solubilization of metal salts and metal colloids. Macromolecules 29(11):3800–3806
23. Li Z, Kesselman E, Talmon Y et al (2004) Multicompartment micelles from ABC miktoarm stars in water. Science 306(5693):98–101
24. Lipshutz BH, Ghorai S, Cortes-Clerget M (2018) The hydrophobic effect applied to organic synthesis: recent synthetic chemistry "in water." Chem Eur J 24(26):6672–6695
25. Arunan E, Desiraju GR, Klein RA et al (2011) Definition of the hydrogen bond (IUPAC recommendations 2011). Pure Appl Chem 83(8):1637–1641
26. Emsley J (1980) Very strong hydrogen bonding. Chem Soc Rev 9(1):91
27. Grabowski SJ (2004) Hydrogen bonding strength—measures based on geometric and topological parameters. J Phys Org Chem 17(1):18–31
28. Watson JD, Crick FHC (1953) Molecular structure of nucleic acids: a structure for deoxyribose nucleic acid. Nature 171(4356):737–738
29. Pauling L, Corey RB, Branson HR (1951) The structure of proteins; two hydrogen-bonded helical configurations of the polypeptide chain. Proc Natl Acad Sci 37(4):205–211
30. Wang Y, Liu M, Gao J (2020) Enhanced receptor binding of SARS-CoV-2 through networks of hydrogen-bonding and hydrophobic interactions. Proc Natl Acad Sci 117(25):13967–13974
31. Brunsveld L, Folmer BJ, Meijer EW et al (2001) Supramolecular polymers. Chem Rev 101(12):4071–4098
32. Schmuck C, Wienand W (2001) Self-complementary quadruple hydrogen-bonding motifs as a functional principle: from dimeric supramolecules to supramolecular polymers. Angew Chem Int Ed 40(23):4363
33. Mazzier D, De S, Wicher B et al (2020) Parallel homochiral and anti-parallel heterochiral hydrogen-bonding interfaces in multi-helical abiotic foldamers. Angew Chem Int Ed 59(4):1606–1610
34. Cao J, Lu C, Zhuang J et al (2017) Multiple hydrogen bonding enables the self-healing of sensors for human-machine interactions. Angew Chem Int Ed 56(30):8795–8800
35. Song Y, Liu Y, Qi T et al (2018) Towards dynamic but supertough healable polymers through biomimetic hierarchical hydrogen-bonding interactions. Angew Chem Int Ed 57(42):13838–13842
36. Tamate R, Hashimoto K, Horii T et al (2018) Self-healing micellar ion gels based on multiple hydrogen bonding. Adv Mater 30(36):e1802792
37. Steed JW, Turner DR, Wallace KJ (2007) Core concepts in supramolecular chemistry and nanochemistry. John Wiley, Chichester, England, Hoboken, NJ
38. Li F, Liu D, Wang T et al (2017) J-aggregation in porphyrin nanoparticles induced by diphenylalanine. J Solid State Chem 252:86–92
39. Maiti NC, Mazumdar S, Periasamy N (1998) J- and H-Aggregates of porphyrin−surfactant complexes: time-resolved fluorescence and other spectroscopic studies. J Phys Chem B 102(9):1528–1538
40. Pasternack RF, Huber PR, Boyd P et al (1972) On the aggregation of meso-substituted water-soluble porphyrins. J Am Chem Soc 94(13):4511–4517
41. Miller ED, Jones ML, Henry MM et al (2018) Optimization and validation of efficient models for predicting polythiophene self-assembly. Polymers 10(12)
42. Surin M, Sonar P, Grimsdale AC et al (2005) Supramolecular organization in fluorene/indenofluorene—oligothiophene alternating conjugated copolymers. Adv Func Mater 15(9):1426–1434
43. Bjørnholm T, Hassenkam T, Greve DR et al (1999) Polythiophene Nanowires. Adv Mater 11(14):1218–1221

44. De Luca G, Liscio A, Maccagnani P et al (2007) Nucleation-governed reversible self-assembly of an organic semiconductor at surfaces: long-range mass transport forming giant functional fibers. Adv Func Mater 17(18):3791–3798
45. Chetia M, Debnath S, Chowdhury S et al (2020) Self-assembly and multifunctionality of peptide organogels: oil spill recovery, dye absorption and synthesis of conducting biomaterials. RSC Adv 10(9):5220–5233
46. Brunsveld L, Vekemans JAJM, Hirschberg JHKK et al (2002) Hierarchical formation of helical supramolecular polymers via stacking of hydrogen-bonded pairs in water. Proc Natl Acad Sci 99(8):4977–4982
47. Hirschberg JH, Brunsveld L, Ramzi A et al (2000) Helical self-assembled polymers from cooperative stacking of hydrogen-bonded pairs. Nature 407(6801):167–170
48. Terashima T, Mes T, de Greef TFA et al (2011) Single-chain folding of polymers for catalytic systems in water. J Am Chem Soc 133(13):4742–4745
49. George SJ, Ajayaghosh A, Jonkheijm P et al (2004) Coiled-Coil gel nanostructures of Oligo(p-phenylenevinylene)s: gelation-induced helix transition in a higher-order supramolecular self-assembly of a rigidπ-conjugated system. Angew Chem 116(26):3504–3507
50. van Gorp JJ, Vekemans JAJM, Meijer EW (2002) C3-symmetrical supramolecular architectures: fibers and organic gels from discotic trisamides and trisureas. J Am Chem Soc 124(49):14759–14769
51. Li M, Zajaczkowski W, Velpula G et al (2020) Transformation from helical to layered supramolecular organization of asymmetric perylene diimides via multiple intermolecular hydrogen bonding. Chem Sci 11(19):4960–4968
52. Kurokawa Y, Shirakawa N, Terada M et al (1980) Formation of polyelectrolyte complex and its adsorption properties. J Appl Polym Sci 25(8):1645–1653
53. Biesheuvel PM, Cohen Stuart MA (2004) Electrostatic free energy of weakly charged macromolecules in solution and intermacromolecular complexes consisting of oppositely charged polymers. Langmuir 20(7):2785–2791
54. Dautzenberg H, Hartmann J, Grunewald S et al (1996) Stoichiometry and structure of polyelectrolyte complex particles in diluted solutions. Ber Bunsenges Phys Chem 100(6):1024–1032
55. Starchenko V, Müller M, Lebovka N (2012) Sizing of PDADMAC/PSS complex aggregates by polyelectrolyte and salt concentration and PSS molecular weight. J Phys Chem B 116(51):14961–14967
56. Tsuchida E (1994) Formation of polyelectrolyte complexes and their structures. J Macromol Sci Part A 31(1):1–15
57. Weber D, Torger B, Richter K et al (2018) Interaction of poly(l-lysine)/polysaccharide complex nanoparticles with human vascular endothelial cells. Nanomaterials 8(6):358
58. Argelles-Monal W, Grciga M, Peniche-Covas C (1990) Study of the stoichiometric polyelectrolyte complex between chitosan and carboxymethyl cellulose. Polym Bull 23(3):307–313
59. Chavasit V, Kienzle-Sterzer C, Antonio Torres J (1988) Formation and characterization of an insoluble polyelectrolyte complex: chitosan-polyacrylic acid. Polym Bull 19(3):223–230
60. Kawashima Y, Handa T, Kasai A et al (1985) Novel method for the preparation of controlled-release theophylline granules coated with a polyelectrolyte complex of sodium polyphosphate-chitosan. J Pharm Sci 74(3):264–268
61. Sato H, Nakajima A (1975) Formation of a polyelectrolyte complex from carboxymethyl cellulose and poly(ethylenimine). Polym J 7(2):241–247
62. Xu Y, Borisov OV, Ballauff M et al (2010) Manipulating the morphologies of cylindrical polyelectrolyte brushes by forming interpolyelectrolyte complexes with oppositely charged linear polyelectrolytes: an AFM study. Langmuir 26(10):6919–6926
63. Kleinen J, Klee A, Richtering W (2010) Influence of architecture on the interaction of negatively charged multisensitive poly(N-isopropylacrylamide)-co-methacrylic acid microgels with oppositely charged polyelectrolyte: absorption vs adsorption. Langmuir 26(13):11258–11265
64. Procházka K, Martin TJ, Webber SE et al (1996) Onion-Type micelles in aqueous media. Macromolecules 29(20):6526–6530

65. Tsitsilianis C, Voulgaris D, Štěpánek M et al (2000) Polystyrene/Poly(2-vinylpyridine) heteroarm star copolymer micelles in aqueous media and onion type micelles stabilized by diblock copolymers. Langmuir 16(17):6868–6876
66. van der Kooij HM, Spruijt E, Voets IK et al (2012) On the stability and morphology of complex coacervate core micelles: from spherical to wormlike micelles. Langmuir 28(40):14180–14191
67. Voets IK, de Keizer A, Cohen Stuart MA et al (2006) Core and corona structure of mixed polymeric micelles. Macromolecules 39(17):5952–5955
68. Voets IK, de Keizer A, de Waard P et al (2006) Double-faced micelles from water-soluble polymers. Angew Chem Int Ed 45(40):6673–6676
69. Voets IK, de Keizer A, Leermakers FAM et al (2009) Electrostatic hierarchical co-assembly in aqueous solutions of two oppositely charged double hydrophilic diblock copolymers. Eur Polymer J 45(10):2913–2925
70. Wang J, de Keizer A, Fokkink R et al (2010) Complex coacervate core micelles from iron-based coordination polymers. J Phys Chem B 114(25):8313–8319
71. Wang J, Sun S, Wu B et al (2019) Processable and luminescent supramolecular hydrogels from complex coacervation of polycations with lanthanide coordination polyanions. Macromolecules 52(22):8643–8650
72. Yan Y, Martens AA, Besseling NAM et al (2008) Nanoribbons self-assembled from triblock peptide polymers and coordination polymers. Angew Chem 120(22):4260–4263
73. Yan Y, de Keizer A, Cohen Stuart MA et al (2008) Stability of complex coacervate core micelles containing metal coordination polymer. J Phys Chem B 112(35):10908–10914
74. Zhou W, Wang J, Ding P et al (2019) Functional polyion complex vesicles enabled by supramolecular reversible coordination polyelectrolytes. Angew Chem Int Ed 58(25):8494–8498
75. Li G, Yang X, Wang J (2008) Raspberry-like polymer composite particles via electrostatic heterocoagulation. Colloids Surf A 322(1–3):192–198
76. Anraku Y, Kishimura A, Oba M et al (2010) Spontaneous formation of nanosized unilamellar polyion complex vesicles with tunable size and properties. J Am Chem Soc 132(5):1631–1636
77. Anraku Y, Kishimura A, Yamasaki Y et al (2013) Living unimodal growth of polyion complex vesicles via two-dimensional supramolecular polymerization. J Am Chem Soc 135(4):1423–1429
78. Kishimura A, Koide A, Osada K et al (2007) Encapsulation of myoglobin in PEGylated polyion complex vesicles made from a pair of oppositely charged block ionomers: a physiologically available oxygen carrier. Angew Chem Int Ed 46(32):6085–6088
79. Koide A, Kishimura A, Osada K et al (2006) Semipermeable polymer vesicle (PICsome) self-assembled in aqueous medium from a pair of oppositely charged block copolymers: physiologically stable micro-/nanocontainers of water-soluble macromolecules. J Am Chem Soc 128(18):5988–5989
80. Sing CE, Zwanikken JW, Olvera de la Cruz M (2014) Electrostatic control of block copolymer morphology. Nat Mater 13(7):694–698
81. Hollingsworth WR, Magnanelli TJ, Segura C et al (2018) Polyion charge ratio determines transition between bright and dark excitons in donor/acceptor-conjugated polyelectrolyte complexes. The Journal of Physical Chemistry C 122(39):22280–22293
82. Gröhn F, Klein K, Brand S (2008) Facile route to supramolecular structures: self-assembly of dendrimers and naphthalene dicarboxylic acids. Chem Eur J 14(23):6866–6869
83. Willerich I, Gröhn F (2008) Switchable nanoassemblies from macroions and multivalent dye counterions. Chem Eur J 14(30):9112–9116
84. Ruthard C, Maskos M, Kolb U et al (2009) Finite-Size networks from cylindrical polyelectrolyte brushes and porphyrins. Macromolecules 42(3):830–840
85. Li Y, Yildiz UH, Müllen K et al (2009) Association of DNA with multivalent organic counterions: from flowers to rods and toroids. Biomacromol 10(3):530–540
86. Mariani G, Kutz A, Di Z et al (2017) Inducing Hetero-aggregation of different Azo dyes through electrostatic self-assembly. Chem Eur J 23(26):6249–6254

87. Willerich I, Ritter H, Gröhn F (2009) Structure and thermodynamics of ionic dendrimer-dye assemblies. J Phys Chem B 113(11):3339–3354
88. Yildiz ÜH, Koynov K, Gröhn F (2009) Fluorescent nanoparticles through self-assembly of linear ionenes and pyrenetetrasulfonate. Macromol Chem Phys 210(20):1678–1690
89. Willerich I, Gröhn F (2010) Photoswitchable nanoassemblies by electrostatic self-assembly. Angew Chem Int Ed 49(44):8104–8108
90. Gröhn F (2010) Soft matter nanoparticles with various shapes and functionalities can form through electrostatic self-assembly. Soft Matter 6(18):4296
91. Gröhn F, Klein K, Koynov K (2010) A novel type of vesicles based on Ionic and π-π Interactions. Macromol Rapid Commun 31(1):75–80
92. Frühbeißer S, Gröhn F (2012) Catalytic activity of macroion-porphyrin nanoassemblies. J Am Chem Soc 134(35):14267–14270
93. Gröhn F (2008) Electrostatic self-assembly as route to supramolecular structures. Macromol Chem Phys 209(22):2295–2301
94. Düring J, Hölzer A, Kolb U et al (2013) Supramolecular organic-inorganic hybrid assemblies with tunable particle size: interplay of three noncovalent interactions. Angew Chem Int Ed 52(33):8742–8745
95. Frühbeißer S, Gröhn F (2017) Porphyrin-Polyelectrolyte nanoassemblies: the role of charge and building block architecture in self-assembly. Macromol Chem Phys 218(17):1600526
96. Kutz A, Mariani G, Schweins R et al (2018) Self-assembled polyoxometalate-dendrimer structures for selective photocatalysis. Nanoscale 10(3):914–920
97. Frühbeißer S, Fuenzalida Werner JP, Gröhn F (2018) Metalloporphyrin–polyelectrolyte assemblies in aqueous solution: Influence of the metal center and the polyelectrolyte architecture. J Polym Sci Part B: Polym Phys 56(6):484–500
98. Krieger A, Fuenzalida Werner JP, Mariani G et al (2017) Functional supramolecular porphyrin-dendrimer assemblies for light harvesting and photocatalysis. Macromolecules 50(9):3464–3475
99. Kutz A, Gröhn F (2017) Improving photocatalytic activity: versatile polyelectrolyte—photosensitizer assemblies for methyl viologen reduction. ChemistrySelect 2(4):1504–1515
100. Mariani G, Krieger A, Moldenhauer D et al (2018) Light-Responsive shape: from micrometer-long nanocylinders to compact particles in electrostatic self-assembly. Macromo Rapid Commun 39(17):e1700860
101. Willerich I, Gröhn F (2011) Thermodynamics of photoresponsive polyelectrolyte-dye assemblies with irradiation wavelength triggered particle size. Macromolecules 44(11):4452–4461
102. Willerich I, Gröhn F (2011) Molecular structure encodes nanoscale assemblies: understanding driving forces in electrostatic self-assembly. J Am Chem Soc 133(50):20341–20356
103. Willerich I, Schindler T, Gröhn F (2011) Effect of polyelectrolyte architecture and size on macroion-dye assemblies. J Phys Chem B 115(32):9710–9719
104. Mariani G, Moldenhauer D, Schweins R et al (2016) Elucidating electrostatic self-assembly: molecular parameters as key to thermodynamics and nanoparticle shape. J Am Chem Soc 138(4):1280–1293
105. Willerich I, Li Y, Gröhn F (2010) Influencing particle size and stability of ionic dendrimer–dye assemblies. J Phys Chem B 114(47):15466–15476
106. Mariani G, Schweins R, Gröhn F (2016) Structure tuning of electrostatically self-assembled nanoparticles through pH. J Phys Chem B 120(7):1380–1389
107. Mariani G, Schweins R, Gröhn F (2017) Structure tuning of electrostatically self-assembled nanoparticles through pH: the role of charge ratio. Macromol Chem Phys 218(24):1700191
108. Mariani G, Schweins R, Gröhn F (2016) Electrostatic self-assembly of dendrimer macroions and multivalent dye counterions: the role of solution ionic strength. Macromolecules 49(22):8661–8671
109. Kutz A, Alex W, Krieger A et al (2017) Hydrogen-Bonded polymer-porphyrin assemblies in water: supramolecular structures for light energy conversion. Macrom Rapid Commun 38(17)
110. Srebnik S, Douglas JF (2011) Self-assembly of charged particles on nanotubes and the emergence of particle rings, chains, ribbons and chiral sheets. Soft Matter 7(15):6897

111. Damasceno PF, Engel M, Glotzer SC (2012) Predictive self-assembly of polyhedra into complex structures. Science 337(6093):453–457
112. Glotzer SC, Solomon MJ (2007) Anisotropy of building blocks and their assembly into complex structures. Nat Mater 6(8):557–562
113. Ruthard C, Maskos M, Kolb U et al (2011) Polystyrene sulfonate-porphyrin assemblies: influence of polyelectrolyte and porphyrin structure. J Phys Chem B 115(19):5716–5729
114. Henle ML, Pincus PA (2005) Equilibrium bundle size of rodlike polyelectrolytes with counterion-induced attractive interactions. Phys Rev E 71(6):60801
115. Kovaric BC, Kokona B, Schwab AD et al (2006) Self-Assembly of peptide porphyrin complexes: toward the development of smart biomaterials. J Am Chem Soc 128(13):4166–4167
116. Kuciauskas D, Caputo GA (2009) Self-assembly of peptide−porphyrin complexes leads to pH-dependent excitonic coupling. J Phys Chem B 113(43):14439–14447
117. Liu Y, Jin J, Deng H et al (2016) Protein-Framed multi-porphyrin micelles for a hybrid natural-artificial light-harvesting nanosystem. Angew Chem Int Ed 55(28):7952–7957
118. Segura CJ, Lucero M, Ayzner AL (2019) Disassembly of an interconjugated polyelectrolyte complex using ionic surfactants. ACS Appl Polym Mater 1(5):1034–1044
119. Kistler ML, Patel KG, Liu T (2009) Accurately tuning the charge on giant polyoxometalate type keplerates through stoichiometric interaction with cationic surfactants. Langmuir 25(13):7328–7334
120. Kutz A, Mariani G, Gröhn F (2016) Ionic dye–surfactant nanoassemblies: interplay of electrostatics, hydrophobic effect, and π–π stacking. Colloid Polym Sci 294(3):591–606
121. Li H, Sun H, Qi W et al (2007) Onionlike hybrid assemblies based on surfactant-encapsulated polyoxometalates. Angew Chem Int Ed 46(8):1300–1303
122. Wald G (1968) The molecular basis of visual excitation. Nature 219(5156):800–807
123. Andrade SM, Costa SMB (2002) Spectroscopic studies on the interaction of a water soluble porphyrin and two drug carrier proteins. Biophys J 82(3):1607–1619
124. Ruthard C, Schmidt M, Gröhn F (2011) Porphyrin-polymer networks, worms, and nanorods: pH-triggerable hierarchical self-assembly. Macromol Rapid Commun 32(9–10):706–711
125. Helseth LE (2012) Pyranine-induced self-assembly of colloidal structures using poly(allylamine-hydrochloride). J Colloid Interface Sci 375(1):23–29
126. Goodman M, Falxa ML (1967) Conformational aspects of polypeptide structure. 23. Photoisomerization of azoaromatic polypeptides. J Am Chem Soc 89(15):3863–3867
127. Koga T, Ushirogochi M, Higashi N (2007) Regulation of self-assembling process of a cationic β-sheet peptide by photoisomerization of an anionic azobenzene derivative. Polym J 39(1):16–17
128. Moldenhauer D, Gröhn F (2013) Nanoassemblies with light-responsive size and density from linear flexible polyelectrolytes. J Polym Sci Part B: Polym Phys 51(10):802–816
129. Guo Y, Gong Y, Yu Z et al (2016) Rational design of photo-responsive supramolecular nanostructures based on an azobenzene-derived surfactant-encapsulated polyoxometalate complex. RSC Adv 6(18):14468–14473
130. Xie X, Wang L, Liu X et al (2020) Light-powered and transient peptide two-dimensional assembly driven by trans-to-cis isomerization of azobenzene side chains. Chem Commun 56(12):1867–1870
131. Markiewicz G, Pakulski D, Galanti A et al (2017) Photoisomerisation and light-induced morphological switching of a polyoxometalate-azobenzene hybrid. Chem Commun 53(53):7278–7281
132. Carl N, Müller W, Schweins R et al (2020) Controlling self-assembly with light and temperature. Langmuir 36(1):223–231
133. Moldenhauer D, Fuenzalida Werner JP, Strassert CA et al (2019) Light-Responsive size of self-assembled spiropyran-lysozyme nanoparticles with enzymatic function. Biomacromol 20(2):979–991
134. Cardenas-Daw C, Gröhn F (2015) Photo-Induced assembly of nanostructures triggered by short-lived proton transfers in the excited-state. J Am Chem Soc 137(27):8660–8663

135. Agmon N, Rettig W, Groth C (2002) Electronic determinants of photoacidity in cyanonaphthols. J Am Chem Soc 124(6):1089–1096
136. Zika A, Bernhardt S, Gröhn F (2020) Photoresponsive photoacid-macroion nano-assemblies. Polymers 12(8)
137. Zika A, Gröhn F (2021) Multiswitchable photoacid-hydroxyflavylium-polyelectrolyte nano-assemblies. Beilstein J Org Chem 17:166–185
138. Feng L, Xu L, Dong S et al (2016) Thermo-reversible capture and release of DNA by zwitterionic surfactants. Soft Matter 12(36):7495–7504
139. Liu F, Wang J, Cao Q et al (2015) One-step synthesis of magnetic hollow mesoporous silica (MHMS) nanospheres for drug delivery nanosystems via electrostatic self-assembly templated approach. Chem Commun 51(12):2357–2360
140. Liu H, Wang L, Wang X et al (2019) Vesicle transition of catanionic redox-switchable surfactants controlled by DNA with different chain lengths. J Colloid Interface Sci 549:89–97
141. Guo Y, Gong Y, Gao Y et al (2016b) Multi-stimuli responsive supramolecular structures based on azobenzene surfactant-encapsulated polyoxometalate. Langmuir 32(36):9293–9300

Chapter 6
Electrostatic Layer-by-Layer Self-Assembly Method: A Physico-Chemical Perspective

Eduardo Guzmán, Ana Mateos-Maroto, Francisco Ortega, and Ramón G. Rubio

Abstract The use of the Layer-by-Layer (LbL) method for the fabrication of structural and functional materials through the alternate deposition of polyelectrolyte bearing opposite charges has undergone a spectacular development due to the numerous avenues that offer for controlling the assembly process by simply tuning some operational parameters or characteristic of the layering molecules. This is only possible by a careful examination of the physicochemical bases underlying the assembly process. This chapter tries to provide a broad physicochemical perspective trying to disentangle some of the most fundamental aspects underlying the exploitation of the LbL electrostatic self-assembly for opening new avenues in the design of novel nanomaterials.

6.1 Introduction

The use of the Layer-by-Layer (LbL) self-assembly methodology has evolved very fast in the last years to become a widespread technology for the assembly of structural and functional materials with applications ranging from the biomedical and pharmaceutical industries (tissue engineering, biomedical devices, wound healing dressing, encapsulation platforms, biosensors, cardiovascular devices or implants) to the fabrication of electronic devices (conductive layers, light-emitting thin films, electrochromic films, photonic systems or non-linear optical devices), and from the fabrication of self-healing materials to preparation of different coatings (antireflection coatings or superhydrophobic surfaces) [1–10]. This has been possible by the

E. Guzmán (✉) · F. Ortega · R. G. Rubio
Departamento de Química Física, Universidad Complutense de Madrid, Ciudad Universitaria s/n, 28040 Madrid, Spain
e-mail: eduardogs@quim.ucm.es

Instituto Pluridisciplinar, Universidad Complutense de Madrid, Paseo Juan XXIII 1, 28040 Madrid, Spain

A. Mateos-Maroto
Max Planck Institute for Polymer Research, Ackermannweg 10, 55128 Mainz, Germany

M. A. Aboudzadeh and A. Frontera (eds.), *Supramolecular Assemblies Based on Electrostatic Interactions*,
https://doi.org/10.1007/978-3-031-00657-9_6

combination of low cost and simplicity with modularity and versatility which have enabled the manufacturing of multi-layered materials with structure and properties tailored almost at will. Thus, the LbL approach makes possible the fabrication of materials including a broad range of structural and functional characteristics [6, 8, 11–13].

The seminal works on the LbL films took advantage of the ability of polyelectrolytes to form self-organized structured when they interact with a polyelectrolyte bearing the opposite charge [14–17]. Thus, it was possible to fabricate polyelectrolyte multi-layered films by the alternate deposition of layers of polyelectrolytes bearing opposite charges onto a solid substrate, i.e., the so-called polyelectrolyte multilayers (PEMs). However, the advances in the implementation of the LbL technology has allowed their use for the self-assembly of hierarchical supramolecular structures by the alternate deposition of a broad range of materials (building blocks), with the only limitation arriving from the need for complementary interactions between the compounds assembled in adjacent layers. This has allowed the fabrication of LbL films including, among other types of compounds, different types of colloidal particles and nano-objects (graphene and graphene oxide nanoplatelets, carbon nanotubes, dendrimers, clays, microgels, polymeric, ceramic, or metallic particles), biomolecules (proteins and peptides, polysaccharides, nucleic acid, lipids), dyes, viruses, synthetic polymers and even in some cases small molecules [1, 3, 18–36]. The extension of the lists of building blocks used in the assembly of LbL materials requires considering the use of interactions with non-electrostatic origin as driving force of the process, and now it is common for the fabrication of multi-layered films using the LbL approach through hydrogen bonding [37, 38], charge transfer interactions [39], molecular recognition [40, 41], coordination interactions [42], chiral recognition [43], host–guest interactions [44], π-π interactions [45], biospecific interactions [46], sol–gel reactions [47], or even covalent bond ("click chemistry" reactions) [48, 49].

Moreover, the possibility to use any kind of substrate, independently of its chemical nature, shape, geometry, or size, as template for the deposition of the multi-layered films has also contributed to the success of the LbL methodology in the fabrication of different types of materials. Nowadays, it is common for the fabrication of LbL in any kind of substrate, with the only requirement being the need that the substrate can be easily accessed for the solution. This has led to LbL films on macroscopic solid charged flat substrates, but also on colloidal micro-and nanoparticles, liposomes or vesicles, micelles, fluid interfaces (floating multilayers), emulsion droplets, or even cells [50–58]. It should be noted that the substrates can play two different roles: (i) define the geometry and morphology of the assembled material, remaining as part of the finalized material, and (ii) be a template during the assembly process, which can be removed from the finalized material using chemical or physical procedures (sacrificial template), and hence allowing the fabrication of free-standing films and hollow capsules [59, 60]. Thus, the LbL method allows controlling almost at will the size, shape, and morphology of the assembled materials, which enable its application in the manufacturing of a broad range of structural and functional materials, e.g., flat films, nano-, and micro-capsules or multi capsules including several hierarchically organized nano-containers [11, 56, 61–66]. Furthermore, the versatility and modularity of the LbL technology allow its use in the fabrication of many sophisticated materials, including particles with onion-like structures, sponges, membranes, or nanotubes [5, 67, 68].

This chapter is intended to present a critical perspective on the most relevant aspects related to the physicochemical bases underlying the assembly of polyelectrolyte films using the LbL technology. It is expected that the discussion in this chapter can help in the understanding of the most relevant theoretical and experimental aspects governing the formation of the LbL films and their properties. This is important to contribute to the current multidisciplinary efforts aimed to exploit the LbL methodology as a key enabling technology for the fabrication of materials with a direct technological output.

6.2 Assembly of LbL Films: Methodological Approaches

The LbL approach is currently established as a simple and inexpensive methodology for manufacturing different types of multi-layered materials through the alternate deposition onto a substrate of building blocks with complementary interactions [69–71]. This is possible using different methodological approaches which may be considered, in most of the cases, as simply extended versions of the seminal approach introduced by Decher et al. [16], including specific modifications to adapt the assembly methodology to the characteristics of the substrate used as template, e.g., size, shape, morphology or chemical nature. It should be noted that the fabrication of LbL materials with a true technological application requires the introduction of strategies for ensuring the scalability of the assembly from laboratory prototypes to the industrial fabrication [72]. This has stimulated the design of different methodological approaches, including some automated deposition machines, enabling the reduction of the deposition time and a fine control of the film properties.

The work by Decher et al. [16] proposed a very simple approach for fabricating LbL films onto flat substrates through the alternate immersion of the substrate in solutions containing the building blocks to be assembled (layering solutions), with intermediate rinsing steps between two consecutive adsorption steps, i.e., deposition by dipping. It should be noted that the rinsing steps are included to remove the material weakly adsorbed to the multilayer. Thus, it is possible to avoid cross-contamination of the assembled films due to the formation of aggregates for interaction in solution of the building blocks and their subsequent precipitation onto the multilayer [73–76]. Figure 6.1 shows a simplified picture of the deposition by dipping of a LbL film formed by polyelectrolytes bearing opposite charges.

The above-discussed approach for the assembly of LbL films presents an important drawback associated with the long time required for the fabrication of each layer. A potential alternative for speeding up the deposition process is to add an organic solution partially miscible with water, e.g., dimethylformamide or dimethylsulfoxide, to the layering solution. Thus, it is possible to induce a dewetting process during the

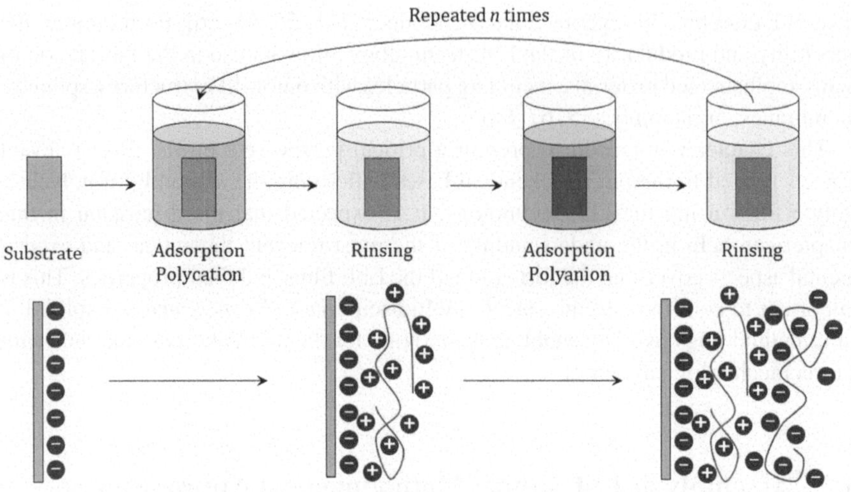

Fig. 6.1 Sketch of the common methodological approach used for the fabrication of LbL multilayers formed by polyelectrolytes bearing oppositely charged on a flat substrate. Reprinted with permission from Mateos-Maroto et al. [76]. Open access under a CC BY 4.0 license, https://creativecommons.org/licenses/by/4.0

deposition of the polyelectrolyte solution, which avoids the rinsing step between the deposition of consecutive layers, resulting in a multilayer assembly with a velocity 30 times higher than when water is used as solvent [77, 78]. The deposition under continuous stirring of the solutions is also a very powerful alternative for fastening the process, allowing a reduction of the deposition time down to 10–20 s per layer [79]. More sophisticated methodological approaches for the reduction of the fabrication time, and increasing the homogeneity of the deposited films, are the spray-assisted and spin-assisted deposition methods (see Fig. 6.2) [80–84].

The fabrication of LbL by spin-assisted deposition relies on the casting of the layering solution onto the surface of the substrate which is going to be used as template for the multilayer assembly. Then, the substrate undergoes a spinning process at a constant velocity for ensuring a homogeneous spreading of the film onto the substrate and the evaporation of the solvent. Once the layer is deposited, the rinsing step of the deposited layer is performed following the same procedure used for the layer deposition, maintaining the substrate under spinning until the deposited film is completely dried. The successive repetition of the deposition and rinsing cycles allows the fabrication of multi-layered films [86, 87]. Spin-coating fabrication of LbL films results in films with lower roughness and better organization than those obtained by dipping approach. Furthermore, the stratification of the films is also enhanced in relation to layers obtained by dipping due to the complex interplay between electrostatic and hydrodynamic interactions (centrifugal, air shear, and viscous forces), with the former controlling the adsorption and rearrangement of the polymer chains, whereas the hydrodynamic one contributes to the control of

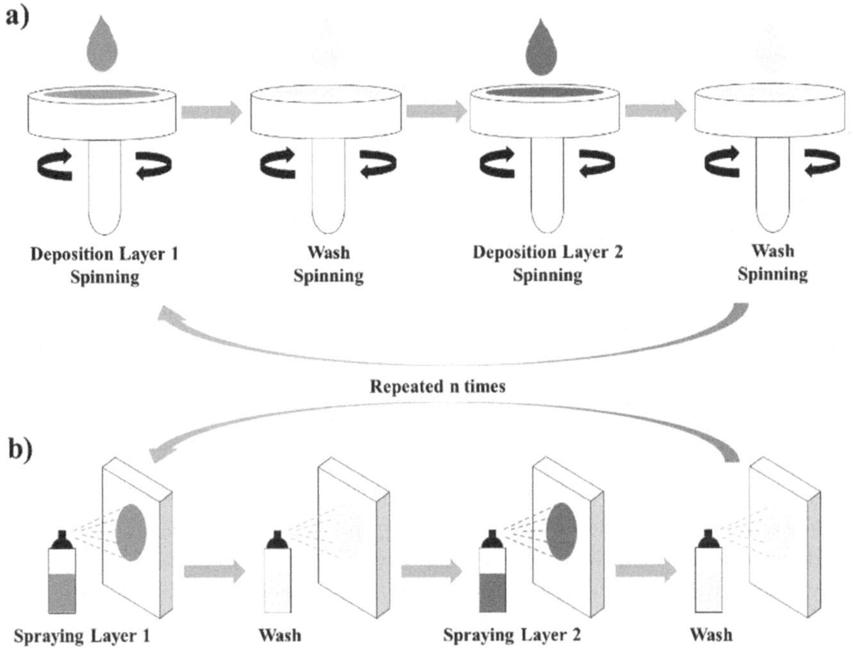

Fig. 6.2 Sketch of the deposition of LbL films through spin-coating **a** and spray-assisted deposition **b**. Reprinted with permission from Guzmán et al. [85]. Copyright (2020) (Elsevier)

the desorption process of the weakly bound material and the dehydration of the films [87]. Furthermore, the combination of electrostatic and hydrodynamic interactions also contributes to the reduction of the time required for the fabrication of films using spin-coating and the smallest thickness of the obtained films, with the latter being governed by the spinning velocity. Another advantage of spin-coating deposition is related to the significant reduction of the amount of material required for the assembly of the multi-layered films [87]. Nevertheless, spin-assisted deposition presents a very important drawback on the low volatility of the water which is the solvent most frequently used for film assembly, which limits its practical application for the fabrication of LbL films.

The deposition under high-gravity fields using a rotating substrate appears as a very attractive alternative for fastening the deposition of LbL films [88], having as its main advantage the hastening of the diffusion process. This reduces the time needed for reaching the adsorption equilibrium, allowing the fabrication of films with a similar quality to those obtained by dipping. This results from the increase in the concentration gradients and the existence of turbulence during the deposition process of the layers.

The spray-assisted deposition is another versatile alternative for the fabrication of LbL films [89], relying on the alternate spraying of solutions containing the different building blocks onto the surface of the substrate, with intermediate rising steps between the deposition of two consecutive layers. It should be noted that the advantages stated above for the spin-assisted deposition are also applicable for the deposition by spraying [89]. Spray-assisted deposition results in the deposition of multilayered films with lower roughness and thickness than those obtained by immersive deposition, i.e., dipping [90–93]. This is explained considering that during spraying, adsorption and drainage occur simultaneously. This makes it necessary to perform the spraying in the direction perpendicular to the surface, which ensures the gravitational drainage and favors a fast removal of the non-adsorbed material. The fabrication of LbL layers using the spray-assisted deposition allows the reduction of the contact time (less than 10 s may be enough), which reduces the interpenetration between adjacent layers [83], that in turn has stimulated the application of spray-assisted deposition at industrial level [10, 94]. Furthermore, the use of the LbL deposition assisted by spraying allows manufacturing of multi-layered films on substrates with a large area [95, 96]. The effectiveness of the layer fabrication using this methodology requires a careful examination of different methodological aspects including the spray-substrate distance, and other operational parameters, e.g., concentration and volume of the sprayed solutions, spraying flow, and waiting times between consecutive deposition steps. Furthermore, the necessity of rinsing steps between consecutive deposition steps and their duration are also very important parameters for controlling the spray-assisted deposition [83, 89, 93, 97–101]. The fabrication of LbL coating by simultaneous spray-assisted of interacting species (SSCIS) based in the spraying on a substrate of solutions containing simultaneously two or more interacting species has become a very promising alternative to the conventional fabrication of LbL films by using a spray-assisted methodology, ensuring a faster interaction between the molecules forming the multi-layered film. This makes possible a continuous fabrication of LbL films with gradual growth. It should be noted that the multi-layered films obtained following this methodology present thicknesses which are similar compared to those obtained by the conventional methodology based on the alternate deposition of oppositely charged layers by spraying. The modulation of the layer thickness can be done by changing the time of each spraying step, the nature of the solvent, and the drainage process of the excess deposited material [98, 102].

The application of external fields, such as electric or magnetic ones, can facilitate the deposition process of LbL multi-layered films [103, 104], leading to the formation of more compact and thicker LbL films than those obtained using any other methodological approach [105, 106]. The deposition assisted by electric field relies commonly on the use of an electrode as template for the film deposition (electrodeposition), which occurs upon the application of a voltage within the formed electrolytic cell. The most frequently used procedure for the electrodeposition of LbL films is based on the immersion of two electrodes in the layering solution, followed by the application of an electric current to drive the layer deposition. Afterward, the electrodes must be rinsed with the solvent, and then they are introduced into a solution containing the compound used for the deposition of the second layer, afterwards the process is repeated as many times as necessary [107]. There are several experimental designs enabling the electrodeposition of LbL films, which can be applied to the multi-layered films onto macroscopic surfaces or colloidal particles [108].

The optimization of the applied voltage and the time of the process are critical for controlling the thickness and roughness of the electrodeposited films [109]. The electrodeposition of LbL films may be also performed by taking advantage of local effects appearing at the electrode surfaces, e.g., redox reactions or pH changes. This may be understood considering that the local change of pH occurring in the vicinity of the electrode in relation to that of the solution may favor the layer deposition [110]. However, the increase of the film thickness reduces the penetration of the current toward the electrode, which reduces the impact of the local pH of the electrode surface on the layer deposition. This limits the maximum number of layers that can be deposited following this approach.

Magnetic fields can be also used for controlling the deposition of LbL films [111]. This is possible by the immersion of the substrate into a solution of magnetic active materials, which can be deposited by conventional dipping. The magnetic fields are applied for controlling the packing between adjacent layers, hence the multilayer thickness [112].

The above discussion was focused on some of the most common methodologies used for the deposition of LbL multi-layered films on flat substrates. However, the LbL method can be used for fabricating multi-layered films on any kind of substrate, which requires a careful examination of the specific characteristic of the used substrates. Traditional dipping can be applied to the deposition of LbL films onto colloidal templates. However, this requires considering that colloidal particles are commonly dissolved or suspended in a solvent, which makes it necessary to consider the introduction of separation steps. The most common methodological approach for depositing multi-layered films onto colloidal templates relies on the pelleting of the decorated colloids using commonly centrifugation/re-dispersion cycles on the washing steps after each layer deposition for removing the excess of unbound materials [50, 51, 113–118]. This means that once a layer is adsorbed, the suspension is centrifuged, separating the sediment containing the polymer-decorated particles and the supernatant where the excess of non-adsorbed material remains. Afterward, the polymer-decorated particles must be re-dispersed in the solvent and the sedimentation/re-dispersion process is repeated, normally a minimum of three times, for ensuring the complete removal of the non-adsorbed material. Thus, it is possible to obtain a clean suspension of polymer-decorated particles in which can be deposited a new layer following a similar methodology [119, 120]. Figure 6.3 shows a scheme of the different steps involved in the fabrication of polyelectrolyte multilayers onto colloidal particles. It should be noted that the above-described methodology allows the fabrication of both core–shell structures and hollow capsules. However, the fabrication of hollows capsules requires the introduction of an additional step of dissolution of the template (sacrificial substrate) [13].

The use of centrifugation for the separation of the supernatant and the polyelectrolyte-decorated particles is a limitation for extending the above-described approach to the coating of nano-sized colloids, affecting also to the possible automatization of the assembly process. The use of the serum replacement method may be a suitable

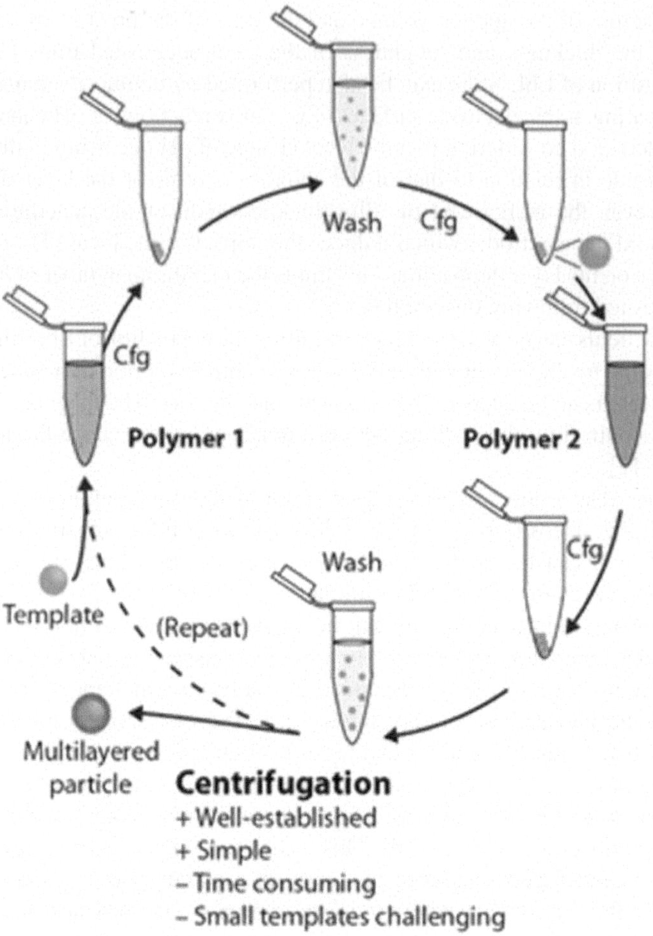

Fig. 6.3 Idealized representation of the most common methodological approach used for the deposition of LbL multi-layered films using colloidal particles as template. Reprinted with permission from Yan et al. [115]. Copyright (2014) (American Chemical Society)

alternative for overcoming the problems associated with the separation by centrifugation [121]. Another alternative for such purpose is the precipitation of the non-adsorbed material by the addition of a bad solvent [122]. Thus, it is possible to speed up the assembly process, increasing the recovery yield. This is a very important issue for a possible industrial scaling up of the assembly process, and for the preparation of concentrated capsule suspensions, which remains one of the most important challenges for the fabrication of LbL systems. The above problems have been partially solved by using tubular flow type reactors. This makes possible the continuous fabrication of capsules with a fixed number of layers. However, this methodology presents a very important drawback related to the possible cross-contamination because a small amount of the last deposited polyelectrolyte always remains in the medium after its deposition step, which drive the formation of inter-polyelectrolyte complexes and their subsequent precipitation on the formed capsules [123]. The application of magnets appears also as a very interesting alternative to the centrifugation for ensuring the separation of the polymer-decorated particles and the excess of non-adsorbed molecules [103]. This requires the use of substrates with magnetic activity [124, 125].

It should be noted that the washing/separation steps can be avoided by adding just the amount of polyelectrolyte necessary for saturating the colloidal particles, which allows increasing the velocity of the assembly by a factor of 3 [50, 126]. This requires a careful evaluation of the change of the zeta-potential of the dispersion with the addition of the adsorbing polymer for avoiding the aggregation. The sonication of the dispersion during the layer deposition can reduce the aggregation of the particles [126–129].

The use of particles lighter than water (e.g., emulsion drops, vesicles, or liposomes) as template requires introducing alternative methodologies for ensuring the separation of the polymer-decorated templates and the excess of non-adsorbed material. For the particular case of the deposition of polyelectrolyte multilayers on droplets of oil in water emulsion, the use of creaming/skimming cycles appears as a very promising alternative for ensuring the separation of the excess of non-adsorbed material and the decorated droplets. This is possible because the lower density of the oil droplets in relation to water leads to their flotation on the top of the polymer solution [130, 131]. It should be noted that the creaming process can be enhanced by centrifugation when the deposition in emulsion droplets is considered [132–134]. On the other side, the deposition of LbL multilayers on vesicles or liposomes is a very complex process involving up to three different steps for each deposited bilayer [135]: (i) addition of the layering solution to a diluted suspension of vesicles or liposomes; (ii) once the first layer is deposited, the solution containing the compound to be assembled as second layer is added to the suspension of polyelectrolyte-decorated vesicle/liposomes, which also contains the excess of non-adsorbed polyelectrolyte during the deposition of the first layer. This leads to the formation of a bilayer onto the template, the formation of inter-polyelectrolyte complexes, and (iii) the inter-polyelectrolyte complexes are settled down by centrifugation for obtaining a clean suspension of polyelectrolyte-decorated vesicles/liposomes. It should be noted that the use of centrifugation is possible because the polyelectrolyte shell ensures the stability of the vesicles/liposomes, avoiding aggregation and fusion phenomena. After the deposition of the first bilayer, it is possible to continue with the assembly of additional layers by repeating the above-discussed procedure. It should be noted that the deposition of polyelectrolyte layers following the above-described procedure is limited to a reduced number of bilayers, about 5–6, because the formation of supramolecular aggregates between the inter-polyelectrolyte complexes and the polyelectrolyte-decorated vesicles/liposomes, as well as the centrifugation process, lead to the reduction of the total number of vesicles/liposomes after the deposition of each pair of layers [135].

The separation steps can be avoided by fixing the colloidal templates using an immobilization agent, e.g., agarose hydrogel. This leads to a situation in which the colloidal particles remain trapped forming a planar substrate, and hence it is possible the use of conventional dipping for the fabrication of LbL films onto colloidal templates. Once the desired number of layers has been built, it is possible to release the decorated particles from the agarose trapping matrix by heating it at 37°C. Afterward, the separation of the agarose and the polyelectrolyte-decorated particles is attained by several cycles of centrifugation/re-dispersion (commonly 3) [136]. This methodology makes possible the use of robotic dipping for the fabrication of LbL multi-layered films on colloidal substrates [136].

The use of microfluidic devices is a very promising alternative for fabricating LbL films on colloidal templates placed in the channels [137–140]. This allows the minimization of the aggregation processes [141–145]. The most common approach to exploiting the use of microfluidic devices to produce LbL materials relies on the application of pressure or vacuum for controlling the sequential transport of the adsorbing species and the washing solutions within the microfluidic chips [146–148]. The assembly of LbL films on particulate substrates is also possible by using fluidized beds [149], taking advantage of the upward force of the washing or polymer solutions that counteracts the gravitational forces which push the particle to the sedimentation. This leads to the lifting of the particles to form a fluidized bed, which enables the fabrication of LbL films using substrates with different sizes (up to 3 μm) and a permeability controlled almost at will [150]. It should be noted that the difficulties in optimizing the methodology and the high cost of the instrumentation have limited the implementation of microfluidic approaches as a common tool for the fabrication of LbL coating on colloidal templates [108].

6.3 Polyelectrolyte Multilayer Growth

The ability of polyelectrolytes to self-organizing in supramolecular structures together with their capacity to form complexes when oppositely charged polyelectrolytes are mixed allows controlling the thickness almost at will, which plays a central role in the control of different properties of LbL materials, e.g., transparency of optical devices, retention of encapsulated drugs or wetting properties of the surfaces and adhesion.

The growth of LbL polyelectrolyte multilayers is commonly defined in terms of the change in amount of adsorbed material (or layer thickness) with the number of deposited bilayers, N. This drives the possible emergence of two different growth mechanisms in LbL materials. The first one, commonly defined as linear growth, is characterized by an almost constant change of the adsorbed amount upon the deposition of each bilayer, which leads to quasi-linear change of the multilayer thickness with the deposition of each bilayer on the number of bilayers. Thus, the multilayer thickness increases after the deposition of each pair of bilayer in such a way that can be considered proportional to the sum of the characteristic sizes of the two molecules forming the multi-layered film, i.e., the size of polycation and polyanion, and consequently the thickness of each bilayer is reduced to a few nanometers. Some classical examples of multilayers displaying linear growth are those formed by poly(allylamine hydrochloride) (PAH) and poly(4-styrenesulfonate of sodium) (PSS), i.e., $(PAH–PSS)_n$ (Note that, in this, the text, the nomenclature of multilayers will include the information of the interacting species defined as A and B, and a subindex, n, indicating the number of deposited bilayer, leading to a representation defined as $(A–B)_n$, or those formed by poly(diallyl-dimethyl-ammonium chloride) (PDADMAC) and PSS assembled under conditions in which both polymer present a high charge density (low ionic strength solutions). There are many other multilayers growing linearly, e.g., $(PM2VP—PSS)_n$ or $(PAH—PAA)_n$ films (with PM2VP and PAA being poly(N-methyl-2-vinyl pyridinium chloride) and poly(acrylic acid), respectively) [151]. The second type of dependence of the adsorbed amount on the number of bilayer is the so-called non-linear growth or most commonly exponential growth. Multilayers growing following this type of dependence undergo a change in their thickness higher than that corresponding to a linear growth upon the deposition of each pair of bilayers. This leads to an increasing change of thickness per bilayer in each deposition cycle, resulting in a thickness per bilayer that is not only associated with the molecular size of the assembled species. Examples of non-linear growth multilayers are $(PDADMAC—PSS)_n$ ones, assembled from conditions in which the effective charge of the polyelectrolyte is highly screened [152–155]. Furthermore, most of the multilayers including biopolymers display non-linear growth, e.g., $(CHI—PAA)_n$, $(PLL—HA)_n$, or $(PLL—PGA)_n$ (with CHI, PLL, HA, and PGA being chitosan, poly(L-lysine), hyaluronic acid and poly(glutamic acid), respectively) [19, 20, 156–158]. In addition to the above-mentioned growth dependences, specific combinations of polyelectrolyte pairs can appear as more exotic dependences of the adsorbed amount on the number of deposited bilayers [159, 160]. Figure 6.4 shows an idealized picture of the two most common growth dependences appearing in LbL films.

Despite the research efforts made for understanding the assembly of polyelectrolyte multilayers obtained following the LbL methodology, it remains a strong controversy about the molecular bases explaining the emergence of different types of growth in polyelectrolyte multilayers. The first work trying to provide a realistic explanation for the appearance of different types of growth in polyelectrolyte multilayers was done by Elbert et al. [161]. They identified the linear growth with an initial deposition of molecular polyelectrolyte layers followed by the diffusion of some chains to the inner region of the multilayer, which agrees with the theoretical description provided by Subbotin and Semenov [162] using a mean-field approach. Therefore, the possibility to diffuse the polyelectrolyte during the deposition process leads to a certain degree of intermingling between adjacent layers. On the other hand, for non-linear growth multilayers, each deposition cycle is associated with a higher adsorption than expected for a single layer. This may be explained considering the existence of diffusion of the polyelectrolyte chains within the multilayer structure, which enables the deposition of an amount of polyelectrolyte higher to that corresponding to a single layer. This occurs for specific polyelectrolyte pairs

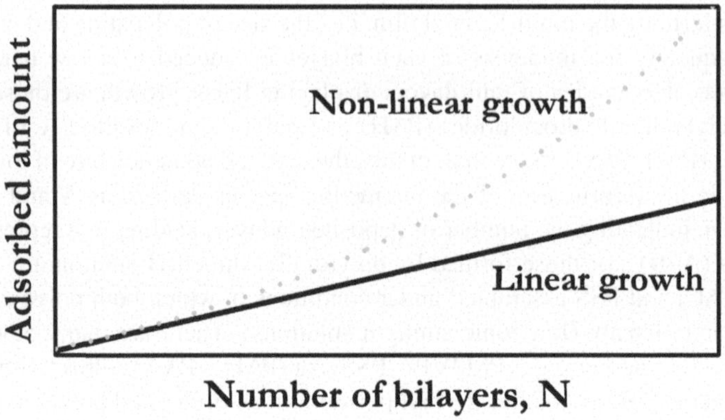

Fig. 6.4 Idealized picture of the linear and non-linear dependences of the adsorbed amount on the number of bilayers, N, appearing in polyelectrolyte multilayers. Reprinted with permission from Mateos-Maroto et al. [76]. Open access under a CC BY 4.0 license, https://creativecommons.org/licenses/by/4.0

and assembly conditions through a process involving the interfacial coacervation at the solution-multilayer interface. This coacervation process occurs between polyelectrolyte chains in solution and the oppositely charged polyelectrolytes diffusing from the inner region multilayer. The coacervates may precipitate on the multilayer surface, which leads to the deposition of a film thicker than a polymer molecular layer [163, 164]. Therefore, the multilayer must be considered as a polymer reservoir, which allows explaining the increase of the thickness with the number of layers as result of the increase in the number of chains that are available for diffusing to the solution-multilayer interface, and the formation of coacervates. This agrees with the theoretical description provided by Tang and Besseling [165], which considers the similarities between the formation process of inter-polyelectrolyte complexes in solution and the assembly of polyelectrolyte multilayers. This suggests that the differences in the growth mechanisms are ascribable to a different dynamics of the polyelectrolyte chains within the multilayers [166].

The scenario proposed by Elbert et al. [161] for describing the non-linear growth of polyelectrolyte multilayers agrees with the results by Picart et al. [156, 157, 167–169]. They explained the non-linear growth of polyelectrolyte multilayers as a result of the existence of an in and out-diffusion of at least one of the polyelectrolytes within the multilayer structure. However, they neglect the role of polyelectrolyte diffusion for multilayers exhibiting linear growth [156, 168, 170]. For $(\text{PLL—HA})_n$ multilayers, an in and out-diffusion of PLL chains within the multilayer structure was found, with this diffusion depending on the nature of the layering solution. Thus, PLL chains can diffuse to the inner region of the multilayer upon the exposure of the film to a solution containing such polymer, whereas the diffusion of PLL to the multilayer surface occurs upon exposure of the film to HA solutions. This leads to the PLL-HA complexation at the solution-multilayer interface, which results in the adsorption of higher amount of material than that expected for a single monolayer, and consequently, the thickness of the film is higher than that expected considering the adsorption of true molecular layers. The use of confocal laser scanning microscopy (CLSM) provided further details on the role of the in and out-diffusion on the emergence of non-linear growth [168]. This technique allows monitoring of the adsorption process within the axis perpendicular to the multilayer by including fluorescently labeled polyelectrolytes at different positions of the multi-layered film. The tracking of fluorescently labeled PLL using CLSM evidenced the ability of PLL for diffusing within the whole multilayer, whereas fluorescently labeled HA chains remain fixed in the positions where they were initially adsorbed. This allows confirming the important role of the interdiffusion of at least one of the polyelectrolytes in the equilibration of non-linear growth multilayers, or both as was reported for $(PLL\text{-}PGA)_n$ multilayers [147]. The in and out-diffusion of the polyelectrolyte chains is possible due to emergence of a Donnan potential within the multilayer, as result of the charge excess associated with the existence of mobile polymer chains. The requirement of equilibration of such Donnan potential leads to the interdiffusion process [167, 171]. The above discussion provides neither a suitable explanation that allows neglecting the role of interdiffusion in multilayers exhibiting linear growth nor a justification for the transition from linear to non-linear growth appearing in some specific polyelectrolyte pairs, e.g., $(PDADMAC—PSS)_n$, as result of the modification of the assembly conditions [152, 153, 172]. The latter may be explained considering that the transition from a linear to a non-linear growth occurs as a consequence of a loose ionic cross-linking. Furthermore, the existence of a strong mutual interaction between the polycation and the polyanion may drive the emergence of the interdiffusion of the polycation chains, and consequently the non-linear growth [173]. This picture is compatible with the enhanced interdiffusion reported by Guzmán et al. [174] in $(PDADMAC\text{-}PSS)_n$ multilayers as result of the increase of the ionic strength of the solution, which drives the transition from linear to non-linear growth, and hence it does not limit the interdiffusion to films exhibiting a non-linear growth.

Therefore, the interdiffusion cannot be considered as the differencing feature between linear and non-linear growth films as was clearly evidenced by Guzmán et al. [174]. They suggested that the film roughness may play a very important role in the emergence of non-linear growth. This may be understood considering that the higher the surface roughness of the multilayer the higher is the area available for the deposition of the following layer, and hence the amount of material deposited in each cycle. This increase in the roughness may be correlated with the coacervation-like process described by Elbert et al. [161], which as matter allows growth faster than linear. On the other side, the deposition of chains respecting their molecular size should lead to a linear growth [152, 172, 174–176]. Haynie et al. [177] discussed furtherly the role of the film roughness on the emergence of non-linear growth. They assume that its emergence results from the formation, propagation, growth, and coalescence of heterogeneous adsorption patterns, e.g., dendritic or isolated aggregates, on the multilayer surface, which increases the film roughness and leads to the emergence of the non-linear growth of the multilayer. This picture was confirmed by Hernández-Montelongo et al. [178] using fractal analysis of Atomic Force Microscopy images.

They found that the adsorption and aggregation processes of linear and non-linear growth multilayers present strong differences. Thus, the assembly of linear growth multilayers occurs by an initial aggregation of the deposited molecules followed by their rearrangement, whereas a diffusion-limited aggregation was found when non-linear growth films are considered. The latter results in diffusion gradients which drive the formation of isolated aggregates and a significant increase in the multilayer roughness. On the other side, linear growth multilayers present relatively smooth and regular surfaces. The existence of different aggregation patterns on polyelectrolyte multilayers was also reported by Naas et al. [179] as a very important driving force in the emergence of different types of growth. It should be noted that Picart et al. [156] in the assembly of (PLL—HA)$_n$ multilayers also found the formation of heterogeneous patterns. However, they did not ascribe to the formation of such pattern any role in the emergence of non-linear growth.

The above discussion points out clearly the absence of a true understanding of the molecular mechanism controlling the assembly of polyelectrolyte layers. However, the current picture suggests that the specific chemical nature of the assembled polymer, and in particular its specific diffusivity, plays an essential role in the control of the polyelectrolyte uptake and consequently the type of growth [170, 173, 180].

6.4 Charge Balance in Polyelectrolyte Multilayers

The direct electrostatic interaction between polyelectrolytes deposited in adjacent layers is not enough for describing the assembly process of LbL films. This can be understood considering that the assembly of polyelectrolyte layers involves at least three different types of interactions: polyelectrolyte—polyelectrolyte, polyelectrolyte—solvent, and polyelectrolyte—template [154, 181].

The emergence of charge inversion is commonly accepted as one of the most important aspects guiding the deposition of charged species on oppositely charged surfaces. This means that the adsorption continues beyond the neutralization of the surface charge, leading to a final charge on the surface with the same sign that the layering molecule [182]. This results from the steric hindrance, which does not allow an effective neutralization of the bare charge of the surface and makes necessary the deposition of a number of molecules higher than that required for a stoichiometric neutralization of the surface charge. This results in the emergence of a charge inversion phenomenon and the formation of a fuzzy layer with charged segments protruding into the solution. The addition of successive layers to the film follows a similar pathway to the above-described, with the charge overcompensation guiding the alternate deposition of the subsequent layers [20, 52, 152, 183]. Evidence of the overcompensation can be obtained for any technique allowing the evaluation of the shifts of the surface charge upon the alternate exposure of a LbL film to layering solutions containing molecules bearing opposite charges, e.g., surface potential, streaming potential, or zeta potential [20, 52, 127, 152–154, 156, 184–186]. Figure 6.5 shows the change of the zeta potential for (PDADMAC-PSS)$_n$ deposited

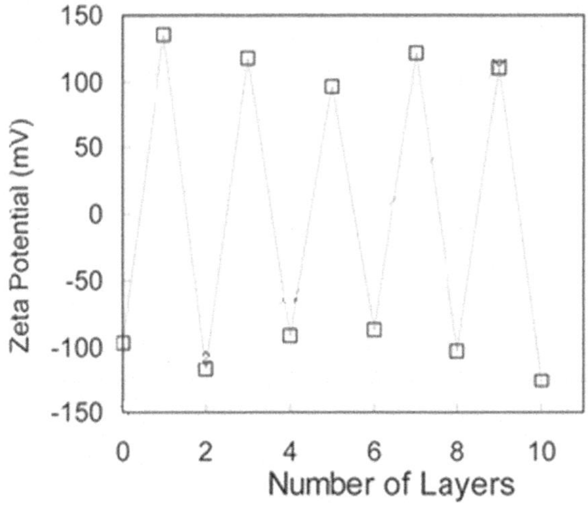

Fig. 6.5 ξ potential shift as result of the alternate deposition of PDADMAC and PSS layers onto silicon wafers from polyelectrolyte solutions with concentration of 10 mM, and ionic strength fixed in 100 mM. Adapted with permission from reference Ferriz-Mañas and Schlenoff [187]. Copyright (2014) (American Chemical Society)

onto silicon wafers [187], with the overcompensation being evidenced by the oscillation of the zeta potential between values about + 100 mV and − 100 mV, upon the deposition of PDADMAC and PSS layers, respectively.

The overcompensation emerges from the absence of a perfect matching of the number of charged monomers in adjacent layers. This results in a charge excess which leads to the instability of the multilayers, and hence it must be minimized to ensure that the electroneutrality boundary conditions are fulfilled [153, 188]. This requires a contribution that enables the counterbalancing of the charge excess. This is possible by the incorporation of small ions into the multilayer, which compensates for the excess charge appearing from the unpaired polyelectrolyte segments [152, 155, 189–194]. The incorporation of small ions for compensating the charge excess leads to the emergence of the so-called extrinsic compensation. It should be noted that for specific polyelectrolyte pairs or assembly conditions, the role of the counterions in the compensation appears negligible, which leads to an intrinsic compensation. This is characterized by a perfect matching between the number of charged monomers in adjacent layers, and hence the ionic equilibrium is established in the multilayers upon the expulsion of the small ions from the film, which leads to multilayers with stoichiometry 1:1 (polycation:polyanion). The formation of these multilayers is driven by a very favorable entropic contribution to the reduction of the free energy due to the release of the counterions from the film to the surrounding media. The picture appears very different in absence of a perfect matching between the charges of the polyelectrolyte in adjacent layers. This requires the incorporation of counterions for ensuring the electroneutrality of the multilayers, which can result in multi-layered films presenting a broad range of different stoichiometries.

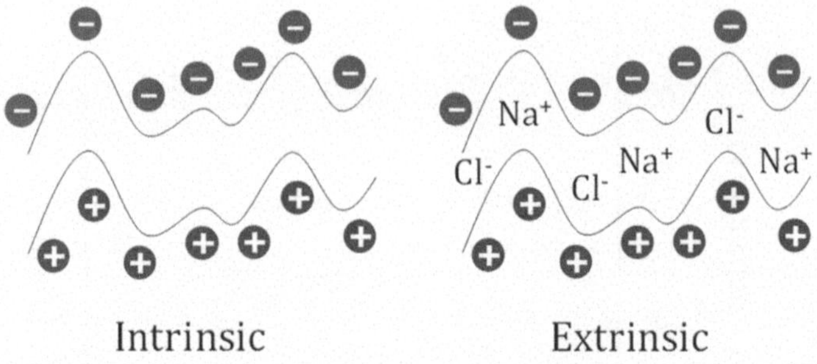

Fig. 6.6 Sketch of the idealized distribution of polyelectrolytechains and counterions in adjacent layers depending on the type of compensation mechanism. Reprinted with permission from Mateos-Maroto et al. [76]. Open access under a CC BY 4.0 license, https://creativecommons.org/licenses/by/4.0

Furthermore, the role of the entropy in the assembly process can be disregarded for extrinsically compensated films. Figure 6.6 shows a scheme representing the idealized distribution of counterions and polyelectrolytes in adjacent layers depending on the type of compensation mechanism.

The compensation mechanism depends on the ionic equilibrium, and hence the modification of any parameter altering it may influence the structure and the physicochemical properties of multi-layered films deposited using the LbL methods [152–154]. The change of the ionic strength is probably the way to induce the strongest modulation of the compensation mechanism of polyelectrolyte multilayers due to the ionic condensation that modifies the effective charge density of the polyelectrolyte chains [195] as was verified by Schlenoff and Dubas [154]. They found that the compensation mechanism of (PDADMAC-PSS)$_n$ may be switched from a mainly intrinsic one to a true extrinsic one. This was rationalized considering that at low ionic strength the assembly of the layers can occur with a release of counterions from the film to the solution, whereas as the ionic strength increases the entropic contribution associated with the release of counterions becomes almost negligible, and most of the counterions remain embedded within the multilayer, leading to the emergence of extrinsic compensation. It should be noted that the different role of the counterions released in the assembly of LbL films leads to a very different energetic landscape on intrinsic and extrinsic compensated films [152, 153]. The type of compensation can be quantified by the introduction of compensation ratio $R_c = \rho^+_{mon}/\rho^-_{mon}$ with ρ^+_{mon} and ρ^-_{mon} being the densities of positively and negatively charged monomers in adjacent layers [18]. Thus, for $R_c \approx 1$, it should be expected a mostly intrinsic compensation, whereas for values for $R_c \neq 1$ the compensation should be considered extrinsic, with an excess of cationic monomers for $R_c > 1$ and of anionic ones for $R_c < 1$. Figure 6.7a displays the dependence of the compensation ratio of the ionic strength, I, for (PDADMAC-PSS)$_n$ multilayers [152]. Independently of the ionic strength, the compensation ratio appears higher than 1, which indicates the existence of excess PDADMAC monomers in relation to PSS ones in adjacent layers. This can be interpreted assuming that the compensation is extrinsic within the whole range of ionic

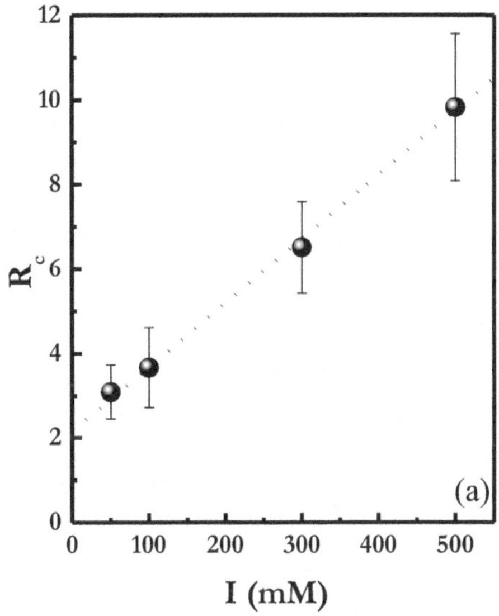

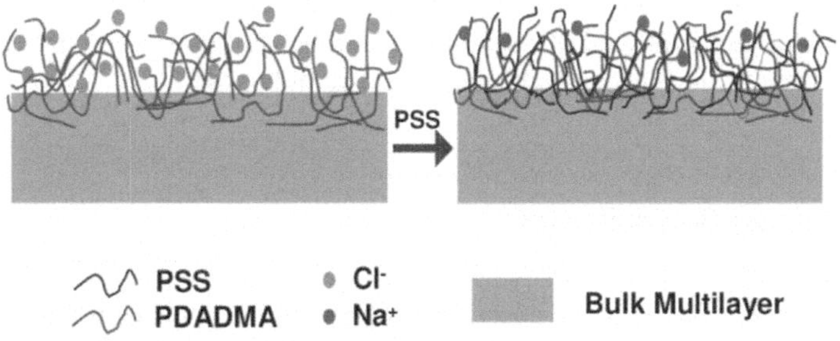

(b)

Fig. 6.7 **a** Compensation ratio R_c in (PDADMAC—PSS)$_n$ multilayers as function of the ionic strength. Adapted with permission from Guzmán et al. [152], Open access under a CC BY 4.0 license, https://creativecommons.org/licenses/by/4.0. **b** Representation of the asymmetrical compensation in polyelectrolyte multilayers as function of the nature of the last deposited layer. Reprinted with permission from Lehaf et al. [196]. Copyright (2012) (American Chemical Society)

strengths. Furthermore, the increase of the compensation ratio with the ionic strength from values slightly higher than one suggests that the change of ionic strength drives a transition from a quasi-intrinsic compensation to a clearly extrinsic compensation. On the other side, the values of R_c also evidences that the compensation is strongly dependent on the specific nature of the capping layer of the multilayer, and hence it presents an asymmetric character [153, 196]. This is clearly assuming that the values of the compensation ratio above 1 are associated with a large excess of PDADMAC monomers in relation to PSS ones, which allows assuming that PDADMAC-capped multilayers present a strongly extrinsic compensation, whereas the compensation becomes intrinsic for multilayers capped with PSS layers. This asymmetry is associated with the specific nature of the polyelectrolyte pair and governs the structure and physicochemical properties of the layers. In particular, for (PDADMAC—PSS)$_n$ multilayers, the distribution of counterions within the multilayers leads to an asymmetric osmotic stress, resulting in the formation of swelled and highly hydrated PDADMAC layers, where PSS ones have a more collapsed conformation. Thus, PDADMAC-capped multilayers present a higher roughness than those capped with a PSS layer [153, 197] (see inset Fig. 6.7b).

The emergence of different compensation mechanisms in the assembly of polyelectrolyte multilayers depends on the specific nature of the assembled pair and the assembly conditions used in the deposition of the layers. This is related to the different impacts of the enthalpic and entropic contributions on the assembly process [198]. The assembly of (PDADMAC-PSS)$_n$ multilayers is governed by a strongly exothermic complexation for films deposited from solutions of low ionic strength, whereas endothermic complexation emerges with the ionic strength increases [85].

It should be noted that the entropic balance on the assembly of LbL multilayers goes beyond the role of the ionic equilibrium, including two additional contributions: (i) release and reorientation of hydration water [199–201], and (ii) reduction of the degrees of freedom of the molecules due to their attachment to the surface [202, 203]. However, the role of two contributions in the assembly is smaller than that associated with the release of counterions, and hence it is possible to neglect their contributions to the entropic balance of the assembly [204].

6.5 Adsorption Kinetics

The timescale involved in the adsorption process of the polyelectrolyte layers plays a critical role in controlling the physicochemical properties and structure of LbL materials [174, 175, 205, 206]. This can be understood assuming that the stratification appearing in different multi-layered films emerges from an arrested adsorption, which leads to the formation of films having a structure reminiscent of a freezing of the deposition far from the steady-state conditions [207–209]. This may be ascribed to the long time required for ensuring a complete reorganization of the polyelectrolyte chains within the whole structure of the multilayers [210], which is compatible with the better stratification obtained of multilayers assembled using methodologies involving a low contact time between the layering solution and the multilayer (spin-coating or spraying), in relation to those obtained by common dipping [91, 211].

The adsorption of polyelectrolyte layers can be considered a quasi-irreversible process, and hence once the polyelectrolyte chains are anchored to the surface, they remain trapped. This may be explained considering that the adsorption of polymer chains on surfaces involves the binding of multiple monomers, and consequently, the reversibility of the adsorption process will require their simultaneous desorption, which is rather improbable because during the desorption of the bound monomers it is possible that other segments can be attached on the surface. This makes almost impossible a complete desorption of a polymer chain from the surface within an accessible time window [212, 213].

A detailed analysis of the kinetics of the layering process of polyelectrolytes on LbL films shows that the deposition consists of at least two steps [20, 174, 175, 205, 214], which can be accounted by a phenomenological model defining the time dependence of the surface concentration $\boldsymbol{\Gamma}(\boldsymbol{t})$ as [215–217]

$$\boldsymbol{\Gamma}(t) = \boldsymbol{A}_1\left(1 - \boldsymbol{e}^{-t/\tau_1}\right) + \boldsymbol{A}_2(1 - \boldsymbol{e}^{-t/\tau_2})^{\boldsymbol{n}}, \tag{6.1}$$

where $\boldsymbol{A}_1$ and $\boldsymbol{A}_2$ define for the amplitudes of the two adsorption steps, the fast and slow adsorption one, respectively and $\boldsymbol{\tau}_1$ and $\boldsymbol{\tau}_2$ indicates the corresponding characteristic times. The first step includes a diffusion-controlled adsorption combined with the polymer adsorption through a barrier (electrostatic or steric) [213, 218]. The second term of the model account for the reorganization of the adsorbed chains occurring after their initial adsorption on the multilayer, and hence it includes the role of the reorganization of the polymer molecules within the surface plane, and the diffusion of the polymer chains toward the inner region of the multilayer [214, 174]. The exponent$\boldsymbol{n}$, having a value close to 1 in most of the for most multilayers, makes it possible to reorganize Eq. (6.1) in terms of the limit surface concentration, $\boldsymbol{\Gamma}_\infty = \boldsymbol{A}_1 + \boldsymbol{A}_2$,

$$\boldsymbol{\Gamma} = \boldsymbol{\Gamma}_\infty - \boldsymbol{A}_1\boldsymbol{e}^{-t/\tau_1} - \boldsymbol{A}_2\boldsymbol{e}^{-t/\tau_2}. \tag{6.2}$$

The above model includes the contribution of two kinetic processes occurring in time scales that emerge well-separated. The first process includes a fast nucleation of polyelectrolyte clusters on the surface, generally occurring within the five initial minutes, whereas the second accounts for the slow reorganization of the material adsorbed initially, and can be extended during a time ranging from few minutes to several hours [222]. Figure 6.8 displays a set of data corresponding to the layering process of a PDADMAC layer corresponding to the assembly of a (PDADMAC—PSS)$_n$ multilayer in which the mathematical procedure allowing the separation of the two adsorption processes is also defined.

Figure 6.8 shows that the above model provides a suitable representation of the adsorption process, with the first step accounting for about the 60–80% of the final Γ [18, 20, 53, 205]. The first adsorption process $\boldsymbol{\tau}_1$ appears commonly independent on the number of deposited layers [20, 53]. However, $\boldsymbol{\tau}_1$ emerges strongly dependent on any variable that affect to the assembly process (pH, ionic strength, T, etc.) [213]. On the other side, for $\boldsymbol{\tau}_2$ it appears very complex dependences on the number of layers, which makes very difficult the prediction of how this parameter change when a specific polyelectrolyte pair is assembled [175, 174, 219]. This may be rationalized considering that $\boldsymbol{\tau}_2$ accounts for any reorganization process occurring in the multilayer plane, i.e., processes occurring within the multilayer surface of the multilayer, in (PAH—PSS)$_n$ or (PDMAEMA—PSS)$_n$ films (PDMAEMA corresponds to poly[2-(*N*, *N*-diethylamino)ethyl methacrylate]), and hence $\boldsymbol{\tau}_2$ remains almost unchanged with the number of deposited layers [18, 175, 174]. However,

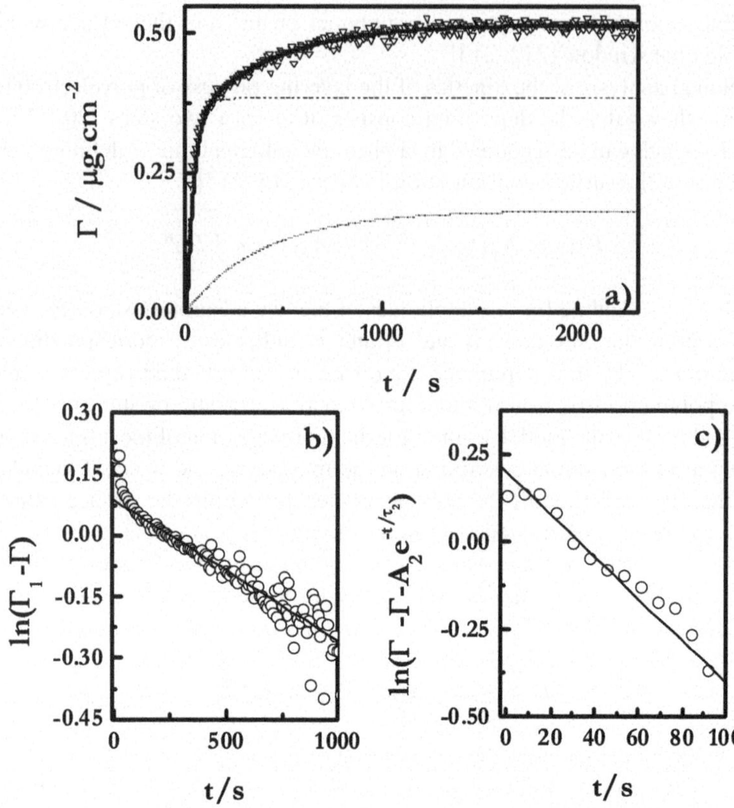

Fig. 6.8 Layering kinetics, and discrimination of the two involved steps for the adsorption of PDADMAC layer on a (PDADMAC—PSS)$_n$ film. **a** Adsorption kinetics. The solid line shows the best fit to Eq. (6.2), with the two exponential steps being evidenced: (–) first fast step and (· · ·) second slow step. **b** Behavior in the long time limit for the adsorption plotted as $\ln(\boldsymbol{\Gamma}_\infty - \boldsymbol{\Gamma})$ versus time, with the solid line representing the fit to a straight line. (c) Behavior in the short time limit for the adsorption kinetics plotted as $\ln(\boldsymbol{\Gamma}_\infty - \boldsymbol{\Gamma} - A_2 e^{-t/\tau_2})$ versus time, with the fit evidenced by a solid line. Adapted with permission from Guzmán et al. [175], Copyright (2011) (Elsevier)

there are other systems, e.g., (PDADMAC—PSS)$_n$ or (CHI—PAA)$_n$ films, in which τ_2 accounts for two different types of processes: (i) in plane reorganizations of the initially adsorbed material, and (ii) the diffusion of the molecules toward the inner region of the 3D structure of the multi-layered films [20, 175, 174]. This leads to a continuous increase of the τ_2 value with the number of deposited bilayers. The above description of the adsorption kinetics can be considered analogous to that proposed by Lane et al. [214], which are included up to three different steps: (i) initial transport of the molecules from the bulk to the vicinity of the surface where they adsorb; (ii) reorganization of the adsorbed molecules within the layer surface, and (iii) diffusion of molecules toward the inner region of the self-assembled multi-layered structure.

6.6 Internal Structure

The extension of the lamellar structure in polyelectrolyte multilayers, i.e., whether adjacent layers are truly independent or are interpenetrated, is one of the most controversial aspects in the study of polyelectrolyte multilayers. This is important because the understanding of this issue presents a fundamental importance for the definition of the stratification degree of LbL multi-layered films as stratified materials [91, 175, 211]. The emergence of lamellar order has been found strongly dependent on the specific polyelectrolyte pair, the conditions used for the layering, and the fabrication protocol (contact time and deposition method) [13, 91, 175, 211].

An oversimplified view of the internal structure of polyelectrolyte multilayers makes a division of 3D structure of the film into three different zones (three-zone model), existing a progressive transition between the structure of adjacent regions [185, 220, 221]. Thus, the zone I define the region closest to the substrate, and it is the first region formed during the assembly of LbL films. This main characteristic of the layers forming this region is the alignment of the chains along the substrate surface, and its thickness remains unchanged during the whole assembly process. Furthermore, the polyelectrolyte chains forming this region remain strongly trapped and their mobility within the rest of the multilayer structure is very limited. The zone II is formed following the zone I, and its thickness increases as the thickness of the whole multilayer increases. This region presents a structure that is reminiscent of that what appears in inter-polyelectrolyte complexes formed in solution [206, 220]. The most external region of the multilayer is the so-called zone III. This region presents an almost constant thickness during the whole fabrication process, and its structure resembles that appearing of free polyelectrolyte chains in solution [221, 222]. Therefore, it is possible to assume that polyelectrolyte multilayers are formed by the following regions: (i) an inhomogeneous region close to the substrate, formed due to the interactions between the surface and the polyelectrolyte, and (ii) a homogeneous region after the deposition of a certain number of layers. Figure 6.9 shows how the different zones of the multilayers evolve as additional layers are deposited.

The above picture presents a general approach to the description of the internal structure of polyelectrolyte multilayers. However, a deeper understanding of the

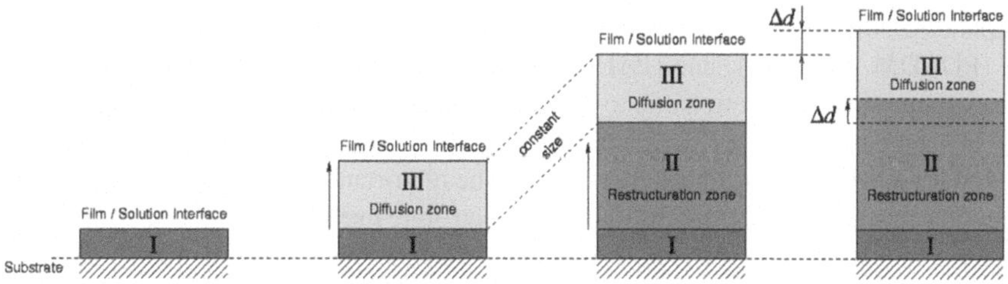

Fig. 6.9 Sketch of the three-zone model for the structure of a polyelectrolyte multilayer structure, and its change as the number of layers increases. Adapted with permission from Porcel et al. [221]. Copyright (2017) (Elsevier)

structure of polyelectrolyte multilayers makes necessary the study of the organization of the layers using neutron (NR) or X-Ray (XRR) reflectometry, or X-Ray photoelectron spectroscopy (XPS) [152, 175, 205, 208, 209]. The combination of NR and XRR allows the first systematic characterization of the stratification of (PAH—PSS)$_n$ multilayers [208, 209]. These multilayers were found to present a certain degree of stratification in the multilayers. Nevertheless, the structure of these multi-layered films was found to be far from a truly lamellar system, existing a certain intermingling between adjacent bilayers (three or more bilayers) [208, 209, 223]. Furthermore, they found that the structural pattern found from the NR and XRR results was compatible with the three-zone model, which enables an appropriate description of the correlations existing between the stratification degree and the thickness of the multilayer. On the other hand, it was reported that the roughness of (PAH—PSS)$_n$ multilayers increases as the number of layers increases, reaching an stationary value due to the densification of the film [208, 209]. This is just the opposite situation to that found for (PDADMAC-PSS)$_n$ multilayers, where the average roughness of the film decreases as the number of layers increases [224]. This difference may be the result of the different mechanisms involved in the assembly of (PAH—PSS)$_n$ and (PDADMAC—PSS)$_n$ multilayers, and the different roles played in the interdiffusion of the polyelectrolyte chains within the film, which is particularly important for (PDADMAC—PSS)$_n$ films [174, 175]. Therefore, it is possible to distinguish between the topological roughness as that found in (PAH—PSS)$_n$ multilayers [208, 209] and the roughness resulting from the assembly process as that appearing in (PDADMAC—PSS)$_n$ films [224].

It should be noted that differences between (PAH—PSS)$_n$ and (PDADMAC—PSS)$_n$ multilayers goes beyond the different impact of the assembly processes on their roughness [152, 174, 175, 205, 208, 209]. Guzmán et al. [152, 175, 205] showed the absence of a lamellar structure on (PDADMAC-PSS)$_n$ films with independence of the conditions used for the film layering, which may be ascribed to the longer time used for the deposition of this multilayers in relation to those used on the deposition of (PAH—PSS)$_n$ ones [208, 209]. This can be an indication of the existence of a dynamic constrain to the equilibration of the film, which leads to the formation of stratified films due to the arresting of the layering process far from the equilibration. This agrees with the results by Panchagnula et al. [207] obtained using molecular dynamics simulation results. Thus, the emergence of a true stratification on the LbL films depends on the adsorption kinetics, which is particularly clear from the comparison of (PDADMAC—PSS)$_n$ and (PAH—PSS)$_n$ multilayers. In the former one, there is no signature of stratification due to the important contribution of the interdiffusion to the layering, whereas the emergence of certain time-dependent lamellar structure was found for (PAH—PSS)$_n$ multilayers [175]. The importance of the contact time on the emergence of a lamellar structure on polyelectrolyte multilayers can be summarized in terms of two main premises as was demonstrated by Selin et al. [225]: (i) short contact times lead to the formation of films where the intermixing occurs only within the most external layers, and (ii) long contact times result in a complete intermingling along with the 3D structure of the multilayer.

6.7 Conclusions

This chapter provides a comprehensive discussion of the current knowledge related to some of the most fundamental aspects underlying the fabrication of polyelectrolyte multi-layered films using the Layer-by-Layer (LbL) method. The simplicity, flexibility, and versatility of the LbL technique have led to a rapid evolution of this methodology as a reference method for the fabrication of novel functional and structural materials. This has been accompanied by a continuous development of new concepts and methodological approaches that make the study of the physicochemical bases underlying the assembly process a key issue for helping the development of new fundamental and applied aspects in the LbL field.

Acknowledgements This work was funded by MICINN under grant PID2019-106557GB-C21, by Banco Santander-Universidad Complutense grant PR87/19-22513 (Spain), and by E.U. on the framework of the European Innovative Training Network-Marie Sklodowska-Curie Action NanoPaint (grant agreement 955612).

References

1. Kumar B, Park YT, Castro M, Grunlan JC, Feller JF (2012) Fine control of carbon nanotubes–polyelectrolyte sensors sensitivity by electrostatic layer by layer assembly (eLbL) for the detection of volatile organic compounds (VOC). Talanta 88:396–402. https://doi.org/10.1016/j.talanta.2011.11.006
2. del Mercato LL, Rivera-Gil P, Abbasi AZ, Ochs M, Ganas C, Zins I et al (2010) LbL multilayer capsules: recent progress and future outlook for their use in life sciences. Nanoscale 2(4):458–467. https://doi.org/10.1039/B9NR00341J
3. del Mercato LL, Ferraro MM, Baldassarre F, Mancarella S, Greco V, Rinaldi R et al (2014) Biological applications of LbL multilayer capsules: from drug delivery to sensing. Adv Colloid Interface Sci 207:139–154. https://doi.org/10.1016/j.cis.2014.02.014
4. Costa RR, Mano JF (2014) Polyelectrolyte multilayered assemblies in biomedical technologies. Chem Soc Rev 43:3453–3479. https://doi.org/10.1039/C3CS60393H
5. Silva JM, Reis RL, Mano JF (2016) Biomimetic extracellular environment based on natural origin polyelectrolyte multilayers. Small 12:4308–4342. https://doi.org/10.1002/smll.201601355
6. Guzmán E, Mateos-Maroto A, Ruano M, Ortega F, Rubio RG (2017) Layer-by-Layer polyelectrolyte assemblies for encapsulation and release of active compounds. Adv Colloid Interface Sci 249:290–307. https://doi.org/10.1016/j.cis.2017.04.009
7. de Villiers MM, Otto DP, Strydom SJ, Lvov YM (2011) Introduction to nanocoatings produced by layer-by-layer (LbL) self-assembly. Adv Drug Deliv Rev 63:701–715. https://doi.org/10.1016/j.addr.2011.05.011
8. Lavalle P, Voegel J-C, Vautier D, Senger B, Schaaf P, Ball V (2011) Dynamic aspects of films prepared by a sequential deposition of species: perspectives for smart and responsive materials. Adv Mat 23:1191–1221. https://doi.org/10.1002/adma.201003309
9. de Villiers MM, Lvov YM (2011) Layer-by-layer self-assembled nanoshells for drug delivery. Adv Drug Deliv Rev 63:699–700. https://doi.org/10.1016/j.addr.2011.06.001
10. Krogman KC, Cohen RE, Hammond PT, Rubner MF, Wang BN (2013) Industrial-scale spray layer-by-layer assembly for production of biomimetic photonic systems. Bioinspir Biomim 8:045005. https://doi.org/10.1088/1748-3182/8/4/

11. Zhao S, Caruso F, Dähne L, Decher G, De Geest BG, Fan J et al (2019) The future of layer-by-layer assembly: a tribute to ACS nano associate editor Helmuth Möhwald. ACS Nano 13:6151–6169. https://doi.org/10.1021/acsnano.9b03326
12. Schlenoff JB (2009) Retrospective on the future of polyelectrolyte multilayers. Langmuir 25:14007–14010. https://doi.org/10.1021/la901950c
13. Decher G, Schlenoff JB (eds) (2003) Multilayer thin films-sequential assembly of nanocomposite materials. Germany Wiley-VCH Verlag, Berlin
14. Decher G, Hong JD (1991) Buildup of ultrathin multilayer films by a self-assembly process 2. Consecutive adsorption of anionic and cationic bipolar amphiphiles and polyelectrolytes on charged surfaces. Ber Bunsen-Ges Phys Chem Chem Phys 95:1430–1434. https://doi.org/10.1002/bbpc.19910951122
15. Decher G, Hong JD (1991) Buildup of ultrathin multilayer films by a self-assembly process 1. Consecutive adsorption of anionic and cationic bipolar amphiphiles on charged surfaces. Makromol Chem, Macromol Symp 46:321–327. https://doi.org/10.1002/masy.19910460145
16. Decher G, Hong JD, Schmitt J (1992) Buildup of ultrathin multilayer films by a self-assembly process. 3. Consecutively alternating adsorption of anionic and cationic polyelectrolytes on charged surfaces. Thin Solid Films 210:831–835. https://doi.org/10.1016/0040-6090(92)90417-a
17. Gucht J, Spruijt E, Lemmers M, Cohen Stuart MA (2011) Polyelectrolyte complexes: bulk phases and colloidal systems. J Colloid Interface Sci 361:407–422. https://doi.org/10.1016/j.jcis.2011.05.080
18. Guzmán E, San Miguel V, Peinado C, Ortega F, Rubio RG (2010) Polyelectrolyte multilayers containing triblock copolymers of different charge ratio. Langmuir 26:11494–11502. https://doi.org/10.1021/la101043z
19. Guzmán E, Chuliá-Jordán R, Ortega F, Rubio RG (2011) Influence of the percentage of acetylation on the assembly of LbL multilayers of poly(acrylic acid) and chitosan. Phys Chem Chem Phys 13:18200–18207. https://doi.org/10.1039/C1CP21609K
20. Guzmán E, Cavallo JA, Chuliá-Jordán R, Gómez C, Strumia MC, Ortega F et al (2011) pH-induced changes in the fabrication of multilayers of poly(acrylic acid) and chitosan: fabrication, properties, and tests as a drug storage and delivery system. Langmuir 27:6836–6845. https://doi.org/10.1021/la200522r
21. Keeney M, Jiang XY, Yamane M, Lee M, Goodman S, Yang F (2015) Nanocoating for biomolecule delivery using layer-by-layer self-assembly. J Mat Chem B 3:8757–8770. https://doi.org/10.1039/C5TB00450K
22. Aggarwal N, Altgärde N, Svedhem S, Zhang K, Fischer S, Groth T (2014) Study on multilayer structures prepared from heparin and semi-synthetic cellulose sulfates as polyanions and their influence on cellular response. Colloids Surf B 116:93–103. https://doi.org/10.1016/j.colsurfb.2013.12.043
23. Hsu BB, Hagerman SR, Hammond PT (2016) Rapid and efficient sprayed multilayer films for controlled drug delivery. J Appl Polym Sci 133:43563. https://doi.org/10.1002/app
24. Zhuk A, Mirza R, Sukhishvili S (2011) Multiresponsive clay-containing layer-by-layer films. ACS Nano 5:8790–8799. https://doi.org/10.1021/nn202812a
25. Li Y-C, Schulz J, Mannen S, Delhom C, Condon B, Chang S et al (2012) Flame retardant behavior of polyelectrolyte-clay thin film assemblies on cotton fabric. ACS Nano 4:3325–3337. https://doi.org/10.1021/nn100467e
26. Priolo MA, Gamboa D, Holder KM, Grunlan JC (2010) Super gas barrier of transparent polymer-clay multilayer ultrathin films. Nano Lett 10:4970–4974. https://doi.org/10.1021/nl103047k
27. Correa-Duarte MA, Kosiorek A, Kandulski W, Giersig M, Liz-MarzÃ¡n LM (2005) Layer-by-layer assembly of multiwall carbon nanotubes on spherical colloids. Chem Mat 17:3268–3272. https://doi.org/10.1021/cm047710e
28. Qin S, Qin D, Ford WT, Herrera JE, Resasco DE (2004) Grafting of poly(4-vinylpyridine) to single-walled carbon nanotubes and assembly of multilayer films. Macromolecules 37:9963–9967. https://doi.org/10.1021/ma048692p

29. Salloum DS, Schlenoff JB (2004) Protein adsorption modalities on polyelectrolyte multilayers. Biomacromol 5:1089–1096. https://doi.org/10.1021/bm034522t
30. Jackler G, Czeslik C, Steitz R, Royer CA (2005) Spatial distribution of protein molecules adsorbed at a polyelectrolyte multilayer. Phys Rev E 71:041912. https://doi.org/10.1103/PhysRevE.71
31. Svensson O, Lindh L, Cárdenas M, Arnebrant T (2006) Layer-by-layer assembly of mucin and chitosan-Influence of surface properties, concentration and type of mucin. J Colloid Interface Sci 299:608–616. https://doi.org/10.1016/j.jcis.2006.02.027
32. Watanabe J, Shen H, Akashi M (2008) Polyelectrolyte droplets facilitate versatile layer-by-layer coating for protein loading interface. Acta Biomater 4:1255–1262. https://doi.org/10.1016/j.actbio.2008.03.009
33. Pedano ML, Martel L, Desbrieres J, Defrancq E, Dumy P, Coche-Guerente L et al (2004) Layer-by-layer deposition of Chitosan derivatives and DNA on gold surfaces for the development of biorecognition layers. Anal Lett 37:2235–2250. https://doi.org/10.1081/AL-200028021
34. He P, Bayachou M (2005) Layer-by-layer fabrication and characterization of DNA-wrapped single-walled carbon nanotube particles. Langmuir 21:6086–6092. https://doi.org/10.1021/la050581b
35. vander Straeten A, Lefèvre D, Demoustier-Champagne S, Dupont-Gillain C (2020) Protein-based polyelectrolyte multilayers. Adv Colloid Interface Sci 280:102161. https://doi.org/10.1016/j.cis.2020
36. Lipton J, Weng G-M, Röhr JA, Wang H, Taylor AD (2020) Layer-by-layer assembly of two-dimensional materials: meticulous control on the nanoscale. Matter 2:1148–1165. https://doi.org/10.1016/j.matt.2020.03.012
37. Stockton WB, Rubner MF (1997) Molecular-level processing of conjugated polymers. 4. Layer-by-layer manipulation of polyaniline via hydrogen-bonding interactions. Macromolecules 30:2717–2725. https://doi.org/10.1021/ma9700486
38. Wang LY, Wang ZQ, Zhang X, Shen JC (1997) A new approach for the fabrication of an alternating multilayer film of poly(4-vinylpyridine) and poly(acrylic acid) based on hydrogen bonding. Macromol Rapid Commun 18:509–514. https://doi.org/10.1002/marc.997.030180609
39. Shimazaki Y, Mitsuishi M, Ito S, Yamamoto M (1997) Preparation of the layer-by-layer deposited ultrathin film based on the charge-transfer interaction. Langmuir 13:1385–1387. https://doi.org/10.1021/la9609579
40. Anzai J, Kobayashi Y, Nakamura N, Nishimura M, Hoshi T (1999) Layer-by-layer construction of multilayer thin films composed of avidin and biotin-labeled Poly(amine)s. Langmuir 15:221–226. https://doi.org/10.1021/la980743m
41. Bourdillon C, Demaille C, Moiroux J, Savéant JM (1994) Step-by-Step immunological construction of a fully active multilayer enzyme electrode. J Am Chem Soc 116:10328–10329. https://doi.org/10.1021/ja00101a074
42. Xiong HM, Cheng MH, Zhou Z, Zhang X, Shen JC (1998) A new approach to the fabrication of a self-organizing film of heterostructured polymer/Cu_2S nanoparticles. Adv Mater 10:529–532. https://doi.org/10.1002/(SICI)521-4095(199805)10:7%3c529::AID-ADMA529%3e3.0.CO;2-E
43. Serizawa T, Hamada K-i, Kitayama T, Fujimoto N, Hatada K, Akashi M (2000) Stepwise stereocomplex assembly of stereoregular poly(methyl methacrylate)s on a substrate. J Am Chem Soc 122(9):1891–1899. https://doi.org/10.1021/ja9913535
44. van der Heyden A, Wilczewski M, Labbé P, Auzély R (2006) Multilayer films based on host–guest interactions between biocompatible polymers. Chem Comm 3220–3222. https://doi.org/10.1039/B604672J
45. Zhu J, Shim BS, Di Prima M, Kotov NA (2011) Transparent conductors from carbon nanotubes LBL-assembled with polymer dopant with $\pi-\pi$ electron transfer. J Am Chem Soc 133:7450–7460. https://doi.org/10.1021/ja111687t

46. Lvov Y, Ariga K, Ichinose I, Kunitake T (1995) Layer-by-layer architectures of concanavalin A by means of electrostatic and biospecific interactions. J Chem Soc Chem Comm 1995:2313–2314. https://doi.org/10.1039/C39950002313
47. Ichinose I, Kawakami T, Kunitake T (1998) Alternate molecular layers of metal oxides and hydroxyl polymers prepared by the surface sol-gel process. Adv Mat 10:535–539. https://doi.org/10.1002/(sici)521-4095(199805)10:7%3c535::Aid-adma535%3e3.0.Co;2-q
48. Such GK, Quinn JF, Quinn A, Tjipto E, Caruso F (2006) Assembly of ultrathin polymer multilayer films by click chemistry. J Am Chem Soc 128:9318–9319. https://doi.org/10.1021/ja063043+
49. An Q, Huang T, Shi F (2018) Covalent layer-by-layer films: chemistry, design, and multidisciplinary applications. Chem Soc Rev 47:5061–5098. https://doi.org/10.1039/C7CS00406K
50. Sukhorukov GB, Donath E, Lichtenfeld H, Knippel E, Knippel M, Budde A et al (1998) Layer-by-layer self assembly of polyelectrolytes on colloidal particles. Colloids Surf A 137:253–266. https://doi.org/10.1016/S0927-7757(98)00213-1
51. Donath E, Sukhorukov GB, Caruso F, Davis SA, Möhwald H (1998) Novel hollow polymer shells by colloid-templated assembly of polyelectrolytes. Angew Chem Int Ed 37:2201–2205. https://doi.org/10.1002/(SICI)521-3773(19980904)37:16%3c2201::AID-ANIE2201%3e3.0.CO;2-E
52. Caruso F, Donath E, Möhwald H (1998) Influence of polyelectrolyte multilayer coatings on Förster resonance energy transfer between 6-carboxyfluorescein and rhodamine B-labeled particles in aqueous solution. J Phys Chem B 102:2011–2016. https://doi.org/10.1021/jp980198y
53. Guzmán E, Ritacco H, Ortega F, Svitova T, Radke CJ, Rubio RG (2009) Adsorption kinetics and mechanical properties of ultrathin polyelectrolyte multilayers: liquid-supported versus solid-supported films. J Phys Chem B 113:7128–7137. https://doi.org/10.1021/jp811178a
54. Ferri JK, Dong W-F, Miller R, Mohwald H (2006) Elastic moduli of asymmetric ultrathin free-standing polyelectrolyte nanocomposites. Macromolecules 39:1532–1537. https://doi.org/10.1021/ma0516485
55. Ferri JK, Dong W-F, Miller R (2005) Ultrathin free-standing polyelectrolyte nanocomposites: a novel method for preparation and characterization of assembly dynamics. J Phys Chem B 109:14764–14768. https://doi.org/10.1021/jp052672n
56. Shchukina EM, Shchukin DG (2012) Layer-by-layer coated emulsion microparticles as storage and delivery tool. Curr Opin Colloid Interface Sci 17:281–289. https://doi.org/10.1016/j.cocis.2012.06.003
57. Cuomo F, Lopez F, Miguel MG, Lindman B (2010) Vesicle-templated layer-by-layer assembly for the production of nanocapsules. Langmuir 26:10555–10560. https://doi.org/10.1021/la100584b
58. Kozlovskaya V, Zavgorodnya O, Chen Y, Ellis K, Tse HM, Cui W et al (2012) Ultrathin polymeric coatings based on hydrogen-bonded polyphenol for protection of pancreatic islet cells. Adv Funct Mat 22:3389–3398. https://doi.org/10.1002/adfm.201200138
59. Sukhishvili SA, Granick S (2002) Layered, erasable polymer multilayers formed by hydrogen-bonded sequential self-assembly. Macromolecules 35:301–310. https://doi.org/10.1021/ma011346c
60. Hiller J, Mendelsohn JD, Rubner MF (2002) Reversibly erasable nanoporous anti-reflection coatings from polyelectrolyte multilayers. Nat Mat 1:59–63. https://doi.org/10.1038/nmat719
61. Denkbas EB, Ottenbrite RM (2006) Perspectives on: Chitosan drug delivery systems based on their geometries. J Bioact Compat Polym 21:351–368. https://doi.org/10.1177/0883911506066930
62. Städler B, Chandrawati R, Goldie K, Caruso F (2009) Capsosomes: subcompartmentalizing polyelectrolyte capsules using liposomes. Langmuir 25:6725–6732. https://doi.org/10.1021/la900213a
63. Städler B, Chandrawati R, Price AD, Chong S-F, Breheney K, Postma A et al (2009) A microreactor with thousands of subcompartments: enzyme-loaded liposomes within polymer capsules. Angew Chem Int Ed 48:4359–4362. https://doi.org/10.1002/anie.200900386

64. Chandrawati R, Hosta-Rigau L, Vanderstraaten D, Lokuliyana SA, Städler B, Albericio F et al (2010) Engineering advanced capsosomes: maximizing the number of subcompartments, cargo retention, and temperature-triggered reaction. ACS Nano 4:1351–1361. https://doi.org/10.1021/nn901843j
65. Hosta-Rigau L, Städler B, Yan Y, Nice EC, Heath JK, Albericio F et al (2010) Capsosomes with multilayered subcompartments: assembly and loading with hydrophobic Cargo. Adv Funct Mat 20:59–66. https://doi.org/10.1002/adfm.200901297
66. Vikulina AS, Skirtach AG, Volodkin D (2019) Hybrids of polymer multilayers, lipids, and nanoparticles: mimicking the cellular microenvironment. Langmuir 35:8565–8573. https://doi.org/10.1021/acs.langmuir.8b04328
67. Johnston APR, Cortez C, Angelatos AS, Caruso F (2006) Layer-by-layer engineered capsules and their applications. Curr Opin Colloid Interface Sci 11:203–209. https://doi.org/10.1016/j.cocis.2006.05.001
68. Tong W, Song X, Gao C (2012) Layer-by-layer assembly of microcapsules and their biomedical applications. Chem Soc Rev 41:6103–6124. https://doi.org/10.1039/C2CS35088B
69. Lvov Y, Essler F, Decher G (1993) Combination of polycation/polyanion self-assembly and Langmuir-Blodgett transfer for the construction of superlattice films. J Phys Chem 97:13773–13777
70. Ariga K, Lvov Y, Kunitake T (1997) Assembling alternate dye-polyion molecular films by electrostatic layer-by-layer adsorption. J Am Chem Soc 119:2224–2231. https://doi.org/10.1021/ja963442c
71. Richardson JJ, Cui J, Björnmalm M, Braunger JA, Ejima H, Caruso F (2016) Innovation in layer-by-layer assembly. Chem Rev 116:14828–14867
72. Hammond PT (2011) Engineering materials layer-by-layer: Challenges and opportunities in multilayer assembly. AIChE J 57:2928–2940
73. Xiang Y, Lu S, Jiang SP (2012) Layer-by-layer self-assembly in the development of electrochemical energy conversion and storage devices from fuel cells to supercapacitors. Chem Soc Rev 41:7291–7321
74. Michel M, Toniazzo V, Ruch D, Ball V (2012) Deposition mechanisms in layer-by-layer or step-by-step deposition methods: from elastic and impermeable films to soft membranes with ion exchange properties. ISRN Mat Sci 2012:701695
75. Iler RK (1966) Multilayers of colloidal particles. J Colloid Interface Sci 21:569–594
76. Mateos-Maroto A, Abelenda-Núñez I, Ortega F, Rubio RG, Guzmán E (2021) Polyelectrolyte multilayers on soft colloidal nanosurfaces: a new life for the layer-by-layer method. Polymers 13:1221
77. Shim BS, Podsiadlo P, Lilly DG, Agarwal A, Lee J, Tang Z et al (2007) Nanostructured thin films made by Dewetting method of layer-by-layer assembly. Nano Lett 7:3266–3273
78. Richardson JJ, Björnmalm M, Caruso F (2015) Technology-driven layer-by-layer assembly of nanofilms. Science 348
79. Fu Y, Li S-J, Xu J, Yang M, Zhang J-D, Jiao Y-H et al (2011) Facile and efficient approach to speed up layer-by-layer assembly: dipping in agitated solutions. Langmuir 27:672–677
80. Li Y, Wang X, Sun J (2012) Layer-by-layer assembly for rapid fabrication of thick polymeric films. Chem Soc Rev 41:5998–6009
81. Lee S-S, Hong J-D, Kim CH, Kim K, Koo JP, Lee K-B (2001) Layer-by-layer deposited multilayer assemblies of ionene-type polyelectrolytes based on the spin-coating method. Macromolecules 34:5358–5360
82. Espinosa-Dzib A, Chen J, Zavgorodnya O, Kozlovskaya V, Liang X, Kharlampieva E (2015) Tuning assembly and enzymatic degradation of silk/poly(*N*-vinylcaprolactam) multilayers via molecular weight and hydrophobicity. Soft Matter 11:5133–5145
83. Izquierdo A, Ono SS, Voegel JC, Schaaf P, Decher G (2005) Dipping versus spraying: exploring the deposition conditions for speeding up layer-by-layer assembly. Langmuir 21:7558–7567

84. Kolansinska M, Krastev R, Gutberlet T, Warszynski P (2009) Layer-by-layer deposition of polyelectrolytes. Dipping versus spraying. Langmuir 25:1224–1232
85. Guzmán E, Rubio RG, Ortega F (2020) A closer physico-chemical look to the layer-by-layer electrostatic self-assembly of polyelectrolyte multilayers. Adv Colloid Interface Sci 282:102197
86. Kiel M, Mitzscherling S, Leitenberger W, Santer S, Tiersch B, Sievers TK et al (2010) Structural characterization of a spin-assisted colloid-polyelectrolyte assembly: stratified multilayer thin films. Langmuir 26:18499–18502. https://doi.org/10.1021/la103609f
87. Cho J, Char K, Hong J-D, Lee K-B (2001) Fabrication of highly ordered multilayer films using a spin self-assembly method. Adv Mat 13:1076–1078. https://doi.org/10.1002/521-4095(200107)13:14%3c1076::AID-ADMA1076%3e3.0.CO;2-M
88. Ma L, Cheng M, Jia G, Wang Y, An Q, Zeng X et al (2012) Layer-by-layer self-assembly under high gravity field. Langmuir 28:9849–9856. https://doi.org/10.1021/la301553w
89. Schlenoff JB, Dubas ST, Farhat T (2000) Sprayed polyelectrolyte multilayers. Langmuir 16:9968–9969. https://doi.org/10.1021/la001312i
90. Elosua C, Lopez-Torres D, Hernaez M, Matias IR, Arregui FJ (2013) Comparative study of layer-by-layer deposition techniques for poly(sodium phosphate) and poly(allylamine hydrochloride). Nanosc Res Lett 8:539. https://doi.org/10.1186/556-276X-8-539
91. Félix O, Zheng Z, Cousin F, Decher G (2009) Are sprayed LbL-films stratified? A first assessment of the nanostructure of spray-assembled multilayers by neutron reflectometry. C R Chim 12:225–234. https://doi.org/10.1016/j.crci.2008.09.009
92. Dierendonck M, De Koker S, De Rycke R, De Geest BG (2014) Just spray it—LbL assembly enters a new age. Soft Matter 10:804–807. https://doi.org/10.1039/C3SM52202D
93. Schaaf P, Voegel J-C, Jierry L, Boulmedais F (2012) Spray-assisted polyelectrolyte multilayer buildup: from step-by-step to single-step polyelectrolyte film constructions. Adv Mat 24:1001–1016. https://doi.org/10.1002/adma.201104227
94. Hammond PT (2012) Building biomedical materials layer-by-layer. Mater Today 15:196–206. https://doi.org/10.1016/S369-7021(12)70090-1
95. Ariga K, Ahn E, Park M, Kim B-S (2019) Layer-by-layer assembly: recent progress from layered assemblies to layered nanoarchitectonics. Chem Asian J 14:2553–2566. https://doi.org/10.1002/asia.201900627
96. Sui Z, Salloum D, Schlenoff JB (2003) Effect of molecular weight on the construction of polyelectrolyte multilayers: stripping versus sticking. Langmuir 19:2491–2495. https://doi.org/10.1021/la026531d
97. Michel M, Izquierdo A, Decher G, Voegel JC, Schaaf P, Ball V (2005) Layer by layer self-assembled polyelectrolyte multilayers with embedded phospholipid vesicles obtained by spraying: integrity of the vesicles. Langmuir 21:7854–7859. https://doi.org/10.1021/la050497w
98. Lefort M, Boulmedais F, Jierry L, Gonthier E, Voegel JC, Hemmerl J et al (2011) Simultaneous spray coating of interacting species: general rules governing the poly(styrene sulfonate)/poly(allylamine) system. Langmuir 27:4653–4660
99. Merrill MH, Sun CT (2009) Fast, simple and efficient assembly of nanolayered materials and devices. Nanotechnology 20:075606. https://doi.org/10.1088/0957-4484/20/7/
100. Mulhearn WD, Kim DD, Gu Y, Lee D (2012) Facilitated transport enhances spray layer-by-layer assembly of oppositely charged nanoparticles. Soft Matter 8:10419–10427. https://doi.org/10.1039/C2SM26456K
101. Alongi J, Carosio F, Frache A, Malucelli G (2013) Layer by Layer coatings assembled through dipping, vertical or horizontal spray for cotton flame retardancy. Carbohydr Polymers 92:114–119. https://doi.org/10.1016/j.carbpol.2012.08.086
102. Lefort M, Popa G, Seyrek E, Szamocki R, Felix O, Hemmerlé J et al (2010) Spray-on organic/inorganic films: a general method for the formation of functional nano- to microscale coatings. Angew Chem Int Ed 49:10110–10113. https://doi.org/10.1002/anie.201002729
103. Hong X, Li J, Wang M, Xu J, Guo W, Li J et al (2004) Fabrication of magnetic luminescent nanocomposites by a layer-by-layer self-assembly approach. Chem Mat 16:4022–4027. https://doi.org/10.1021/cm049422o

104. Sun J, Gao M, Feldmann J (2001) Electric field directed layer-by-layer assembly of highly fluorescent CdTe nanoparticles. J Nanosci Nanotech 1:133–136. https://doi.org/10.1166/jnn.2001.029
105. Shi L, Lu Y, Sun J, Zhang J, Sun C, Liu J et al (2003) Site-Selective lateral multilayer assembly of bienzyme with polyelectrolyte on ITO electrode based on electric field-induced directly layer-by-layer deposition. Biomacromol 4:1161–1167. https://doi.org/10.1021/bm030003e
106. Richardson JJ, Ejima H, Lörcher SL, Liang K, Senn P, Cui J et al (2013) Preparation of nano- and microcapsules by electrophoretic polymer assembly. Angew Chem Int Ed 52:6455–6458. https://doi.org/10.1002/anie.201302092
107. Van Tassel PR (2012) Polyelectrolyte adsorption and layer-by-layer assembly: electrochemical control. Curr Opin Colloid Interface Sci 17:106–113. https://doi.org/10.1016/j.cocis.2011.08.008
108. Richardson JJ, Björnmalm M, Caruso F (2015) Technology-driven layer-by-layer assembly of nanofilms. Science 348:aaa2491. https://doi.org/10.1126/science.aaa2491
109. Ko YH, Kim YH, Park J, Nam KT, Park JH, Yoo PJ (2011) Electric-field-assisted layer-by-layer assembly of weakly charged polyelectrolyte multilayers. Macromolecules 44:2866–2872. https://doi.org/10.1021/ma102112a
110. Wang Z, Zhang X, Gu J, Yang H, Nie J, Ma G (2014) Electrodeposition of alginate/chitosan layer-by-layer composite coatings on titanium substrates. Carbohydr Polymers 103:38–45. https://doi.org/10.1016/j.carbpol.2013.12.007
111. Mu B, Liu P, Du P, Dong Y, Lu C (2011) Magnetic-targeted pH-responsive drug delivery system via layer-by-layer self-assembly of polyelectrolytes onto drug-containing emulsion droplets and its controlled release. J Polymer Sci A Polymer Chem 49:1969–1976. https://doi.org/10.1002/pola.24623
112. Dey S, Mohanta K, Pal AJ (2010) Magnetic-field-assisted layer-by-layer electrostatic assembly of ferromagnetic nanoparticles. Langmuir 26:9627–9631. https://doi.org/10.1021/la101132z
113. Caruso F, Caruso RA, Möhwald H (1998) Nanoengineering of inorganic and hybrid hollow spheres by colloidal templating. Science 282:1111–1114
114. Bertrand P, Jonas A, Laschewsky A, Legras R (2000) Ultrathin polymer coatings by complexation of polyelectrolytes at interfaces: suitable materials, structure and properties. Macromol Rapid Commun 21:319–348
115. Yan Y, Björnmalm M, Caruso F (2014) Assembly of layer-by-layer particles and their interactions with biological systems. Chem Mat 26:452–460
116. Donath E, Walther D, Shilov VN, Knippel E, Budde A, Lowack K et al (1997) Nonlinear hairy layer theory of electrophoretic fingerprinting applied to consecutive layer by layer polyelectrolyte adsorption onto charged polystyrene latex particles. Langmuir 13:5294–5305
117. Sukhorukov GB, Donath E, Davis S, Lichtenfeld H, Caruso F, Popov VI et al (1998) Stepwise polyelectrolyte assembly on particle surfaces: a novel approach to colloid design. Polym Adv Technol 9:759–767
118. Caruso F, Lichtenfeld H, Giersig M, Möhwald H (1998) Electrostatic self-assembly of silica nanoparticle–polyelectrolyte multilayers on polystyrene latex particles. J Am Chem Soc 120(33):8523–8524
119. Bagaria HG, Wong MS (2011) Polyamine-salt aggregate assembly of capsules as responsive drug delivery vehicles. J Mat Chem 21:9454–9466
120. Tong W, Gao C (2008) Multilayer microcapsules with tailored structures for bio-related applications. J Mat Chem 18:3799–3812
121. Voigt A, Lichtenfeld H, Sukhorukov GB, Zastrow H, Donath E, Baumler H et al (1999) Membrane filtration for microencapsulation and microcapsules fabrication by layer-by-layer polyelectrolyte adsorption. Ind Eng Chem Res 38:4037–4043
122. Nagaraja AT, You Y-H, Choi J-W, Hwang J-H, Meissner KE, McShane MJ (2016) Layer-by-layer modification of high surface curvature nanoparticles with weak polyelectrolytes using a multiphase solvent precipitation process. J Colloid Interface Sci 466:432–441

123. Elizarova IS, Luckham PF (2016) Fabrication of polyelectrolyte multilayered nano-capsules using a continuous layer-by-layer approach. J Colloid Interface Sci 470:92–99
124. Mu B, Liu P, Du P, Dong Y, Lu C (2011) Magnetic-targeted pH-responsive drug delivery system via layer-by-layer self-assembly of polyelectrolytes onto drug-containing emulsion droplets and its controlled release. J Polym Sci A: Polym Chem 49:1969–1976
125. Wilson R, Spiller DG, Prior IA, Bhatt R, Hutchinson A (2007) Magnetic microspheres encoded with photoluminescent quantum dots for multiplexed detection. J Mat Chem 17:4400–4406
126. Szczepanowicz K, Hoel HJ, Szyk-Warszynska L, Bielańska E, Bouzga AM, Gaudernack G et al (2010) Formation of biocompatible nanocapsules with emulsion core and Pegylated shell by polyelectrolyte multilayer adsorption. Langmuir 26:12592–12597. https://doi.org/10.1021/la102061s
127. Hoogeveen NG, Cohen Stuart MA, Fleer GJ, Böhmer MR (1996) Formation and stability of multilayers of polyelectrolytes. Langmuir 12:3675–3681
128. Bantchev G, Lu Z, Lvov Y (2009) Layer-by-layer nanoshell assembly on colloids through simplified washless process. J Nanosci Nanotechnol 9:396–403
129. Szczepanowicz K, Dronka-Góra D, Para G, Warszyński P (2010) Encapsulation of liquid cores by layer-by-layer adsorption of polyelectrolytes. J Microencapsul 27:198–204
130. Grigoriev DO, Bukreeva T, Möhwald H, Shchukin DG (2008) New method for fabrication of loaded micro- and nanocontainers: emulsion encapsulation by polyelectrolyte layer-by-layer deposition on the liquid core. Langmuir 24:999–1004
131. Thanasukarn P, Pongsawatmanit R, McClements D (2006) Utilization of layer-by-layer interfacial deposition technique to improve freeze–thaw stability of oil-in-water emulsions. Food Res Int 39:721–729
132. Li J, Stöver HDH (2010) Pickering emulsion templated layer-by-layer assembly for making microcapsules. Langmuir 26:15554–15560
133. Rossier-Miranda FJ, Schroën K, Boom R (2012) Microcapsule production by an hybrid colloidosome-layer-by-layer technique. Food Hydrocolloids 27:119–125
134. Liu H, Gu X, Hu M, Hu Y, Wang C (2014) Facile fabrication of nanocomposite microcapsules by combining layer-by-layer self-assembly and Pickering emulsion templating. RSC Adv 4:16751–16758
135. Guzmán E, Ruano M, Ortega F, Rubio RG (2014) Stratified interpolyelectrolyte complexes: fabrication, structure and properties. In: Visakh PM, Bayraktar O, Picó GA (eds) Polyelectrolytes. Springer International Publishing, Cham (Switzerland), pp 299–347
136. Richardson JJ, Liang K, Kempe K, Ejima H, Cui J, Caruso F (2013) Immersive polymer assembly on immobilized particles for automated capsule preparation. Adv Mat 25:6874–6878
137. Wang Y, Zhou J, Guo X, Hu Q, Qin C, Liu H et al (2017) Layer-by-layer assembled biopolymer microcapsule with separate layer cavities generated by gas-liquid microfluidic approach. Mater Sci Eng C Mater Biol Appl 81:13–19
138. Wang Y, Liu Y, Cheng Y, Kim E, Rubloff GW, Bentley WE et al (2011) Coupling electrodeposition with layer-by-layer assembly to address proteins within microfluidic channels. Adv Mater 23:5817–5821
139. Lee UN, Day JH, Haack AJ, Bretherton RC, Lu W, DeForest CA et al (2020) Layer-by-layer fabrication of 3D hydrogel structures using open microfluidics. Lab Chip 20:525–536
140. Alkekhia D, Hammond PT, Shukla A (2020) Layer-by-layer biomaterials for drug delivery. Ann Rev Biomed Eng. 22:1–24
141. Björnmalm M, Yan Y, Caruso F (2014) Engineering and evaluating drug delivery particles in microfluidic devices. J Control Rel 190:139–149
142. Priest C, Quinn A, Postma A, Zelikin AN, Ralston J, Caruso F (2008) Microfluidic polymer multilayer adsorption on liquid crystal droplets for microcapsule synthesis. Lab Chip 8:2182–2187
143. Matosevic S, Paegel BM (2013) Layer-by-layer cell membrane assembly. Nat Chem 5:958–963
144. Mets JM, Wilson JT, Cui W, Chaikof EL (2013) An automated process for layer-by-layer assembly of polyelectrolyte multilayer thin films on viable cell aggregates. Adv Healthc Mater 2:266–270

145. Kantak C, Beyer S, Yobas L, Bansal T, Trau D (2011) A 'microfluidic pinball' for on-chip generation of layer-by-layer polyelectrolyte microcapsules. Lab Chip 11:1030–1035
146. Raman N, Lee M-R, Palecek SP, Lynn DM (2014) Polymer multilayers loaded with antifungal β-peptides kill planktonic *Candida albicans* and reduce formation of fungal biofilms on the surfaces of flexible catheter tubes. J Control Release 191:54–62
147. Madaboosi N, Uhlig K, Jäger MS, Möhwald H, Duschl C, Volodkin DV (2012) Microfluidics as a tool to understand the build-up mechanism of exponential-like growing films. Macromol Rapid Comm 33:1775–1779
148. Kim H-J, Lee K, Kumar S, Kim J (2005) Dynamic sequential layer-by-layer deposition method for fast and region-selective multilayer thin film fabrication. Langmuir 18:8532–8538
149. Richardson JJ, Teng D, Björnmalm M, Gunawan ST, Guo J, Cui J et al (2014) Fluidized bed layer-by-layer microcapsule formation. Langmuir 30:10028–10034
150. Noi KF, Roozmand A, Björnmalm M, Richardson JJ, Franks GV, Caruso F (2015) Assembly-controlled permeability of layer-by-layer polymeric microcapsules using a tapered fluidized bed. ACS Appl Mater Interfaces 7(50):27940–27947
151. Johansson E, Blomberg E, LingstroÌ^m R, Wägberg L (2009) Adhesive interaction between polyelectrolyte multilayers of polyallylamine hydrochloride and polyacrylic acid studied using atomic force microscopy and surface force apparatus. Langmuir 25:2887–2894
152. Guzmán E, Maestro A, Llamas S, Álvarez-Rodríguez J, Ortega F, Maroto-Valiente Á et al (2016) 3D solid supported inter-polyelectrolyte complexes obtained by the alternate deposition of poly(diallyldimethylammonium chloride) and poly(sodium 4-styrenesulfonate). Beilstein J Nanotech 7:197–208
153. Guzmán E, Ritacco H, Rubio JEF, Rubio RG, Ortega F (2009) Salt-induced changes in the growth of polyelectrolyte layers of poly(diallyl-dimethylammonium chloride) and poly(4-styrene sulfonate of sodium). Soft Matter 5:2130–2142
154. Schlenoff JB, Dubas ST (2001) Mechanism of polyelectrolyte multilayer growth: charge overcompensation and distribution. Macromolecules 34:592–598
155. Dubas ST, Schlenoff JB (1999) Factors controlling the growth of polyelectrolyte multilayers. Macromolecules 32:8153–8160
156. Picart C, Lavalle P, Hubert P, Cuisinier FJG, Decher G, Schaaf P et al (2001) Buildup mechanism for poly(L-lysine)/hyaluronic acid films onto a solid surface. Langmuir 17:7414–7424
157. Lavalle P, Gergely C, Cuisinier FJG, Decher G, Schaaf P, Voegel JC et al (2002) Comparison of the structure of polyelectrolyte multilayer films exhibiting a linear and an exponential growth regime: an in situ atomic force microscopy study. Macromolecules 35:4458–4465
158. Schneider A, Richert L, Francius G, Voegel J-C, Picart C (2007) Elasticity, biodegradability and cell adhesive properties of chitosan/hyaluronan multilayer films. Biomed Mater 2:S45–S51
159. Cini N, Tulun T, Decher G, Ball V (2010) Step-by-step assembly of self-patterning polyelectrolyte films violating (almost) all rules of layer-by-layer deposition. J Am Chem Soc 132 8264–8265
160. Cini N, Tulun T, Blanck C, Toniazzo V, Ruch D, Decher G et al (2012) Slow complexation dynamics between linear short polyphosphates and polyallylamines: analogies with '"layer-by-layer"' deposition. Phys Chem Chem Phys 14:3048–3056
161. Elbert DL, Herbert CB, Hubbel JA (1999) Thin polymer layers formed by polyelectrolyte multilayer techniques on biological surfaces. Langmuir 15:5355–5362
162. Subbotin AV, Semenov AN (2021) The structure of polyelectrolyte complex coacervates and multilayers. Macromolecules 54:1314–1328
163. Guzmán E, Fernández-Peña L, Ortega F, Rubio RG (2020) Equilibrium and kinetically-trapped aggregates in polyelectrolyte-oppositely charged surfactant mixtures. Curr Opin Colloid Interface Sci 48:91–108
164. Llamas S, Guzmán E, Baghdadli N, Ortega F, Cazeneuve C, Rubio RG et al (2016) Adsorption of poly(diallyldimethylammonium chloride)—sodium methyl-cocoyl-taurate complexes onto solid surfaces. Colloids Surf A 505:150–157

165. Tang K, Besseling NAM (2016) Formation of polyelectrolyte multilayers: ionic strengths and growth regimes. Soft Matter 12:1032–1040
166. Xu L, Pristinski D, Zhuk A, Stoddart C, Ankner JF, Sukhishvili SA (2012) Linear versus exponential growth of weak polyelectrolyte multilayers: correlation with polyelectrolyte complexes. Macromolecules 45:3892–3901
167. Lavalle P, Picart C, Mutterer J, Gergely C, Reiss H, Voegel J-C et al (2003) Modeling the buildup of polyelectrolyte multilayer films having exponential growth. J Phys Chem B 108:635–648
168. Picart C, Mutterer J, Richert L, Luo Y, Prestwich GD, Schaaf P et al (2002) Molecular basis for the explanation of the exponential growth of polyelectrolyte multilayers. Proc Nat Acad Sci USA 99:12531–12535
169. Jourdainne L, Lecuyer S, Arntz Y, Picart C, Schaaf P, Senger B, et al (2008) Dynamics of poly(l-lysine) in hyaluronic acid/poly(l-lysine) multilayer films studied by fluorescence recovery after pattern photobleaching. Langmuir 24:7842–7847
170. Sustr D, Hlaváček A, Duschl C, Volodkin D (2018) Multi-fractional analysis of molecular diffusion in polymer multilayers by FRAP: a new simulation-based approach. J Phys Chem B 122(3):1323–1333
171. Hoda N, Larson RG (2009) Modeling the buildup of exponentially growing polyelectrolyte multilayer films. J Phys Chem B 113:4232–4241
172. Guzmán E, Ortega F, Rubio RG (2016) Comment on "Formation of polyelectrolyte multilayers: ionic strengths and growth regimes" by K Tang and A. M. Besseling. Soft Matter 12:1032 (12:8460–8463)
173. Yuan W, Weng G-M, Lipton J, Li CM, Van Tassel PR, Taylor AD (2020) Weak polyelectrolyte-based multilayers via layer-by-layer assembly: approaches, properties, and applications. Adv Colloid Interface Sci 282:102200
174. Guzmán E, Ritacco HA, Ortega F, Rubio RG (2012) Growth of polyelectrolyte layers formed by poly(4-styrenesulfonate sodium salt) and two different polycations: new insights from study of adsorption kinetics. J Phys Chem C 116:15474–15483
175. Guzmán E, Ritacco H, Ortega F, Rubio RG (2011) Evidence of the influence of adsorption kinetics on the internal reorganization of polyelectrolyte multilayers. Colloids Surf A 384:274–281
176. McAloney RA, Sinyor M, Dudnik V, Goh MC (2001) Atomic force microscopy studies of salt effects on polyelectrolyte multilayer film morphology. Langmuir 17:6655–6663
177. Haynie DT, Cho E, Waduge P (2011) "In and out diffusion" hypothesis of exponential multilayer film buildup revisited. Langmuir 27:5700–5704
178. Hernandez-Montelongo J, Nascimento VF, Hernández-Montelongo R, Beppu MM, Cotta MA (2020) Fractal analysis of the formation process and morphologies of hyaluronan/chitosan nanofilms in layer-by-layer assembly. Polymer 191:122283
179. Naas C, Scheler U, Lappan U (2021) Influence of pH on the growth and the local dynamics of polyelectrolyte multilayers. Macromolecules 54:1043–1051
180. Selin V, Ankner JF, Sukhishvili SA (2017) Nonlinear layer-by-layer films: effects of chain diffusivity on film structure and swelling. Macromolecules 50(16):6192–6201
181. Joanny JF (1999) Polyelectrolyte adsorption and charge inversion. Eur Phys J B 117–122
182. Berndt P, Kurihara K, Kunitake T (1992) Adsorption of poly(styrenesulfonate) onto an ammonium monolayer on mica: a surface forces study. Langmuir 8:2486–2490
183. Schwarz S, Eichhorn KJ, Wischerhoff E, Laschewsky A (1999) Polyelectrolyte adsorption onto planar surfaces: a study by streaming potential and ellipsometry measurements. Colloids Surf A 159:491–501
184. Ringwald C, Ball V (2015) Shear induced changes in the streaming potential of polyelectrolyte multilayer films. Colloids Surf A 464:41–45
185. Ladam G, Schaad P, Voegel JC, Schaaf P, Decher G, Cuisinier F (2000) In situ determination of the structural properties of initially deposited polyelectrolyte multilayers. Langmuir 16:1249–1255

186. Adamczyk Z, Zembala M, Kolasińska M, Warszyński P (2007) Characterization of polyelectrolyte multilayers on mica and oxidized titanium by streaming potential and wetting angle measurements. Colloids Surf A 302:455–460
187. Ferriz-Mañas M, Schlenoff JB (2014) Zeta potential of polyelectrolyte multilayers using the spinning disk method. Langmuir 30:8776–8783
188. Fares HM, Schlenoff JB (2017) Equilibrium overcompensation in polyelectrolyte complexes. Macromolecules 50:3968–3978
189. von Klitzing R, Moehwald H (1995) Proton concentration profile in ultrathin polyelectrolyte films. Langmuir 11:3554–3559
190. Fernández-Peña L, Guzmán E, Ortega F, Bureau L, Leonforte F, Velasco D et al (2021) Physico-chemical study of polymer mixtures formed by a polycation and a zwitterionic copolymer in aqueous solution and upon adsorption onto negatively charged surfaces. Polymer 217:123442
191. Rmaile HH, Schlenoff JB (2002) "Internal pKa's" in polyelectrolyte multilayers: coupling protons and salt. Langmuir 18:8263–8265
192. Lourenço JMC, Ribeiro PA, Botelho do Rego AM, Raposo M (2007) Counterions in layer-by-layer films—influence of the drying process. J Colloid Interface Sci. 313:26–33
193. Lourenço JMC, Ribeiro PA, Botelho do Rego AM, Braz Fernandes FM, Moutinho AMC, Raposo M (2004) Counterions in poly(allylamine hydrochloride) and poly(styrene sulfonate) layer-by-layer films. Langmuir 20:8103–8109
194. Llamas S, Guzmán E, Ortega F, Rubio RG (2020) Adsorption of mixtures of a Pegylated lipid with anionic and Zwitterionic surfactants at solid/liquid. Colloids Interfaces 4:47
195. Ghostine RA, Markarian MZ, Schlenoff JB (2013) Asymmetric growth in polyelectrolyte multilayers. J Am Chem Soc 135:7636–7646
196. Lehaf AM, Hariri HH, Schlenoff JB (2012) Homogeneity, modulus, and viscoelasticity of polyelectrolyte multilayers by nanoindentation: refining the buildup mechanism. Langmuir 28:6348–6355
197. Ghostine RA, Jisr RM, Lehaf A, Schlenoff JB (2013) Roughness and salt annealing in a polyelectrolyte multilayer. Langmuir 11,742−11,750
198. Volodkin D, von Klitzing R (2014) Competing mechanisms in polyelectrolyte multilayer formation and swelling: Polycation–polyanion pairing versus polyelectrolyte–ion pairing. Curr Opin Colloid Interface Sci 19:25–31
199. Guzmán E, Ortega F, Prolongo MG, Starov VM, Rubio RG (2011) Influence of the molecular architecture on the adsorption onto solid surfaces: comb-like polymers. Phys Chem Chem Phys 13:16416–16423
200. Maestro A, Jones D, Sánchez de Rojas Candela C, Guzman E, Duits MHG, Cicuta P (2018) Tuning interfacial properties and processes by controlling the rheology and structure of poly(n-isopropylacrylamide) particles at air/water interfaces. Langmuir 34:7067–7076
201. Llamas S, Mendoza AJ, Guzmán E, Ortega F, Rubio RG (2013) Salt effects on the air/solution interfacial properties of PEO-containing copolymers: equilibrium, adsorption kinetics and surface rheological behavior. J Colloid Interface Sci 400:49–58
202. Guzmán E, Ortega F, Baghdadli N, Cazeneuve C, Luengo GS, Rubio RG (2011) Adsorption of conditioning polymers on solid substrates with different charge density. ACS Appl Mat Interfaces. 3:3181–3188
203. Guzmán E, Ortega F, Baghdadli N, Luengo GS, Rubio RG (2011) Effect of the molecular structure on the adsorption of conditioning polyelectrolytes on solid substrates. Colloids Surf A 375:209–218
204. Kotov NA (1999) Layer-by-layer self-assembly: the contribution of hydrophobic interactions. Nanostr Mat 12:789–796
205. Guzmán E, Ritacco H, Rubio JEF, Rubio RG, Ortega F (2009) Salt-induced changes in the growth of polyelectrolyte layers of poly(diallyldimethylammoniumchloride) and poly(4-styrene sulfonate of sodium). Soft Matter 5:2130–2142. https://doi.org/10.1039/B901193E
206. Fares HM, Schlenoff JB (2017) Equilibrium Overcompensation in Polyelectrolyte Complexes. Macromolecules 50(10):3968–3978

207. Panchagnula V, Jeon J, Dobrynin AV (2004) Molecular dynamics simulations of electrostatic layer-by-layer self-assembly. Phys Rev Lett 93(3):037801. https://doi.org/10.1103/PhysRevLett.93
208. Lösche M, Schmitt J, Decher G, Bouwman WG, Kjaer K (1998) Detailed structure of molecularly thin polyelectrolyte multilayer films on solid substrates as revealed by neutron reflectometry. Macromolecules 31:8893–8906. https://doi.org/10.1021/ma980910p
209. Schmitt J, Griinewald T, Decher G, Pershan PS, Kjaer K, Losche M (1993) Internal structure of layer-by-layer adsorbed polyelectrolyte films: a neutron and X-ray reflectivity study. Macromolecules 26:7058–7063. https://doi.org/10.1021/ma00077a052
210. Rmaile HH, Schlenoff JB (2002) Internal pKa's in polyelectrolyte multilayers: coupling protons and salt. Langmuir 18:8263–8265. https://doi.org/10.1021/la025624s
211. Kharlampieva E, Kozlovskaya V, Chan J, Ankner JF, Tsukruk VV (2009) Spin-assisted layer-by-layer assembly: variation of stratification as studied with neutron reflectivity. Langmuir 25:14,017–14,024. https://doi.org/10.1021/la9014042
212. Holmberg K, Jönsson B, Kronberg B, Lindman B (2002) Surfactants and polymers in aqueous solution. Wiley, Chichester, United Kingdom
213. Cohen-Stuart MA, Hoogendam CW, de Keizer A (1997) Kinetics of polyelectrolyte adsorption. J Phys Cond Matter 9:7767–7783. https://doi.org/10.1088/0953-8984/9/37/009
214. Lane TJ, Fletcher WR, Gormally MV, Johal MS (2008) Dual-beam polarization interferometry resolves mechanistic aspects of polyelectrolyte adsorption. Langmuir 24:10633–10636. https://doi.org/10.1021/la802496h
215. Raposo M, Oliveira ON (2000) Energies of adsorption of poly(o-methoxyaniline) layer-by-layer films. Langmuir 16:2839–2844. https://doi.org/10.1021/la990945y
216. Raposo M, Oliveira ON (1998) Adsorption mechanisms in layer-by-layer films. Braz J Phys 28:392–404. https://doi.org/10.1590/S0103-97331998000400014
217. Raposo M, Pontes RS, Mattoso LHC, Oliveira ON (1997) Kinetics of adsorption of poly(o-methoxyaniline) self-assembled films Macromolecules 30:6095–6101. https://doi.org/10.1021/ma970228f
218. Linse P, Källrot N (2010) Polymer adsorption from bulk solution onto planar surfaces: effect of polymer flexibility and surface attraction in good solvent. Macromolecules 43:2054–2068. https://doi.org/10.1021/ma902338m
219. Kienle DF, Schwartz DK (2019) Complex salt dependence of polymer diffusion in polyelectrolyte multilayers. J Phys Chem Lett 10:987–992. https://doi.org/10.1021/acs.jpclett.9b00004
220. Das BP, Tsianou M (2017) From polyelectrolyte complexes to polyelectrolyte multilayers: electrostatic assembly, nanostructure, dynamics, and functional properties. Adv Colloid Interface Sci 244:71–89. https://doi.org/10.1016/j.cis.2016.12.004
221. Porcel C, Lavalle P, Decher G, Senger B, Voegel JC, Schaaf P (2007) Influence of the polyelectrolyte molecular weight on exponentially growing multilayer films in the linear regime. Langmuir 23:1898–1904. https://doi.org/10.1021/la062728k
222. Nazaran P, Bosio V, Jaeger W, Anghel DF, von Klitzing R (2007) Lateral mobility of polyelectrolyte chains in multilayers. J Phys Chem B 111:8572–8581. https://doi.org/10.1021/jp068768e
223. Korneev D, Lvov Y, Decher G, Schmitt J, Yaradaikin S (1995) Neutron reflectivity analysis of self assembled film superlattices with alternate layers of deuterated and hydrogenated polysterenesulfonate and polyallylamine. Physica B 213:954–956. https://doi.org/10.1016/0921-4526(95)00333-5
224. Soltwedel O, Ivanova O, Nestler P, Müler M, Köhler R, Helm CA (2010) Interdiffusion in polyelectrolyte multilayers. Macromolecules 43:7288–7293. https://doi.org/10.1021/ma101279q
225. Selin V, Ankner JF, Sukhishvili SA (2017) Nonlinear layer-by-layer films: effects of chain diffusivity on film structure and swelling. Macromolecules 50:6192–6201. https://doi.org/10.1021/acs.macromol.7b01218

Chapter 7
Supramolecular Assemblies Based on σ-hole Interactions

Antonio Bauzá and Antonio Frontera

Abstract Elements belonging to Groups 14–17 and Periods 3–6 frequently act as Lewis acids which are able to establish directional noncovalent interactions (NCI) with a variety of Lewis bases (lone pair donors), π-systems (aromatic rings, triple and double bonds) and non-nucleophilic anions (BF_4^-, PF_6^-, ClO_4^-, etc.). These promising NCIs are named in general as σ-hole interactions that are subdivided as tetrel bonds for elements belonging to group 14, pnictogen bonding for group 15, chalcogen bonding for group 16, and halogen bonding for group 17. In general, σ-hole interactions offer differentiating features when moving down in the same group (larger and more positive σ-holes) or moving left in the same row (number of available σ-holes and directionality) of the periodic table. This chapter shows that Molecular Electrostatic Potential (MEP) surface calculation is a powerful tool to explain the solid-state architecture of many X-ray structures. This is exemplified by using many examples retrieved from the Cambridge Structural Database (CSD), especially focused on σ-hole interactions.

7.1 Introduction

Atoms of Groups (Gs.) 13–18 bonded to electron-withdrawing groups (EWG) form favorable supramolecular complexes with electron-rich atoms, anions, and π-systems [1–13]. The electrophilic atom was used to designate the ubiquitous hydrogen bonding, consequently, the scientific community is employing nowadays the name of the group of the Periodic Table (PT) the electrophilic atom belongs to refer to donor–acceptor noncovalent interactions (NCIs) [14, 15]. Actually, the "International Union of Pure and Applied Chemistry (IUPAC)" has defined the terms "halogen bond (HaB)" for NCIs of G–17 [16] and "chalcogen bond (ChB)" [17] for G. 18. Moreover, other terms are commonly used for other groups of the p-block, which are

A. Bauzá · A. Frontera (✉)
Universitat de Les Illes Balears, Crta de Valldemossa km 7.5, 07122 Palma de Mallorca (Baleares), Spain
e-mail: toni.frontera@uib.es

M. A. Aboudzadeh and A. Frontera (eds.), *Supramolecular Assemblies Based on Electrostatic Interactions*,
https://doi.org/10.1007/978-3-031-00657-9_7

"aerogen or noble gas bonding" (NgB, G–18) [12], "pnictogen bonding" (PnB, G–15)" [18, 19], "tetrel bonding" (TtB, G–14) [20], and "triel bonding" (TrB, G–13) [7]. "Spodium bonding" (SpB, G–12) [21–23] and "coinage or regium bonding" (CiB, G–11) are also utilized to name to NCIs where post-transition atoms act as electrophiles and to discriminate this type of NCIs from classical coordination bonds [24, 25]. Most recently, "matere" and "osme" bonding terms have been proposed to name interactions involving groups 7 [26] and 8 [27], respectively. Several investigations have shown that this family of NCIs can be used as a real alternative to HB in several fields related to supramolecular chemistry. Excellent book chapters and review articles describe interesting applications and fundamental aspects of this class of NCIs [28–45]. Furthermore, interesting investigations have compared the σ-hole interactions with the ubiquitous HB [46–51] in terms of geometric and energetic features.

The attractive interaction (EWG–X$\cdots$A) between the p-block element (Lewis acid, X) and σ-hole acceptor (A, Lewis base, anion, π-system) is the sum of different contributions (see Fig. 7.1 for a representation of a prototypical $CF_3I\cdots NH_3$ complex): electrostatics, charge transfer, orbital mixing, polarization and dispersion forces [52, 53]. Typically, the electrostatic term is rationalized by the attraction between the electron-rich nucleophile (A) and the region of positive electrostatic potential located at the prolongation of the EWG–X bond, known as σ-hole. Dispersion and polarization terms are important contributors in those NCIs involving the heavier elements of the groups, which are easily polarized [45, 54–58]. The orbital term is rationalized in σ-hole NCIs as the global stabilization of the system due to the electron donation from the electron-rich atom or group of atoms to the antibonding EWG–X orbital [LP(A) or $\pi(A) \rightarrow \sigma^*$(EWG–Y)]. That is, the lone pair (LP) or π-type orbitals (double/triple bond, aromatic ring, etc.) belonging to A interact with the antibonding σ^* orbital of the EWG–X bond. As commented above, the polarizability of the atoms increases by descending the group of the periodic table. The positive potential of the σ-hole consequently increases if the EWG–X bond is more polarized and the strength of the NCI increases. Another way to polarize the EWG–X bond

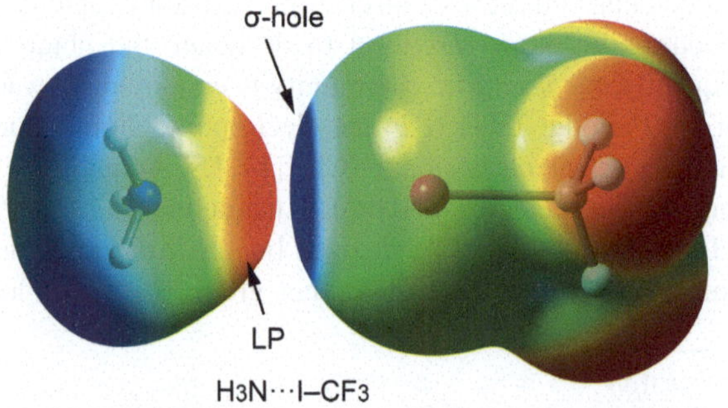

Fig. 7.1 Supramolecular complex of NH_3 and CF_3I as σ-hole acceptor and donor, respectively. Color code: blue positive and red negative

is by increasing the electron-withdrawing ability of EWG. Therefore, the combination of heavy elements and strong EWG augments the positive potential and size of the σ-hole and the NCI strengthens (the electrostatic term increases). Moreover, the utilization of heavier elements in combination with strong EWGs decreases the energy level of the σ*(EWG–X) orbitals, thus favoring the NCI by increasing the stabilization via LP(A) or $\pi(A) \rightarrow \sigma^*$(EWG–Y) donation [53] and also by increasing the contribution of the favorable dispersion term [59, 60].

Studies of NCIs [61, 62] involving elements of G–14 to G–17 are growing very fast [14, 15, 63]. The distribution of electron density is anisotropic in halogen, chalcogen, and pnictogen atoms due to the concurrence of σ-lumps and σ–holes on the same element as shown by several investigations [36–46, 61, 62]. In monovalent G–17 (one covalent bond), divalent G–16 (two covalent bonds), trivalent G–15 (three covalent bonds), and tetravalent G–14 (four covalent bonds) elements, the number of σ-lumps (LPs) goes from three to none and the number of σ-holes from one to four (ongoing from G–17 to G–14).

In this chapter, MEP surfaces and X-ray structures retrieved from the CSD are used to (i) illustrate the ability of p-block elements to participate in σ-hole interactions; (ii) the utility of MEP surfaces to rationalize most of the solid-state architectures observed in the X-ray structures and (iii) the superior ability of heavier elements to establish structure-directing σ-hole NCIs.

7.2 Results and Discussion

7.2.1 Tetrel Bonding

Table 7.1 summarizes the polarizabilities (α) and van der Waals (vdR) radii of the tetrel elements from periods 2 to 5. The atomic α value increases from 9.0 a.u. in C to 37.3 a.u. in Sn. Remarkably, the difference in the atomic polarizability between C and Si is quite large (~ ×3) and conversely, it is small between Si and Ge and modest between Ge and Sn. The molecular electrostatic potential values at the σ-holes of the fluoride derivatives of the four Tt elements studied in this section are also given

Table 7.1 Atomic polarizabilities (α, a.u.) of tetrel (Tt) elements, van der Waals radii (in Å), and σ-hole MEP values (in kcal/mol) of their tetrafluoride derivatives

Tt	α	R_{vdW}	MEP (TtF_4)
C	9.0	1.70	18.6
Si	26.1	2.10	39.0
Ge	28.4	2.11	50.2
Sn	37.3	2.17	66.5

Adapted from Ref. [1]

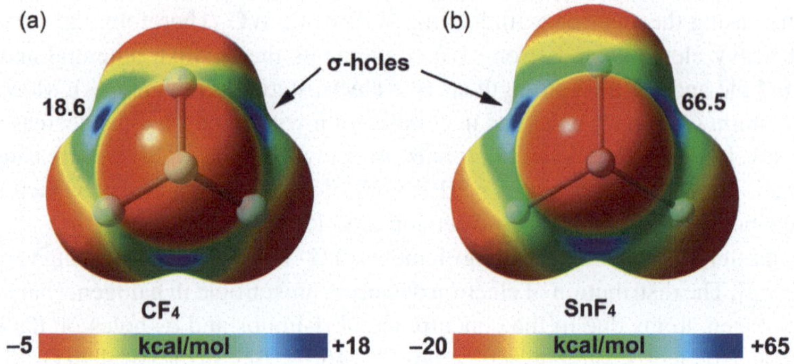

Fig. 7.2 MEP surfaces of CF_4 (**a**) and SnF_4 (**b**) using the 0.001 a.u. isosurface

in Table 7.1 and the MEP surfaces of two of them are represented in Fig. 7.2. The energetic data evidence that the MEP values increase when descending in the group, parallel to the behavior observed for the atomic polarizability.

Figure 7.2 shows the MEP surfaces of CF_4 and SnF_4 as illustrative examples of the whole series. Both present similar distribution of the electron-rich and poor regions and they show that the σ-holes are small, thus anticipating a strong directionality of the TtB interaction. As aforementioned, the MEP value measured at the σ-hole is significantly larger in SnF_4 than in CF_4. Moreover, the MEP values at the F-atoms are also more negative in SnF_4 than in CF_4 (see MEP scale), thus confirming that the Sn–F is significantly more polarized than the C–F bond.

G–14 interactions are quite different compared to G15–G17 interactions in terms of steric requirements. The approaching Lewis Base in an HaB complex is opposite to the EWG that is bonded to the G–17 elements. Consequently, the bulkiness of the EWG does not affect the halogen bonding NCI. However, steric effects intensify on going from G–16 to G–14 NCIs, being the situation of ChBs rather similar to HaBs. For PnBs the Lewis base is able to approach the G–15 elements without feeling substantial repulsion. In G–14 σ-hole NCIs, the tetrahedral arrangement of the four groups increases the steric demands that are more critical in the lighter elements of the group [63].

7.2.1.1 Carbon

The behavior of the heavier elements of G–14 functioning as Lewis acids has been investigated and reported for decades [20, 36, 64–67]. However, investigations dedicated to study the carbon atom [68] are much more recent, in spite of it is the most abundant Tt atom. Some recent investigations have evidenced that 1,1,2,2-tetracyanocyclopropane [69–71] is a convenient σ-hole donor (both σ-holes merge in the same spatial region) that establishes strong TtBs with electron-rich entities. In fact, the intense σ-hole located in the middle of the $(CN)_2C–C(CN)_2$ bond has been used in crystal engineering to construct highly ordered co-crystals. Figure 7.3

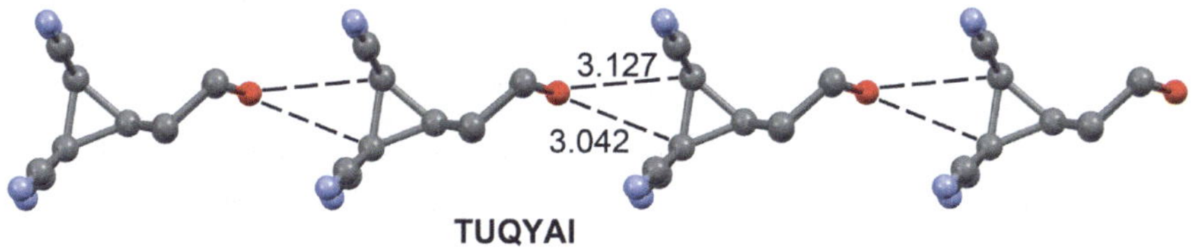

Fig. 7.3 TtBs in the solid-state of CSD refcode "TUQYAI" (represented in ball and stick). Distances in Å. H-atoms omitted

shows the X-ray packing of "TUQYAI" [72] that was designed to combine the σ-hole donor and acceptor within the same molecule. The packing of this molecule forms 1D infinite chains in the solid-state that propagate via C··· O TtB interactions.

Moreover, Mani and Arunan [73] have described the ability of sp^3 hybridized carbon atoms in methyl group or aliphatic chains to participate in σ-hole interactions leading to noncovalent "carbon bonding". The term "carbon bonding" is not recommended since it is widely used in covalent chemistry and it should be replaced by tetrel bonding following the nomenclature recommended by the IUPAC for halogen and chalcogen bonds [16, 17].

The importance of TtB as structure-guiding force in crystalline solids where the σ-hole donor is an sp^3 hybridized carbon atom has been recently reviewed [2]. The functional groups that are well-suited to form TtBs are basically methyl and methylene groups bound to cations: ammonium, pyridinium, and sulfonium. These strong EWGs foster quite short and directional TtBs. In a lesser extent, neutral EWGs like fluorine, nitro, and cyano substituents are also able to promote the formation of TtBs in the solid-state. Interestingly, it has been demonstrated that this type of TtBs involving sp^3-hybridized carbon atom is also relevant in enzymatic chemistry, in particular the inhibition of serine proteases and antagonists of the muscarinic acetylcholine receptor [74], as represented in Fig. 7.4.

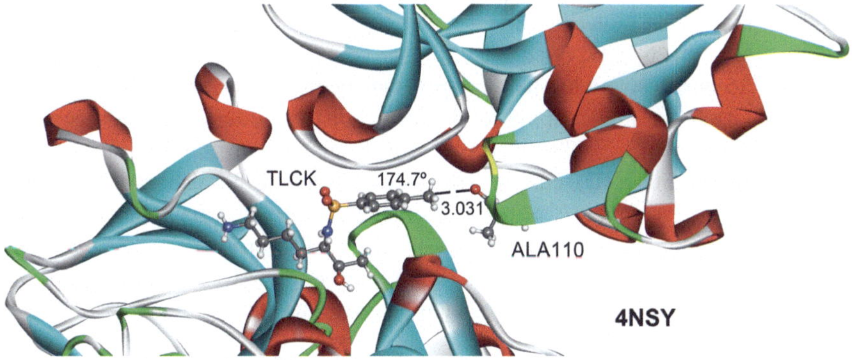

Fig. 7.4 Partial view of the X-ray structure of 4NSY with indication of the carbon TtB (distance in Å) between the alanine 110 and the TLCK ligand, both represented in ball and stick

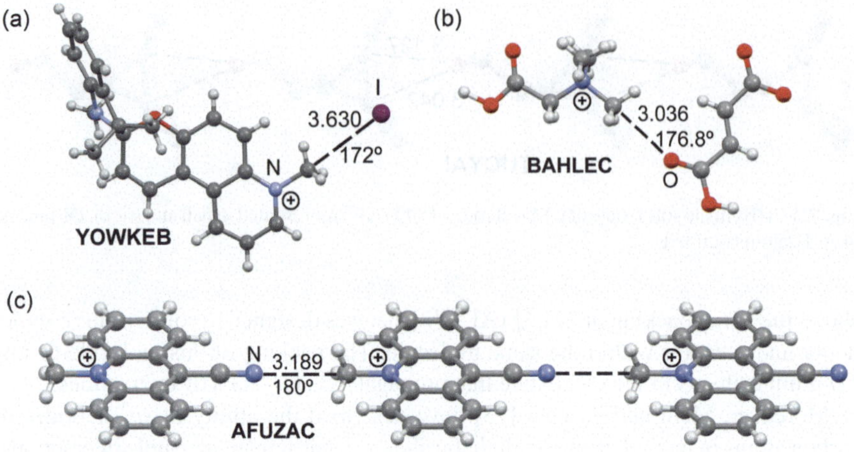

Fig. 7.5 TtBs in the solid-state of CSD refcodes "YOWKEB" (**a**), "BAHLEC" (**b**), and "AFUZAC" (**c**). Distances in Å

X-ray solid-state structures of several exemplifying compounds are shown in Fig. 7.5 where the TtBs involving carbon are relevant. The interactions are very directional (angles very close to linearity) thus ruling out the formation of H-bonds. The ammonium cation highly polarizes the C–N bond, thus promoting the formation of the charge-assisted TtBs. For instance, the methyl-quinolinium moiety in "YOWKEB" [75] interacts with the iodide counter-anion by means of a short and directional TtB (172°), see Fig. 7.5a. The methyl group of *N*,*N*,*N*-trimethyl glycine fumarate [76] establishes a quite short and directional N–C···O TtB (Fig. 7.5b) with a C···O separation of 3.036 Å and N–C···O angle of 176.8° ("BAHLEC" structure). Finally, the 9-cyano-10-methylacridinium [77] in "AFUZAC" (Fig. 7.5c) propagates in the solid-state X-ray structure forming infinite 1D supramolecular assemblies where the methyl group establishes a TtB with the N-atom of the cyano group of the adjacent molecule (3.189 Å, see Fig. 7.5c) that is located exactly at the prolongation of the N–CH_3 bond.

Figure 7.6 shows two additional examples with neutral instead of charged systems. Directional TtBs are crucial in the solid-state of [bis(trifluoroacetoxy)iodo]benzene (refcode "CEZBEO01") [78] directing the formation of infinite supramolecular one-dimensional (1D) chains. The second example corresponds to chlorotricyanomethane ("CTCYME") [79] where three directional TtBs guide the formation of a trimer with the lone pair donor atoms located opposite the C–Cl bonds (see Fig. 7.6b). Curiously, the N-atoms are located opposite the Cl atom, instead of the more EWG cyano group. This unexpected behavior has been previously described in chalcogen bonding [80].

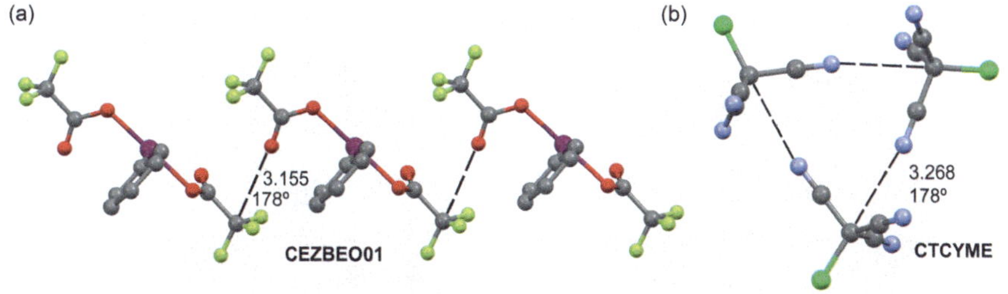

Fig. 7.6 TtBs in the solid-state of CSD refcodes "CEZBEO01" (**a**) and "CTCYME" (**b**). Distances in Å. H-atoms omitted

7.2.1.2 Silicon

Silicon is probably more prone to participate in noncovalent TtBs than the heavier tetrels. This is due to the fact that tetravalent Ge and Sn can easily expand their valence, thus engaging in covalent/coordination chemistry. Moreover, tetrahedral lead(IV) compounds are quite uncommon and lead(II) has a rich coordination chemistry [81–84] and is considered a metal. Nonetheless, hypervalent species of Si are also known [85–95] but are less abundant. NCIs involving silicon were originally studied by Alkorta et al. [96] firstly in tetrahalosilanes theoretically and secondly in aminopropylsilanes [97] combining theory and experiment. The energetic and geometric features of TtBs in neutral and protonated $RTtF_3$ (R = pyridinyl and furanyl) systems have been analyzed by Scheiner et al. [98] for C, Si, and Ge, using NH_3 as LP donor.

Figure 7.7 depicts two X-ray structures where relevant TtBs interactions involving anions (chloride and fluoride) are established and the Si-atom has not expanded valence. In "QOMBID" [99] structure (Fig. 7.7a), the bis(1-((1,1-dimethyl-2-benzoylhydrazonio)methyl-C,O)-methyl)disiloxane molecule is able to recognize chloride anions by means of two short Si···Cl TtBs contacts. It is interesting that

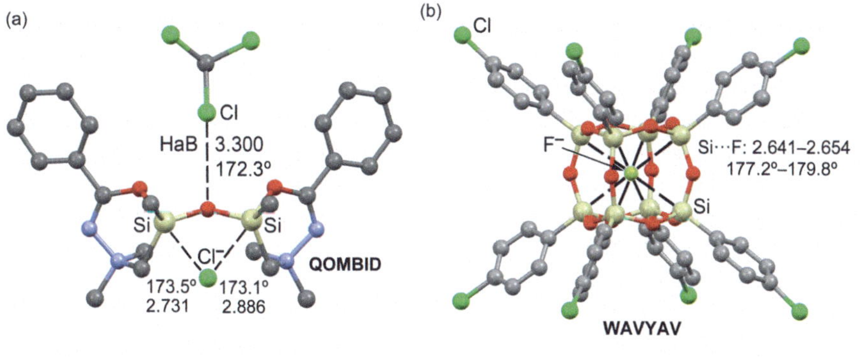

Fig. 7.7 TtBs in the solid-state of CSD refcodes "QOMBID" (**a**) and "WAVYAV" (**b**). Distances in Å. H-atoms omitted

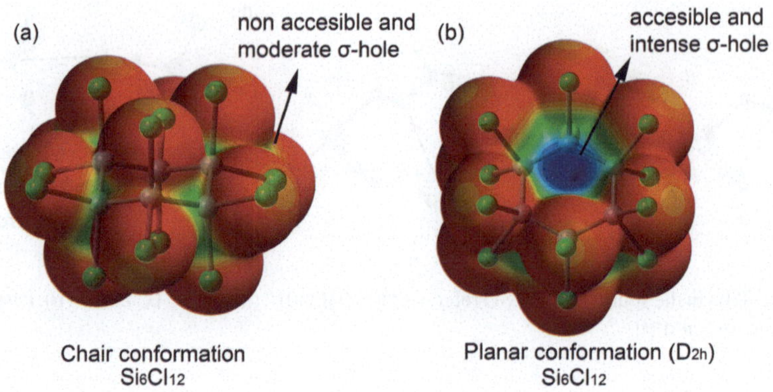

Fig. 7.8 MEP surface (0.001 a.u.) of dodecachlorohexasilinane in a chair conformation. (d) MEP surface (0.001 a.u.) of dodecachlorohexasilinane in a planar conformation (D_{6h}). Color code for MEP surfaces: red negative and blue positive

the bridging O-atom also establishes a directional Cl· · · O HaB with a co-crystallized solvent molecule ($CHCl_3$). Both σ-hole interactions likely influence each other leading to a mutual reinforcement. That is, the formation of the HaB enhances the σ-hole at the Si-atom thus reinforcing the TtB. Moreover, the formation of the TtB increases the nucleophilicity of the O-atom, thus reinforcing the HaB. The other selected structure is a spherosilicate ("WAVYAV", see Fig. 7.7b) where interesting TtBs have been described. Experimentally, it has been observed that the utilization of fluoride salts significantly increases the reaction yield in the synthesis of spherosilicates [100]. In fact, a fluoride anion is incorporated in the X-ray structure, perfectly centered within the octasiloxane cage, where eight Si· · · F TtBs [101] are established.

As aforementioned, steric effect is likely the major drawback for the utilization of TtBs as an alternative to other interactions like HB or HaB for the design and synthesis of functional systems. In Fig. 7.8a the MEP surface of dodecachlorohexasilinane in the chair-like confirmation is shown. This conformation corresponds to the minimum energy structure and evidence that the σ-holes are not accessible, and the MEP values are very small (green color). However, if a planar the D_{6h} conformation is imposed, two intense and accessible positive regions appear above and below the Si_6 ring (see Fig. 7.8b) because the σ-holes of the Si-atoms merge in the same region.

Figure 7.9a shows the X-ray solid-state structure of Si_6Cl_{12} that presents the expected chair-like confirmation in line with the theoretical calculations (refcode "AZAZUX") [102]. Notably, the addition of chloride or iodide anions induces planarity in the dodecachlorohexasilinane ring, as shown in Fig. 7.9b, c, refcodes "LECXIC" [103] and "AZEBAJ" [102]. Parallel behavior is observed in dodecabromohexasilinane that becomes planar upon addition of bromide anion (see Fig. 7.9d) refcode "AZEBEN" [103]. The unexpected formation of these "inverse sandwich" complexes [103, 104] upon addition of two equivalents of halide anions is due to the concurrent formation of six TtBs. In fact, the formation of these TtBs strongly agrees with the MEP surface shown in Fig. 7.8b and clearly compensates for the

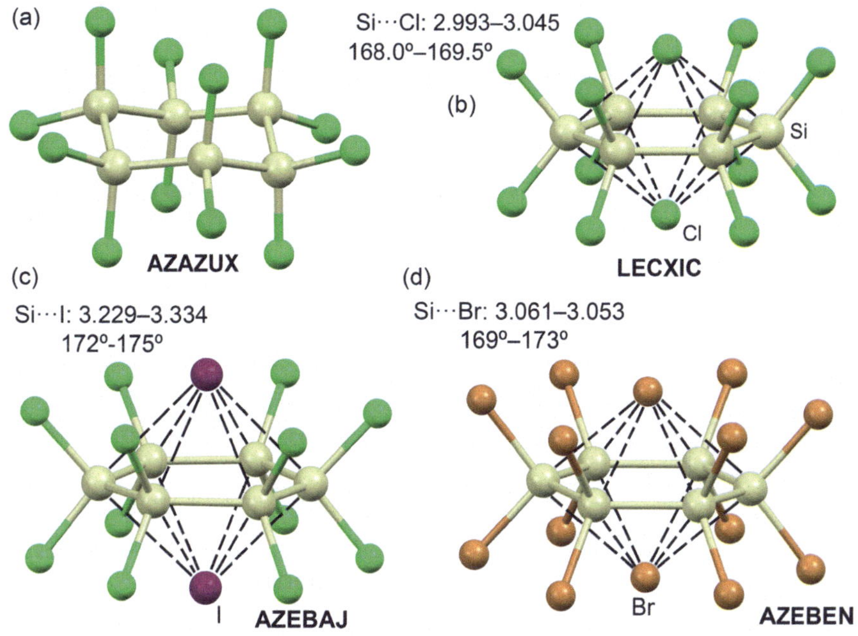

Fig. 7.9 **a** X-ray structure of dodecachlorohexasilinane (refcode "AZAZUX"). **b** TtBs in the solid-state of CSD refcode "LECXIC". **c** TtBs in the solid-state of CSD refcode "AZEBAJ". **d** TtBs in the solid-state of CSD refcode "AZEBEN". Distances in Å

energy difference between the chair and planar conformation of the six-membered dodecachlorohexasilinane or dodecabromohexasilinane rings. The TtB distances are longer and more directional (172°–175°) for the iodide anion ("AZEBAJ" structure) compared to the bromide ("AZEBEN") and chloride ("LECXIC") derivatives.

7.2.1.3 Germanium and Tin

The investigation on TtBs [2, 20, 96, 97] is concentrated in the lighter elements and those of the heavier elements are rare [105]. However, two reviews have been published gathering much structural information demonstrating the tendency of Ge, Sn to form very directional NCIs [106, 107] with a structural guiding role in crystal structures of Ge and Sn. These reviews evidence that G–14 NCIs are more directional than G–15 and G–16 NCIs and equivalent to G–17 NCIs.

Figure 7.10 shows two X-ray structures showing directional Ge··· O and Ge··· F σ-hole G–14 NCIs. The first one propagates in the solid-state forming infinite 1D assemblies via directional σ-hole bonds (refcode "QAHXIG") [108] exhibiting a TtB distance of 3.520 Å and strong linearity (179.8°). The X-ray solid-state structure of "VUZVUH" [109] is represented in Fig. 7.10b, showing the cuboidal $Ge_8O_{12}(OH)_8$

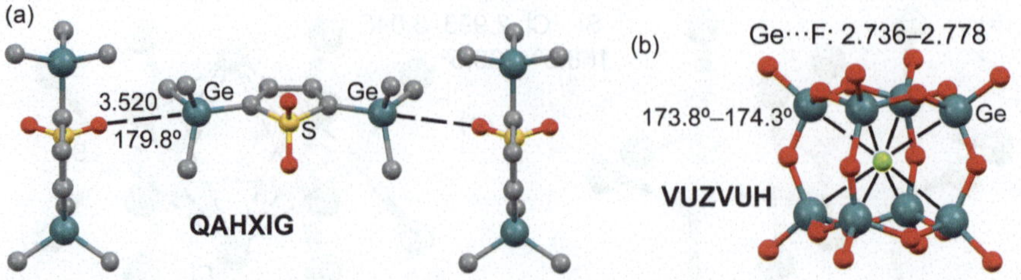

Fig. 7.10 TtBs in CSD refcodes "QAHXIG" (**a**) and "VUZVUH" (**b**). Distances in Å. H-atoms omitted. Adapted from Ref. [1]

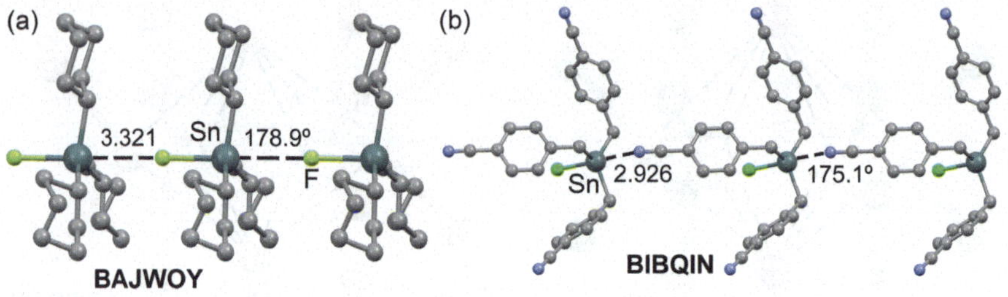

Fig. 7.11 TtBs in CSD refcodes "BAJWOY" (**a**) and BIBQIN (**b**). Distances in Å. H-atoms omitted. Adapted from Ref. [1]

moiety. Interestingly, the fluoride anion is encapsulated in the center, similar to the silicon cuboid structure commented in Fig. 7.7b (refcode "WAVYAV").

For Tin, two X-ray structures have been selected characterized by the formation of infinite 1D assemblies. The σ-lump donor atom approaches the tin element opposite to the Sn–F bond in "BAJWOY" [110] and Sn–Cl in "BIBQIN" [111]. In the latter, the Sn element establishes a quite short Sn$\cdots$N σ-hole tetrel bond taking into consideration the low basicity of the sp-hybridized N-atom. Similarly, the fluorine atom in "BAJWOY" (Fig. 7.11a) forms a short σ-hole TtB opposite to the covalent Sn–F bond, thus proliferating the 1D assembly, in spite of the poor basicity of fluorine. The short σ-hole bonds involving tin are due to its strong Lewis acidity, in line with the MEP results (see Fig. 7.2).

7.2.2 Pnictogen Bonding

Many works describing G–15 NCIs theoretically are available in the literature [43, 44, 112–119], but those describing experimental evidence are less frequent [120–123]. It has been recently used in supramolecular catalysis [124] and crystal engineering [43, 44, 125]. Some examples are highlighted in this section.

Table 7.2 gathers the atomic polarizabilities (α) and vdW radii of G–15 elements. The polarizability augments from 5.3 a.u. in nitrogen to 30.8 a.u. in antimony. As

Table 7.2 Atomic polarizabilities (α, a.u.) of pnictogen (Pn)elements, van der Waals radii (R_{vdW}, Å), and σ-hole MEP values (kcal/mol) of their tetrafluoride derivatives

Pn	α	R_{vdW}	MEP (PnF_3)
N	5.3	1.55	15.9
P	16.9	1.80	27.4
As	21.6	1.85	38.5
Sb	30.8	2.06	46.7

Adapted from Ref. [1]

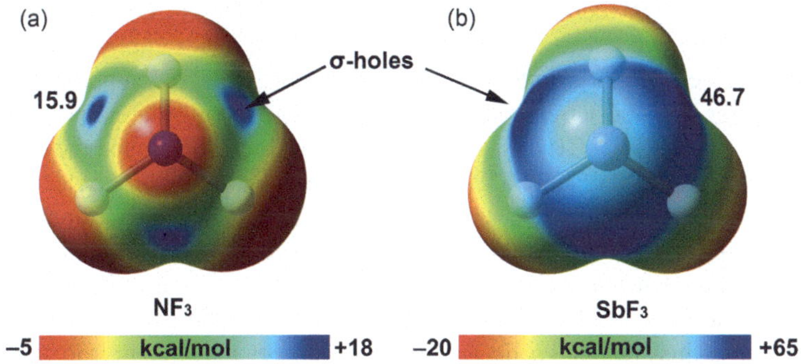

Fig. 7.12 MEP surfaces of NF_3 (**a**) and SbF_3 (**b**) using the 0.001 a.u. isosurface

described in the tetrel atoms, the α difference between the elements of periods 2 and 3 is large (~ ×3) and small elements of periods 3 and 4. The difference is again important between As and Sb. Compared to tetrel atoms, the pnictogen atoms are less polarizable and exhibit smaller van der Waals radii. The molecular electrostatic potential values at the σ-holes of the fluoride derivatives of the four Pn elements studied in this section are also given in Table 7.2 and two of them are represented in Fig. 7.12. The energetic data evidence that the MEP values at the σ-hole become more positive when descending in the group, parallel to the behavior observed for the tetrel atoms (see Sect. 7.2.1).

MEP surfaces of NF_3 and SnF_4 are shown in Fig. 7.12 as examples of the whole series. The NF_3 molecule presents well-defined σ-holes opposite to the NF covalent bonds. In contrast, the SbF_3 molecule presents a wide region of positive potential where the global maximum is located approximately opposite to the Sb–F bonds. Therefore, it is expected that PnBs involving the heavier Pn-atoms are less directional than TtBs. The MEP value is significantly larger in SbF_3 than in NF_3. As also observed for the TtF_4 molecules, the MEP values at the F-atoms are more negative in the SbF_3 than in NF_3, thus revealing that the Sb–F is significantly more polarized than the N–F bond. The MEP values at the σ-holes are greater for tetrel atoms than for pnictogen ones, in agreement with the atomic polarizabilities.

7.2.2.1 Nitrogen

As far as our knowledge extends, there is no experimental evidence of σ-hole interactions involving nitrogen acting as Lewis acid (σ-hole donor). In sharp contrast, there are numerous examples where the N-atom participates in π-hole interactions, particularly in nitro-derivatives (R-NO_2) [126–128] and azide metal complexes (M-N_3) [129]. Nitrate anions can even act as π-holes donors in crystal structures when the negative charge is partially transferred to adjacent molecules by means of noncovalent interactions such as hydrogen bonding [130] or coordination bonds [131]. Moreover, rotational spectroscopy experiments revealed that π-hole bonding geometry prevails in the gas phase for the complex between Me_3N and $MeNO_2$ [132].

In this section, we have explored the CSD to search for examples of pnictogen bonding considering charged species, and in particular R–N^+H_3 compounds. Similarly to charge-assisted tetrel bonding interactions involving the carbon atom in complexes of type R_3N^+–$CH_3\cdots$A (see Fig. 7.5), it can be envisaged the existence of pnictogen bonding of type R–$N^+H_3\cdots$A. In this type of complex, the electron-rich atom must approach the N along with the R–N bond, thus preventing the formation of H-bonds. In Fig. 7.13, a prototypical example is shown where the ammonia as Lewis base approaches methylammonium cation, establishing a PnB, as shown by the QTAIM analysis [133] of bond critical points (red spheres). Interestingly, no bond critical points and bond paths are obtained between the N-atom from ammonia and the H-atoms of the ammonium cation, thus suggesting the true PnB nature of this interaction. The interaction energy is very large (–18.9 kcal/mol) due to the cationic nature of the pnictogen bond donor atom.

X-ray solid-state structures of three exemplifying compounds are shown in Fig. 7.14 where the C–$N^+H_3\cdots$O PnBs are important. The interactions are very directional (angles very close to linearity, C–N$\cdots$O > 176°) thus preventing the formation of H-bonds. The 2-carboxyanilinium moiety in DISGEU [134] forms self-assembled dimers where two symmetrically equivalent PnBs are established (see Fig. 7.14a). The O-atoms of the carboxy groups approximate the RN^+H_3 groups by means of very short (2.738 Å) and directional (176.5°) C–N$\cdots$O PnBs. In the ENOBUF [135] structure, the benzylammonium cation and the adjacent benzoate form a highly directional PnB (Fig. 7.14b) with a N$\cdots$O separation of 3.023 Å

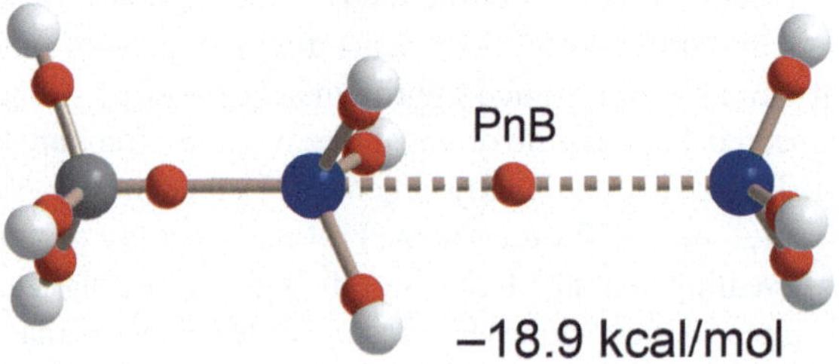

Fig. 7.13 QTAIM analysis of bond critical points and bond path in the prototypical $CH_3N^+H_3\cdots NH_3$ σ-hole complex

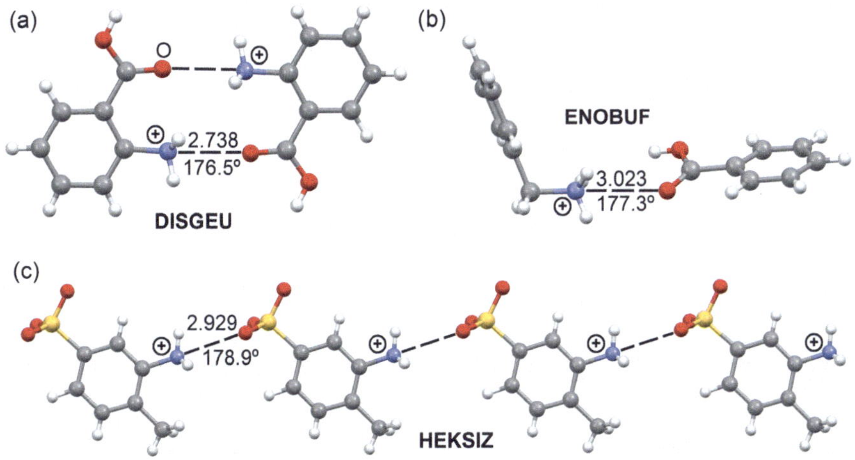

Fig. 7.14 PnBs in the solid-state of CSD refcodes DISGEU (**a**), ENOBUF (**b**) and HEKSIX (**c**). Distances in Å

and N–C· · · O angle of 177.3° The third example corresponds the zwitterionic form of 2-aminotoluene-4-sulfonic acid (refcode HEZSIZ) [136] that propagates in the solid-state X-ray structure forming infinite 1D supramolecular assemblies where the ammonium group establishes a PnB with the O-atom of the sulfonate group of the adjacent molecule (3.929 Å, see Fig. 7.14c) that is located at the prolongation of the C–NH_3 bond (178.9°).

7.2.2.2 Phosphorus

The ability of phosphorous to participate in PnBs has been studied theoretically [44] and experimentally [122, 137, 138]. Recently, it has been demonstrated the formation of P· · · O donor–acceptor PnBs in PCl_3-CH_3OH adducts using matrix isolation infrared spectroscopy [122]. Similarly, experimental results have proved the formation of P· · · O and P· · · π PnBs in PCl_3-H_2O and PCl_3-C_6H_6 systems, where PnBs dominate over conventional hydrogen bonds [137, 138].

There are many works in the literature where PnBs are recognized as important interactions in the solid-state of arsenic [139–141], antimony [142–146], or bismuth [147–155] compounds. However, those involving phosphorus are quite uncommon [156]. Figure 7.15 shows four X-ray structures involving tertiary amines and PBr_3 where highly directional PnBs are formed [157]. For some amines like trimethylamine (“MEWXOB”) or tetramethylethylenediamine (“MEWXUH” and “MEWYAO”), one or two Br atoms are substituted by the nucleophilic N-atoms (covalent P–N bonds are formed) while the bromide leaving group remains close to the P-atom establishing P· · · Br–PnBs (see Fig. 7.15). In contrast, in the X-ray co-crystal of 1,4-dimethyl pyperazine and PBr_3 (“MEWYES”), the substitution is not

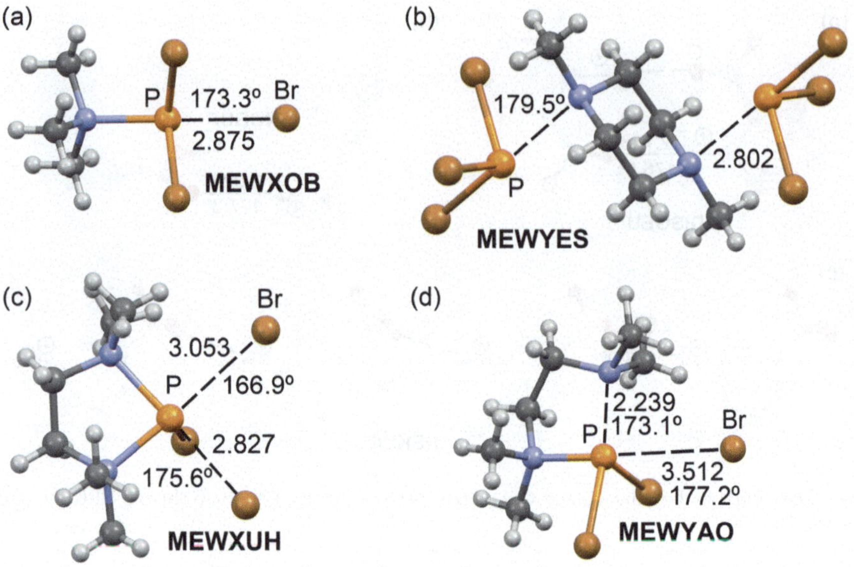

Fig. 7.15 PnBs in the solid-state of CSD refcodes "MEWXOB" (**a**), "MEWYES" (**b**) "MEWXUH" (**c**) and "MEWYAO" (**d**). Distances in Å

observed and trimeric adducts are formed where two very directional P···N bonds (179.5°) are formed with the axial lone pairs pointing exactly to the elongation of the Br–P bonds [157].

An additional example of a solid-state structure governed by PnBs is given in Fig. 7.16. In the "ICOJAO" structure [158], two different P···O PnBs involving two O-atoms from trifluoromethanesulfonate anion as donor

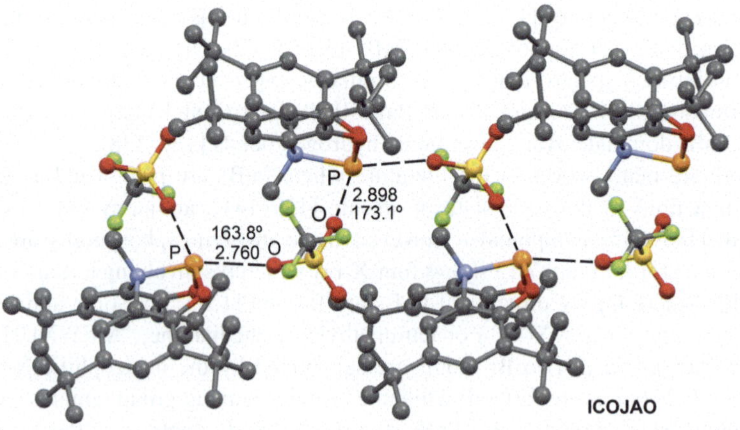

Fig. 7.16 PnBs in the solid-state of CSD refcode "ICOJAO". Distances in Å. H-atoms omitted

are observed. These interconnect two molecules of 2,4,8,10-tetra-t-butyl-12-methyl[1,3,2]benzoxazaphospholo[2,3-b][1,3,2]benzoxazaphosphol-12-ium cations. The structure forms a supramolecular polymer (Fig. 7.16) guided by these PnBs, where the shortest one (2.760 Å) corresponds to the electrostatically enhanced N^+–P$\cdots$O contact involving the most polarized bond. The other one O–P$\cdots$O is longer but more directional (173.1°).

7.2.2.3 Arsenic and Antimony

As aforesaid, As and Sb form stronger σ-hole PnBs than nitrogen and phosphorous. In fact, it has recently shown its utility in catalysis [159] and molecular recognition of anions [160]. Sb is the most promising element to construct supramolecular assemblies based on PnBs and receptors for the trapping of anions [161, 162].

Figure 7.17 shows the packing $As(N_3)_3$ (ICSD refcode "413360") and $Sb(N_3)_3$ (ICSD refcode "413359") structures [163], to exemplify the ability of trivalent arsenic and antimony to form concurrent σ-hole G–15 NCIs. In the crystal packing, each $Pn(N_3)$ molecule establishes three PnBs, thus determining their solid-state architecture. The three σ-hole interactions are symmetrically equivalent in the antimony derivative.

$As(CN)_3$ molecule in the solid-state establishes three concurrent PnBs that determine its crystal packing, as shown in Fig. 7.18a, refcode "USEPUF" [164]. The As$\cdots$N distances are shorter than the sum of van der Waals radii and the angles characterizing the PnBs range from 161° to 162°. In "UROZIL" [165] structure (Fig. 7.18b) the As atom establishes three contacts with the O-atoms of the phosphine oxide groups with PnB angles ranging 167° to 177°. In "UROYUW" [165] structure (Fig. 7.18b) the SbF_3 units interact with two electron-rich O-atoms (PnB contacts) generating a supramolecular ring. In case of "USEQAM" structure [164] the $Sb(CN)_3$ and 2,2′-bipyridine form a co-crystal (Fig. 7.18d). Two of the σ-hole

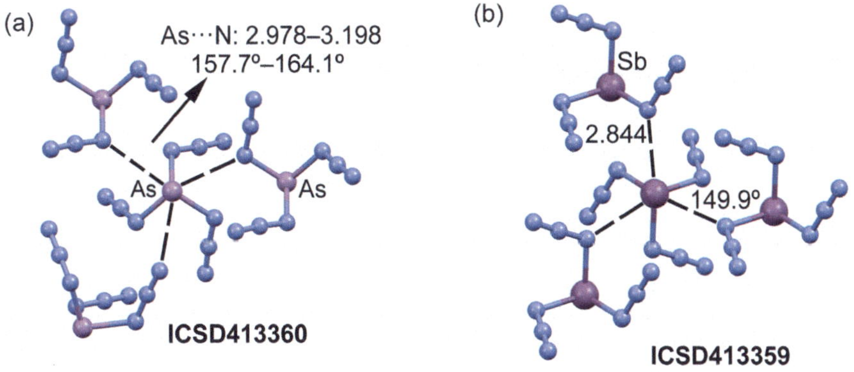

Fig. 7.17 PnBs in the solid-state of ICSD refcodes 413,360 (**a**) and 413,359 (**b**). Distances in Å. Adapted from Ref. [1]

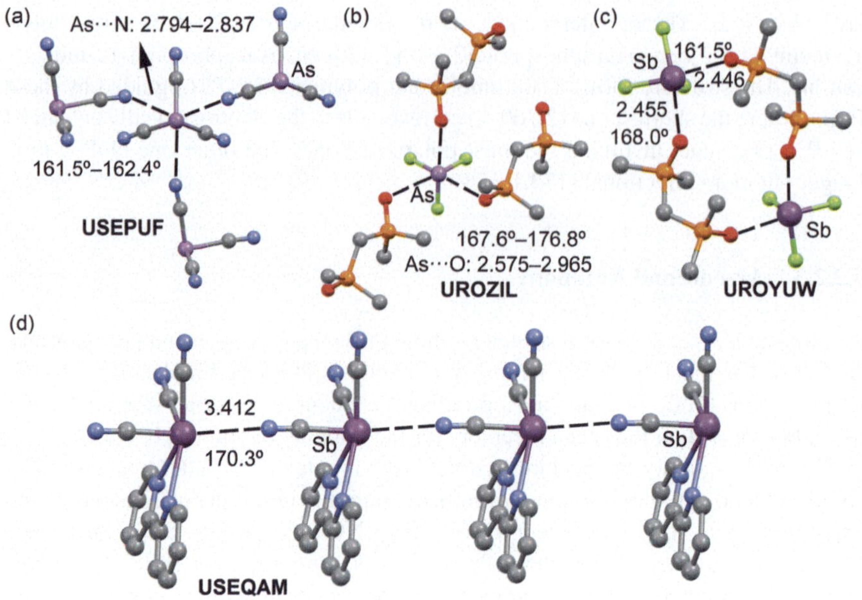

Fig. 7.18 PnBs in the solid-state of CSD refcodes "USEPUF" (**a**), "UROZIL" (**b**), "UROYUW" (**c**), and "USEQAM" (**d**). Distances in Å. Adapted from Ref. [1]

contacts are very short (with bipyridine) and the other one is longer (established with the N-atom of the cyano group) in line with the low basicity of the latter. These PnBs govern the X-ray packing of the co-crystal.

The angles observed for the PnBs NCIs shown in Figs. 7.17 and 7.18 are less linear than those described for G–14 NCIs (see previous section) in agreement with the MEP analysis (see Figs. 7.2 and 7.12) evidencing more intense and accessible σ-holes in the Pn-derivatives.

7.2.3 Chalcogen Bonding

A significant advancement has been accomplished in the field of ChB applications, especially in supramolecular chemistry, crystal engineering, and supramolecular catalysis [3, 9, 10, 42].

Table 7.3 summarizes the atomic α values and vdW radii of the G–16 atoms from periods 2 to 5. The polarizability of the Ch-atoms value augments from 3.0 a.u. in oxygen atom to 25.9 a.u. in tellurium atom. Remarkably, the increment in the α value on going from oxygen atom to sulfur atom is very significant (~ ×4), similarly to previous observations for G–14 and G–15 elements. The increments are smaller on going from S to Te. The maximum (σ-hole) molecular electrostatic potential values are indicated in Table 7.3 and two of them are depicted in Fig. 7.19. The MEP values

Table 7.3 Atomic polarizabilities (α, a.u.) of G–16 atoms, vdW radii (R_{vdW}, Å), and σ-hole MEP values (kcal/mol) of ChF_2

Ch	α	R_{vdW}	MEP (ChF_2)
O	3.0	1.52	16.8
S	11.8	1.80	35.6
Se	17.5	1.90	44.9
Te	25.9	2.06	52.6

Values are taken from Ref. [1]

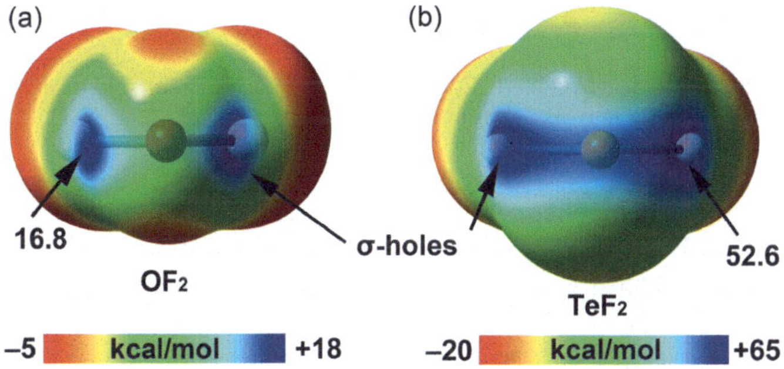

Fig. 7.19 MEP plots of OF_2 (**a**) and TeF_2 (**b**) using the 0.001 a.u. isosurface

at the σ-hole augment from 16.8 kcal/mol in OF_2 to 52.6 kcal/mol in TeF_2. The MEP values at the σ-holes of the chalcogen fluoride are smaller than those of pnictogen and tetrel fluorides, revealing that the electrophilicity decreases when moving from G–14 to G–16.

Figure 7.19 depicts the MEP surfaces of OF_2 and TeF_2 as illustrative examples of the whole series. In OF_2 the σ-holes opposite to the F–O bonds are small and clearly defined. In contrast, TeF_2 shows a large and positive MEP region along the molecular plane connecting both σ-hole. Therefore, the formation of linear ChBs is not expected in the heavier Ch-atoms, in line with the heavier Pn-atoms (see Fig. 7.12). In contrast, all G–14 atoms exhibit small σ-hole (see Fig. 7.2 for CF_4 and SnF_4) explaining their linear TtBs in the X-ray structures.

7.2.3.1 Oxygen

The oxygen atom is a very poor Ch-bond donor due to its low polarizability (lower than C and N, see Tables 7.1, 7.2 and 7.3) and large electronegativity. Several theoretical investigations have demonstrated the existence of attractive chalcogen bonds in $O(CN)_2$ and OF_2 compounds, among others [166–168] with a variety of electron donors.

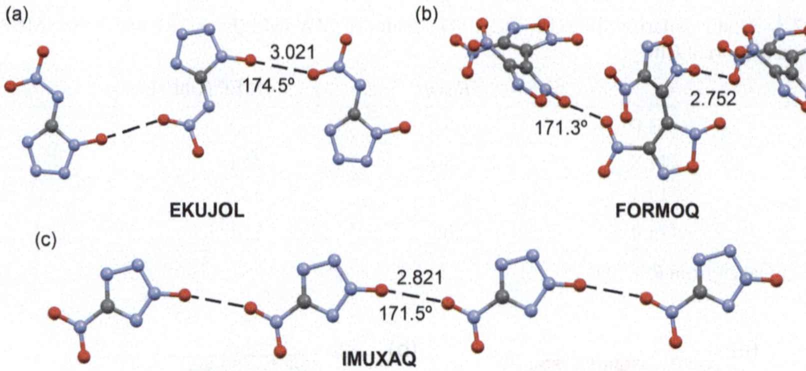

Fig. 7.20 ChBs in the solid-state of CSD refcodes "EKUJOL" (**a**), "FORMOQ" (**b**) and "IMUXAQ" (**c**). Distances in Å. H-atoms omitted

To date, there is not any conclusive experimental evidence for the existence of σ-hole ChBs involving O-atom as Lewis acid. However, it should be mentioned that Scilabra et al. [169] have proposed their participation in the crystal packing of 2-methylsulfonyl-3-(2-chloro-5-nitrophenyl)oxaziridine and several oxynitrobenzodifuroxan derivatives. In Fig. 7.20, three examples where directional O$\cdots$O ChBs dictate the formation of 1D supramolecular polymers are shown. In case of "EKUJOL" [170] and "IMUXAC" structures [171] the monomers are anionic (tetrazolate rings) thus forming anion$\cdots$anion ChBs. In contrast "FORMOQ" [172] structure corresponds to the neutral 4,4′-dinitro-3,3′-bi-1,2,5-oxadiazole 2,2′-dioxide. It should be emphasized the very short O$\cdots$O distance observed in this X-ray structure that rules out the possibility that the position of the O-atoms is simply due to packing effects in the crystal state, where the O-atoms are located in the least repulsive position.

7.2.3.2 Sulfur

ChBs involving sulfur is not very strong compared to the heavier chalcogens and consequently, their utilization in solution is not frequent. In contrast, ChBs involving sulfur in the solid-state are commonly described in sulfides, disulfides, and trisulfides derivatives as typical σ-hole donors [169]. Moreover, sulfur-containing heteroaromatics [169] are also excellent synthons frequently used in crystal engineering. As example of sulfide, we have selected the crystal structure of 2,2,4,4-tetrafluoro-1,3-dithietane (refcode "FAZVUA" [173], see Fig. 7.21a). Single crystal X-ray analysis shows that intermolecular C–S$\cdots$F chalcogen bonds are the shortest contacts in its solid-state structure. Analysis of the MEP surface (Fig. 7.21b) indicates that these contacts result from the attraction between the negative regions of fluorine atoms and the positive regions resulting from the merging of positive σ-holes located at sulfur and carbon atoms. This merging of both σ-holes causes the small C–S$\cdots$F angles in

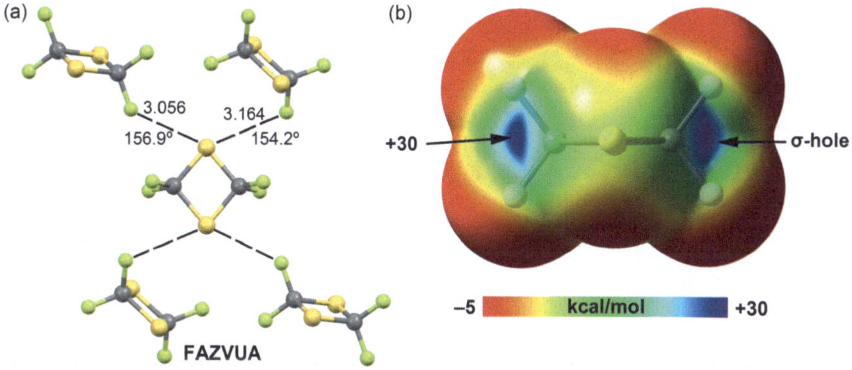

Fig. 7.21 **a** ChBs in the solid-state of CSD refcode "FAZVUA". Distances in Å. H-atoms omitted. **b** MEP surface of perfluoro-1,3-dithietane using the 0.001 a.u. isosurface

the ChB contacts that are far from linearity (154.2° and 156.9°), where the F-atoms are displaced toward the C-atoms.

Figure 7.22 shows two representative examples of ChB in disulfides and trisulfides. The X-ray structure of the 4-(4-nitrophenyl)-1,2,3,5-dithiadiazolyl radical (refcode "ASOFAR") shows that the O-atoms of the nitro group are located opposite the S–N bonds establishing two concurrent and directional S···O ChBs, thus promoting the formation of 1D supramolecular polymers. Similarly, in the selected example to illustrate ChBs in trisulfides (refcode "DAHDOF" [174], Fig. 7.22b), the 6-trifluoromethyl-4-nitrobenzotrithiole propagates in the solid-state forming 1D

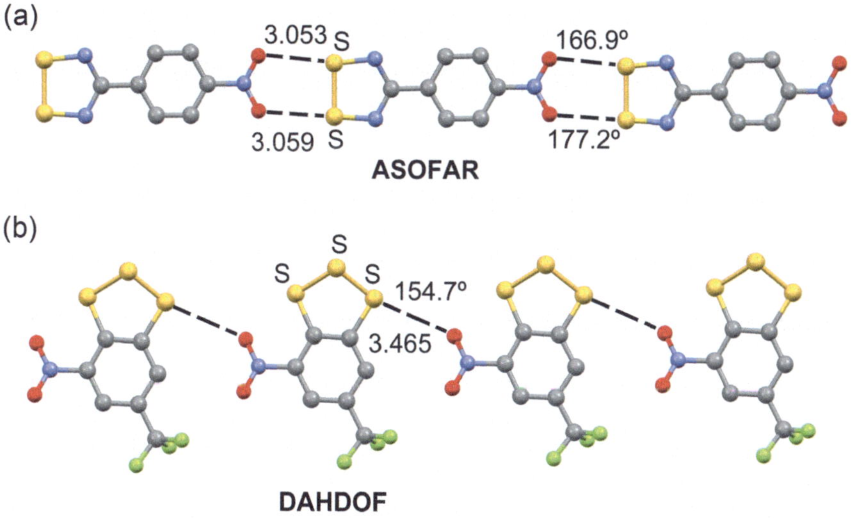

Fig. 7.22 ChBs in the solid-state of CSD refcodes "ASOFAR" (**a**) and "DAHDOF" (**b**). Distances in Å

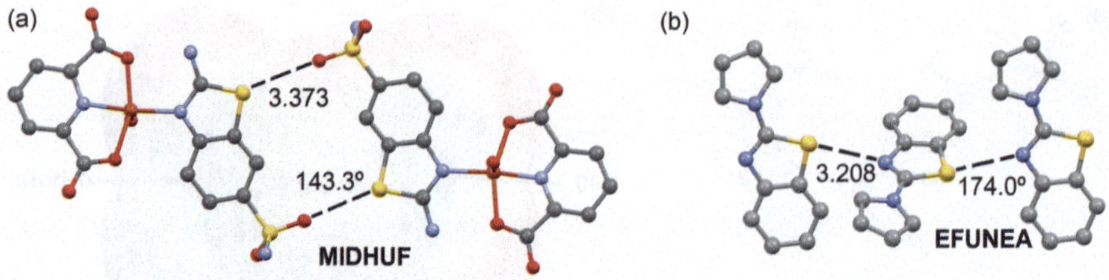

Fig. 7.23 ChBs in the solid-state of CSD refcodes "MIDHUF" (**a**) and "EFUNEA" (**b**). Distances in Å. H-atoms omitted

supramolecular polymers guided by ChBs established between the nitro group as donor and the trisulfide as acceptor. In this example the directionality is modest (154.7°) thus further confirming that ChBs are less directional than TtBs.

Finally, "MIDHUF" [175] and "EFUNEA" [176] structures exemplify ChBs in heteroaromatic rings. Figure 7.23a shows the X-ray structure of (2-amino-1,3-benzothiazole-6-sulfonamide)-aqua-(pyridine-2,6-dicarboxylato)-Cu(II) complex. This molecule forms self-assembled dimers in the solid-state by means of two S···O contacts that are quite short (3.373 Å). The coordination of the thiazole ring to the Cu(II) increases the Lewis acidity at the S-atom, thus reinforcing the ChB interaction. The "EFUNEA" (Fig. 7.23b) solid-state structure forms 1D supramolecular chains in the solid-state where the sp^2-hybridized lone pair is located opposite to the C–S bond, thus establishing short and quite directional ChB (174°).

7.2.3.3 Selenium and Tellurium

Selenium is a vital trace element [177, 178] that is continuously under investigation. Its derivatives are important in the pharmaceutical industry. Several reviews [9, 40, 41] are available in the literature devoted to the study of organoselenium compounds and their interactions with electron-rich species. Many examples in the literature demonstrate the ability of organic diselenides to establish both intermolecular and intramolecular ChB with electron-rich atoms, which are stronger and more directional in the prolongation of the Se–Se bond [9]. Two examples are shown in Fig. 7.24, one illustrating ChB interactions in the prolongation of the diselenide bond (refcode "FEYBAP" [179]) and the other illustrating ChBs in the middle of the Se–Se bond, as a consequence of the overlapping of both σ-holes (refcode "HOGBOW" [180]).

Two examples of σ-hole ChBs in selenodiazoles are depicted in Fig. 7.25. The "SAJPAY" and "SAJQED" [181] are co-crystals. The coformers of "SAJPAY" structure are 3,4-dicyano-1,2,5-selenodiazole and p-methoxypyridine-N-oxide and those of "SAJQED" structure are 1,2,5-selenodiazole and p-phenylpyridine-N-oxide. They exhibit the same motif (four-membered supramolecular rings) governed by the formation of two σ-hole chalcogen bonding interactions (Fig. 7.25a, b). To illustrate the relevance of selenium as Lewis acid, the structure of refcode "GEFVOC10" [182] is

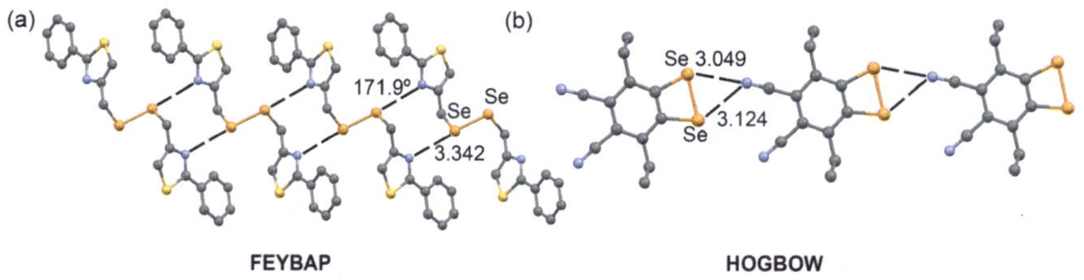

Fig. 7.24 ChBs in CSD refcodes "FEYBAP" (**a**) and "HOGBOW" (**b**). Distances in Å. H-atoms omitted. Adapted from Ref. [1]

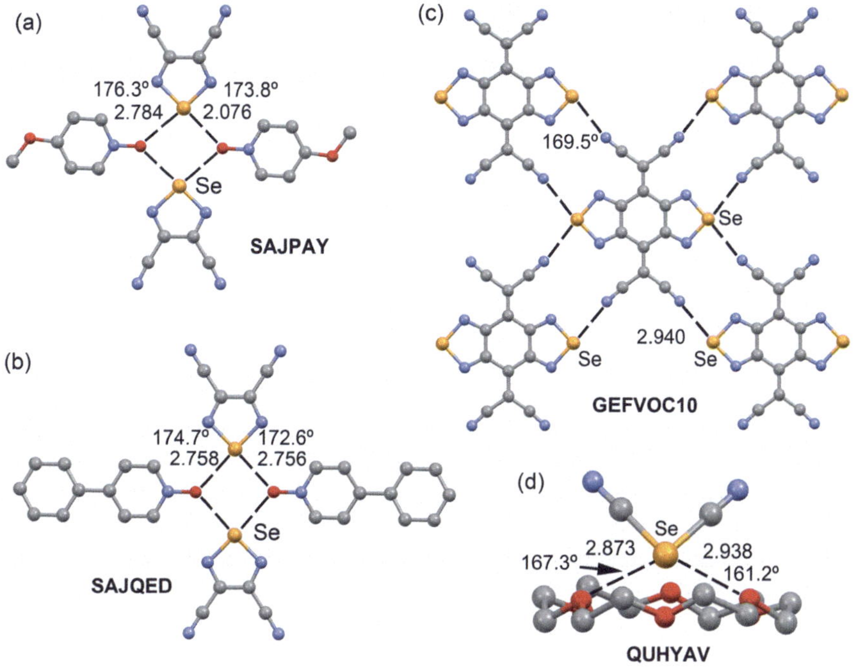

Fig. 7.25 ChBs in the solid-state of CSD refcodes "SAJPAY" (**a**), "SAJQED" (**b**), "GEFVOC10" (**c**), and "QUHYAV" (**d**). Distances in Å. H-atoms omitted. Adapted from Ref. [1]

represented in Fig. 7.25c. It is flat and the selenium atoms act as double σ-hole donors toward the nitrogen atoms of the CN groups with an angle of 169.5°. These short ChBs connect the molecule with four neighbors (Fig. 7.25c), dictating the formation of infinite 2D supramolecular assemblies. The co-crystal of 18-crown-6 with selenium dicyanide (refcode "QUHYAV" [183]) is represented in Fig. 7.25d. It shows two short and moderately directional ChBs that are established opposite to both electron-withdrawing cyano groups, confirming the σ-hole nature of the interaction.

Organotellurium compounds usually exhibit intense σ-holes without being bonded to strong electron-withdrawing groups. Consequently, their utilization in

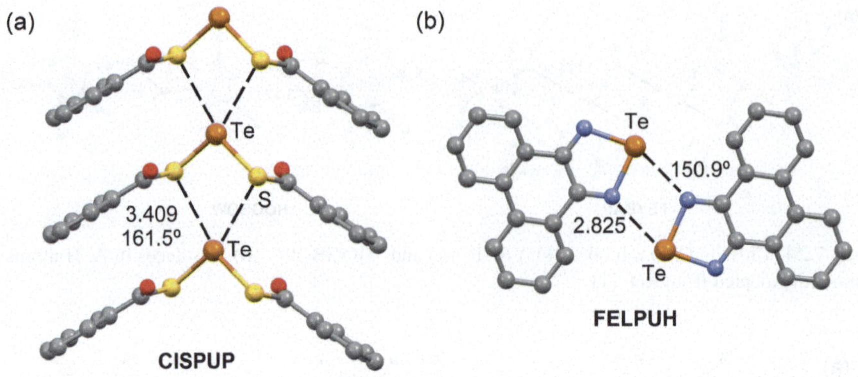

Fig. 7.26 ChBs in CSD refcodes "CISPUP" (**a**) and "FELPUH" (**b**). Distances in Å. Adapted from Ref. [1]

host–guest chemistry and catalysis [11, 184, 185]. An excellent review highlighting ChBs involving tellurium derivatives has been reported [186]. In addition, theoretical works have demonstrated that the heavier G–16 atoms form strong chalcogen bonds [187, 188].

Figure 7.26 represents two tellurium derivatives exhibiting ChBs. In the refcode "CISPUP" [189], an interesting head-to-tail 1D polymer is observed in the solid-state. That is, the Te atom establishes double chalcogen bonds with the sulfur atoms of the next molecule (Fig. 7.26a) propagating the 1D polymer. Refcode "FELPUH" [190] self-assembles forming dimers in the crystal packing via two short and symmetrically equivalent Te···N ChBs. The directionality of the ChBs is in general moderate in the two structures represented in Fig. 7.26. This fact agrees well with the MEP surface of TeF_2 that shows extended and deep σ-holes, thus allowing larger rearrangements and readjustments of the geometries in these molecules due to the influence of secondary interactions and packing effects without a significant weakening of the ChB.

7.2.4 Halogen Bonding

After hydrogen bonding, the HaB interaction is likely the most popular σ-hole based NCI [29–39] and has inspired the investigation and generalization of σ-hole NCIs in G–14 to G–16 elements. The HaB interaction is briefly described herein since several book chapters and reviews are available in the literature [29–39].

Table 7.4 gathers the α-values and vdW radii of the G–17 elements (periods 2–5). The polarizability of halogen atoms augments from 1.8 a.u. in fluorine atoms to 22.6 a.u. in iodine atoms. As previously observed for G–16 atoms, the atomic polarizability of chlorine atom is four times greater than that of fluorine atom while the differences for the rest of G–17 elements are smaller. The molecular electrostatic potential values gathered in Table 7.4 (see also Fig. 7.27) show the classical behavior.

Table 7.4 Atomic polarizabilities (α, a.u.) of halogen (Ha) elements, van der Waals radii (R_{vdW}, Å), and σ-hole MEP values (kcal/mol) of their tetrafluoride derivatives

Ha	α	R_{vdW}	MEP (HaF)
F	1.8	1.47	15.9
Cl	8.4	1.75	42.3
Br	14.0	1.89	49.7
I	22.6	1.98	55.8

Values are taken from Ref. [1]

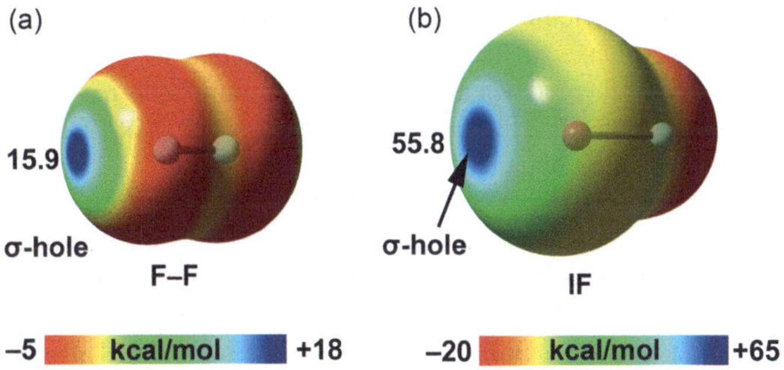

Fig. 7.27 MEP plots of F_2 (**a**) and IF (**b**) using the 0.001 a.u. isosurface

That is, the MEP values increase from 15.9 kcal/mol in F_2 to 55.8 kcal/mol in IF, parallel to the behavior observed for the atomic polarizability values.

Figure 7.27 shows the MEP surfaces of F_2 and IF as illustrative examples of the whole series. In both molecules the σ-holes are small and well-defined opposite to the F(I)–F bonds thus predicting stronger directionality in HaB in comparison to ChB or PnB and similar to TtBs that also showed well-defined σ-hole regions (see Fig. 7.2).

7.2.4.1 Fluorine

The participation of fluorine in donor–acceptor interactions as acceptor is rather uncommon [191]. In the literature, it has occasionally termed fluorine bond [192–194]. Several theoretical studies have demonstrated that fluorine has the capability of forming HaBs if the EWG bonded to F is sufficiently electron-withdrawing. [194–198]. In 2011 Hardegger et al. [199] analyzed both the CSD and PDB databases and suggested that it is possible to form an XB between organofluorine compounds and electronegative atoms. In fact, the experimental evidence for the existence of a HaB arises from crystal structure analyses where an F atom sits in fairly close proximity to an electron-rich atom (closer than the sum of their vdW radii). A number of crystals [200–204] have shown close contact of F with electron donor atoms. However, it is not obvious to distinguish an interaction attractive enough to affect the structure as opposed to one in which the atoms are close to each other simply because they do not repel one another [205].

7.2.4.2 Chlorine

Chlorine is less used in molecular recognition, crystal engineering, and catalysis than the heavier halogens bromine and iodine. It is because the HaB interactions involving chlorine as σ-hole donor are significantly weaker and less competitive in polar media than those of bromine and iodine. In contrast, HaBs involving chlorine are abundant in crystal structures (solvates). It has been demonstrated that the co-crystallization of halogenated solvents is favored whenever the crystal components have free available lone pairs to participate in HaB interactions. Actually, solvates of CCl_4, $CHCl_3$, and CH_2Cl_2 [39] are frequently obtained.

Figure 7.28 represents the X-ray structures of three solvates as representative examples. These are refcodes "CABZOU11" [206], "FIYLEG" [207], and "NAXYIW" [208] corresponding to $CHCl_3$, CH_2Cl_2, and CCl_4 co-crystals, respectively. In all cases, the HaB interaction with the solvent molecule is very directional (>173°) as common in HaBs and in agreement with the MEP surface analysis commented above. It is worthy to comment the X-ray structure of the tetrachloromethane solvate of 3,5-dichloro-4-cyanobenzoic acid (Fig. 7.28c) where the 3,5-dichloro-4-cyanobenzoic acid forms self-assembled dimers in the solid-state stabilized by two symmetrically equivalent Cl···N HaBs, where the electron donor is likely the π-system of the cyano group. Furthermore, the self-assembled dimer

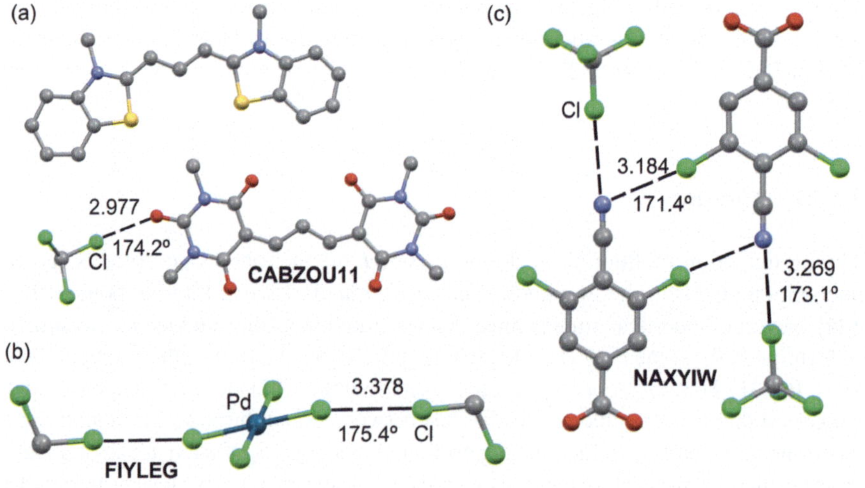

Fig. 7.28 HaBs in the solid-state of three solvates, CSD refcodes "CABZOU11" (**a**), "FIYLEG" (**b**), and "NAXYIW" (**c**). Distances in Å

interacts with two adjacent CCl_4 molecules forming two additional and directional Cl$\cdots$N contacts.

7.2.4.3 Bromine and Iodine

HaBs involving bromine and iodine as σ-hole donors are commonly used in crystal engineering, catalysis, supramolecular chemistry, molecular recognition of anions, and membrane transport and plenty of information is available in the literature [28–39, 47–51]. Therefore, this section is devoted to discussing more unconventional and recent investigations on halogen bonding where the Lewis base is a transition metal complex, that is the d_z2[M] acts as electron donor instead of a more conventional Lewis base (see Fig. 7.29).

Conventional Lewis bases are LP-bearing atoms (N, P, O, S, Ha) that are typically used in halogen bonding. Furthermore, electron-rich carbon atoms in neutral molecules (CO), anions (CN^-, RCC^-), and π-systems (double and triple bonds) are used as typical σ-hole acceptors. Much uncommon and counterintuitive is the utilization of a lone pair located in a filled d_z2-orbital as halogen bond acceptor. In fact, a metal cationic center is able to act as σ-hole acceptor if it contains one LP in a filled d-orbital. Several works report the utility of such uncommon halogen bonds in supramolecular chemistry and crystal engineering [209, 210].

Nitroguanidate Ni(II) complex is represented in Fig. 7.3, refcode "MEBXID" [210], that is co-crystallized with 1,3,5-triiodotrifluorobenzene (FIB), a good halogen bond donor. It establishes two symmetrically related Ni$\cdots$I interactions with FIB. The Ni$\cdots$I distance is shorter than the ΣR_{vdW}(Ni + I = 3.61 Å). It can be observed that the directionality of the interaction does not correspond to the expected for a

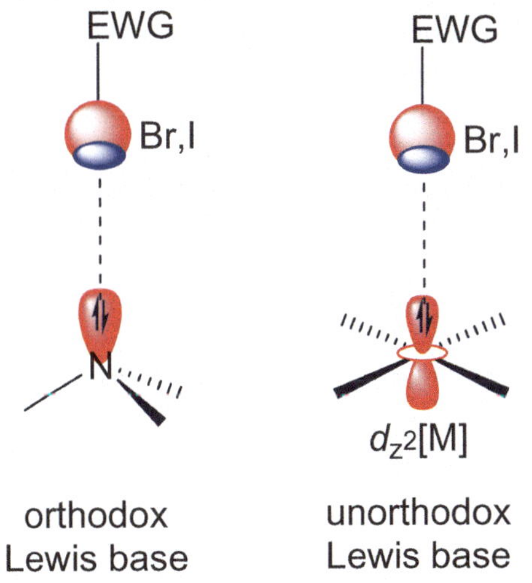

Fig. 7.29 Schematic representation of conventional and unconventional HaBs

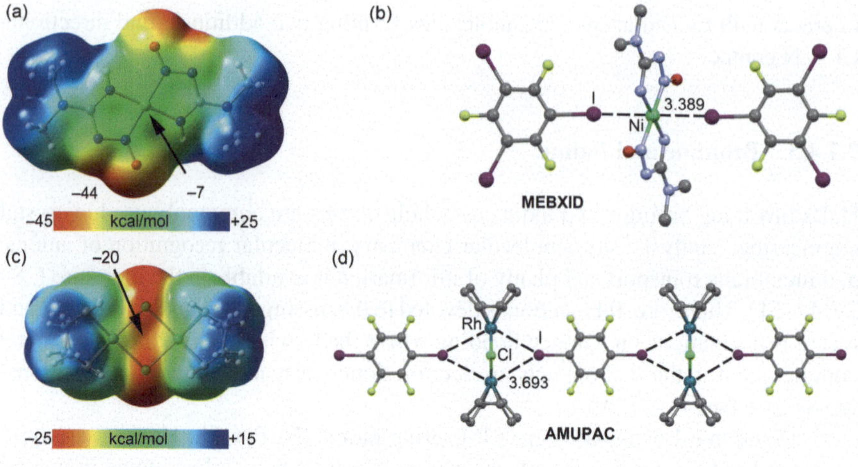

Fig. 7.30 MEP surfaces of Ni(nitroguanidate)$_2$ (**a**) and $[RhCl_2(COD)]_2$ (**c**) using the 0.001 a.u. isosurface (energies in kcal/mol). HaBs in the solid-state of three solvates, CSD refcodes "MEBXID" (**b**) and "AMUPAC" (**d**). Distances in Å. Adapted from Ref. [1]

semicoordination bond, since the C–I$\cdots$Ni angle is 142.5°. The molecular electrostatic potential plot (see Fig. 7.30a) evidences that the potential value is negative over the nickel atom. The MEP the minimum is located between the nitrogen and oxygen atoms (in the molecular plane of the nitroguanidate moiety). The potential value over Ni atoms suggests that this contact can be defined as HaB, which is also sustained by a recent work that has proved that the FIB molecule is able to establish HaBs with C–I$\cdots$LP angles far from linearity (up to 105°) [211].

Rh(I) complex of formula $[RhCl_2(COD)]_2$ (COD = 1,5-cyclooctadiene) has been recently used in crystal engineering as σ-hole acceptor molecule [212]. Combined theoretical and experimental investigations showed that the d_z2-orbitals of rhodium(I) metals deliver enough nucleophilicity to form three-center halogen bonds with iodine derivatives as σ-hole donors. Both Rh(I) metal centers act concurrently as HaB acceptors, as observed in refcode "AMUPAC" structure (see Fig. 7.30d). The molecular electrostatic potential plot is represented in Fig. 7.30c evidencing the nucleophilicity of the Rh_2Cl_2 core (–20 kcal/mol). This double σ-hole halogen bonding interaction is responsible for the formation of an infinite 1D assembly in the X-ray structure (see Fig. 7.30d).

Remarkably, square planar Pd(II) and Pt(II) halide complexes have the remarkable ability to establish halogen bonds where the M(II) center (M = Pd and Pt) acts as σ-hole acceptor [213–217]. The pioneering work using Br,I$\cdots d_z2$[Pd] halogen bonding interactions was published in 2018 by Boyarskiy and co-workers [214]. Several crystal structures using Pd and Pt as metals of formulas *trans*-$[MCl_2(NCNMe_2)_2]\cdot 2CHX_3$ were reported (X = Br, I). Two of them are represented in Fig. 7.31a, b where it can be observed that the C–Ha bonds ("LIHMIB", Ha = Br) and ("LIHMEX", Ha = I) point to the metal center. Theoretical and experimental

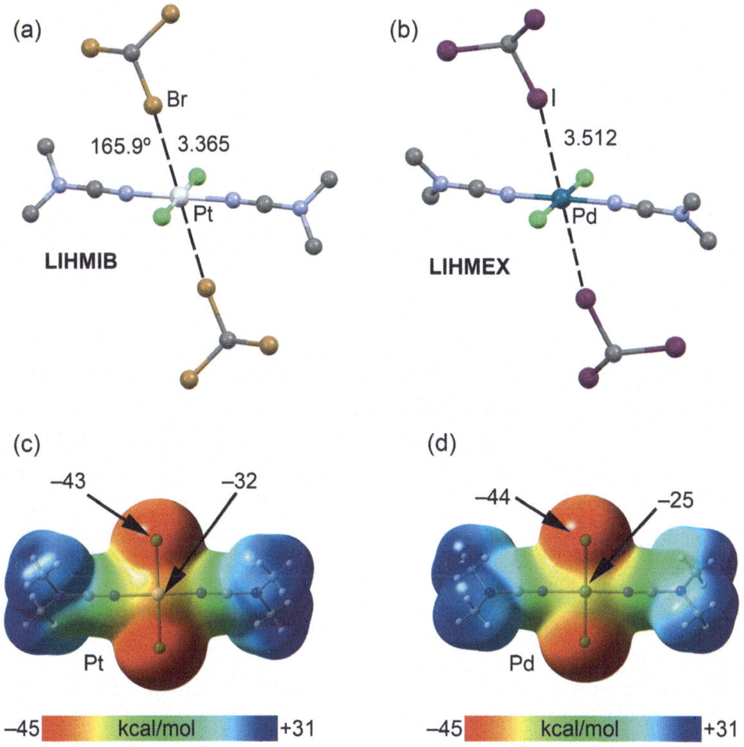

Fig. 7.31 HaBs in the solid-state of three solvates, CSD refcodes LIHMIB (**a**) and AMUPAC (**b**). Distances in Å. MEP surfaces of $Pt(NCNMe_2)_2Cl_2$ (**c**) and $Pt(NCNMe_2)_2Cl_2$ (**d**) using the 0.001 a.u. isosurface (energies in kcal/mol). Adapted from Ref. [1]

results corroborate that these contacts are halogen bonding interactions in nature. Indeed, the molecular electrostatic potential surfaces represented in Fig. 7.31c, d evidence that the potential values above the metal centers are negative (–32 and –25 kcal/mol for Pt and Pd, respectively), thus adequate for acting as Lewis bases.

7.3 Concluding Remarks

The data gathered in Tables 7.1, 7.2 and 7.3 shows that the polarizability of the elements belonging to G–14 to G–17 increases going from period 2 to 5. There is a large gap between elements of periods 2 and 3 that explain the very low ability of the lightest elements of the groups to participate in σ-hole interactions (apart from carbon). The molecular electrostatic potential values obtained for the fluoride derivatives of those elements at the σ-holes augments when descending in the group, parallel to the atomic polarizability. G–14 and G–17 elements exhibit more intense

σ-holes than G–15 and G–16 elements. Conversely, the size of σ-holes is more extended for the heavier elements of groups 15 and 16 than those of groups 14 and 17, in line with the stronger linearity of PnBs and ChBs.

Sections 7.2.1–7.2.4 provide significant support to the strong structure-directing role of σ-hole G–14 to G–17 interactions. The solid-state architecture of the examples provided in those sections demonstrates that these interactions are very directional and can be used to design new crystalline solids and other materials like supramolecular polymers. This type of contact is not always identified by the original authors or described as secondary interactions. However, they are predictable and general in X-ray structures of G–14 to G–17 derivatives. Hence, they should be incorporated into the toolbox of scientist working in crystal growth and crystal prediction. Moreover, their future in other fields like supramolecular catalysis and material science is promising. Finally, the existence of unconventional σ-hole HaB interactions has been highlighted in the last part of the chapter, where the σ-lump is located at the d_z2[M] orbital. Such unexplored interactions are expected to occur also for the elements belonging to G–14 to G–16.

References

1. Frontera A, Bauzá A (2021) On the importance of σ–hole interactions in crystal structures. Crystals 11:1205. https://doi.org/10.3390/cryst11101205
2. Daolio A, Scilabra P, Terraneo G, Resnati GC (2020) (sp^3) atoms as tetrel bond donors: a crystallographic survey. Coord Chem Rev 413:213265. https://doi.org/10.1016/j.ccr.2020.213265
3. von der Heiden D, Vanderkooy A, Erdélyi M (2020) Halogen bonding in solution: NMR spectroscopic approaches. Coord Chem Rev 407:213147. https://doi.org/10.1016/j.ccr.2019.213147
4. Gill H, Gokel MR, McKeever M, Negin, S Patel MB, Yin S, Gokel GW (2020) Supramolecular pore formation as an antimicrobial strategy. Coord Chem Rev 412:213264. https://doi.org/10.1016/j.ccr.2020.213264
5. Xu Y, Szell PMJ, Kumar V, Bryce DL (2020) Solid-state NMR spectroscopy for the analysis of element-based non-covalent interactions. Coord Chem Rev 411:213237. https://doi.org/10.1016/j.ccr.2020.213237
6. Wang W, Zhang Y, Jin WJ (2020) Halogen bonding in room-temperature phosphorescent materials. Coord Chem Rev 404:213107. https://doi.org/10.1016/j.ccr.2019.213107
7. Grabowski SJ (2020) Triel bond and coordination of triel centres – Comparison with hydrogen bond interaction. Coord Chem Rev 407:213171. https://doi.org/10.1016/j.ccr.2019.213171
8. Fromm KM (2020) Chemistry of alkaline earth metals: It is not all ionic and definitely not boring! Coord Chem Rev 408:213192. https://doi.org/10.1016/j.ccr.2020.213193
9. Fourmigué M, Dhaka A (2020) Chalcogen bonding in crystalline diselenides and selenocyanates: from molecules of pharmaceutical interest to conducting materials. Coord Chem Rev 403:213084. https://doi.org/10.1016/j.ccr.2019.213084
10. Scheiner S, Michalczyk M, Zierkiewicz W (2020) Coordination of anions by noncovalently bonded σ-hole ligands. Coord Chem Rev 405:213136. https://doi.org/10.1016/j.ccr.2019.213136
11. Taylor MS (2020) Anion recognition based on halogen, chalcogen, pnictogen and tetrel bonding. Coord Chem Rev 413:213270. https://doi.org/10.1016/j.ccr.2020.213270

12. Bauzá A, Frontera A (2020) σ/π-Hole noble gas bonding interactions: Insights from theory and experiment. Coord Chem Rev 404:213112. https://doi.org/10.1016/j.ccr.2019.213112
13. Biot N, Bonifazi D (2020) Chalcogen-bond driven molecular recognition at work. Coord Chem Rev 413:213243. https://doi.org/10.1016/j.ccr.2020.213243
14. Cavallo G, Metrangolo P, Pilati T, Resnati G, Terraneo G (2014) Naming interactions from the electrophilic site. Cryst Growth Des 14:2697–2702. https://doi.org/10.1021/cg5001717
15. Terraneo G, Resnati G (2017) Bonding matters. Cryst Growth Des 17:1439–1440. https://doi.org/10.1021/acs.cgd.7b00309
16. Desiraju GR, Ho PS, Kloo L, Legon AC, Marquardt R, Metrangolo P, Politzer P, Resnati G, Rissanen K (2013) Definition of the halogen bond (IUPAC recommendations 2013). Pure Appl Chem 85:1711–1713. https://doi.org/10.1351/PAC-REC-12-05-10
17. Aakeroy CB, Bryce DL, Desiraju GR, Frontera A, Legon AC, Nicotra F, Rissanen K, Scheiner S, Terraneo G, Metrangolo P, Resnati G (2019) Definition of the chalcogen bond (IUPAC recommendations 2019). Pure Appl Chem 91:1889–1892. https://doi.org/10.1515/pac-2018-0713
18. Zahn S, Frank R, Hey-Hawkins E, Kirchner B (2011) Pnicogen bonds: a new molecular linker? Chem Eur J 17:6034–6038. https://doi.org/10.1002/chem.201002146
19. Girolami GS (2009) Origin of the terms pnictogen and pnictide. J Chem Educ 86:1200–1201. https://doi.org/10.1021/ed086p1200
20. Bauzá A, Mooibroek TJ, Frontera A (2013) Tetrel bonding interaction: rediscovered supramolecular force? Angew Chem Int Ed 52:12317–12321. https://doi.org/10.1002/anie.201306501
21. Bauza A, Alkorta I, Elguero J, Mooibroek TJ, Frontera A (2020) Spodium bonds: noncovalent interactions involving group 12 elements. Angew Chem Int Ed 59:17482–17487. https://doi.org/10.1002/anie.202007814
22. Alkorta I, Elguero J, Frontera A (2020) 17. Not only hydrogen bonds: other noncovalent interactions. Crystals 10 180. https://doi.org/10.3390/cryst10030180
23. Gomila RM, Bauza A, Mooibroek TJ, Frontera A (2021) Spodium bonding in five coordinated Zn(II): a new player in crystal engineering? CrystEngComm 23:3084–3093. https://doi.org/10.1039/D1CE00221J
24. Daolio A, Pizzi A, Terraneo G, Ursini M, Frontera A, Resnati G (2021) Anion··· anion coinage bonds: the case of tetrachloridoaurate. Angew Chem Int Ed. 60:14385–14389. https://doi.org/10.1002/anie.202104592
25. Frontera A, Bauzà A (2018) Regium–π bonds: an unexplored link between noble metal nanoparticles and aromatic surfaces. Chem Eur J 24:7228–7234. https://doi.org/10.1002/chem.201800820
26. Daolio A, Pizzi, A Terraneo, G, Frontera A, Resnati, G (2021) σ-holes allow for attractive anion··· anion interactions involving perrhenate, permanganate, and pertechnetate anions. ChemPhysChem 22:2281–2285. https://doi.org/10.1002/cphc.202100681
27. Daolio A, Pizzi A, Calabrese M, Terraneo G, Bordignon S, Frontera A, Resnati G (2021) Molecular electrostatic potential and noncovalent interactions in derivatives of group 8 elements. Angew Chem Int Ed 60:20723–20727. https://doi.org/10.1002/anie.202107978
28. Politzer P, Lane P, Concha MC, Ma Y, Murray JS (2007) An overview of halogen bonding. J Mol Model 13:305–311. https://doi.org/10.1007/s00894-006-0154-7
29. Metrangolo P, Resnati G (2001) Halogen bonding: a paradigm in supramolecular chemistry. Chem Eur J 7:2511–2519. https://doi.org/10.1002/1521-3765(20010618)7:12%3c2511::AID-CHEM25110%3e3.0.CO;2-T
30. Metrangolo P, Neukirch H, Pilati T, Resnati G (2005) Halogen bonding based recognition processes: a world parallel to hydrogen bonding. Acc Chem Res 38:386–395. https://doi.org/10.1021/ar0400995
31. Metrangolo P, Meyer F, Pilati T, Resnati G, Terraneo G (2008) Halogen bonding in supramolecular chemistry. Angew Chem Int Ed 47:6114–6127. https://doi.org/10.1002/anie.200800128

32. Metrangolo P, Resnati G (eds) (2008) Halogen bonding: fundamentals and applications. Springer, Berlin
33. Politzer P, Murray JS, Clark T (2010) Halogen bonding: an electrostatically-driven highly directional noncovalent interaction. Phys Chem Chem Phys 12:7748–7757. https://doi.org/10.1039/C004189K
34. Erdélyi M (2012) Halogen bonding in solution. Chem Soc Rev 41:3547–3557. https://doi.org/10.1039/C2CS15292D
35. Beale TM, Chudzinski MG, Sarwar MG, Taylor MS (2013) Halogen bonding in solution: thermodynamics and applications. Chem Soc Rev 42:1667–1680. https://doi.org/10.1039/C2CS35213C
36. Politzer P, Murray JS, Clark T (2013) Halogen bonding and other σ-hole interactions: a perspective. Phys Chem Chem Phys 15:11178–11189. https://doi.org/10.1039/C3CP00054K
37. Metrangolo P, Resnati G (eds) (2015) Halogen bonding II Impact on materials chemistry and life sciences. Springer International Publishing AG, Heidelberg
38. Gilday LC, Robinson SW, Barendt TA, Langton, MJ Mullaney BR, Beer PD (2015) Halogen bonding in supramolecular chemistry. Chem Rev 115:7118–7195. https://doi.org/10.1021/cr500674c
39. Cavallo G, Metrangolo P, Milani R, Pilati T, Priimagi A, Resnati G, Terraneo G (2016) The halogen bond. Chem Rev 116:2478–2601. https://doi.org/10.1021/acs.chemrev.5b00484
40. Mahmudov KT, Kopylovich MN, Guedes da Silva MFC, Pombeiro AJL (2017) Chalcogen bonding in synthesis, catalysis and design of materials. Dalton Trans 46:10121–10138. https://doi.org/10.1039/C7DT01685A
41. Gleiter R, Haberhauer G, Werz DB, Rominger F, Bleiholder C (2018) From noncovalent chalcogen–chalcogen interactions to supramolecular aggregates: experiments and calculations. Chem Rev 118:2010–2041. https://doi.org/10.1021/acs.chemrev.7b00449
42. Vogel L, Wonner P, Huber SM (2019) Chalcogen bonding: an overview. Angew Chem Int Ed 58:1880–1891. https://doi.org/10.1002/anie.201809432
43. Scheiner S (2013) The pnicogen bond: its relation to hydrogen, halogen, and other noncovalent bonds. Acc Chem Res 46:280–288. https://doi.org/10.1021/ar3001316
44. del Bene JE, Alkorta I, Elguero J (2015) The pnicogen bond in review: structures, binding energies, bonding properties, and spin–spin coupling constants of complexes stabilized by pnicogen bonds. In: Scheiner S (ed) Noncovalent forces. Springer, Heidelberg, pp 191–264. https://doi.org/10.1007/978-3-319-14163-3_8
45. Brammer L (2017) Halogen bonding, chalcogen bonding, pnictogen bonding, tetrel bonding: origins, current status and discussion. Faraday Discuss 203:485–507. https://doi.org/10.1039/C7FD00199A
46. Politzer P, Murray JS, Lane P (2007) σ-Hole bonding and hydrogen bonding: competitive interactions. Int J Quantum Chem 107:3046–3052. https://doi.org/10.1002/qua.21419
47. Zhu S, Xing C, Xu, W Jin G, Li Z (2004) Halogen bonding and hydrogen bonding coexist in driving self-assembly process. Cryst Growth Des 4:53–56. https://doi.org/10.1021/cg0300275
48. Priimagi A, Cavallo G, Forni A, Gorynsztejn-Leben M, Kaivola M, Metrangolo P, Milani R, Shishido A, Pilati T, Resnati G, Terraneo G (2012) Halogen bonding versus hydrogen bonding in driving self-assembly and performance of light-responsive supramolecular polymers. Adv Funct Mater 22:2572–2579. https://doi.org/10.1002/adfm.201200135
49. Mukherjee A, Tothadi S, Desiraju GR (2014) Halogen bonds in crystal engineering: like hydrogen bonds yet different. Acc Chem Res 47:2514–2524. https://doi.org/10.1021/ar5001555
50. Scheiner S (2019) Forty years of progress in the study of the hydrogen bond. Struct Chem 30:1119–1128. https://doi.org/10.1007/s11224-019-01357-2
51. Corradi E, Meille SV, Messina MT, Metrangolo P, Resnati G (2000) Halogen bonding versus hydrogen bonding in driving self-assembly processes. Angew Chem Int Ed 39:1782–1786. https://doi.org/10.1002/(SICI)1521-3773(20000515)39:10<1782::AID-ANIE1782>3.0.CO;2-5

52. Pascoe DJ, Ling KB, Cockroft SL (2017) The origin of chalcogen-bonding interactions. J Am Chem Soc 139:15160–15167. https://doi.org/10.1021/jacs.7b08511
53. Bora PL, Novák M, Novotný J, Foroutan-Nejad C, Marek R (2017) Supramolecular covalence in bifurcated chalcogen bonding. Chem Eur J 23:7315–7323. https://doi.org/10.1002/chem.201700179
54. Bhandary S, Sirohiwal A, Kadu R, Kumar S, Chopra D (2018) Dispersion stabilized Se/Te···π double chalcogen bonding synthons in in situ cryocrystallized divalent organochalcogen liquids. Cryst Growth Des 18:3734–3739. https://doi.org/10.1021/acs.cgd.8b00585
55. Ams MR, Trapp N, Schwab A, Milić JV, Diederich F (2019) Chalcogen bonding "2S–2N Squares" versus competing interactions: exploring the recognition properties of sulfur. Chem Eur J 25:323–333. https://doi.org/10.1002/chem.201804261
56. Politzer P, Murray JS, Concha MC (2008) σ-hole bonding between like atoms; a fallacy of atomic charges. J Mol Model 14:659–665. https://doi.org/10.1007/s00894-008-0280-5
57. Cozzolino AF, Vargas-Baca I, Mansour S, Mahmoudkhani AH (2005) The nature of the supramolecular association of 1, 2, 5-chalcogenadiazoles. J Am Chem Soc 127:3184–3190. https://doi.org/10.1021/ja044005y
58. De Vleeschouwer F, Denayer M, Pinter B, Geerlings P, De Proft F (2018) Characterization of chalcogen bonding interactions via an in-depth conceptual quantum chemical analysis. J Comput Chem 39:557–572. https://doi.org/10.1002/jcc.25099
59. Bleiholder C, Gleiter R, Werz DB, Köppel H (2007) Theoretical investigations on heteronuclear chalcogen−chalcogen interactions: on the nature of weak bonds between chalcogen centers. Inorg Chem 46:2249–2260. https://doi.org/10.1021/ic062110y
60. Bleiholder C, Werz DB, Köppel H, Gleiter R (2006) Theoretical investigations on chalcogen−chalcogen interactions: what makes these nonbonded interactions bonding? J Am Chem Soc 128:2666–2674. https://doi.org/10.1021/ja056827g
61. Politzer P, Murray JS (2019) An overview of strengths and directionalities of noncovalent interactions: σ-holes and π-holes. Curr Comput-Aided Drug Des 9:165. https://doi.org/10.3390/cryst9030165
62. Politzer P, Murray JS, Clark T, Resnati G (2017) The σ-hole revisited. Phys Chem Chem Phys 19:32166–32178. https://doi.org/10.1039/C7CP06793C
63. Scheiner S (2018) Steric crowding in tetrel bonds. J Phys Chem A 122:2550–2562. https://doi.org/10.1021/acs.jpca.7b12357
64. Murray JS, Riley KE, Politzer P, Clark T (2010) Directional weak intermolecular interactions: σ-hole bonding. Aust J Chem 63:1598–1607. https://doi.org/10.1071/CH10259
65. Clark T (2013) σ-holes. Wiley Interdiscip Rev Comput Mol Sci 3:13–20. https://doi.org/10.1002/wcms.1113
66. Bundhun A, Ramasami P, Murray JS, Politzer P (2013) Trends in σ-hole strengths and interactions of F_3MX molecules (M = C, Si, Ge and X = F, Cl, Br, I) J Mol Model 19:2739–2746. https://doi.org/10.1007/s00894-012-1571-4
67. Grabowski SJ (2014) Tetrel bond–σ-hole bond as a preliminary stage of the SN2 reaction. Phys Chem Chem Phys 16:1824–1834. https://doi.org/10.1039/C3CP53369G
68. Thomas SP, Pavan MS, Row TNG (2014) Experimental evidence for "carbon bonding" in the solid state from charge density analysis. Chem Commun 50:49–51. https://doi.org/10.1039/C3CC47226D
69. Bauzá A, Mooibroek TJ, Frontera A (2014) Small cycloalkane $(CN)_2C$–$C(CN)_2$ structures are highly directional non-covalent carbon-bond donors. Chem Eur J 20:10245–10248. https://doi.org/10.1002/chem.201403680
70. Bauzá A, Mooibroek TJ, Frontera A (2016) 1,1,2,2-tetracyanocyclopropane (TCCP) as supramolecular synthon. Phys Chem Chem Phys 18:1693–1698. https://doi.org/10.1039/C5CP06350G
71. Escudero-Adán EC, Bauzá A, Frontera A, Ballester P (2015) Nature of noncovalent carbon-bonding interactions derived from experimental charge-density analysis. ChemPhysChem 16:2530–2533. https://doi.org/10.1002/cphc.201500437

72. Roeleveld JJ, Lekanne Deprez SJ, Verhoofstad A, Frontera A, van der Vlugt JI, Mooibroek TJ (2020) Engineering crystals using sp^3-C centred tetrel bonding interactions. Chem Eur J 26:10126–10132. https://doi.org/10.1002/chem.202002613
73. Mani D, Arunan E (2013) The X–C$\cdots$Y (X= O/F, Y= O/S/F/Cl/Br/N/P) 'carbon bond' and hydrophobic interactions. Phys Chem Chem Phys **15**:14377–14383. https://doi.org/10.1039/C3CP51658J
74. Bauza A (2016) Frontera A $RCH_3\cdots O$ interactions in biological systems: are they trifurcated H-bonds or noncovalent carbon bonds? Curr Comput-Aided Drug Des 6:26. https://doi.org/10.3390/cryst6030026
75. Morgunov RB, Mushenok FB, Aldoshin SM, Sanina NA, Yur'eva EA, Shilov GV, Tkachev VV (2009) Thermally-induced paramagnetism of spiropyrane iodides. New J Chem 33:1374. https://doi.org/10.1039/b822567b
76. Haussuhl E, Schreuer J (2016) Crystal structure and elastic properties of betaine fumarate and of betaine maleate, two isomers of $((CH_3)_3NCH_2COOH$. $OOC(CH)_2COOH$. Z Kristallogr 216:616–622. https://doi.org/10.1524/zkri.216.12.616.22492
77. Huta OM, Patsaj IO, Konitz A, Meszko J, Blazejowski J (2002) 9-Cyano-10-methylacridinium hydrogen dinitrate. Acta Cryst C58:o295–o297
78. Alcock NW, Harrison WD, Howes C (1984) Secondary bonding. Part 13. Aryl-tellurium(IV) and -iodine(III) acetates and trifluoroacetates. The crystal and molecular structures of bis-(p-methoxyphenyl)tellurium diacetate, μ-oxo-bis[diphenyltrifluoroacetoxytellurium] hydrate, and [bis(trifluoroacetoxy)iodo]benzene. J Chem Soc, Dalton Trans 1709. https://doi.org/10.1039/dt9840001709
79. Witt JR, Britton D, Mahon C (1972) The crystal structures of $BrC(CN)_3$, $ClC(CN)_3$, and $CH_3C(CN)_3$. Acta Cryst B28:950. https://doi.org/10.1107/S0567740872003450
80. Franconetti A, Quiñonero D, Frontera, A, Resnati G (2019) Unexpected chalcogen bonds in tetravalent sulfur compounds. Phys Chem Chem Phys 21:11313–11319. https://doi.org/10.1039/C9CP01033E
81. Patia S, Rappoport Z (eds) (1995) Chemistry of Functional Groups: the chemistry of organic germanium, tin and lead compounds, vol 19. Wiley
82. Parr J (2004) Comprehensive coordination chemistry II. In: McCleverty JA, Meye TJ (eds), vol 3. Elsevier Pergamon, Oxford, p 545
83. Sato T (1995) Comprehensive organometallic chemistry II. In: Abel EW, Stone FGA, Wilkinson G (eds), vol 11. Pergamon Press, Oxford, p 389
84. Pinhey JT (1995) Comprehensive organometallic chemistry II. In: Abel EW, Stone FGA, Wilkinson G (eds), vol 11. Pergamon Press, Oxford, p 461
85. Greenberg A, Wu G (1990) Structural relationships in silatrane molecules. Struct Chem 1:79–85. https://doi.org/10.1007/BF00675787
86. Hencsei P (1991) Evaluation of silatrane structures by correlation relationships. Struct Chem 2:21–26. https://doi.org/10.1007/BF00673485
87. Voronkov MG, Barishok VP, Petukhov LP, Rahklin RG, Pestunovich VA (1988) 1-Halosilatranes. J Organomet Chem 358:39–55. https://doi.org/10.1016/0022-328X(88)87069-4
88. Lukevics E, Dimens V, Pokrovska N, Zicmane I, Popelis J, Kemme A (1999) Addition of nitrile oxides to 2,3-dihydrofurylsilanes. Crystal and molecular structure of tetrahydrofuro-[2,3-d]-isoxazolylsilanes. J Organomet Chem 586:200–207. https://doi.org/10.1016/S0022-328X(99)00266-1
89. Corriu RJP (1990) Hypervalent species of silicon: structure and reactivity. J Organomet Chem 400:81–106. https://doi.org/10.1016/0022-328X(90)83007-7
90. Shen Q, Hilderbrandt RL (1980) The structure of methyl silatrane (1-methyl-2,8,9-trioxa-5-aza-1-silabicyclo (3.3.3) undecane) as determined by gas phase electron diffraction. J Mol Struct 64:257–262. https://doi.org/10.1016/0022-2860(80)80136-0
91. Förgics G, Kolonits M, Hargittai I (1990) The gas-phase molecular structure of 1-fluorosilatrane from electron diffraction. Struct Chem 1:245–250. https://doi.org/10.1007/BF00674268

92. Eujen R, Petrauskas E, Roth A, Brauer DJ (2000) The structures of 1-chlorogermatrane and of 1-fluorogermatrane, revisited. J Organomet Chem 613:86–92. https://doi.org/10.1016/S0022-328X(00)00514-3
93. Lukevics E, Ignatovich L, Beliakov S (1999) Synthesis and molecular structure of phenyl and tolylgermatranes. J Organomet Chem 588:222–230. https://doi.org/10.1016/S0022-328X(99)00378-2
94. Livant P, Northcott J, Webb TR (2001) Structure of an oxo-bridged germatrane dimer. J Organomet Chem 620:133–138. https://doi.org/10.1016/S0022-328X(00)00807-X
95. Karlov SS, Shutov PL, Churakov AV, Lorberth J, Zaitseva GS (2001) New approach to 1-(phenylethynyl)germatranes and 1-(phenylethynyl)-3,7,10-trimethylgermatrane. Reactions of 1-(phenylethynyl)germatrane with N-bromosuccinimide and bromine. J Organomet Chem 627:1–5. https://doi.org/10.1016/S0022-328X(01)00708-2
96. Alkorta I, Rozas I (2001) Elguero J Molecular complexes between silicon derivatives and electron-rich groups. J Phys Chem A 105:743–749. https://doi.org/10.1021/jp002808b
97. Alkorta I, Elguero J, Fruchier A, Macquarrie DJ, Virgili A (2001) Aminopropylsilanes versus silatranes: an experimental and theoretical study. J Organomet Chem 625:148–153. https://doi.org/10.1016/S0022-328X(00)00808-1
98. Liu M, Li Q, Scheiner S (2017) Comparison of tetrel bonds in neutral and protonated complexes of pyridineTF_3 and furanTF_3 (T = C, Si, and Ge) with NH_3. Phys Chem Chem Phys 19:5550–5559. https://doi.org/10.1039/C6CP07531B
99. Kalikhman I, Girshberg O, Lameyer L, Stalke D, Kost D (2001) Tautomeric equilibrium between penta-and hexacoordinate silicon chelates. A chloride bridge between two pentacoordinate silicons. J Am Chem Soc 123:4709–4716. https://doi.org/10.1021/ja004118r
100. Taylor PG, Bassindale AR, El Aziz Y, Pourny M, Stevenson R, Hursthouse MB, Coles SJ (2012) Further studies of fluoride ion entrapment in octasilsesquioxane cages; X-ray crystal structure studies and factors that affect their formation. Dalton Trans 41:2048–2059. https://doi.org/10.1039/C1DT11340B
101. Bassindale AR, Pourny M, Taylor PG, Hursthouse MB, Light ME (2003) Fluoride-ion encapsulation within a Silsesquioxane cage. Angew Chem Int Ed 42:3488–3490. https://doi.org/10.1002/ange.200351249
102. Dai X, Choi S-B, Braun CW, Vaidya P, Kilina S, Ugrinov A, Schulz DL, Boudjouk P (2011) Halide coordination of perhalocyclohexasilane Si_6X_{12} (X = Cl or Br) cage. Inorg Chem 50:4047–4053. https://doi.org/10.1021/ic102535n
103. Tillmann J, Meyer-Wegner F, Nadj A, Becker-Baldus J, Sinke T, Bolte M, Holthausen MC, Wagner M, Lerner H-W (2012) Unexpected disproportionation of tetramethylethylenediamine-supported perchlorodisilane $Cl_3SiSiCl_3$. Inorg Chem 51:8599–8606. https://doi.org/10.1021/ic301283m
104. Dai X, Schulz D L, Braun CW, Ugrinov A, Boudjouk P (2010) "Inverse Sandwich" complexes of perhalogenated cyclohexasilane. Organometallics **29**:2203–2205. https://doi.org/10.1021/om1000716
105. Grabowski SJ (2017) Lewis acid properties of tetrel tetrafluorides—the coincidence of the σ-hole concept with the QTAIM approach. Curr Comput-Aided Drug Des 7:43. https://doi.org/10.3390/cryst7020043
106. Scilabra P, Kumar V, Ursini M, Resnati G (2018) Close contacts involving germanium and tin in crystal structures: experimental evidence of tetrel bonds. J Mol Model 24:37. https://doi.org/10.1007/s00894-017-3573-8
107. Bauza A, Seth SK, Frontera A (2019) Tetrel bonding interactions at work: impact on tin and lead coordination compounds. Coord Chem Rev 384:107–125. https://doi.org/10.1016/j.ccr.2019.01.003
108. Lukevics E, Arsenyan P, Belyakov S, Popelis J, Pudova O (1999) Cycloaddition reactions of nitrile oxides to silyl-and germyl-substituted thiophene-1,1-dioxides. Organometallics 18:3187–3193. https://doi.org/10.1021/om9902129
109. Villaescusa LA, Lightfoot P, Morris RE (2002) Synthesis and structure of fluoride-containing GeO_2 analogues of zeolite double four-ring building units. Chem Commun:2220–2221. https://doi.org/10.1039/B207374A

110. Calogero S, Ganis P, Peruzzo V, Tagliavini G, Valle G (1981) X-ray and Mössbauer studies of tricyclohexyltin (IV) halides. The crystal structures of (cyclo-$C_6H_{11})_3SnX$ (X = F, Br and I) J Organomet Chem 220:11–20. https://doi.org/10.1016/S0022-328X(00)83099-5
111. Wang J-Q, Kuang D-Z, Zhang F-X, Feng Y-L, Xu Z-F (2004) CCDC 211279: Experimental crystal structure determination. https://doi.org/10.5517/cc72vgl
112. Trujillo C, Sanchez-Sanz G, Alkorta I, Elguero J (2015) Halogen, chalcogen and pnictogen interactions in (XNO_2) 2 homodimers (X = F, Cl, Br, I). New J Chem 39:6791–6802. https://doi.org/10.1039/C5NJ00600G
113. Alkorta I, Elguero J, Grabowski SJ (2015) Pnicogen and hydrogen bonds: complexes between PH3X+ and PH2X systems. Phys Chem Chem Phys 17:3261–3272. https://doi.org/10.1039/C4CP04840G
114. Esrafili MD, Mohammadian-Sabet F, Baneshi MM (2015) The dual role of halogen, chalcogen, and pnictogen atoms as Lewis acid and base: triangular XBr:SHX:PH2X complexes (X = F, Cl, Br, CN, NC, OH, NH_2, and OCH_3). Int J Quantum Chem 115:1580–1586. https://doi.org/10.1002/qua.24987
115. Adhikari U, Scheiner S (2012) Sensitivity of pnicogen, chalcogen, halogen and H-bonds to angular distortions. Chem Phys Lett 532:31–35. https://doi.org/10.1016/j.cplett.2012.02.064
116. Bauzà A, Quinonero D, Deya PM, Frontera A (2013) Halogen bonding versus chalcogen and pnicogen bonding: a combined Cambridge structural database and theoretical study. CrystEngComm 15:3137–3144. https://doi.org/10.1039/C2CE26741A
117. Murray JS, Lane P, Politzer P (2007) A predicted new type of directional noncovalent interaction. Int J Quantum Chem 107:2286–2292. https://doi.org/10.1002/qua.21352
118. Alkorta I, Sánchez-Sanz G, Elguero J (2014) Pnicogen bonds between X-PH_3 (X = O, S, NH, CH2) and phosphorus and nitrogen bases. J Phys Chem A 118:1527–1537. https://doi.org/10.1021/jp411623h
119. Shukla R, Chopra D (2016) Characterization of N$\cdots$O non-covalent interactions involving σ-holes: "electrostatics" or "dispersion." Phys Chem Chem Phys 18:29946–29954. https://doi.org/10.1039/C6CP05899J
120. Mokrai R, Jaie Barrett J, Apperley DC, Benkő Z, Heift D (2020) Tweaking the charge transfer: bonding analysis of Bismuth(III) complexes with a flexidentate phosphane ligand. Inorg Chem 59:8916–8924. https://doi.org/10.1021/acs.inorgchem.0c00734
121. Sarkar S, Pavan MS, Guru Row TN (2015) Experimental validation of "pnicogen bonding" in nitrogen by charge density analysis. Phys Chem Chem Phys 17:2330–2334. https://doi.org/10.1039/C4CP04690K
122. Joshi PR, Ramanathan N, Sundararajan K, Sankaran K (2015) Evidence for phosphorus bonding in phosphorus trichloride–methanol adduct: a matrix isolation infrared and ab initio computational study. J Phys Chem A 119:3440–3451. https://doi.org/10.1021/jp511156d
123. Nelyubina YV, Korlyukov AA, Lyssenko KA (2015) Experimental charge density evidence for pnicogen bonding in a crystal of ammonium chloride. ChemPhysChem 16:676–681. https://doi.org/10.1002/cphc.201402673
124. Andrea Gini A, Paraja M, Galmés B, Besnard C, Poblador-Bahamonde AI, Sakai N, Frontera A, Matile S (2020) Pnictogen-bonding catalysis: brevetoxin-type polyether cyclizations. Chem Sci 11:7086–7091. https://doi.org/10.1039/D0SC02551H
125. Scilabra P, Terraneo G, Resnati G (2017) Fluorinated elements of Group 15 as pnictogen bond donor sites. J Fluor Chem 203:62–67. https://doi.org/10.1016/j.jfluchem.2017.10.002
126. Bauza A, Mooibroek TJ, Frontera A (2015) Directionality of π-holes in nitro compounds. Chem Commun 51:1491–1493. https://doi.org/10.1039/C4CC09132A
127. Bauza A, Frontera A, Mooibroek TJ (2016) π-hole interactions involving nitro compounds: directionality of nitrate esters. Cryst Growth Des 16:5520–5524. https://doi.org/10.1021/acs.cgd.6b00989
128. Bauza A, Sharko AV, Senchyk GA, Rusanov EB, Frontera A, Domasevitch KV (2017) π–hole interactions at work: crystal engineering with nitro-derivatives. CrystEngComm 19:1933–1937. https://doi.org/10.1039/C7CE00267J

129. Bauza A, Frontera A, Mooibroek TJ (2019) Chem Eur J 25:13436–13443. Hazari A, Das LK, Kadam RM, Bauzá A, Frontera A, Ghosh A (2015) Dalton Trans 44:3862–3876. https://doi.org/10.1039/C7CE00267J
130. Bauza A, Frontera A, Mooibroek TJ (2017) NO_3^- anions can act as Lewis acid in the solid state. Nat Commun 8:14522. https://doi.org/10.1038/ncomms14522
131. Mooibroek TJ (2017) Coordinated nitrate anions can be directional π-hole donors in the solid state: a CSD study. CrystEngComm 19:4485–4488. https://doi.org/10.1039/C7CE01266G
132. Li W, Spada L, Tasinato N, Rampino S, Evangelisti L, Gualandi A, Cozzi PG, Melandri S, Barone V, Puzzarini C (2018) Theory meets experiment for noncovalent complexes: the puzzling case of pnicogen interactions. Angew Chem Int Ed 57:13853–13857. https://doi.org/10.1002/anie.201807751
133. Bader RFW (1991) A quantum theory of molecular structure and its applications. Chem Rev 91:893–928. https://doi.org/10.1021/cr00005a013
134. Allouche F, Selmi W, Zid MF, Benlecheb T (2019) Theoretical and experimental study of new hybrid compound rich in hydrogen bonding: 2-carboxyanilinium hypophosphite. J Mol Struct 1179:756–763. https://doi.org/10.1016/j.molstruc.2018.11.069
135. Lemmerer A (2011) Six two- and three-component ammonium carboxylate salt structures with a ladder-type hydrogen-bonding motif, three incorporating neutral carboxylic acid molecules. Acta Cryst C67:o92–o99. https://doi.org/10.1107/S0108270111003714
136. Shubnell AJ, Squattrito PJ (1994) Monohydrates of two isomers of aminotoluenesulfonic acid. Acta Cryst C50:1296–1299. https://doi.org/10.1107/S0108270193012077
137. Joshi PR, Ramanathan N, Sundararajan K, Sankaran K (2017) Phosphorous bonding in PCl_3: H_2O adducts: a matrix isolation infrared and ab initio computational studies. J Mol Spectrosc 331:44–52. https://doi.org/10.1016/j.jms.2016.11.005
138. Ramanathan N, Sankaran K, Sundararajan K (2016) PCl3–C6H6 heterodimers: evidence for P···π phosphorus bonding at low temperatures. Phys Chem Chem Phys 18:19350–19358. https://doi.org/10.1039/C6CP03825E
139. Carter TG, Healey ER, Pitt MA, Johnson DW (2005) Secondary bonding interactions observed in two arsenic thiolate complexes. Inorg Chem 44:9634–9636. https://doi.org/10.1021/ic051522o
140. Carter TG, Vickaryous WJ, Cangelosi VM, Johnson DW (2007) Supramolecular arsenic coordination chemistry. Comments Inorg Chem 28:97–122. https://doi.org/10.1080/02603590701560994
141. Vickaryous WJ, Herges R, Johnson DW (2004) Arsenic-π Interactions stabilize a self-sssembled As2L3 supramolecular complex. Angew Chem Int Ed 43:5831–5833. https://doi.org/10.1002/ange.200461011
142. Moaven S, Yu J, Yasin J, Unruh DK, Cozzolino AF (2017) Precise Steric control over 2D versus 3D self-assembly of Antimony (III) alkoxide cages through strong secondary bonding interactions. Inorg Chem 56:8372–8380. https://doi.org/10.1021/acs.inorgchem.7b01049
143. Leroy C, Johannson R, Bryce DL (2019) $^{121/123}$Sb nuclear quadrupole resonance spectroscopy: characterization of non-covalent pnictogen bonds and NQR crystallography. J Phys Chem A 123:1030–1043. https://doi.org/10.1021/acs.jpca.8b11490
144. Cangelosi VM, Pitt MA, Vickaryous WJ, Allen CA, Zakharov LN, Johnson DW (2010) Design considerations for the group 15 elements: the pnictogen···π interaction as a complementary component in supramolecular assembly design. Cryst Growth Des 10:3531–3536. https://doi.org/10.1021/cg100444n
145. Benjamin SL, Levason W, Reid G, Warr RP (2012) Halostibines $SbMeX_2$ and $SbMe_2X$: Lewis acids or Lewis bases? Organometallics 31:1025–1034. https://doi.org/10.1021/om2010996
146. Diel BN, Huber TL, Ambacher WG (1999) Synthesis and characterization of the first examples of 1,3,2-diazastibole and 1,3,2-diazabismole ring compounds. Heteroat Chem 10:423–429. https://doi.org/10.1002/(SICI)1098-1071(1999)10:5%3c423::AID-HC13%3e3.0.CO;2-B
147. Anderson KM, Baylies CJ, Jahan AHMM, Norman NC, Orpen AG, Starbuck J (2003) Coordination complexes of the bismuth(iii) thiolates Bi(SC6F5)3 and Bi(SC6Cl5)3 with pyridineligands. Dalton Trans 3270–3277. https://doi.org/10.1039/B305711A

148. Charmant JPH, Jahan AHMM, Norman NC, Orpen AG, Podesta TJ (2004) Coordination polymers formed by Bi (SC6F5)3 with multidentate polypyridyl ligands. CrystEngComm 6:29–33. https://doi.org/10.1039/B315176J
149. Murafuji T, Nagasue M, Tashiro Y, Sugihara Y, Azuma N (2000) Structural characteristics of aryloxybismuthanes stabilized by hypervalent bond formation. Synthesis, incorporation of 4-methoxyphenol through hydrogen bonding, and crystal supramolecularity. Organometallics 19:1003–1007. https://doi.org/10.1021/om9908534
150. Heift D, Mokrai R, Barrett J, Apperley D, Batsanov A, Benkö Z (2019) Weak pnictogen bond with bismuth: experimental evidence based on Bi–P through-space coupling. Chem Eur J 25:4017–4024. https://doi.org/10.1002/chem.201900266
151. Silvestru C, Breunig HJ, Althaus H (1999) Structural chemistry of bismuth compounds. I. Organobismuth derivatives. Chem Rev 99:3277–3328. https://doi.org/10.1021/cr980083q
152. Nekoueishahraki B, Samuel PP, Roesky HW, Stern D, Matussek J, Stalke D (2012) Organobismuth (III) and dibismuthine complexes bearing N, N'-Disubstituted 1, 8-diaminonaphthalene ligand: synthesis, structure, and reactivity. Organometallics 31:6697–6703. https://doi.org/10.1021/om300758s
153. Briand GG, Burford N (2000) Coordination complexes of bismuth (III) involving organic ligands with pnictogen or chalcogen donors. Adv Inorg Chem 50:285–357. https://doi.org/10.1016/S0898-8838(00)50007-5
154. Agocs L, Burford N, Cameron TS, Curtis JM, Richardson JF, Robertson KN, Yhard GB (1996) Spectroscopic, Structural, and Mass spectrometric studies on two systematic series of dithiabismuth(III) heterocycles: identification of bismuthenium cations and their solvent complexes. J Am Chem Soc 118:3225–3232. https://doi.org/10.1021/ja9539756
155. Benjamin SL, Levason W, Reid G, Rogers MC, Warr RP (2012) Lewis base complexes of methyldihalobismuthines BiMeX2 (X = Cl or Br). J Organomet Chem 708–709:106–111. https://doi.org/10.1016/j.jorganchem.2012.02.030
156. Trubenstein HJ, Moaven S, Vega M, Unruh DK, Cozzolino AF (2019) Pnictogen bonding with alkoxide cages: which pnictogen is best? New J Chem 43:14305–14312. https://doi.org/10.1039/C9NJ03648B
157. Muller G, Brand J, Jetter SE (2001) Donor-acceptor complexes between organoamines and phosphorus tribromide. Z Naturforschung, B: Chem Sci 56:1163–1171. https://doi.org/10.1515/znb-2001-1111
158. Robinson TP, Lo S-K, De Rosa D, Aldridge S, Goicoechea JM (2016) On the ambiphilic reactivity of geometrically constrained Phosphorus(III) and Arsenic(III) compounds: insights into their interaction with ionic substrates. Chem Eur J 22:15712–15724. https://doi.org/10.1002/chem.201603135
159. Benz S, Poblador-Bahamonde AI, Low-Ders N, Matile S (2018) Catalysis with pnictogen, chalcogen, and halogen bonds. Angew Chem Int Ed 57:5408–5412. https://doi.org/10.1002/anie.201801452
160. Lim JYC, Beer PD (2018) Sigma-hole interactions in anion recognition. Chem 4:731–783. https://doi.org/10.1016/j.chempr.2018.02.022
161. Hirai M, Cho J, Gabbaï FP (2016) Promoting the hydrosilylation of benzaldehyde by using a dicationic antimony-based Lewis acid: evidence for the double electrophilic activation of the carbonyl substrate. Chem Eur J 22:6537–6541. https://doi.org/10.1002/chem.201600971
162. Qiu J, Unruh DK, Cozzolino AF (2016) Design, synthesis, and structural characterization of a Bisantimony(III) compound for anion binding and the density functional theory evaluation of halide binding through antimony secondary bonding interactions. J Phys Chem A 120:9257–9269. https://doi.org/10.1021/acs.jpca.6b08170
163. Haiges R, Vij A, Boatz JA, Schneider S, Schroer T, Gerken M, Christe KO (2004) First structural characterization of binary AsIII and SbIII azides. Chem Eur J 10:508–517. https://doi.org/10.1002/chem.200305482
164. Deokar P, Leitz D, Stein TH, Vasiliu M, Dixon DA, Christe KO, Haiges R (2016) Preparation and characterization of antimony and arsenic tricyanide and their 2,2′-bipyridine adducts. Chem Eur J 22:13251–13257. https://doi.org/10.1002/chem.201602436

165. Levason W, Light ME, Maheshwari S, Reid G, Zhang W (2011) Unusual neutral ligand coordination to arsenic and antimony trifluoride. Dalton Trans 40:5291–5297. https://doi.org/10.1039/C1DT10113G
166. Murray JS, Lane P, Clark T, Politzer P (2007) σ-hole bonding: molecules containing group VI atoms. J Mol Model 13:1033–1038. https://doi.org/10.1007/s00894-007-0225-4
167. Varadwaj PR, Varadwaj A, Marques HM, MacDougall PJ (2019) The chalcogen bond: can it be formed by oxygen? Phys Chem Chem Phys 21:19969–19986. https://doi.org/10.1039/C9CP03783G
168. Varadwaj PR (2019) Does oxygen feature chalcogen bonding? Molecules 24:3166. https://doi.org/10.3390/molecules24173166
169. Scilabra P, Terraneo G, Resnati G (2019) The chalcogen bond in crystalline solids: a world parallel to halogen bond. Acc Chem Res 52:1313–1324. https://doi.org/10.1021/acs.accounts.9b00037
170. Fischer D, Klapötke TM, Stierstorfer J (2015) 5-Nitriminotetrazole 1-oxide: an exciting oxygen- and nitrogen-rich heterocycle. Eur J Inorg Chem 2015:4628–4632. https://doi.org/10.1002/ejic.201500944
171. Göbel M, Karaghiosoff K, Klapötke TM, Piercey DG, Stierstorfer J (2010) Nitrotetrazolate-2N-oxides and the strategy of N-oxide introduction. J Am Chem Soc 132:17216–17226. https://doi.org/10.1021/ja106892a
172. Fischer D, Klapötke TM, Stierstorfer J (2014) Synthesis and characterization of diaminobisfuroxane. Eur J Inorg Chem 2014:5808–5811. https://doi.org/10.1002/ejic.201402960
173. Nayak SK, Kumar V, Murray JS, Politzer P, Terraneo G, Pilati T, Metrangolo P, Resnati G (2017) Fluorination promotes chalcogen bonding in crystalline solids. CrystEngComm 19:4955–4959. https://doi.org/10.1039/C7CE01070B
174. Chenard BL, Harlow RL, Johnson AL, Vladuchick SA (1985) Synthesis, structure, and properties of pentathiepins. J Am Chem Soc 107:3871–3879. https://doi.org/10.1021/ja00299a019
175. Alkaya ZA, İlkimen H, Yenikaya C, Tunca E, Bülbül M, Tunç T, Sarı M (2018) Synthesis and characterization of Cu(II) complexes of 2-amino-6-sulfamoylbenzothiazole and their inhibition studies on carbonic anhydrase isoenzymes. Polyhedron 151:199–205. https://doi.org/10.1016/j.poly.2018.05.015
176. Hedidi M, Bentabed-Ababsa G, Derdour A, Roisnel T, Dorcet V, Chevallier F, Picot L, Thiéry V, Mongin F (2014) Synthesis of C, N′-linked bis-heterocycles using a deprotometalation–iodination–N-arylation sequence and evaluation of their antiproliferative activity in melanoma cells. Bioorg Med Chem 22:3498–3507. https://doi.org/10.1016/j.bmc.2014.04.028
177. Burk RF (ed) (1994) Selenium in biology and human health. Springer, New York
178. Saha UK (2017) Selenium: a vital element in soil-pant-animal/human continuum. J Agric Sci Bot 1:1–3. https://doi.org/10.35841/2591-7897.1.1.1-3
179. Popa RA, Licarete E, Banciu M, Silvestru A (2018) Organoselenium compounds containing pyrazole or phenylthiazole groups: synthesis, structure, tin(IV) complexes and antiproliferative activity. Appl Organomet Chem 32:e4252. https://doi.org/10.1002/aoc.4252
180. Kimura T, Nakahodo T, Fujihara H, Suzuki E (2014) 4,5-Dicyano-3,6-diethylbenzo-1,2-diselenete, a Highly Stable 1,2-Diselenete: its preparation, structural characterization, calculated molecular orbitals, and complexation with tetrakis(triphenylphosphine)palladium. Inorg Chem 53:4411–4417. https://doi.org/10.1021/ic5000765
181. Xu Y, Kumar V, Bradshaw MJZ, Bryce DL (2020) Chalcogen-bonded cocrystals of substituted pyridine N-oxides and chalcogenodiazoles: an X-ray diffraction and solid-state NMR investigation. Cryst Growth Des 20:7910–7920. https://doi.org/10.1021/acs.cgd.0c01173
182. Suzuki T, Fujii H, Yamashita Y, Kabuto C, Tanaka S, Harasawa M, Mukai T, Miyashi T (1992) Clathrate formation and molecular recognition by novel chalcogen-cyano interactions in tetracyanoquinodimethanes fused with thiadiazole and selenadiazole rings. J Am Chem Soc 114:3034–3043. https://doi.org/10.1021/ja00034a041
183. Fritz S, Ehm C, Lentz D (2015) Structure and chemistry of $SeF_x(CN)_{4-x}$ compounds. Inorg Chem 54:5220–5231. https://doi.org/10.1021/acs.inorgchem.5b00107

184. Garrett GE, Gibson GL, Straus RN, Seferos DS, Taylor MS (2015) Chalcogen bonding in solution: interactions of benzotelluradiazoles with anionic and uncharged Lewis bases. J Am Chem Soc 137:4126–4133. https://doi.org/10.1021/ja512183e
185. Navarro-García E, Galmés B, Velasco MD, Frontera A, Caballero A (2020) Anion recognition by neutral chalcogen bonding receptors: experimental and theoretical investigations. Chem Eur J 26:4706–4713. https://doi.org/10.1002/chem.201905786
186. Cozzolino AF, Elder PJW, Vargas-Vaca I (2011) A survey of tellurium-centered secondary-bonding supramolecular synthons. Coord Chem Rev 255:1426–1438. https://doi.org/10.1016/j.ccr.2010.12.015
187. Scheiner S (2013) Detailed comparison of the pnicogen bond with chalcogen, halogen, and hydrogen bonds. Int J Quantum Chem 113:1609–1620. https://doi.org/10.1002/qua.24357
188. Nziko VPN, Scheiner S (2014) Chalcogen bonding between tetravalent SF_4 and amines. J Phys Chem A 118:10849–10856. https://doi.org/10.1021/jp509212t
189. Subrahmanyan I, Aravamudan G, Rout GC, Seshasayee M (1984) Synthesis, properties, and molecular structure of bis(thiobenzoato-S)tellurium(II), $C_14H_{10}O_2S_2Te(II)$. J Crystal Spect Res 14:239–248. https://doi.org/10.1007/BF01161161
190. Neidlein R, Knecht D, Gieren A, Ruiz-Perez C (1987) Synthese und Röntgenstrukturanalyse des Phenanthro[9,10-c]-l,2,5-telluradiazols. Z Naturforschung, B: Chem Sci 42:84–90
191. Eskandari K, Lesani M (2015) Does fluorine participate in halogen bonding? Chem Eur J 21:4739–4746. https://doi.org/10.1002/chem.201405054
192. Chopra D, Row TNG (2011) Role of organic fluorine in crystal engineering. CrystEngComm 13:2175–2186. https://doi.org/10.1039/C0CE00538J
193. Laurence C, Graton J, Gal JF (2011) An overview of Lewis basicity and affinity scales. J Chem Educ 88:1651–1657. https://doi.org/10.1021/ed200057b
194. Lu YX, Zou JW, Yu QS, Jiang YJ, Zhao WN (2007) Ab initio investigation of halogen bonding interactions involving fluorine as an electron acceptor. Chem Phys Lett 449:6–10. https://doi.org/10.1016/j.cplett.2007.09.087
195. Murray JS, Lane P, Politzer P (2009) Expansion of the σ-hole concept. J Mol Model 15:723–729. https://doi.org/10.1007/s00894-008-0386-9
196. Murray JS, Politzer P (2009) Molecular surfaces, van der Waals radii and electrostatic potentials in relation to noncovalent interactions. Croat Chem Acta 82:267–275
197. Li W, Zeng Y, Zhang X, Zheng S, Meng L (2014) The enhancing effects of group V σ-hole interactions on the F···O halogen bond. Phys Chem Chem Phys 16:19282–19289. https://doi.org/10.1039/C4CP02430C
198. Metrangolo P, Murray JS, Pilati T, Politzer P, Resnati G, Terraneo G (2011) The fluorine atom as a halogen bond donor, viz. a positive site CrystEngComm 13:6593–6596. https://doi.org/10.1039/C1CE05554B
199. Hardegger LA, Kuhn B, Spinnler B, Anselm L, Ecabert R, Stihle M, Gsell B, Thoma R, Diez J, Benz J, Plancher JM, Hartmann G, Isshiki Y, Morikami K, Shimma N, Haap W, Banner DW, Diederich F (2011) Halogen bonding at the active sites of human cathepsin L and MEK1 kinase: efficient interactions in different environments ChemMedChem 6:2048–2054. https://doi.org/10.1002/cmdc.201100353
200. Dikundwar AG, Row TNG (2012) Evidence for the "Amphoteric" nature of fluorine in halogen bonds: an instance of Cl···F contact. Cryst Growth Des 12:1713–1716. https://doi.org/10.1021/cg3002477
201. Pavan MS, Durga Prasad K, Guru Row TN (2013) Halogen bonding in fluorine: experimental charge density study on intermolecular F···F and F···S donor–acceptor contacts. Chem Commun 49:7558–7560. https://doi.org/10.1039/c3cc43513j
202. Hathwar VR, Chopra D, Panini P, Guru Row TN (2014) Revealing the polarizability of organic fluorine in the trifluoromethyl group: implications in supramolecular chemistry. Cryst Growth Des 14:5366–5369. https://doi.org/10.1021/cg501240r
203. Sirohiwal A, Hathwar VR, Dey D, Regunathan R, Chopra D. Characterization of fluorine-centred 'F···O' [sigma]-hole interactions in the solid state. Acta Crystallogr Sect B Struct Sci Cryst Eng Mater 73:140–152. https://doi.org/10.1107/s2052520616017492

204. Jelsch C, Guillot B (2017) Directional O··· F halogen bonds. Acta Crystallogr Sect B: Struct Sci Cryst Eng Mater 73:136–137. https://doi.org/10.1107/s205252061700467x
205. Scheiner S (2020) J F-halogen bond: conditions for its existence. J Phys Chem A 124:7290–7299. https://doi.org/10.1021/acs.jpca.0c06803
206. Etter MC, Kress RB, Bernstein J, Cash DJ (1984) Solid-state chemistry and structures of a new class of mixed dyes. Cyanine-oxonol. J Am Chem Soc 106:6921–6927. https://doi.org/10.1021/ja00335a009
207. Klapp LRR, Bruhn C, Leibold M, Siemeling U (2013) Ferrocene-based bis (guanidines): superbases for tridentate N, Fe, N-coordination. Organometallics 32:5862–5872. https://doi.org/10.1021/om400454v
208. Britton D (2012) 3,5-Dichloro-4-cyanobenzoic acid co-crystals with carbon tetrachloride, naphthalene, and anthracene. J Chem Crystallogr 42:851–855. https://doi.org/10.1007/s10870-012-0324-7
209. Katlenok EA, Haukka M, Levin OV, Frontera A, Kukushkin VY (2020) Supramolecular assembly of metal complexes by (Aryl) I··· dz2 [PtII] halogen bond. Chem Eur J 26:7692–7701. https://doi.org/10.1002/chem.202001196
210. Bikbaeva ZM, Ivanov DM, Novikov AS, Ananyev IV, Bokach NA, Kukushkin VY (2017) Electrophilic–nucleophilic dualism of Nickel(II) toward Ni··· I noncovalent interactions: semicoordination of iodine centers via electron belt and halogen bonding via σ-hole. Inorg Chem 56:13562–13578. https://doi.org/10.1021/acsinorgchem7b02224
211. Rozhkov AV, Eliseeva AA, Baykov SV, Galmés B, Frontera A, Kukushkin VY (2020) One-Pot Route to X-perfluoroarenes (X = Br, I) based on Fe^{III}-assisted C-F Functionalization and utilization of these arenes as building blocks for crystal engineering involving halogen bonding. Cryst Growth Des 20:5908–5921. https://doi.org/10.1021/acs.cgd.0c00606
212. Eliseeva AA, Ivanov DM, Rozhkov AV, Ananyev IV, Frontera A, Kukushkin VY (2021) Bifurcated halogen bonding involving two Rhodium (I) centers as an integrated σ-hole acceptor. JACS Au 1:354–361. https://doi.org/10.1021/jacsau.1c00012
213. Gossage RA, Ryabov AD, Spek AL, Stufkens DJ, van Beek JAM, van Eldik R, van Koten G (1999) Models for the initial stages of oxidative addition. Synthesis, characterization, and mechanistic investigation of η^1-I_2 organometallic "Pincer" complexes of platinum. X-ray crystal structures of $[PtI(C_6H_3\{CH_2NMe_2\}_2\text{-}2,6)(\eta^1\text{-}I_2)]$ and exo-meso-$[Pt(\eta^1\text{-}I_3)(\eta 1\text{-}I_2)(C_6H_3\{CH_2N(t\text{-}Bu)Me\}_2\text{-}2,6)]$. J Am Chem Soc 121:2488–2497. https://doi.org/10.1021/ja982095z
214. Ivanov DM, Kirina YV, Novikov AS, Starova GL, Kukushkin VY (2016) Efficient π-stacking with benzene provides 2D assembly of trans-$[PtCl_2\ (p\text{-}CF_3C_6H_4CN)_2]$. J Mol Struct 1104:19–23. https://doi.org/10.1016/j.molstruc.2015.09.027
215. Baykov SV, Dabranskaya U, Ivanov DM, Novikov AS, Boyarskiy VP (2018) Pt/Pd and I/Br isostructural exchange provides formation of C–I··· Pd, C–Br··· Pt, and C–Br··· Pd metal-involving halogen bonding. Cryst Growth Des 18:5973–5980. https://doi.org/10.1021/acs.cgd.8b00762
216. Bulatova M, Ivanov DM, Haukka M (2021) Classics meet classics: theoretical and experimental studies of halogen bonding in adducts of Platinum(II) 1,5-cyclooctadiene halide complexes with diiodine, iodoform, and 1,4-diiodotetrafluorobenzene. Cryst Growth Des 21:974–987. https://doi.org/10.1021/acs.cgd.0c01314
217. Ivanov DM, Bokach NA, Kukushkin VY, Frontera A (2021) Metal centers as nucleophiles: oxymoron of halogen bond-involving crystal engineering. Chem Eur J 27. https://doi.org/10.1002/chem.202103173

Chapter 8
Regium Bonds: A Bridge Between Coordination and Supramolecular Chemistry

Antonio Frontera and Antonio Bauzá

Abstract During the past century, the coordination chemistry of coinage metals (elements of Group 11) has been extensively studied owing to their strong implications in crystal engineering, protein-drug chemistry, or catalysis. Very recently, their role has been expanded to the "noncovalent realm" by providing evidence of their ability to behave as Lewis acids and undergo noncovalent binding. More precisely, Cu, Ag, and Au are able to undertake noncovalent interactions (NCI) involving Lewis bases of different nature (e.g., lone pair donors, π-systems) or even small charged anions (e.g., Cl^-). This has been mainly attributed to presence of positive (σ-hole) and negative (σ-lump) electrostatic potential regions, which mimic in some way the electrophilic (σ-hole) and nucleophilic (belt) regions present in aerogens (Group 18), halogens (Group 17), chalcogens (Group 16) and pnictogens (Group 15). Group 11 noncovalent bonds have been named "regium bonds (RgB)" owing to the noble metal character of Cu, Ag, and Au elements. This chapter encompasses a series of both theoretical and experimental examples of RgBs to provide a general picture of the promising features of the interaction in crystal engineering, biological systems, and surface absorption processes as well as interplay and cooperativity between RgBs and other noncovalent forces.

8.1 Introduction

At present, elements of groups 13–18 covalently bound to electron-withdrawing groups (EWG) are well known to favorably interact with Lewis bases, anions, or even π-systems [1–13]. During the past century, the electrophilic site was used to define the ubiquitous hydrogen bonding (HB) interaction, which has led to utilizing the name of the group of the Periodic Table where the electrophilic atom belongs to name the noncovalent interactions (NCIs) involving electrophilic and nucleophilic

A. Frontera · A. Bauzá (✉)
Universitat de les Illes Balears, Crta de Valldemossa km 7.5, 07122 Palma de Mallorca (Baleares), Spain
e-mail: antonio.bauza@uib.es

M. A. Aboudzadeh and A. Frontera (eds.), *Supramolecular Assemblies Based on Electrostatic Interactions*,
https://doi.org/10.1007/978-3-031-00657-9_8

species [14, 15]. For instance, the International Union of Pure and Applied Chemistry (IUPAC) has already recommended the terms halogen bond (HaB) [16] and chalcogen bond (ChB) [17] to define NCIs of groups 17 and 18, respectively. Moreover, other terms are widely used for defining other groups of the p-block, which are aerogen or noble gas bonding (NgB, group 18) [12], pnictogen bonding (PnB, group 15), [18, 19] tetrel bonding (TtB, group 14) [20], and triel bonding [TrB, group 13] [7]. In addition, spodium bonding (SpB) [21–23] has been recently proposed to refer to NCIs where elements of group 12 act as Lewis acids and to differentiate this type of interaction from coordination bonds. Moreover, the terms Alkali bond [24–26], Alkaline earth bond [27–30], and Osme bond [31] have been coined to describe the NCIs involving groups 1, 2, and 8 of the periodic table, respectively. Lastly, the term regium bond (RgB) has been proposed to describe the NCI involving elements from group 11 (Cu, Ag, and Au) [32]. The term coinage bond (CiB) has been also proposed and used in the literature [33]. In this chapter, we have adopted the RgB terminology simply because it is found in a major number of manuscripts. It should be mentioned that the IUPAC has started a project for naming this interaction (https://iupac.org/project/2021-006-2-100) without providing any specific recommendation yet (Fig. 8.1).

This chapter is focused on the RgB interaction, starting from its physical nature and moving to a collection of several selected computational and experimental examples. To understand the physical nature of this novel NCI, a series of Molecular Electrostatic Potential (MEP) surfaces are shown in Fig. 8.2. Concretely, in Fig. 8.2a the MEP surfaces of Cu_2, Ag_2, and Au_2 are represented. As noted, a positive electrostatic potential region (blue region) is located on the extension of the Rg–Rg bond (Rg = Cu, Ag, and Au) due to the valence configuration ($d^{10}s^1$) of Rg (e.g., [Xe5$d^{10}6s^1$] for Au), which results in a singly occupied s-orbital. On the other hand, the Rg–Rg bond displays a negative potential (red region) owing to the polarization of the Rg_2

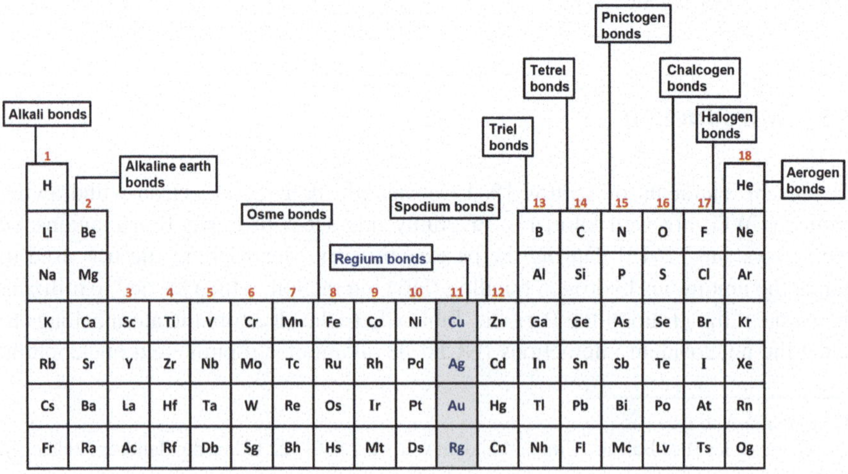

Fig. 8.1 The different noncovalent interactions undertaken by the elements of the periodic table

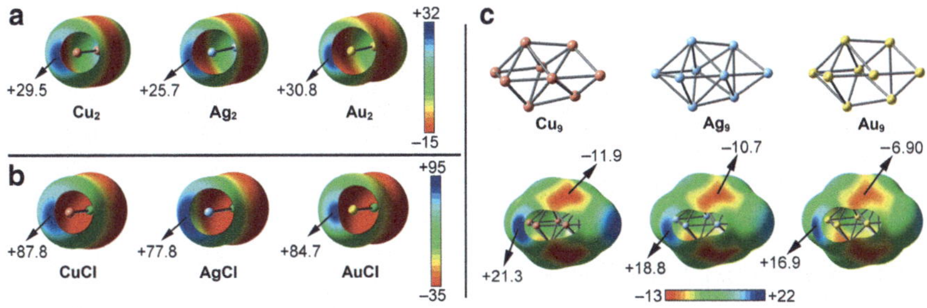

Fig. 8.2 Molecular electrostatic potential (MEP) surfaces of Rg_2 **a** RgCl, **b** and Rg_9, **c** molecules (Rg = Cu, Ag and Au). Energy values at selected points on the surface are given in kcal/mol (0.001 a.u.)

σ–orbital toward the bond. Contrary to the common behavior exhibited by other σ–hole based interactions [34], the Rg_2 σ–hole values become more positive in the case of Cu_2 (+29.5 kcal/mol) and Au_2 (+30.8 kcal/mol) molecules than for Ag_2 (+25.7 kcal/mol).

A similar tendency can be observed in Fig. 8.2b, where the MEP surfaces of CuCl, AgCl, and AuCl are shown. In this case, the σ–hole MEP values obtained dramatically increased owing to the Rg^{+1} oxidation state (e. g. + 30.8 and +84.7 kcal/mol for Au_2 and AuCl, respectively). This fact is interesting since + 1 is the most common oxidation state for noble metals and it critically enhances the Cu, Ag, and Au σ–hole donor ability.

Finally, in Fig. 8.2c the MEP surfaces of three Rg_9 clusters are shown, and the results point in the same way as for the Rg_2 molecules. That is, the overlap of singly occupied s-orbitals between Rg atoms leads to areas of positive potential (σ-holes in blue) at the low-coordinated sites of the atoms and negative potential (σ–lumps in red) in the bonding regions.

From these series of MEP analyses, it is clear that the σ-holes present in these moieties make them suitable for acting as electrophilic sites upon their interaction with electron-rich species (e.g., a lone pair, π–system, or an anion) from an electrostatic viewpoint.

In this chapter, both theoretical and experimental studies are gathered to (i) illustrate the ability of group 11 of elements to participate in σ-hole interactions; (ii) demonstrate the interplay between RgBs and other NCIs (e.g., TrBs and NgBs), and (iii) show the potential applications of the interaction in protein-ligand chemistry, crystal engineering, and surface absorption mechanisms. It is also important to note that although elements of group 10 (Ni, Pd, and Pt) are usually considered part of the noble metal family [35], the examples gathered in this chapter only involve atoms belonging to group 11 to stay true to the definition of the interaction [32].

In the following sections, several selected theoretical and experimental studies involving RgBs as well as their interplay with other NCIs are described. This chapter is intended to serve as a multidisciplinary guide, hence, the studies gathered herein are described with the intention of providing a general vision of the work from the original authors. For a more specific reading or details, see the original article.

8.2 Results and Discussion

8.2.1 Theoretical Studies

Understanding the nature and strength of Regium bonds

Several theoretical studies have been devoted to the study of the physical nature and strength features of RgBs. For instance, Zheng and collaborators [36] studied the ability of phosphine oxide (H_3PO), the *trans* phosphinuous acid (T-PH_2OH), the *cis* phosphinuous acid (C-PH_2OH) and a series of RgX (Rg = Cu, Ag, Au; X = F, Cl, Br) molecules to undergo RgBs at the CCSD(T) [37] and MP2 [38]/aug-cc-pVTZ-PP [39] level of theory.

They evaluated the stability and directionality of both O– and P–shared complexes, which exhibited interaction energy values ranging from −72.1 to −16.8 kcal/mol (CCSD(T)/aug-cc-pVTZ-PP) and P–O–Rg and O–P–Rg angles comprised between 112 and 131° (See Fig. 8.3). The authors found that the P–shared complexes exhibited larger interaction energy values than their O–shared analogous for both *trans* and *cis*

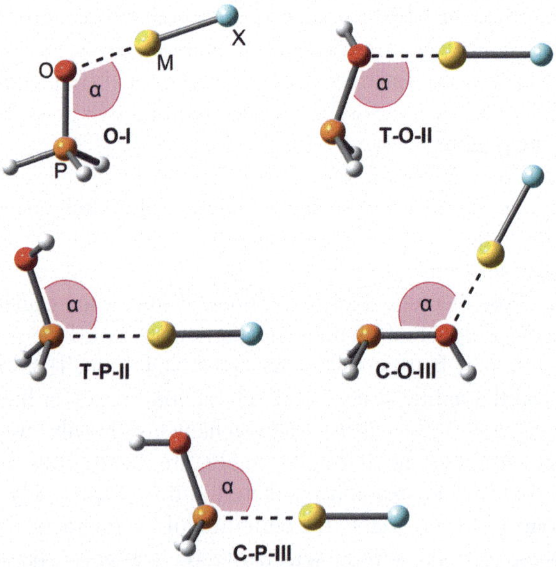

Fig. 8.3 Optimized geometries of several representative complexes $H_3PO\cdots RgX$ (O–I), *trans*-$PH_2OH\cdots RgX$ (T-O-II, T-P-II), and *cis*-$PH_2OH\cdots RgX$ (C-O-III, C-P-III) at the MP2/aug-cc-pVTZ-PP level of theory. α denotes the P–O–Rg or O–P–Rg angles.

geometries and that the RgB strength followed the tendency Au > Cu > Ag, F > Cl > Br, and P··· RgX > O··· RgX.

The authors also analyzed the *trans–cis* equilibrium, concluding that both structures were easily converted into each other due to a low energy barrier. Lastly, the orbital analyses revealed two resonance-type structures of P:Rg-X (ωI) ↔ P–Rg:X (ωII), O:Rg-X (ωI) ↔ O–Rg:X (ωII). The authors attributed the competition between ωI ↔ ωII resonance structures to hyperconjugation interactions present in these complexes.

In a posterior study, Zhou and coworkers [40] evaluated the RgBs established in AuX (X = F, Cl, Br)···PHOHR (R = CH_3, F, CF_3, NH_2, CN) complexes at the CCSD(T) and MP2/aug-cc-pVTZ-PP level of theory. To achieve this goal, the authors used four types of different dispositions, named *trans*-P, *trans*-O, *cis*-P, and *cis*-O (some representative examples are shown in Fig. 8.4). The authors concluded that the strength of the RgBs studied (with interaction energy values ranging from −76.6 to −20.4 kcal/mol at the CCSD(T)/aug-cc-pVTZ-PP level of theory) decreased in the order NH_2 > CH_3 > F > CF_3 > CN, F > Cl > Br, being the nature of the substituent crucial in regulating the strength of the RgB.

Furthermore, the authors also analyzed the *cis/trans* conformation barrier, concluding that for the O–shared complexes (Fig. 8.4 top) the *cis* tautomers were more stable in the case of F, CF_3, and CN–substituted complexes while in NH_2– and CH_3–substituted dimers the *trans* conformation exhibited more stability than the *cis* conformation. On the other hand, among the P–shared complexes (Fig. 8.4 bottom)

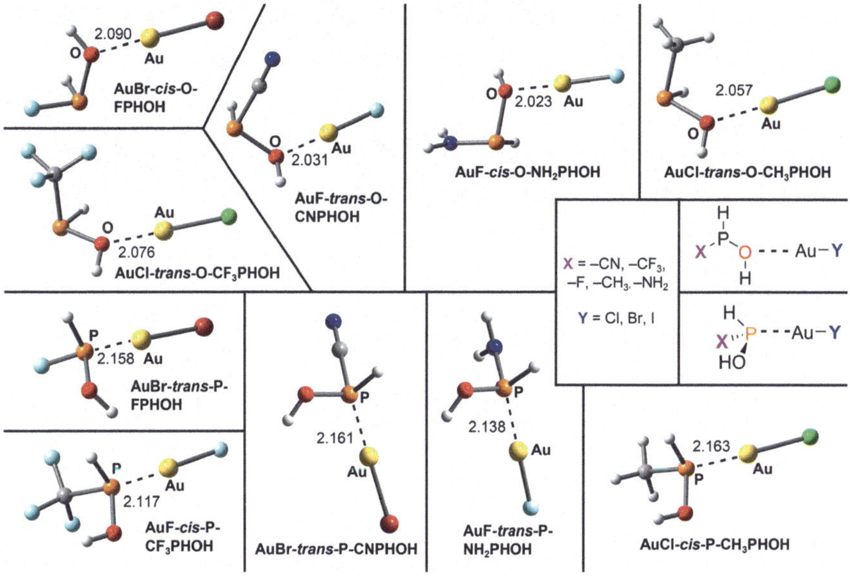

Fig. 8.4 Representative set of optimized complexes at the MP2/aug-cc-pVTZ-PP level of theory. Distances in Å

the *trans* tautomers were more stable than their *cis* analogous. Finally, orbital analysis suggested that dispersion energy dominates the formation of the Au···O RgBs, while for Au···P complexes the interaction was dominated by electrostatics.

Alkorta et al. have studied the recognition and binding stoichiometry of halide anions by pyrazolate in a series of theoretical studies [41, 42], while Ulloa and coworkers [43] have theoretically analyzed the NCIs responsible for the formation of $[Rg_3(3,5\text{-}(CF_3)_2Pz)_3] - C_{60}$ cocrystals in a 4:1 ratio. In the study from Alkorta and corwokers [42], the authors evaluated the strength, geometrical features, and electron density properties of the 1:1, 1:2, and 1:3 complexes involving cyclic $(Py\text{-}Rg)_3$ (Rg = Au, Ag, and Cu) units and halide ions (F^-, Cl^- and Br^-) at the MP2/aug-cc-pVTZ level of theory (see Fig. 8.5a).

To unveil potential areas of anion binding the authors carried out the MEP analysis of three pyrazolate moieties bearing Cu $(Pz\text{-}Cu)_3$, Ag $(Pz\text{-}Ag)_3$, and Au $(Pz\text{-}Au)_3$ atoms (see Fig. 8.5b). As noticed, the MEP surface exhibited both negative regions (highlighted in red) placed above and below the pyrazole moiety and positive regions (highlighted in blue) on the H atoms in the periphery. Interestingly, the core region

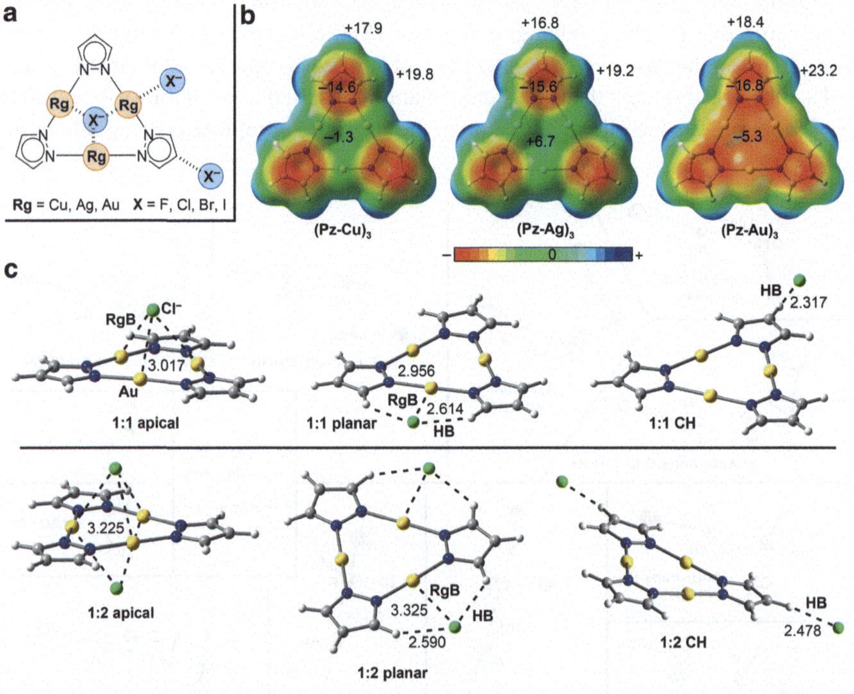

Fig. 8.5 **a** Schematic view of the possible complexes with X^- (X = F, Cl, and Br). The scheme shows the possible interaction sites. **b** MEP surfaces of $(Pz\text{-}Cu)_3$, $(Pz\text{-}Ag)_3$, and $(Pz\text{-}Au)_3$. Energy values at selected points on the surface are given in kcal/mol (0.001 a.u.). **c** Optimized geometries of several $Pz\text{-}Au_3$ dimers (1:1 apical, 1:1 planar, and 1:1CH) and trimers (1:2 apical, 1:2 planar, and 1:2 CH) at the MP2/aug-cc-pVTZ level of theory

of the pyrazolate showed a negative electrostatic potential value in the case of Pz-Cu_3 (−1.3 kcal/mol), which became positive in Pz-Ag_3 (+6.7 kcal/mol) and turned negative again in the case of Pz-Au_3 (−5.3 kcal/mol). On the other hand, all CH groups from the pyrazolate moiety exhibited positive MEP values (ranging from +16.8 to +23.2 kcal/mol) being those associated with the side CH groups from the pyrazole ring more positive than those from the central CH group (e.g., +19.8 and +17.9 kcal/mol in Pz-Cu_3). Finally, in Fig. 8.5c some representative optimized dimers and trimers are shown, thus establishing (i) multiple RgBs (in 1:1 apical and 1:2 apical), (ii) combined hydrogen bonding (HB) and RgBs (in 1:1 planar and 1:2 planar) and (iii) HBs (in 1:1 CH and 1:2 CH).

Among the 1:1 complexes, the binding energies resulted in large and attractive values, as expected for the interaction between an anion and a metal ion, ranging between −48.8 and −8.4 kcal/mol. The authors concluded that the most stable complex of each configuration highly depended on the metal atom and on the anion considered. For example, in 1:1 apical complexes, those involving Ag and Cu derivatives were found to be the most favorable, while for Au derivatives 1:1 planar complexes exhibited the largest stability. Finally, those complexes involving the interaction with the central pyrazole CH group were the less stable of the set.

On the other hand, in case of 1:2 complexes, the interaction energies ranged from –19.6 to +33.5 kcal/mol, strongly depending on the metal and interacting anion. In the case of complexes involving Br^- and Cl^-, the binding energies were found to be repulsive upon formation of the complex. In general, the authors found that for each anion and configuration, the most stable complexes implicated the Ag pyrazolate unit.

Lastly, the authors also carried out calculations using 1:3 HB complexes. However, similarly than that to 1:2 complexes, the 1:3 complexes exhibit positive binding energies ranging between +66.2 and +38.2 kcal/mol, much larger than those for the 1:2 complexes, although the obtained structures represented a minima in the potential energy surface.

Finally, to complement their results with experimental evidences, the authors carried out a search in the Cambridge Structural Database (CSD) looking for complexes where $(Pz\text{-}Rg)_3$ interacts with halides. The search showed a total of 13 X-ray crystal structures with the presence of halogen atoms in apical disposition interacting with Cu(II) atoms bridging the pyrazole rings. Three examples retrieved from the search are shown in Fig. 8.6; RUYGUN [44] and OBOQAY [45] involving

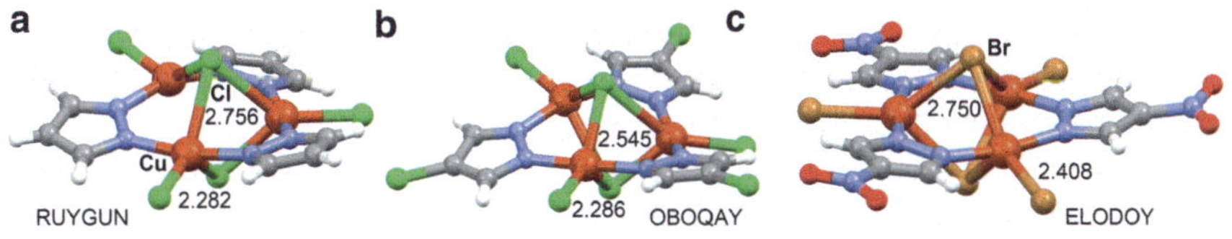

Fig. 8.6 Partial views of the X-ray crystal structures RUYGUN **a** OBOQAY, **b** and ELODOY, **c.** Distances in Å

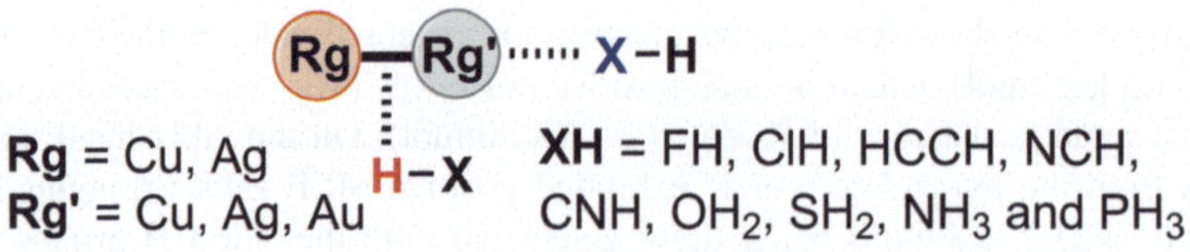

Fig. 8.7 Schematic representation of the HB and RgB complexes studied by Sanchez-Sanz and collaborators

a Cl^- and ELODOY [46] involving a Br^- atom. It is also worthy to emphasize that the global charge was compensated by tetra-n-butylammonium cations.

Other interesting examples of the combination of RgBs and HBs are the studies from Sánchez-Sanz and coworkers [47, 48]. In their study [48] the authors evaluated the formation of RgB and HB complexes formed by the combination of RgRg' species (Rg = Cu, Ag and Rg' = Cu, Ag and Au), which interestingly are able to behave as both HB acceptor (Lewis base) or RgB donor (Lewis acid) and several XH molecules (FH, ClH, OH_2, SH_2, HCN, HNC, HCCH, NH, and PH_3) at the CCSD(T)/CBS [49] level of theory (see Fig. 8.7).

The authors performed all possible combinations of RgB and HB dimers, concluding that in most of the cases, the RgBs established are much stronger than the HBs. The authors also studied the transition barriers for the interconversion between the RgB and HB, which showed that this process was disfavored from a thermodynamic point of view. On the other hand, the inverse process (from HB to RgB) exhibited very small energy barriers, indicating an almost barrierless process. Finally, when mixed binary coinage molecules are used (e.g., CuAg), the authors observed that AgCu···XH complexes behave similarly to Ag_2···XH and Cu_2···XH complexes with small variations of the interaction energy. Interestingly, the large electronegativity difference between Au and Ag (or Cu) created large dipole moments in the RgRg' molecules which dramatically affected the formation of the complexes.

Two interesting works involving RgBs with aromatic surfaces were carried out by us [50, 51]. In our former study [50], we analyzed the energetic and geometric features of a series of regium–π (Rg–π) complexes involving Cu_9, Ag_9, and Au_9 clusters and several aromatic compounds, including benzene, naphthalene, and anthracene (as well as their perfluorinated derivatives) and trifluorobenzene at the PBE0 [52, 53]-D3 [54]/def2-TZVP [55] level of theory. In Fig. 8.8a some representative complexes are shown where the Rg_9 moiety can either interact through the σ–hole (in Cu_9-B, Ag_9–TFB, and Ag_9–NT complexes) or the σ–lump (in Au_9–HFB and Au_9–AT complexes) depending on the electronic nature of the aromatic system (see Fig. 8.8b, c). The interaction energies obtained ranged between −19.7 to −7.7 kcal/mol, which are attractive and moderately strong values.

To conclude this section is also worthy to emphasize the studies from Alkorta [56], Cui [57], Larrañaga [58], and Zierkiewicz [59] and collaborators, where they analyze the strength, directionality, and charge-density properties of RgBs involving both simple molecules (e.g., Cu_2, AgCl) as well as small metal clusters (e.g., Au_6) as RgB donors with neutral Lewis bases.

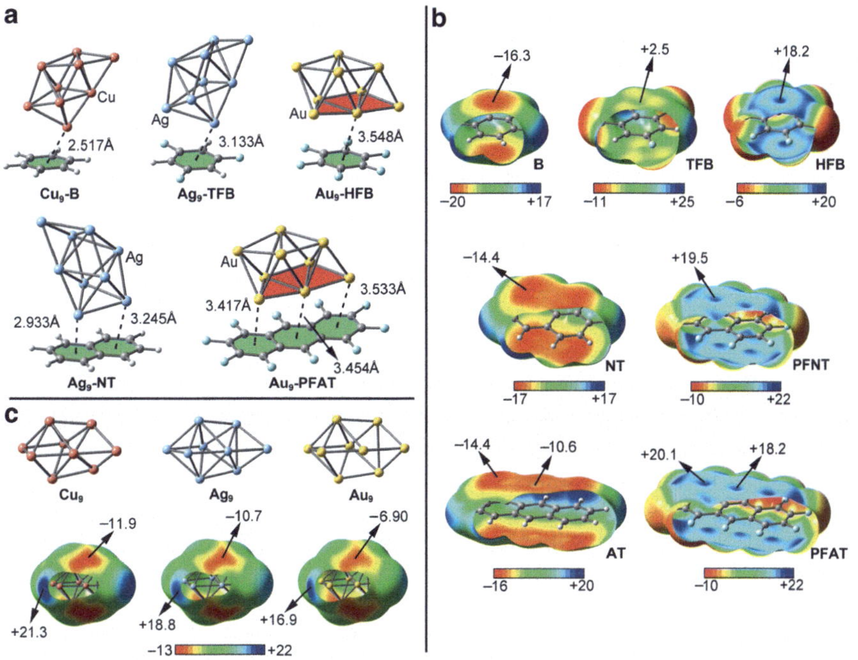

Fig. 8.8 **a** Optimized Rg–π complexes at the PBE0-D3/def2-TZVP level of theory. **b** MEP surfaces of benzene (B), trifluorobenzene (TFB), hexafluorobenzene (HFB), naphthalene (NT), perfluoronaphtalene (PFNT), anthracene (AT) and perfluoroanthracene (PFAT). **c** MEP surfaces of Cu_9, Ag_9, and Au_9 metal clusters. Energy values at selected points on the surface are given in kcal/mol (0.001 a.u.)

Interplay between Regium bonds and other NCIs

Two studies were selected where cooperative effects are present between RgB and TrB and NgB interactions. In the first one, Zhang and collaborators [60] evaluated cooperativity between RgB and TrB interactions at the MP2/aug-cc-pVTZ-PP level of theory. More, in particular, the authors computed the interaction energy of binary TrB complexes of $RgCN\cdots ZX_3$ (Rg = Cu, Ag, Au; Z = B, Al; X = H, F) and $C_2H_4\cdots ZX_3$ as well as binary RgB $C_2H_4\cdots RgCN$ complexes (see Fig. 8.9). The energetic results obtained ranged from −55.5 to −20.1 kcal/mol and the authors observed a dependency of the strength of the TrB on the nature of Tr and Rg atoms involved as well as in the Tr substituents (either H or F). The authors also performed orbital analyses, finding an orbital interaction involving the donation either from a lone pair of the N atom or from an occupied C = C π-orbital to an empty p orbital (p*) of the Tr atom (B or Al). On the other hand, in the case of the RgB complexes two main orbital contributions were unveiled; i) from an occupied C = C π-orbital to an empty σ orbital (σ*) from the metal atom and ii) from a d occupied orbital of the metal atom to the C = C antibonding π* orbital.

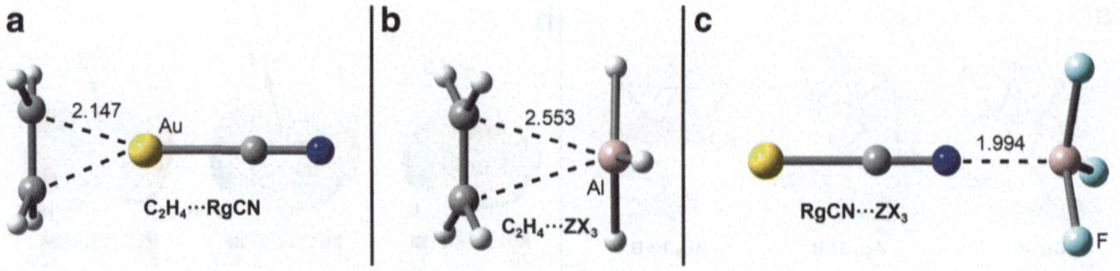

Fig. 8.9 Representative optimized dimers exhibiting regium bonding (RgB) (**a**, $C_2H_4\cdots RgCN$) and triel bonding (TrB) (**b**, $C_2H_4\cdots ZX_3$ and **c**, $RgCN\cdots ZX_3$) interactions at the MP2/aug-cc-pVTZ level of theory. Distances in Å

The authors also carried out an energy decomposition analysis of the TrBs and RgBs studied herein into electrostatic energy (E_{ele}), exchange energy (E_{ex}), repulsion energy (E_{rep}), polarization energy (E_{pol}), and dispersion energy (E_{disp}) to understand the origin of both NCIs. Results showed that among RgB complexes (Fig. 8.9a), the authors found that E_{ele} was larger than E_{pol} and E_{disp} in $C_2H_4\cdots CuCN$ and $C_2H_4\cdots AgCN$ complexes, while the former played a crucial role in $C_2H_4\cdots AuCN$ complexes. In addition, in N···Z TrB complexes (Fig. 8.9c), E_{pol} dominates the interaction beyond E_{ele}, while the latter plays a more prominent role in those complexes involving BF_3 as TrB donor. Furthermore, the F substituents decrease the values of both E_{ele} and E_{pol} in N···B TrBs, while an increased effect on both terms was found in N···Al TrB complexes. Moreover, E_{disp} played a minor role in $RgCN\cdots ZH_3$ complexes. The authors arrived at a similar conclusion by inspecting the TrBs involving $C_2H_4\cdots AlX_3$ system (Fig. 8.9b). Lastly, the weight of each energy term decreased in the order AuCN > CuCN > AgCN.

In a final stage of their study, the authors computed the trimers involving both RgBs and TrBs (see Fig. 8.10 for three representative cases) to unveil favorable/unfavorable cooperative effects between both interactions. They obtained favorable cooperative values in $C_2H_4\cdots RgCN\cdots ZX_3$ (Fig. 8.10a) trimers but unfavorable in NCRg·· $C_2H_4\cdots ZX_3$ and $C_2(CN)_4\cdots RgCN\cdots ZH_3$ trimers (Fig. 8.10b, c, respectively). Surprisingly, in the $C_2H_4\cdots RgCN\cdots ZX_3$ complexes, the TrB exhibited a shorter binding distance compared to its corresponding dimer, while the RgB possessed a longer binding distance than their analogous dimer.

In the second selected example, a computational study by Wang and coworkers at the MP2/aug-cc-pVTZ level of theory analysed the interplay between RgBs and NgBs [61]. The authors started by computing the NgB dimers in $HCN\cdots XeF_2O$ and $C_2H_4\cdots XeF_2O$ complexes. They obtained larger interaction energy values for those complexes involving HCN, in agreement with its stronger Lewis base character compared to C_2H_4. Furthermore, they substituted the H atom in HCN for Cu, Ag, and Au and evaluated its influence on the formation of the NgB, finding that the metal substitution strengthens the NgB in $RgCN\cdots XeF_2O$ (Rg = Cu, Ag, and Au) and its enhancing effect follows the Au < Cu < Ag pattern. Particularly, the authors obtained a NgB interaction strength ranging between −13.7 and −6.5 kcal/mol.

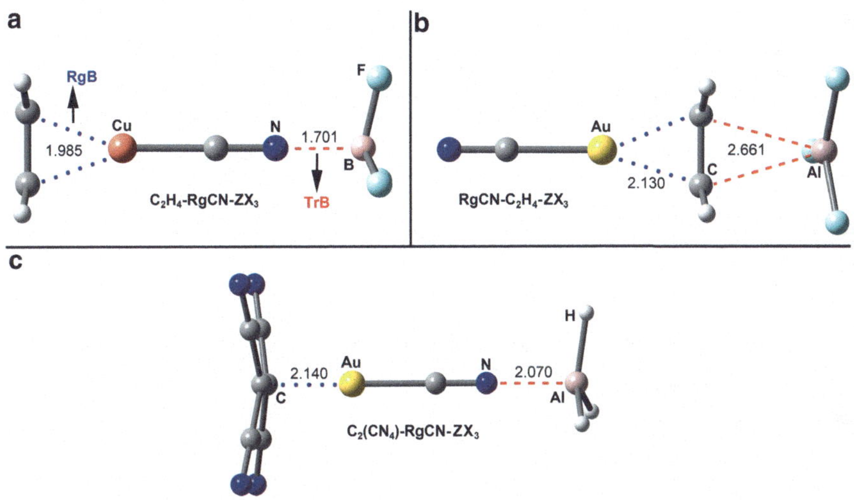

Fig. 8.10 Representative optimized trimers exhibiting cooperativity between Regium (RgB) and Triel (TrB) bonding interactions at the MP2/aug-cc-pVTZ level of theory. **a** C_2H_4-RgCN-ZX_3, **b** NCRg-C_2H_4-ZX_3 and **c** $C_2(CN)_4$-RgCN-ZH_3. Distances in Å

The authors also characterized the NgBs studied herein by an orbital analysis, which showed that for whether RgCN···$XeFO_2$ or C_2H_4···$XeOF_2$ complexes there were two prominent orbital interactions; i) the interaction between a lone pair from the N atom to an antibonding Xe–O σ* orbital and ii) from a lone pair of the Xe atom to an antibonding C–N σ* orbital in RgCN···$XeFO_2$ complexes, iii) the interaction between a C = C π bonding orbital to an antibonding Xe–O σ* orbital and iv) from a lone pair of the Xe atom to an antibonding C = C π* orbital in C_2H_4···$XeOF_2$ systems. To conclude this section, the authors performed an energy decomposition analysis into five physical terms electrostatic, exchange, repulsion, polarization, and dispersion energies. They found that the exchange energy was the largest attractive term followed by electrostatics in all the complexes studied. Thus likely pointing to a major predominance of electrostatics in the formation of the NgB complexes analyzed herein, although with some degree of covalency.

Lastly, the authors studied the cooperative effects between both interactions by computing trimers (some representative examples are shown in Fig. 8.11). They found favorable cooperativity values in C_2H_2-RgCN-XeF_2O and C_2H_4-RgCN-XeF_2O trimers (Fig. 8.11a), where both NgB and RgB are reinforced. On the contrary, both types of interactions are weakened in RgCN-C_2H_4-XeF_2O and $C_2(CN)_4$-RgCN-XeF_2O systems (Fig. 8.11b, c respectively) indicating an unfavorable cooperativity between both interactions.

To conclude this section, it is also worthy to mention the study by Zhang and collaborators [62] where they computationally studied cooperativity effects between RgBs and pnicogen bonds (PnBs) at the MP2/aug-cc-pVTZ-PP level of theory.

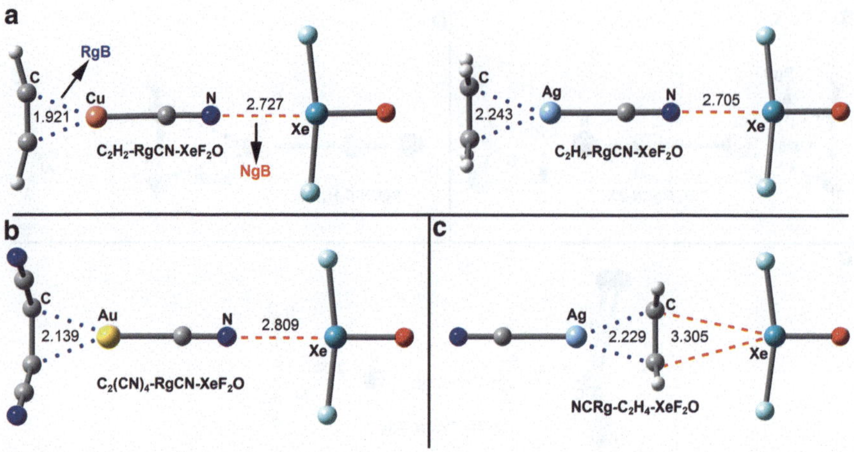

Fig. 8.11 Representative optimized trimers showing cooperative effects between Regium (RgB) and Noble gas bonding interactions (NgB). **a** C_2H_2-RgCN-XeF_2O systems, **b** C_2H_4-RgCN-XeF_2O and **c** NCRg-C_2H_4-XeF_2O. Distances in Å

8.2.2 Experimental Studies

The following selected experimental studies involve RgBs in the fields of crystal engineering, biological systems, and metal surface absorption phenomena.

Regium bonds as a novel tool in crystal engineering

The first selected study is the work of Daolio and collaborators [63], where they synthesized and characterized two tetrachloridoaurates of acetylcholine and dimethylpropiothetine by combining crystallography and computations at the PBE0-D3/def2-TZVP level of theory. More specifically, the authors used $AuCl_4^-$ anions, which act as self-complementary moieties, where the Au and Cl atoms play the role of RgB donor and acceptor entities, respectively. In Fig. 8.12, the X-ray structures from the two newly synthesized compounds are shown with indication of both Au···Cl (in

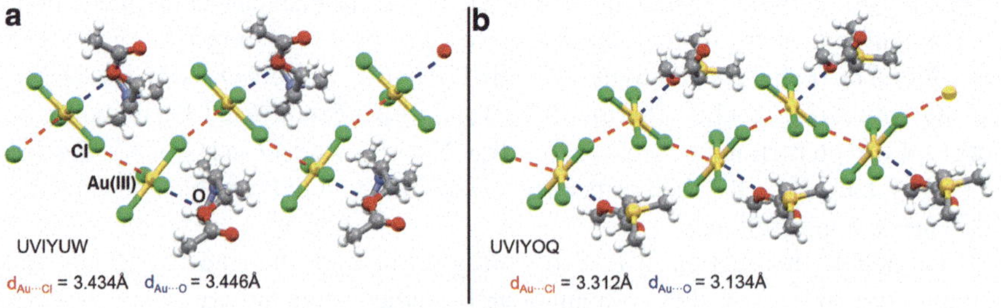

Fig. 8.12 Partial views of the X-ray crystal structures UVIYUM (**a**) and UVIYOQ (**b**) exhibiting Au···Cl (highlighted in red) and Au···O RgBs (highlighted in blue) contacts

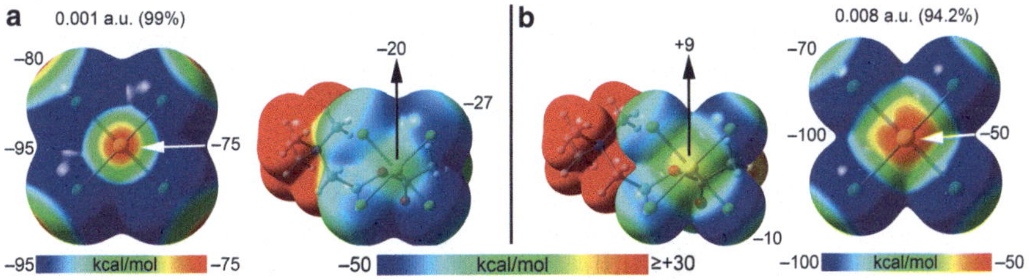

Fig. 8.13 MEP surfaces of the $AuCl_4$ surface unit and $AuCl_4$-acetylcholine complex using 0.001 a.u. **a** and 0.008 a.u. **b** isovalues at the PBE0-D3/def2-TZVP level of theory. Energy values are given in kcal/mol at several points of the surface

red) and Au···O (in blue) RgB contacts. As noted, the Au···Cl contacts involving Au and Cl atoms from different units drive the formation of a supramolecular anionic polymer, wherein the metal centers established a second RgB with the sp^3 hybridized oxygen atom belonging to the ester moieties.

To demonstrate the electrophilic role of Au as well as the attractive nature of both the Au···Cl and Au···O interactions, the authors performed theoretical calculations (PBE0-D3/def2-TZVP level of theory). Concretely, the authors carried out a MEP analysis (see Fig. 8.13) of the isolated anion (which resulted negative over the entire surface owing to its net negative charge and locating the global minimum along the bisector of the Cl-Au-Cl angle). On the other hand, the MEP surface analysis of the salt showed that the value of the potential of the Au center, while remaining negative, was reduced to −20.1 kcal/mol. However, this turned into positive (+9 kcal/mol) while adjusting the MEP surface to the intermolecular Au···O/Cl distances, which were shorter than the sum of the corresponding Au···O and Au···Cl van der Waals radii. This study might pave the way for the design of new materials where RgBs involving Au(III) centers dictate the formation of attractive anion···anion and anion···neutral nucleophile interactions.

In a different study, Priola and coworkers [64] synthesized and characterized the structure of the $\{Cu(NH_3)_4[Au(CN)_2]\}^+[Au(CN)_2]^-$ salt, which exhibited an unprecedented $[d^9]$Cu···$[d^{10}]$Au RgB. In their work, the authors reported first evidence of RgBs between two cationic species, where Cu(II) and Au(I) act as electron acceptor and donor moieties, respectively. This counterintuitive RgB involving Cu and Au atoms played a prominent role in the formation of a supramolecular cationic polymer (see Fig. 8.14). Theoretical calculations (PBE0-D3/def2-TZVP level of theory) provided support for the electrophilic role of Cu and nucleophilic role of Au as well as the attractive nature of the $[d^9]$Cu···$[d^{10}]$Au interaction. More specifically, the authors carried out a MEP analysis of the $\{Cu(NH_3)_4[Au(CN)_2]\}^+$ cationic unit and the ion pair (see Fig. 8.15).

As noted the MEP values are positive over the entire surface of the cationic molecule due to its electronpoor nature, with the exception of the N atom belonging to cyano ligand, which is not coordinated to a metal atom and exhibited a negative potential (−29 kcal/mol). The most positive MEP values were found at the Cu(II)

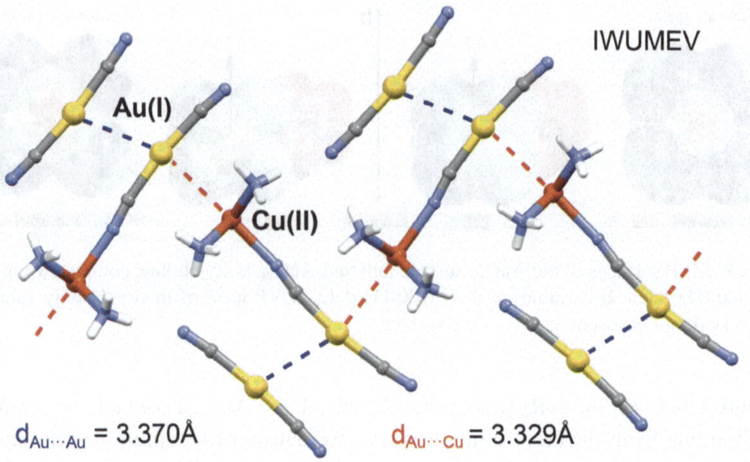

Fig. 8.14 X-ray architecture IWUMEV including Au···Au (in blue) and Au···Cu (in red) contacts

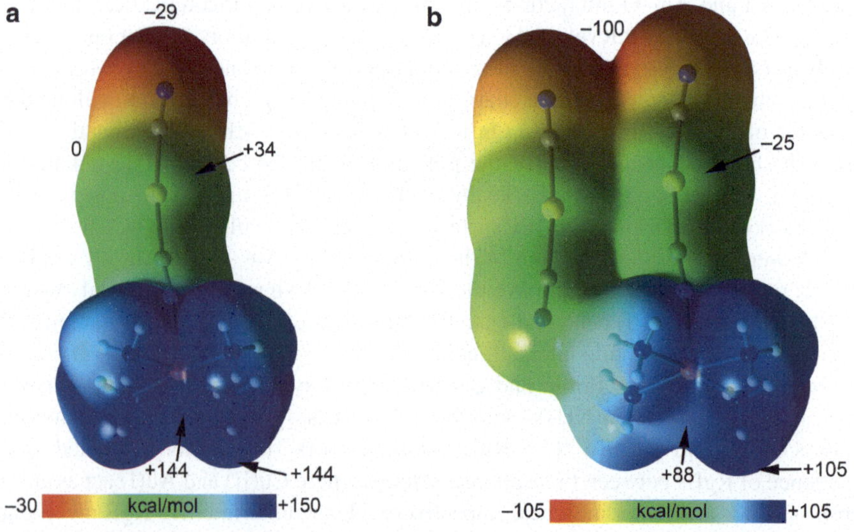

Fig. 8.15 MEP surfaces of the $\{Cu(NH_3)_4[Au(CN)_2]\}^+$ cationic unit (**a**) and the ion pair (**b**) at the PBE0-D3/def2-TZVP level of theory. The values are given in kcal/mol at several points of the surface using 0.001 a.u. isovalues

atom located on the opposite site to the Au(I) coordinated –CN group (exhibiting a σ-hole) as well as the H atoms (+144 kcal/mol) belonging to the Cu(II) coordinated ammonia molecules, thus being useful for understanding the location of the counterions in this particular X-ray structure.

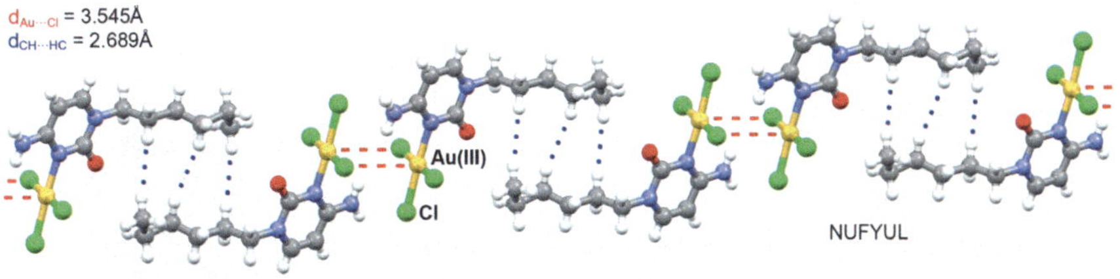

Fig. 8.16 Partial view of the X-ray crystal structure NUFYUL exhibiting Au···Cl RgBs (highlighted in red) and CH···HC interactions (highlighted in blue)

When using the closest anion to perform the MEP surface analysis of the salt, the σ-hole MEP value at the Cu(I) center decreased to +88 kcal/mol while the MEP maxima were located at the H atoms of the coordinated ammonia molecules (+105 kcal/mol). Lastly, the MEP value of the region between the terminal CN groups of both metal coordinated and non-coordinated $[Au(CN)_2]^-$species exhibited a very negative electrostatic potential (–100 kcal/mol). It is also worthy to emphasize that the Au atoms MEP values are negative, thus being adequate for understanding their role as electron donor species.

These findings show that RgBs involving two noble metal atoms (one acting as donor and another as acceptor) are strong enough to drive the formation of attractive cation···cation NCIs, which ultimately is key for determining the crystal packing of the salt.

Another interesting example of a solid-state architecture governed by RgBs is the study by Terrón and collaborators [65]. Concretely, the authors reported the synthesis and X-ray characterization of an Au(III) complex of 1-hexylcytosine via N(3). As noted in Fig. 8.16, the solid-state architecture of this compound is formed by 2D layers where molecules self-assemble by means of Au···Cl contacts (highlighted in red) to form dimers that further interact with each other through hydrophobic C–H···H–C interactions (highlighted in blue) involving the aliphatic hexyl chains.

To verify the presence of a region of positive electrostatic potential over the $AuCl_3N$ moiety that helped to understand the crystal packing pattern exhibited by this compound, the authors carried out a MEP analysis and the result is shown in Fig. 8.17. As it can be observed, the MEP surface exhibits a region of positive potential above and below the Au atom (+11 kcal/mol). The most electropositive region was located on the NH_2 H atoms (+25 kcal/mol), while the most negative MEP values were found at the $AuCl_3N$ plane between two chlorido ligands (−41 kcal/mol). Finally, it is interesting to highlight that the MEP value at the belts of the Cl atoms is negative, thus being suitable to participate as RgB electron donor species (Lewis base).

In addition, a posterior orbital analysis revealed a significant orbital contribution from an electron donation from a lone pair belonging to the chlorine atom to an empty p orbital (p*) of Au (basically constituted by the 6pz atomic orbital), thus remarking the importance of orbital interactions in the formation of this noncovalent complex.

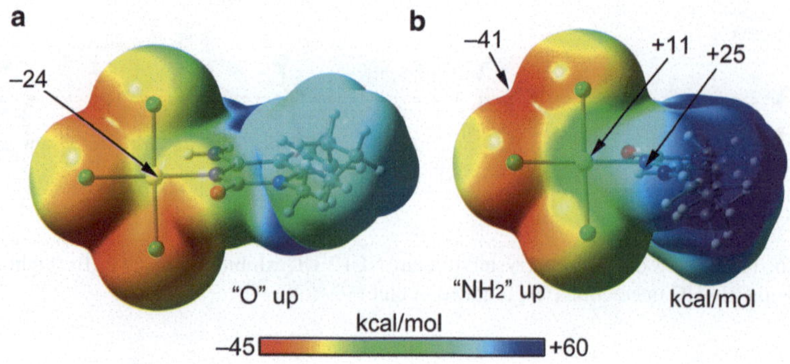

Fig. 8.17 MEP surfaces (using an isovalue of 0.001 a.u.) of the X-ray structure synthesized by Terrón and coworkers at the PBE0-D3/def2-TZVP level of theory, using two different "on-top" orientations ("O" up in A and "NH_2" up in B). Energies at selected points are given in kcal/mol

Regium bonds in biological systems

RgBs have been also spotted in biological systems, playing important roles in protein conformation and protein-drug interactions. Some examples are given below while some others can be found in our work [66]. In this chapter four selected protein–ligand complexes are shown in Fig. 8.18, corresponding to PDB structures 1UT7 [67], 1A8D [68] 4ZFP [69], and 5Y0Y [70].

The first structure (PDB code 1UT7, Fig. 8.18a) encompasses the DNA binding domain of Arabidopsis ANAC (abscisic acid-responsive NAC) protein. This protein family is widespread in plants and their main function is related to the deregulation of crucial biological processes (e.g., transcriptional activity [71], hormone inducible processes [72], and protein complexation routes [73]). Ersnt et al. [67] reported

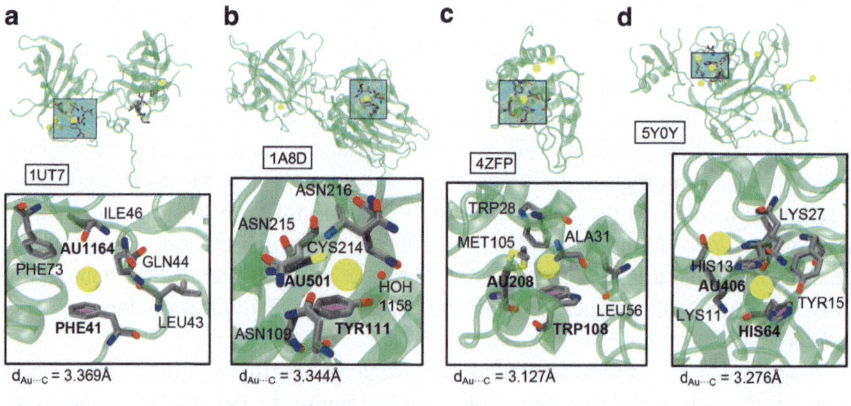

Fig. 8.18 Partial views of the X-ray structures of PDB entries **a** 1UT7, **b** 1A8D, **c** 4ZFP, and **d** 5Y0Y. The regium−π (Rg−π) interaction is magnified in the bottom part of the figure. The Au···C distances are also indicated (measured to the C atoms of the ring in 1A8D and to the ring centroid in 1UT7, 4ZFP, and 5Y0Y)

structural insights of the protein−protein and protein–DNA interactions involving NAC. To achieve this, the authors used dicyanoaurate to yield a high-resolution X-ray crystal structure and identified four Au(I) binding sites (three on one chain of the protein (named A) and one on the other (named B)). In chain A, a hydrophobic pocket formed by Phe73, Ile46, Leu43, and Phe41 residues accommodated an Au(I) ion through the formation of a Rg−π bond (Fig. 8.18a). This finding might lead to structural implications because the corresponding region in monomer B (where no Rg–π bonds were undertaken) exhibited a different structural pattern.

The second example PDB entry 1A8D (Fig. 8.18b) belongs to the Tetanus toxin, an enzyme present in *Clostridium tetani* that produces an exotoxin that can potentially cause paralytic death. To design therapeutic agents capable of effectively inhibiting this enzyme the structure belonging to the ganglioside binding region was solved at 1.6 Å resolution by Knapp and coworkers [68]. Concretely, the authors used Au atoms to solve the X-ray crystal structure and one Au(I) ion coordinated to CYS214 is interacting with a neighboring TYR111 residue through a Rg−π bond (Fig. 8.18b). Thus, Rg–π interactions might become a novel tool for rational drug design as well as design novel therapeutic agents for the Tetanus toxin enzyme.

The third example involves the work of Ferraro and collaborators, [69] who used lysozymes (concretely HEWL) as an enzyme model to further understand protein−Au interactions (Fig. 8.18c). More precisely, in the structure of PDB entry 4ZFP they used AuSac2, a medicinal Au(I) compound, to shed light on how Au anchors inside the protein hydrophobic core. Importantly, an Au(I)−Met105 coordination complex undertook a Rg−π interaction with the Trp108 π–system, which belongs to a hydrophobic pocket composed of Trp28, Ala31, and Leu56 residues. Owing to the important HEWL fluorophore properties of Trp108, the authors carried out fluorescence spectroscopy assays (both in the presence and absence of AuSac2) to evaluate the implications of these hydrophobic residues in the recognition and binding phenomena of Au. They observed an abrupt decrease in the fluorescence intensity of HEWL upon the addition of AuSac2 and proposed an energy transfer mechanism from Trp108 to the Au(I)−Met105 complex, resulting in a noticeable quenching of the protein fluorescence, thus enhancing the visible emission from the ligand. Hence, the presence of a Rg−π bond might play an important role in facilitating the charge transfer process between both counterparts and might be responsible for the observed quenching phenomena.

As a final example, the study from Wu and collaborators [70] sheds light on the structural features of phlebovirus glycoprotein Gn, being one of the envelope proteins on the surface of Severe fever with thrombocytopenia syndrome virus (SFTSV) and Rift Valley fever virus (RVFV), also considered as a major antigenic component. Despite its importance for virus entry and fusion, the molecular features of the phlebovirus Gn are mostly unknown until date. The authors used Au derivatives to solve the X-ray crystal structure of these proteins. In addition, they used HIS tags owing to their convenience for purification and detection assays since these residues are known to effectively bind several types of metal ions under specific buffer conditions. In this case (see Fig. 8.18d), an Au atom (AU406) is trapped between two HIS residues (HIS13 and HIS64) and exhibits both a coordination bond with HIS13

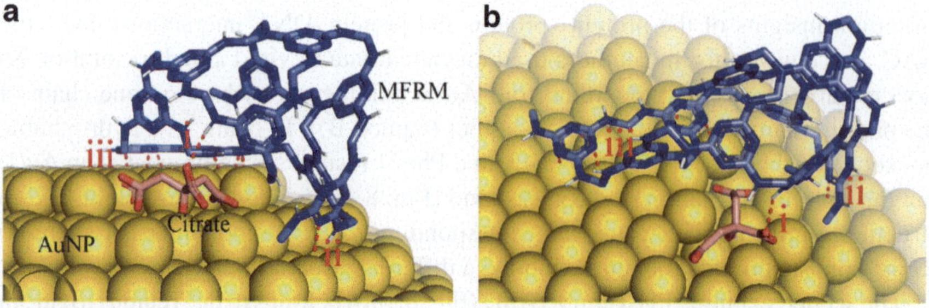

Fig. 8.19 Side (**a**) and zenithal (**b**) views of a molecular docking study between the monomeric unit of MFRM and citrate-AuNP. i, ii, and iii denote electrostatics, hydrogen bonds (HB), and regium–π (Rg–π) interactions, respectively. Reprinted (adapted) with permission from reference [74]. Copyright 2021 American Chemical Society

($d_{Au\cdots N}$ = 2.209 Å) and a Rg–π interaction with HIS64 ($d_{Au\cdots C}$ = 3.276 Å). This example showcases the ability of HIS residues to covalently and noncovalently trap Au(I) ions, which might be important in the field of protein engineering.

Regium bonds are involved in metal surface absorption phenomena

As the last selected example of this chapter, it is worth highlighting the study from Li and coworkers [74], which deals with the synthesis characterization of melamine–formaldehyde resin microspheres (MFRM) as active absorpters of Au nanoparticles (AuNPs). AuNPs possess hazardous effects on the environment, animal species, and human health as well as being potential pollution agents on the environment [75]. Particularly, they are known to exert cytotoxicity and genotoxicity on marine organisms, mammals, including human cells [76]. Therefore, there is a growing interest in the design of new materials capable of binding and removing AuNPs from the ecosystem. During their study, the authors elucidated the molecular recognition mechanism of the MFRM-AuNP absorption by means of docking simulations (see Fig. 8.19). They concluded that the MFRM compound interacts with the AuNP through the combination of i) electrostatics, ii) metal–ligand coordination bonds, and iii) Rg–π interactions. This study might open new ways to design novel adsorbents for AuNP removal, thus finding new ways to noticeably minimize the impact of metal nanoparticles in the environment and living ecosystems.

8.3 Concluding Remarks

We are at the mere naissance of RgBs as novel tools in several fields such as crystal engineering, biology, and environmental chemistry. The X-ray structures and theoretical studies shown in Sects. 2.1 and 2.2 of this chapter, although not intended to be comprehensive, provide theoretical and experimental support to the fact that the elements from 11 are able to undergo noncovalent binding with Lewis bases.

Furthermore, RgBs play a relevant role in governing the solid-state architecture of the crystalline solids. Usually, this interaction is often overlooked or simply identified as a short contact and therefore, it should be considered and used by the scientific community. To date, few works have used this interaction on purpose to design and construct supramolecular assemblies, and thus, it is expected that these interactions will be increasingly used in crystal engineering as well as in supramolecular chemistry and biological systems.

References

1. Pizzi A, Pigliacelli C, Bergamaschi G, Gori A, Metrangolo P (2020) Biomimetic engineering of the molecular recognition and self-assembly of peptides and proteins via halogenation. Coord Chem Rev 411:213242.https://doi.org/10.1016/j.ccr.2020.213242
2. Daolio A, Scilabra P, Terraneo G, Resnati, G (2020) C(sp^3) atoms as tetrel bond donors: a crystallographic survey. Coord Chem Rev 413:213265.https://doi.org/10.1016/j.ccr.2020.213265
3. von der Heiden D, Vanderkooy A, Erdélyi M (2020) Halogen bonding in solution: NMR spectroscopic approaches. Coord Chem Rev 407:213147.https://doi.org/10.1016/j.ccr.2019.213147
4. Gill H, Gokel MR, McKeever M, Negin, S Patel MB, Yin S, Gokel GW (2020) Supramolecular pore formation as an antimicrobial strategy. Coord Chem Rev 412:213264. https://doi.org/10.1016/j.ccr.2020.213264
5. Xu Y, Szell PMJ, Kumar V, Bryce DL (2020) Solid-state NMR spectroscopy for the analysis of element-based non-covalent interactions. Coord Chem Rev 411:213237.https://doi.org/10.1016/j.ccr.2020.213237
6. Wang W, Zhang Y, Jin WJ (2020) Halogen bonding in room-temperature phosphorescent materials. Coord Chem Rev 404:213107.https://doi.org/10.1016/j.ccr.2019.213107
7. Grabowski SJ (2020) Triel bond and coordination of triel centres – Comparison with hydrogen bond interaction. Coord Chem Rev 407:213171.https://doi.org/10.1016/j.ccr.2019.213171
8. Fromm, KM (2020) Chemistry of alkaline earth metals: It is not all ionic and definitely not boring! Coord Chem Rev 408:213192.https://doi.org/10.1016/j.ccr.2020.213193
9. Fourmigué M, Dhaka A (2020) Chalcogen bonding in crystalline diselenides and selenocyanates: from molecules of pharmaceutical interest to conducting materials. Coord Chem Rev 403:213084.https://doi.org/10.1016/j.ccr.2019.213084
10. Scheiner S, Michalczyk M, Zierkiewicz W (2020) Coordination of anions by noncovalently bonded σ-hole ligands. Coord Chem Rev 405:213136.https://doi.org/10.1016/j.ccr.2019.213136
11. Taylor MS (2020) Anion recognition based on halogen, chalcogen, pnictogen and tetrel bonding. Coord Chem Rev 413:213270. https://doi.org/10.1016/j.ccr.2020.213270
12. Bauzá A, Frontera A (2020) σ/π-Hole noble gas bonding interactions: Insights from theory and experiment. Coord Chem Rev 404:213112.https://doi.org/10.1016/j.ccr.2019.213112
13. Biot N, Bonifazi D (2020) Chalcogen-bond driven molecular recognition at work. Coord Chem Rev 413:213243.https://doi.org/10.1016/j.ccr.2020.213243
14. Cavallo G, Metrangolo P, Pilati T, Resnati G, Terraneo G (2014) Naming interactions from the electrophilic site. Cryst Growth Des 14 2697–2702. https://doi.org/10.1021/cg5001717
15. Terraneo G, Resnati G (2017) Bonding matters. Cryst Growth Des 17:1439–1440. https://doi.org/10.1021/acs.cgd.7b00309
16. Desiraju GR, Ho PS, Kloo L, Legon AC, Marquardt R, Metrangolo P, Politzer P, Resnati G, Rissanen K (2013) Definition of the halogen bond (IUPAC Recommendations 2013). Pure Appl Chem 85:1711–1713. https://doi.org/10.1351/PAC-REC-12-05-10

17. Aakeroy CB, Bryce DL, Desiraju GR, Frontera A, Legon AC, Nicotra F, Rissanen K, Scheiner S, Terraneo G, Metrangolo P, Resnati G (2019) Definition of the chalcogen bond (IUPAC Recommendations 2019). Pure Appl Chem 91:1889–1892. https://doi.org/10.1515/pac-2018-0713
18. Zahn S, Frank R, Hey-Hawkins E, Kirchner B (2011) Pnicogen bonds: a new molecular linker? Chem Eur J 17:6034–6038. https://doi.org/10.1002/chem.201002146
19. Girolami GS (2009) Origin of the terms pnictogen and pnictide. J Chem Educ 86:1200–1201. https://doi.org/10.1021/ed086p1200
20. Bauzá A, Mooibroek TJ, Frontera A (2013) Tetrel Bonding Interaction: rediscovered supramolecular force? Angew Chem Int Ed 52:12317–12321. https://doi.org/10.1002/anie.201306501
21. Bauza A, Alkorta I, Elguero J, Mooibroek TJ, Frontera A (2020) Spodium bonds: noncovalent interactions involving group 12 elements. Angew Chem Int Ed 59:17482–17487. https://doi.org/10.1002/anie.202007814
22. Alkorta I, Elguero J, Frontera A (2020) Not only hydrogen bonds: other noncovalent interactions. Crystals 10:180. https://doi.org/10.3390/cryst10030180
23. Gomila RM, Bauza A, Mooibroek TJ, Frontera A (2021) Spodium bonding in five coordinated Zn(II): a new player in crystal engineering? Cryst Eng Comm 23:3084–3093. https://doi.org/10.1039/D1CE00221J
24. Kollman PA, Liebman JF, Allen LC (1970) The lithium bond. J Am Chem Soc 92:1142–1150
25. Esrafili MD, Mohammadirad N (2014) Halogen bond interactions enhanced by sodium bonds–Theoretical evidence for cooperative and substitution effects in NCX···NCNa···NCY complexes (X = F, Cl, Br, I; Y = H, F, OH). Can J Chem 92:653–658. https://doi.org/10.1139/cjc-2014-0163
26. Del Bene JE, Alkorta I, Elguero J (2009) Characterizing complexes with F-Li^+–F lithium bonds: structures, binding energies, and spin-spin coupling constants. J Phys Chem A 113:8359–8365. https://doi.org/10.1021/jp9020917
27. Brea O, Alkorta I, Corral I, Mó O, Yáñez M, Elguero J (2018) Intramolecular beryllium Bonds. Further insights into resonance assistance phenomena. In: Novoa JJ (ed) Intermolecular interactions in crystals; The Royal Society of Chemistry, London, UK, Chapter 15
28. Montero-Campillo MM, Mó O, Yáñez M, Alkorta I, Elguero J (2019) The beryllium bond. Adv Inorg Chem 73:73–121. https://doi.org/10.1016/bs.adioch.2018.10.003
29. Yang X, Li Q, Cheng J, Li W (2013) A new interaction mechanism of $LiNH_2$ with MgH_2: magnesium bond. J Mol Model 19:247–253. https://doi.org/10.1007/s00894-012-1538-5
30. Montero-Campillo MM, Sanz P, Mó O, Yáñez M, Alkorta I, Elguero J (2018) Alkaline-earth (Be, Mg and Ca) bonds at the origin of huge acidity enhancements. Phys Chem Chem Phys 20:2413–2420. https://doi.org/10.1039/C7CP07891A
31. Daolio A, Pizzi A, Calabrese M, Terraneo G, Bordignon S, Frontera A, Resnati G, Molecular electrostatic potential and noncovalent interactions in derivatives of group 8 elements. Angew Chem Int Ed. https://doi.org/10.1002/anie.202107978
32. Stenlid JH, Johansson AJ, Brinck T (2018) σ-Holes and σ-lumps direct the Lewis basic and acidic interactions of noble metal nanoparticles: introducing regium bonds. Phys Chem Chem Phys 20:2676–2692. https://doi.org/10.1039/C7CP06259A
33. Legon AC, Walker NR (2018) What's in a name? "Coinage-metal" non-covalent bonds and their definition. Phys Chem Chem Phys 20:19332–19338. https://doi.org/10.1039/C8CP03432J
34. Bauzá A, Mooibroek TJ, Frontera A (2015) The bright future of unconventional σ/π-Hole interactions. ChemPhysChem 16:2496–2517. https://doi.org/10.1002/cphc.201500314
35. Stenlid JH, Brinck T (2017) Extending the σ-Hole concept to metals: an electrostatic interpretation of the effects of nanostructure in gold and platinum catalysis. J Am Chem Soc 139:11012–11015. https://doi.org/10.1021/jacs.7b05987
36. Zheng B, Liu Y, Wang Z, Zhou F, Liu Y, Ding X, Lu T (2019) Regium bonds formed by MX (M = Cu, Ag, Au; X=F, Cl, Br) with phosphine-oxide/phosphinous acid: comparisons between oxygen-shared and phosphine-shared complexes. Mol Phys 117:2443–2455. https://doi.org/10.1080/00268976.2019.1567847

37. Watts JD, Gauss J, Bartlett RJ (1993) Coupled-cluster methods with noniterative triple excitations for restricted open-shell Hartree-Fock and other general single determinant reference functions. Energies and analytical gradients. J Chem Phys 98. https://doi.org/10.1063/1.464480
38. Weigend F, Häser M (1997) RI-MP2: first derivatives and global consistency. Theor Chem Acc 97:331–340. https://doi.org/10.1007/s002140050269
39. Wilson AK, Woon DE, Peterson KA, Dunning Jr TH. Gaussian basis sets for use in correlated molecular calculations. IX. The atoms gallium through krypton. J Chem Phys 1999; 110:7667. DOI: doi.org/https://doi.org/10.1063/1.478678
40. Zhou F, Liu Y, Wang Z, Yang Q, Zheng B (2020) Probing Au···O and Au···P regium bonding interaction in AuX (X = F, Cl, Br)···RPHOH (R = CH_3, F, CF_3, NH_2, CN) complexes. Comp Theor Chem 1179:112800.https://doi.org/10.1016/j.comptc.2020.112800
41. Alkorta I, Trujillo C, Sánchez-Sanz G, Elguero J (2020) Regium bonds between Silver(I) Pyrazolates Dinuclear complexes and Lewis bases (N_2, OH_2, NCH, SH_2, NH_3, PH_3, CO and CNH) Crystals 10:137. https://doi.org/10.3390/cryst10020137
42. Alkorta I, Elguero J, Trujillo C, Sánchez-Sanz G (2020) Interaction between trinuclear regium complexes of pyrazolate and anions, a computational study. Int J Mol Sci 21:8036. https://doi.org/10.3390/ijms21218036
43. Ulloa CO, Ponce-Vargas M, Muñoz-Castro A (2018) formation of coinage-metal···fullerene adducts. evaluation of the interaction nature between triangular coinage metal complexes (M_3 = Cu, Ag, and Au) and C_{60} through relativistic density functional theory calculations. J Phys Chem C 122:25110–25117. https://doi.org/10.1021/acs.jpcc.8b08417
44. Angaridis PA, Baran P, Boča R, Cervantes-Lee F, Haase W, Mezei G, Raptis, RG, Werner R (2002) Synthesis and structural characterization of trinuclear CuII−Pyrazolato complexes containing μ_3-OH, μ_3-O, and μ_3-Cl Ligands. Magnetic susceptibility study of $[PPN]_2[(\mu_3\text{-}O)Cu_3(\mu\text{-}pz)_3Cl_3]$. Inorg Chem 41:2219–2228. https://doi.org/10.1021/ic010670l
45. Mezei G, Raptis RG (2004) Effect of pyrazole-substitution on the structure and nuclearity of Cu(II)-pyrazolato complexes. Inorg Chim Acta 357:3279–3288. https://doi.org/10.1016/j.ica.2004.03.016
46. Boˇca R, Dlháˇn L, Mezei G, Ortiz-Pérez T, Raptis RG, Telser J (2003) Triangular, ferromagnetically-coupled CuII3−pyrazolato complexes as possible models of particulate methane monooxygenase (pMMO). Inorg Chem 42:5801–5803. https://doi.org/10.1021/ic0344416
47. Sánchez-Sanz G, Trujillo C, Alkorta I, Elguero J (2019) Understanding regium bonds and their competition with hydrogen bonds in Au_2:HX complexes. Chem Phys Chem 20:1572–1580. https://doi.org/10.1002/cphc.201900553
48. Sánchez-Sanz G, Trujillo C, Alkorta I, Elguero J (2020) Rivalry between regium and hydrogen bonds established within diatomic coinage molecules and lewis acids/bases. ChemPhysChem 21:2557–2563. https://doi.org/10.1002/cphc.202000704
49. de Lara-Castells MP, Krems RV, Buchachenko AA, Delgado-Barrio G, Villarreal P (2001) Complete basis set extrapolation limit for electronic structure calculations: energetic and nonenergetic properties of HeBr and $HeBr_2$ van der Waals dimers. J Chem Phys 115:10438.https://doi.org/10.1063/1.1415078
50. Frontera A, Bauzá A (2018) Regium–π bonds: an unexplored link between noble metal nanoparticles and aromatic surfaces. Chem Eur J 24:7228–7234. https://doi.org/10.1002/chem.201800820
51. Bauzá A, Frontera A (2018) Regium-π *vs* Cation-π Interactions in M_2 and MCl (M = Cu, Ag and Au) complexes with small aromatic systems: an ab initio study. Inorganics 6:64. https://doi.org/10.3390/inorganics6030064
52. Adamo C, Barone V (1999) Toward reliable density functional methods without adjustable parameters: The PBE0 model. J Chem Phys 110:6158–6169. https://doi.org/10.1063/1.478522
53. Ernzerhof M, Scuseria GE (1999) Assessment of the Perdew–Burke–Ernzerhof exchange-correlation functional. J Chem Phys 110:5029–5036. https://doi.org/10.1063/1.478401
54. Grimme S, Antony J, Ehrlich S, Krieg H (2010) A consistent and accurate ab initio parametrization of density functional dispersion correction (DFT-D) for the 94 elements H-Pu. J Chem Phys 132:154104–154119. https://doi.org/10.1063/1.3382344

55. Weigend F, Ahlrichs R (2005) Balanced basis sets of split valence, triple zeta valence and quadruple zeta valence quality for H to Rn: design and assessment of accuracy. Phys Chem Chem Phys 7:3297–3305. https://doi.org/10.1039/B508541A
56. Alkorta I, Walker NR, Legon AC (2021) Non-Covalent Interactions of the Lewis Acids Cu–X, Ag–X, and Au–X (X = F and Cl) with nine simple lewis bases B: a systematic investigation of coinage-metal bonds by ab initio calculations. Inorganics 9:13. https://doi.org/10.3390/inorganics9020013
57. Cui J, Zhang X, Meng L, Li Q, Zeng Y (2019) Coinage metal dimers as the noncovalent interaction acceptors: study of the σ–lump interactions. Phys Chem Chem Phys 21:21152–21161. https://doi.org/10.1039/C9CP03686E
58. Larrañaga O, Arrieta A, Guerra CF, Bickelhaupt FM, de Cózar A (2021) Nature of Alkali- and coinage-metal bonds *versus* hydrogen bonds. Chem Asian J 16:315–321. https://doi.org/10.1002/asia.202001201
59. Zierkiewicz W, Michalczyk M, Scheiner S (2018) Regium bonds between Mn clusters (M = Cu, Ag, Au and n = 2–6) and nucleophiles NH_3 and HCN. Phys Chem Chem Phys 20:22498–22509. https://doi.org/10.1039/C8CP03883J
60. Zhang J, Wang Z, Liu S, Cheng J, Li W, Li Q (2019) Synergistic and diminutive effects between triel bond and regium bond: Attractive interactions between π-hole and σ-hole. App Organometal Chem 33: e4806https://doi.org/10.1002/aoc.4806
61. Wang R, Wang Z, Yu X, Li Q (2020) Synergistic and diminutive effects between regium and aerogen bonds. ChemPhysChem 21:2426–2431. https://doi.org/10.1002/cphc.202000720
62. Zhang Z, Lu T, Ding L, Wang G, Wang Z, Zheng Z, Zheng B, Liu Y, Ding, X. L. Cooperativity effects between regium-bonding and pnicogen-bonding interactions in ternary MF···PH_3O···MF (M = Cu, Ag, Au): an *ab initio* study. Mol Phys 2020; e1784478. DOI: https://doi.org/10.1080/00268976.2020.1784478
63. Daolio A, Pizzi A, Terraneo G, Ursini M, Frontera A, Resnati G (2021) Anion···anion coinage bonds: the case of tetrachloridoaurate. Angew Chem Int Ed 60:14385–14389. https://doi.org/10.1002/anie.202104592
64. Priola E, Mahmoudi G, Andreo J, Frontera A (2021) Unprecedented $[d^9]$Cu···$[d^{10}]$Au coinage bonding interactions in $\{Cu(NH_3)_4[Au(CN)_2]\}^+[Au(CN)_2]^-$ salt. Chem Commun 57:7268–7271. https://doi.org/10.1039/D1CC02709C
65. Terrón A, Buils J, Mooibroek TJ, Barceló-Oliver M, García-Raso A, Fiol JJ, Frontera A (2020) Synthesis, X-ray characterization and regium bonding interactions of a trichlorido(1-hexylcytosine)gold(III) complex. Chem Commun 56:3524–3527. https://doi.org/10.1039/d0cc00505c
66. Piña MN, Frontera A, Bauzá A (2020) Regium−π bonds are involved in protein-gold binding. J Phys Chem Lett 11:8259–8263. https://doi.org/10.1021/acs.jpclett.0c02295
67. Ernst HA, Olsen AN, Skriver K, Larsen S, Leggio LL (2004) Structure of the conserved domain of ANAC, a member of the NAC family of transcription factors. EMBO Rep 5:297–303. https://doi.org/10.1038/sj.embor.7400093
68. Knapp M, Segelke B, Rupp B (1998) AM Cryst Assoc Abstr Papers (Annual Meeting) 25:90
69. Ferraro G, Massai L, Messori L, Cinellu MA, Merlino A (2015) Structural evidences for a secondary gold binding site in the hydrophobic box of lysozyme. Biometals 28:745–754. https://doi.org/10.1007/s10534-015-9863-7
70. Wu Y, Zhu Y, Gao F, Yongjun J, Oladejo BO, Chai Y, Bi Y, Lu S, Dong M, Zhang C, Huang G, Wong G, Li N, Zhang Y, Li Y, Feng WH, Shi Y, Liang M, Zhang R, Qi J, Gao GF (2017) Structures of phlebovirus glycoprotein Gn and identification of a neutralizing antibody epitope. Proc Nat Acad Sci 114:E7564–E7573. https://doi.org/10.1073/pnas.1705176114
71. Ren T, Qu F, Morris TJ (2000) HRT gene function requires Interaction between a NAC protein and viral capsid protein to confer resistance to turnip crinkle virus. Plant Cell 12:1917–1925. https://doi.org/10.1105/tpc.12.10.1917
72. Xie Q, Sanz-Burgos AP, Guo H, García JA, Gutiérrez C (1999) GRAB proteins, novel members of the NAC domain family, isolated by their interaction with a geminivirus protein. Plant Mol Biol 39:647–656. https://doi.org/10.1023/a:1006138221874

73. Greve K, La Cour T, Jensen MK, Poulsen FM, Skriver K (2003) Interactions between plant RING-H2 and plant-specific NAC (NAM/ATAF1/2/CUC2) proteins: RING-H2 molecular specificity and cellular localization. Biochem J 371:97–108. https://doi.org/10.1042/BJ20021123
74. Li Q, Liu JJ, Sun X, Xu L (2019) Hierarchically porous melamine-formaldehyde resin microspheres for the removal of nanoparticles and simultaneously as the nanoparticle immobilized carrier for catalysis. ACS Sustain Chem Eng 7:867–876. https://doi.org/10.1021/acssuschemeng.8b04490
75. Navarro E, Baun A, Behra R, Hartmann NB, Filser J, Miao AJ, Quigg A, Santschi PH, Sigg L (2008) Environmental behavior and ecotoxicity of engineered nanoparticles to algae, plants, and fungi. Ecotoxicology 17:372–386. https://doi.org/10.1007/s10646-008-0214-0
76. Sajid M, Ilyas M, Basheer C, Tariq M, Daud M, Baig N, Shehzad F (2015) Impact of nanoparticles on human and environment: review of toxicity factors, exposures, control strategies, and future prospects. Environ Sci Pollut Res 22:4122–4143. https://doi.org/10.1007/s11356-014-3994-1

Chapter 9
Aqueous Supramolecular Assemblies of Photocontrolled Molecular Amphiphiles

Franco King-Chi Leung

Abstract Amphiphilic molecules, are composed of hydrophobic and hydrophilic parts and the intrinsic tendence to assemble in aqueous conditions, producing numerous supramolecular assembled structures and functional systems. Some of the recent challenges in the design of adaptive, responsive, far-from-equilibrium functional systems in aqueous environments, the proper design of photo-controlled moieties in intrinsic charged amphiphilic molecular structures offers fruitful opportunities to create supramolecular assembly systems, based on electrostatic interaction, with response to light in aqueous environment. In this chapter, we discuss the design strategy of photo-controlled molecular amphiphiles, the supramolecular assembled structures in aqueous environment and at air–water interfaces, as well as different strategies for producing dynamic functions in both isotropic and anisotropic supramolecular assembled materials. The motions at air–water interface, foam formation, reversible supramolecular assembly at nanometer length-scale, and life-like artificial muscle function are discussed. Manipulating the molecular structural design, supramolecular assembling conditions, and external stimulation, the photo-controlled molecular amphiphiles open directions toward applications ranging from controlled bio-target delivery to soft robotic.

9.1 Introduction

Amphiphile, this word originated from the Greek *amphis*: both, and *philia*: love, is referring to a chemical structure consisting a polar and a non-polar part with intrinsic hydrophobic (lipid-loving) and hydrophilic (water-loving) properties [1–4]. Most common amphiphiles including surfactants and diglycerides [5], the amphiphilic nature, which means a dual hydrophobic and hydrophilic nature in a small molecular

F. K.-C. Leung (✉)
State Key Laboratory of Chemical Biology and Drug Discovery, Department of Applied Biology and Chemical Technology, The Hong Kong Polytechnic University, Hung Hom, Kowloon, Hong Kong SAR, China
e-mail: kingchifranco.leung@polyu.edu.hk

M. A. Aboudzadeh and A. Frontera (eds.), *Supramolecular Assemblies Based on Electrostatic Interactions*,
https://doi.org/10.1007/978-3-031-00657-9_9

structure, determines the spontaneous supramolecular assembly into a well-defined structure in aqueous environment via noncovalent interactions [6–14], including electrostatic interaction. Bottom-up approach provides natural and synthetic supramolecular assembled structures of molecular amphiphiles across multiple length-scale [15–37].

Nature has created the most beautiful and sophisticated examples of supramolecular assembled systems derived from amphiphilic molecules. Take phospholipids as an example, which is a typical class of natural amphiphiles, supramolecular assembling into biological bilayer membranes, i.e., cell membrane, in living organisms to maintain stable biological conditions and compartmentation [38, 39]. In the presence of various proteins in the supramolecular bilayered structure of cell membrane, it allows the membrane to respond automatically to subtle variations in the surrounding environment and enables living organisms with correct biological functions [30, 40]. Nature provides unlimited inspiration for designing novel artificial mimetic systems. In the past decades, various synthetic supramolecular assembled systems are designed and equipped with stimuli responsiveness and functional tunability, enabling complicated supramolecular assembling pathways and energy landscapes [41–49]. Numerous synthetic stimuli-responsive amphiphiles have been developed to provide structurally well-defined supramolecular assemblies, in which the supramolecular structures can be fine controlled by external stimulation, e.g., gases, heat, small molecules, pH, and light [12, 50–69]. Light is regarded as a noninvasive stimulation to allow high temporal and spatial control of responsive systems as well as the tunable wavelength and intensity [70, 71]. Advanced synthetic strategies allow structural modifications of photoresponsive amphiphilic molecules, including (1) photoresponsive molecular amphiphile, composed of a covalently linked photoresponsive unit, and (2) photoresponsive supramolecular amphiphile, composed of a noncovalently connected photoresponsive unit [11, 53, 72]. Supramolecular assemblies of photoresponsive amphiphilic molecules controlled by light offer ample opportunities toward adaptive and smart soft materials, as well as stimuli-responsive systems in research fields ranging from chemical biology, nanotechnology to materials in robotic science [5, 71, 73, 74].

Naturally, almost all supramolecular assembled structures and their assembling processes with functional capabilities found in living systems occur in aqueous environment [75, 76]. Water is a basic but important and essential medium for supramolecular assembly in biological systems, allowing those systems with advanced complexity, robustness, and adaptability. The supramolecular assemblies and hierarchical supramolecular assemblies of photoresponsive molecular amphiphiles in aqueous environments and conditions offer excellent opportunities to produce biocompatible and adaptable artificial soft functional systems. In this current chapter, we discuss and demonstrate the recent research progress in the area of supramolecular assembled and hierarchical supramolecular structures of photoresponsive charged small molecular amphiphiles in aqueous environments and conditions for both at air–water interfaces and in aqueous solutions, including criteria of the molecular structures, intrinsic hierarchical supramolecular organized structures, novel, and important recognition, and their assembly parameters with their responsive behavior.

The selected areas are discussed: (1) molecular design strategies of photoresponsive charged molecular amphiphiles, (2) intrinsic dynamic or static functions of photoresponsive charged molecular amphiphiles at air–water interfaces, and (3) functional supramolecular self-assembled structures of photoresponsive charged molecular amphiphiles in solutions, including one-dimensional (1D) supramolecular structures to randomly arranged and organized three-dimensional (3D) structures and unidirectionally aligned hierarchical structures. With the focused discussion on the molecular and supramolecular structures and photoresponsive molecular amphiphile's intrinsic functions and the correlated supramolecular assembled structures in aqueous environments, we herein highlight and focus on the recent progress of dynamic systems in water and amplification of molecular motion of molecular machineries to sustain macroscopic life-like robotic functions.

9.2 Photoresponsive Charged Molecular Amphiphiles

To provide delicate design of photoresponsive charged molecular amphiphiles achieving proper supramolecular assembly structures controlled by light in aqueous environments, the variety of molecular structures of amphiphiles should be considered [77, 78]. In consideration of the chemical nature of hydrophilic motif, molecular amphiphiles are classified as ionic (i.e., anionic, cationic), zwitterionic, and nonionic amphiphiles [14, 77, 78]. Based on the mode and number of connecting points of both hydrophilic and hydrophobic moieties, molecular amphiphiles can be classified into head/single tail and double tail amphiphiles, Gemini amphiphiles, and bola-amphiphiles (Fig. 9.1a) [3, 5, 79].

Most commonly found in natural and synthetic amphiphiles types are head/single tail, Gemini amphiphiles, and bola-amphiphiles. Gemini amphiphile is composed of two hydrocarbon tails with two ionic groups linked by a single spacer, allowing aggregation at very low amphiphile concentration and reducing surface tension significantly [5, 30, 80]. Bola-amphiphile is composed of two hydrophilic head groups linked with a single aliphatic chain, which is commonly found in cell membranes of thermophilic bacteria to show high-thermal resistance [81, 82]. The structural properties and order of supramolecular assembly of molecular amphiphiles are highly dependent on the design of its molecular structures. In this connection, not only the type of molecular amphiphile but also the nature and position of the photoresponsive unit (e.g., structural rigidity, geometrical change, change in polarity, etc.) have to be considered to determine the resulting supramolecular assembly and advanced functions. In order to strengthen and maximize the intermolecular interaction of the resulting supramolecular assembly, this chapter focuses on the discussion of anionic, cationic, and zwitterionic amphiphiles, i.e., charged molecular amphiphiles.

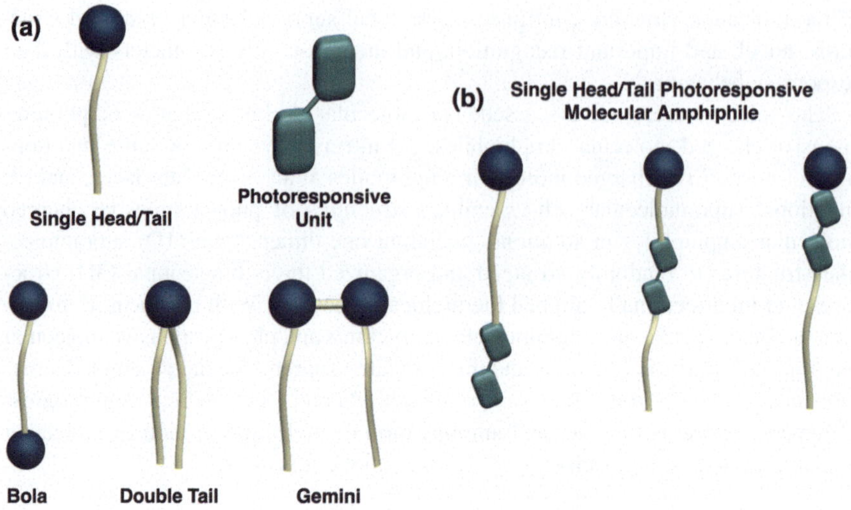

Fig. 9.1 Schematic diagram of **a** common molecular amphiphiles and **b** photoresponsive molecular amphiphiles illustration (single head–tail amphiphiles are employed as example)

By implementation of photoresponsive units either at the charged head group or at different positions of the hydrophobic chain, it is regarded as the typical method to design photoresponsive molecular amphiphiles (Fig. 9.1b). Numerous photoresponsive molecules have been well-developed in past few decades to provide applications ranging from smart responsive systems to material science. Photoresponsive units, such as stilbenes, dithienylethenes, spiropyrans, azobenzenes, donor–acceptor Stenhouse adducts, molecular motors, are commonly applied to the design of photoresponsive molecular amphiphiles [83, 84]. Some representative molecular structures and the corresponding photoisomerization processes of the photoresponsive units have been shown in Scheme 9.1.

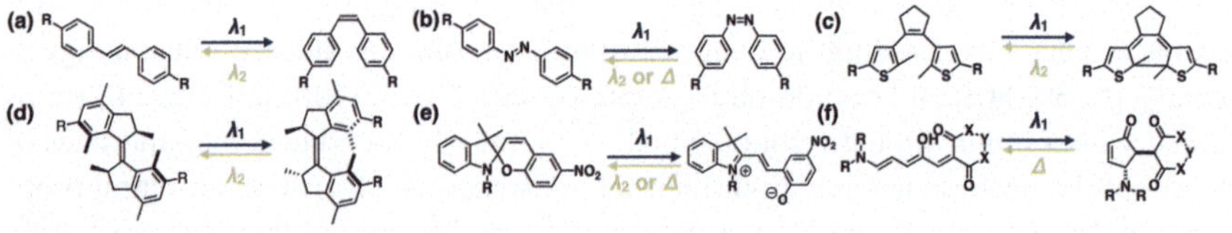

Scheme 9.1 Major photoresponsive molecular structures and the corresponding photoisomerization processes of **a** stilbene, **b** azobenzene, **c** dithienylethene, **d** molecular motor, **e** spiropyrans, and **f** donor–acceptor Stenhouse adduct. (λ: Photoirradiation and Δ: Heating)

9.3 Dynamic Functions of Photoresponsive Charged Molecular Amphiphiles at Air–Water Interfaces

The introduction of molecular amphiphiles into aqueous environments allows molecules to spread over the entire interface. The charged hydrophilic heads of amphiphiles orient toward the polar phase, while the hydrophobic tails orient toward the non-polar phase to produce self-assembled monolayers at the air–water interfaces. On the basis of the solubility of molecular amphiphiles in aqueous environment, self-assembled monolayers can be classified into the Langmuir monolayers and the Gibbs monolayers (adsorption monolayers) [85–87]. Langmuir monolayers are composed of either insoluble or sparingly soluble amphiphiles, providing well-organized layered films. Photoresponsive molecular amphiphiles, such as spiropyrans, azobenzenes, molecular motors, and dithienylethenes, have been applied in Langmuir monolayers in controls of surface pressure/tension reversibly. The supramolecular structure, prepared by packing tight thin films, can be prepared systematically through layer-by-layer Langmuir–Blodgett methodology, indicating the important prospects for materials in sensing technologies and optoelectronic materials [88–105].

Gibbs monolayers (adsorption monolayers) are constructed with molecular amphiphiles with excellent aqueous solubility to provide direct distribution over the interfaces between air and water without involvement of common solvents. Gibbs monolayers demonstrate close correlations between supramolecular assembled structures in aqueous environments and at interfaces between air and water. The pioneer studies in the area of the photoresponsive molecular amphiphiles based Gibbs monolayers were using physical parameters with crucial equilibrium, including critical micelle concentration (CMC) and surface tension under equilibrium, to determine supramolecular assembled structural transformations in aqueous solutions and at air–water interfaces. In the current section, we focus on the discussion of dynamic/static interfacial functions for the photoresponsive molecular amphiphiles [106–114].

Shin and Abbott reported an actively controlled surface tension with dynamic properties of a mixture solution of photoresponsive azobenzene-based bolaamphiphile **1** with sodium dodecyl sulfate (SDS) (Fig. 9.2a) [115]. Through the combination of positive charged **1** with negatively charged SDS, this complementary electrostatic interaction allows a closely packed supramolecular co-assembled structure at the interface between air and water (Gibbs monolayer). Large geometric supramolecular transformations, from *trans* to *cis*, of the photoisomerization of azobenzenes **1** induced a significant variation of surface tension around 25 mN/m dynamically (below CAC, Fig. 9.2b). This molecular design of photoresponsive molecular amphiphile demonstrated a novel strategy for dynamically controlled surface tension of the aqueous solutions and provided an alternative method to an innovative transformation in surface tension under equilibrium conditions. Notably, the large geometrical transformation of azobenzene-based amphiphiles provides significant transformation of dipole moment (commonly around 3 D) [115, 116].

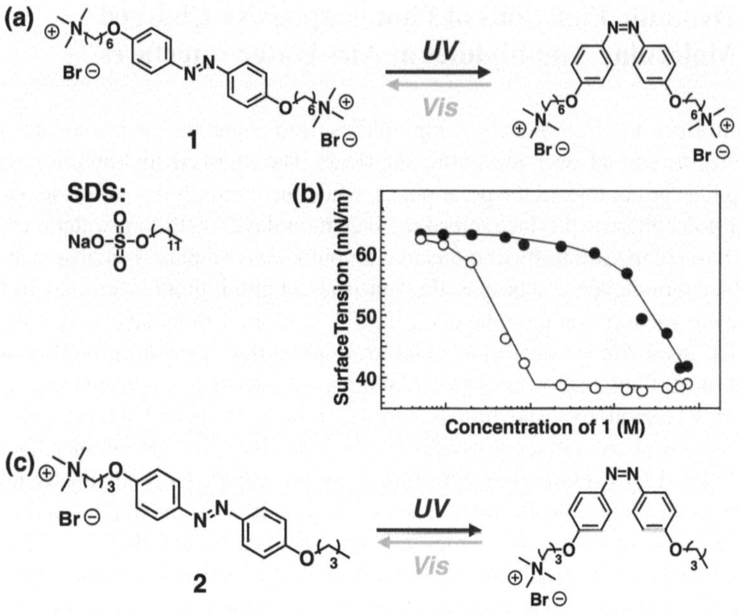

Fig. 9.2 **a** Photoisomerization process and molecular design of **1** with SDS. **b** Surface tension change of **1** with SDS before (filled circles) and after (open-circles) UV photoirradiation. **c** Molecular design of **2** and its photoisomerization process. Adapted image with permission from Ref. [115]. Copyright (1999) (American Chemical Society)

A photoresponsive molecular amphiphile **2**, designed by Monteux and co-workers, is composed of a cationic ammonium head group with an azobenzene-based hydrophobic tail (Fig. 9.2c), to show tunable dynamic surface tension by photo-controlled adsorption/desorption processes of **2** [117]. The electrostatic barrier to adsorption was studied by kinetically limited model. The *cis*-**2** isomer showed 10 times faster than that *trans*-**2** isomer in adsorption process and desorbed 300 times faster. As a result, the monolayers were packed with almost only *trans*-**2** in-equilibrium conditions. Because of the adsorption and desorption processes of *trans*- and *cis*-isomers of **2**, instant variation of assembly occurred to trigger rapid changes in interfacial properties. A solution of **2** and talc particles was observed under optical microscopy to monitor surface flux process. The surface of solution was exposed under UV-light spot ($\lambda = 365$ nm), the majority of *trans*-**2** isomer adsorbed outside of the irradiated spot, meanwhile significant desorption of **2** occurred in the irradiated part because of the *trans*-to-*cis* isomerization. As a result, the surface tension of the solution was rapidly increased to generate surface tension gradient between irradiated and non-irradiated parts. The generated Marangoni flow from outside to inside of the light spot, it induced movements of talc particles toward the light spot. Overall, the particle concentrating phenomenon of the light spot was observed instantly under UV-light (Fig. 9.3) [117].

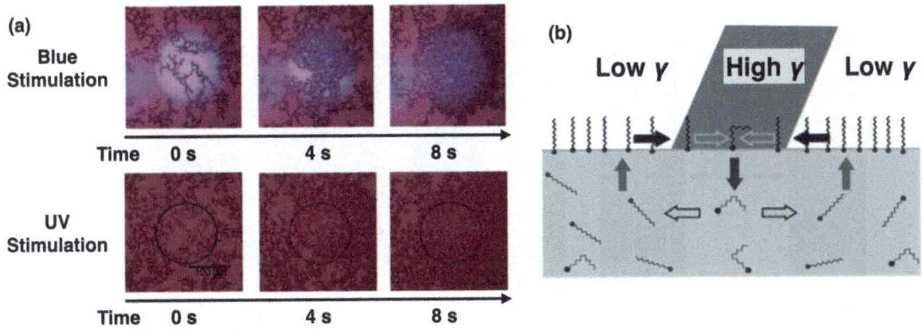

Fig. 9.3 **a** Microscopic images of phenomenon of particle concentrating with UV and blue-light irradiating spot. **b** Schematic diagram of adsorption–desorption with its Marangoni flow of photoresponsive molecular amphiphiles. Adapted images with permission from Ref. [117]. Copyright (2011) (Royal Society of Chemistry)

Nonetheless, blue-light irradiation (λ = 436 nm) can induce a similar particle concentrating phenomenon. The distinct properties of adsorption and desorption processes of *trans*-**2** and *cis*-**2** isomers were demonstrated to provide a novel strategy to control surface gradients at air–water interface by azobenzene-based amphiphile solution with light-induced Marangoni flow.

In using the dynamic and tunable interfacial properties of aqueous solutions of azobenzene-based photoresponsive molecular amphiphile **2**, an in-situ controlled stability and breakage of photoresponsive foams were reported by Monteux and co-workers. The photoresponsive foams were prepared with the aqueous solution of **2** to control stability of foams remotely and non-invasively (Fig. 9.4a) [118]. The majority of *trans*-**2** maintained stable foams, while the foams ruptured upon UV-light irradiation within 100 s (Fig. 9.4b, c). The obtained solution was able to prepare stable foams again after blue-light irradiation for a few minutes, indicating a reversible control of foam formation by UV- and blue-light sequential irradiation. *Trans*-**2** converts to *cis*-**2** upon UV-light irradiation of the stable foam of **2**, inducing amphiphiles desorbed from the air–water interfaces into the bulk solutions. As a result, dynamic adsorption and desorption interfaces was achieved. Possibly due to the dynamic adsorption/desorption phenomenon of **2**, an out-of-equilibrium surface tension gradient can induce Marangoni flows at the interfaces, leading to the breakage of foams upon UV-light irradiation.

Later Monteux and co-workers reported the correlation of the light-induced flow and foam destabilization mechanism, they investigated the dependence of light-induced flow of **2** and its foam breakage at thin-liquid film of bubbles and macroscopic foams [119]. A tunable velocity of Marangoni flow was observed at thin-liquid films by controlling photoisomerization of **2** with various UV-light intensities. The light-induced flow in macroscopic foams is able to reduce foam drainage rate at early stage of UV-light irradiation, along with a subsequent rapid foam rupture. The effect of light irradiation on the disjoining pressure of thin-liquid films showed that UV-light induced foam rupture was attributed to the decrease of electrostatic repulsion

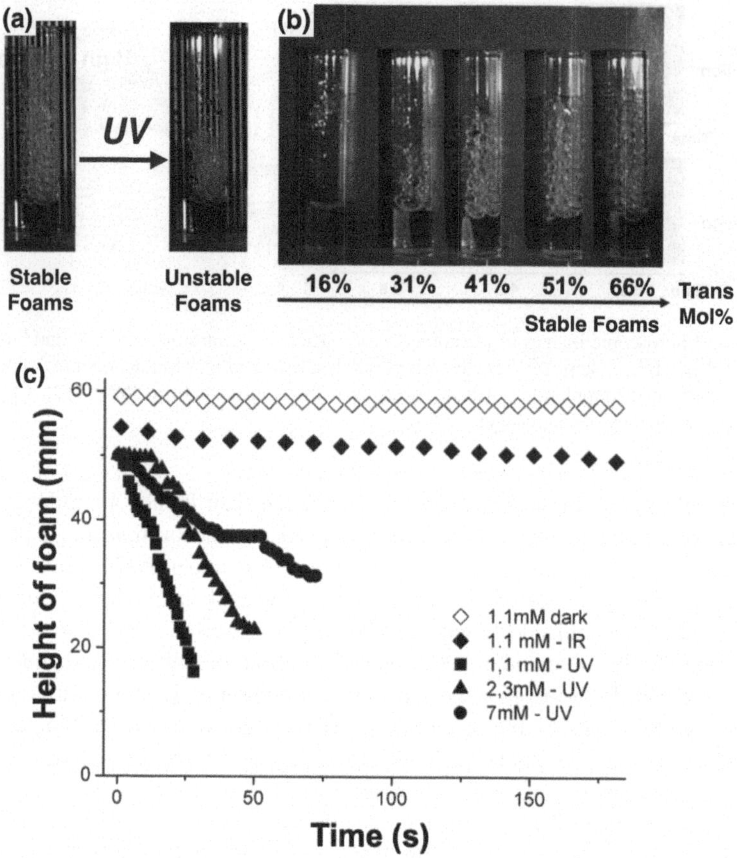

Fig. 9.4 **a** Schematic diagram of photo-controlled foams of **2**. **b** Stable foams of *trans*-**2**. **c** Destabilized foams prepared by UV photoirradiation. Adapted images with permission from Ref. [118]. Copyright (2012) (American Chemical Society)

in the thin foam film and oscillation of the disjoining pressure isotherm [120]. This photoresponsive molecular amphiphile system at the air–water interface shows the potential prospects for applications in remote controls of foam stability.

Jiang and co-workers reported a dual stimuli-responsive molecular amphiphile system, which was synthetically modified on the basis of the azobenzene amphiphile **2** by implementing a tertiary amine head group **3** (Fig. 9.5) [121]. In the conductivity measurements, amphiphile **3** is able to be reversibly transformed between the form of hydrophobic tertiary amine and the form of amphiphilic ammonium salt by purging nitrogen and carbon dioxide alternately. Upon 365 nm light irradiation, the photostationary state (PSS) was attained with a *trans*-**3**: *cis*-**3** ratio 4: 96, while the backward isomerization was controlled by blue-light irradiation. By controlling both gases purging (nitrogen/carbon dioxide) and light (UV/blue light) irradiation, the significant geometrical structural transformation of **3** is achieved to generate dual

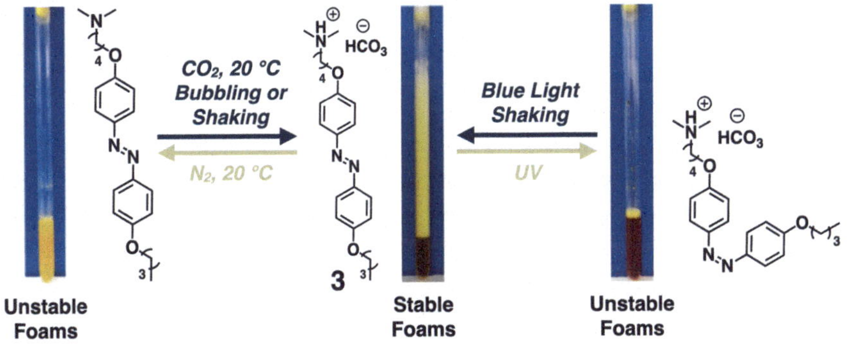

Fig. 9.5 Schematic diagram of the gases and photo-controlled foams of **3**. Adapted image with permission from Ref. [121]. Copyright (2007) (Elsevier)

stimuli-responsive foams. Illusively, the stable foams, prepared from the solutions of **3** by bubbling with carbon dioxide, were ruptured after nitrogen purging for 10 min, or foam rupture was induced by UV-light irradiation. The current approach provides a simple and promising strategy for future multiple stimulation responsive systems.

In addition to the structural modification of the hydrophilic parts of azobenzene amphiphiles, structural rigidity improved hydrophobic part was designed in the azobenzene-based amphiphile **4** with carboxylate charged end-group, to provide significant enhancement of foam stability (Fig. 9.6) [122]. The photoisomerization of **4** was found to attain the PSS with 95% of *cis*-**4**, confirmed by absorption spectroscopy and proton NMR. The critical micellar concentration (CMC) was changed

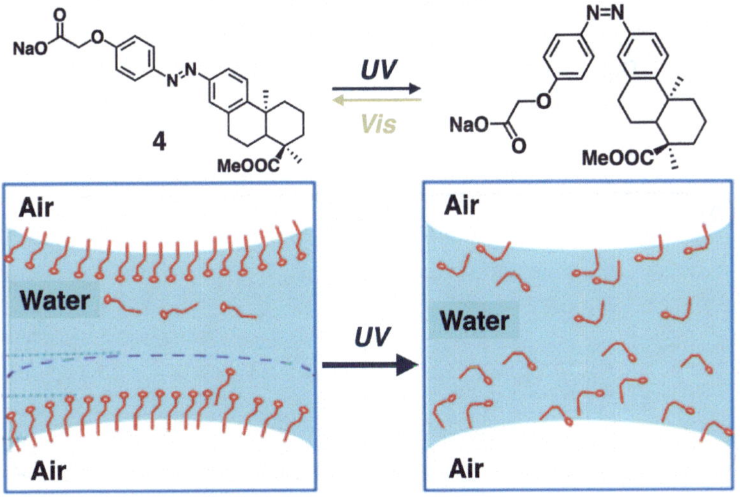

Fig. 9.6 Schematic diagram of photoisomerization of **4** and its photo-controlled foams. Adapted image with permission from Ref. [122]. Copyright (2017) (American Chemical Society)

drastically upon photoirradiation, possibly due to the large geometrical transformation from *trans*-**4** to *cis*-**4**. Stable foams were obtained from a solution of *trans*-**4** (2.0 mM) with a stability half-life of ~16 h. In contrast, all the stable foams were collapsed in 4 min photoirradiation with UV-light, affording photoresponsive foams with significant variation of foam stability triggered by UV-light.

By substituting the bromide counterion of amphiphile **2** with a bis-(trifluoromethanesulfonimide) (BTF) anion, it provided a strategy for controlling foam stability by binding to cucurbit[7]uril (**CB7**) with spermine moieties (Fig. 9.7) [122]. The azobenzene-based amphiphile **5**, composed of the amphiphile part and counterion BTF, was improved the foamability after addition of **CB7** due to closer packing at air–water interface. As confirmed by proton NMR and conductivity measurements, the host–guest complex of **5** ⊂ **CB7** might shield the electrostatic interaction of amphiphile **5** and its counterion BTF. The free BTF is desorbed from the air–water interface to allow a closer packing at the interface. Given that stronger host–guest complex forms between spermine with **CB7**, free amphiphile **5** becomes available for binding with BTF at air–water interface. The decreased foamability was observed because of the loosely packed monolayers of **5** at the interface. The results

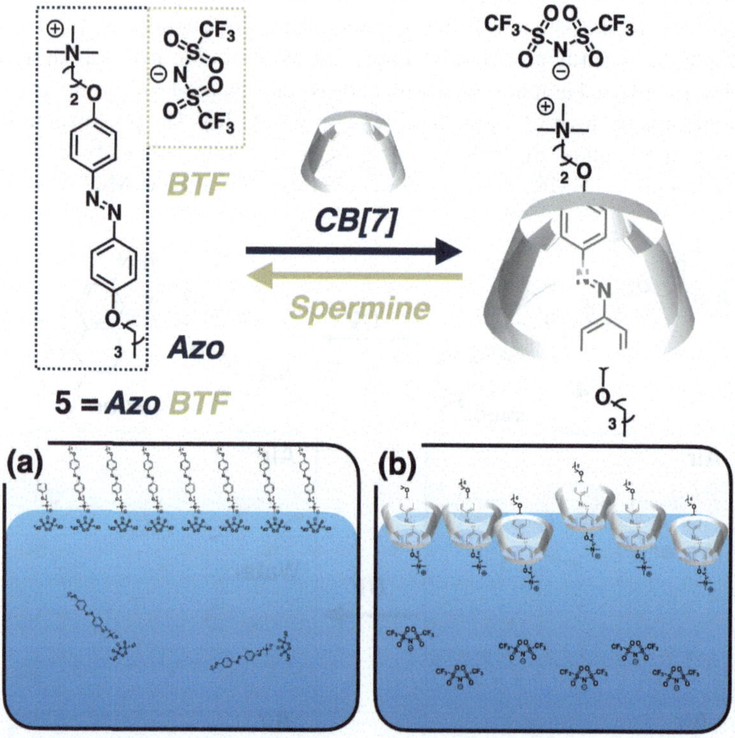

Fig. 9.7 Supramolecular structural transformations at air–water interface and in aqueous solution of **a 5** and **b 5** ⊂ **CB7**

clearly demonstrated that a strategy of supramolecular host–guest interaction allows to reversibly control the foamability by photoresponsive molecular amphiphile.

Chen et al. has synthesized a class of nonionic and cationic azobenzene-based amphiphiles through the modifications of hydrophobic chains and hydrophilic part. These amphiphiles allowed the preparation of photoresponsive foams for textile foam coloring systems with reduced pollutants discharge and less energy consumption than that of the traditional coloring processes [123–128]. By the reversibly tuned light stimulation, the foamability and stability of colored photoresponsive foams could be systematically controlled. The colored stable foams were applied in textile coloring process, after photo-controlled foam rupture, the unused colored photoresponsive amphiphiles were easily recycled, achieving near zero-pollutant discharge coloring process (Fig. 9.8) [128]. Another application of photoresponsive foams prepared from azobenzene-based amphiphiles is involving the flotation processes of quartz particles [129, 130]. All these applicational results demonstrated the feasibility of using photoresponsive molecular amphiphiles in more environmental-friendly industrial processes.

In using isomerization of molecular amphiphiles, the dynamic photoresponsive air–water interfaces provide promising opportunities to control optical particle depositions with defined patterns [131, 132], macroscopic liquid droplet motions [133, 134], and liquid marble transport [135]. A novel strategy for accumulating particles in predefined locations with complicated patterns was reported, which shared similarities to that of observed particle concentrating phenomenon of amphiphile **2** [117]. A cationic azobenzene-based molecular amphiphile **6** was employed in the photoinduced Marangoni flow in evaporating droplets (Fig. 9.9a) [131]. Various complicated patterns were formed by using photomasks, showing patterns with model suspension to complex formulations, e.g., commercial coffee suspensions. Marangoni flows at air–water interfaces were observed by the photoinduced transportation of floating liquid marbles by the solution of molecular amphiphile **6**. (Fig. 9.9b) [135]. By controlling the thickness of liquid substrates, two motion directions of liquid marbles were observed with Marangoni and anti-Marangoni motions, providing complicated dynamic function of photoresponsive molecular amphiphiles at air–water interfaces.

A photoresponsive dissipative assembly system of molecular amphiphile **6** in suspension of anionic polystyrene microparticles was reported by Baigl and coworkers. A reversible and dynamic control of particle arrangements between a highly crystalline assembly and disordered states was observed by photoirradiation at air–water interfaces (Fig. 9.10) [136]. The particle arrangement states were controlled by the photoinduced dynamic adsorption/desorption processes of amphiphile **6** at air–water interfaces. In dark, the suspension was mainly composed of *trans*-**6**, which the particles showed disordered state. At PSS, the suspension was composed of *cis*-**6** either 95% (photo-irradiated with UV for 1 min) or 45% (photo-irradiated with blue-light for 1 min), affording similar colloidal assembly with highly crystalline assembly state. It was proposed that the continuous photoinduced desorption of amphiphile **6** maintained out-of-equilibrium conditions to allow crystallization process. The crystals are transferred to disordered state when the light (energy source) is off.

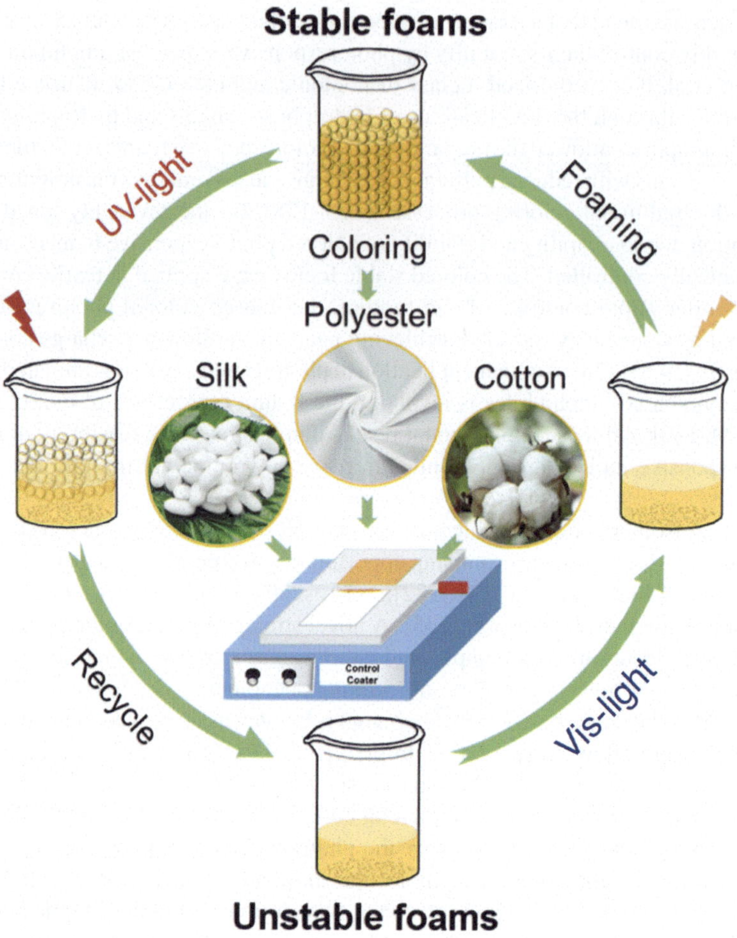

Fig. 9.8 Schematic diagram of foam coloring process with excellent recyclability. Adapted image with permission from Ref. [128]. Copyright (2020) (Elsevier)

The system showed the controlled transformation of colloidal assembly at air–water interfaces to expand the realm of currently known dissipative systems.

Azobenzene-derived photoresponsive molecular amphiphiles **7** and **8** were reported independently by Ravoo and Braunschweig (Fig. 9.11) to provide studies of responsive adsorption properties by vibrational sum-frequency generation and dynamic surface tension, as well as resulting macroscopic functions, such as responsive foams and particle motions. Molecular amphiphile **7**, which was designed with a carboxylate charged group connected with an ethylene glycol-linker, has been demonstrated with controllable surface tension and foam stability by pH and light (Fig. 9.11a) [137]. The most foam stability of solution **7** was observed at pH 7.1 after

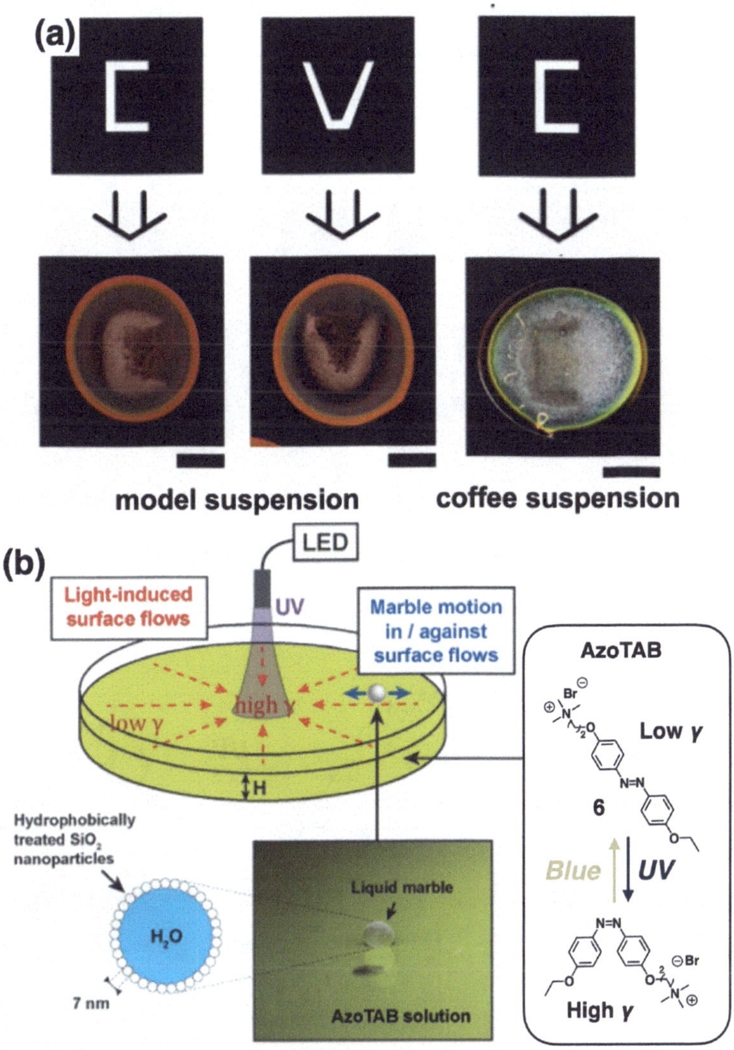

Fig. 9.9 Schematic diagram of **a** particles deposition process and **b** photo-controlled liquid floating marbles transport. Adapted **a** with permission from Ref. [133]. Copyright (2016) (American Chemical Society). Adapted **b** with permission from Ref. [135]. Copyright (2016) (John Wiley and Sons)

green-light irradiation. Molecular amphiphile **8**, which was designed with sulfonate-charged group connected by an alkyl-linker, can provide higher foam stability than that of amphiphile **7**, because of the improved micro-phase separation by hydrophobic part (Fig. 9.11b) [138]. The results showed a class of photoresponsive molecular amphiphiles able to control foam stability and photoinduced particle motions via dynamic adsorption/desorption at Gibbs monolayers.

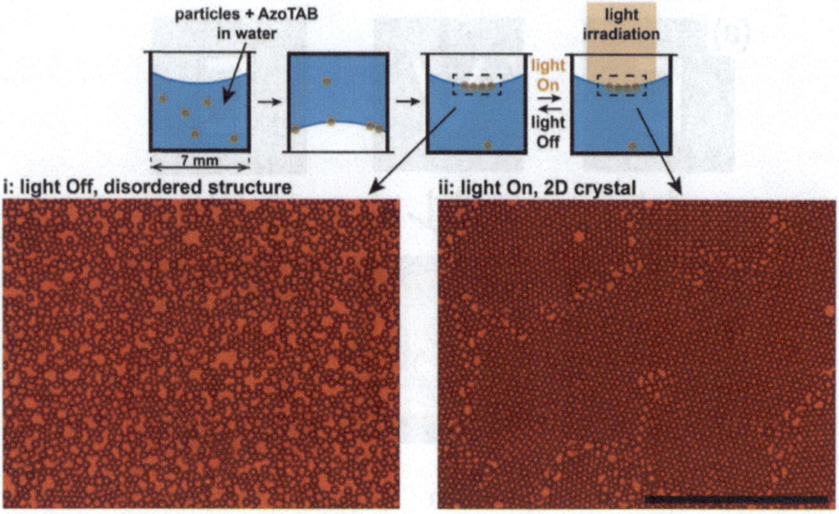

Fig. 9.10 Microscopy figures of photo-controlled (i) disassembling and (ii) colloidal crystallizing processes at air–water interfaces. Scale bar: 100 nm. Adapted image with permission from Ref. [136]. Copyright (2019) (Elsevier)

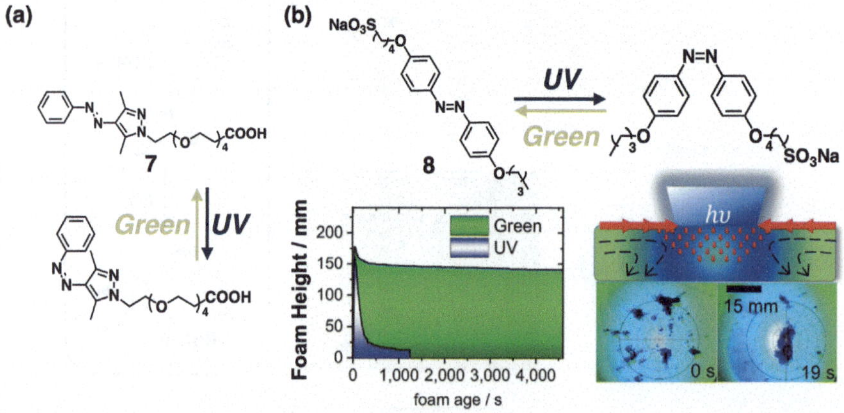

Fig. 9.11 **a** Photoisomerization process of **7**. **b** Photo-controlled foams and photo-controlled particle movements of **8**. Adapted image with permission from Ref. [138]. Open access under a CC BY 4.0 license, https://creativecommons.org/licenses/by/4.0/

Spiropyran-based molecular amphiphiles, as an alternative to azobenzene-base molecular amphiphiles, could provide dynamic photoresponsive Gibbs monolayers for responsive foams and photoinduced motions applications [139–141]. Molecular amphiphile **9**, which was designed with a sulfonate-charged group connected with alkyl-linker to spiropyran core, is able to provide photo- and pH-responsive foams

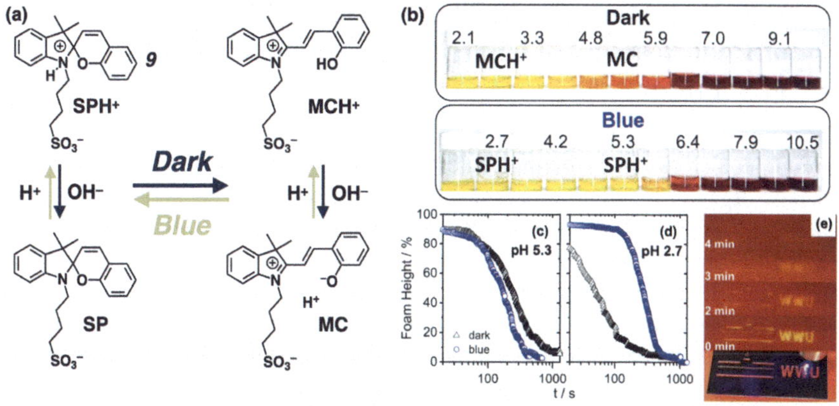

Fig. 9.12 Schematic diagram of **a** photoisomerization of **9** and **b** pH and photo-controlled solution color change. Photo-controlled foams of **9** at **c** pH 5.3 and **d** 2.7. **e** Writing and erasing processes with light of solution **9** (pH 5.3) upon photoirradiation under a photomask. Adapted image with permission from Ref. [141]. Copyright (2020) (American Chemical Society)

(Fig. 9.12) [141]. The color change (from yellow to dark red) of amphiphile **9** can be finely tuned by pH (2.1–10.5). In pH range of 4.8–5.9, a reversible color change (between orange and yellow) was observed by alternating irradiation and kept in dark. In using the advanced photo-controlled color change, photo-writing of information and subsequent self-erasing could be done by photoirradiation with a photomask on top of the solution of amphiphile **9**.

The discussion of the controlled macroscopic functions on the basis of photoresponsive molecular amphiphiles with excellent aqueous solubility (below 1.0 wt%) supramolecularly assembled as Gibbs monolayers have clearly shown different promising application opportunities in the novel development of different environmental-friendly production and industrial processing methodologies [142, 143]. However, currently, in use of photoresponsive molecular amphiphiles at interfaces between air and water for industrial-scale instead of laboratory demonstration remained highly largely unexplored. We believe that the identification of the supramolecular assembly mechanisms and energy in molecular isomerization amplified along multiple length-scale to macroscopic dynamic functions, including macroscopic particle motion and foams with photoswitchability, is able to provide hints in design of photoresponsive molecular amphiphiles for industrial-scale applications. Leung and Feringa reported recently the unique supramolecular assembly transformations of an amphiphile based on molecular motor applied in solution and at interfaces, to afford multiple heat/photo-controlled switching of macroscopic foams. This work might provide the key design strategies for the mechanism of motion amplification from nanoscale to assemblies micrometer-scale, and even responsive foam properties at centimeter-scale (full discussion provided in Sect. 9.4) [144].

9.4 Functional Supramolecular Self-assembly of Photoresponsive Charged Molecular Amphiphiles in Solution

Supramolecular assembly of molecules with amphiphilic properties in no matter aqueous, organic, or mixture organic/aqueous media has been known to produce numerous nanostructures, such as bilayers structures (vesicles, tubes, lamellae) and micelles (worm-like, rod-like spherical,). Because of the biocompatibility and potential bio-related applications, we herein focus on the discussion of the supramolecular assembled structures of molecular amphiphile and their structural transformations in aqueous media. The resulting supramolecular assembled structure showed high dependence of the molecular structural design and experimental conditions, e.g., concentration, temperature, pH, light source, ionic strength [72, 145–152]. To predict the shape and size of supramolecular assembled structures of molecular amphiphiles in aqueous media, according to Israelachvili et al., [14] packing parameters (P) can be employed, which was defined as: $P = v/a_0 l_0$, where v is the volume of the amphiphile tail, a_o and l_0 are the area of the hydrophilic groups and the length of tail in the amphiphile, respectively. The molecular geometrical transformation of molecular amphiphile change the packing parameters to induce supramolecular assembly transformation [14, 29, 145, 153, 154]. In the current section, we focus on the discussion of assembly structure of photoresponsive molecular amphiphiles in aqueous media, ranging from one-dimensional (1D) nanostructures, isotropic entangled three-dimensional (3D) networks to anisotropic 3D structures.

9.4.1 *Isotropic Self-assembly of Photoresponsive Charged Molecular Amphiphiles*

Supramolecular assembly of photoresponsive molecular amphiphiles into 1D nanostructures provides deep insight into achieving precise controls of supramolecular organization and potential bio-related applications. An asymmetric supramolecular helical structure has commonly recognized as attractive target for biomaterials and supramolecular chirality [155–161]. Stupp and co-workers reported a supramolecular structure transformation of a photoresponsive amphiphile **10a** from quadruple helical nanofibers to single strand nanofibers (Fig. 9.13) [162]. By photocaging design, the amphiphile **10a** was composed of 2-nitrobezyl group with short peptide segment $GV_3A_3E_3$ (Gly-Val-Val-Val-Ala-Ala-Ala-Glu-Glu-Glu). Upon 350 nm irradiation, the 2-nitrobenzyl group can be cleaved to afford **10b** (Fig. 9.13). After photocaging, the supramolecular interactions become favorable to form supramolecular structure of **10b**, revealed by transmission electron microscopy (TEM). Atomic force microscopy (AFM) and TEM images of **10a** showed quadruple helical assemblies with uniform pitch and width 92 ± 4 nm and 33 ± 2, respectively (Fig. 9.13b–d). Upon 350 nm irradiation, the dissociation of quadruple-helix assemblies into single

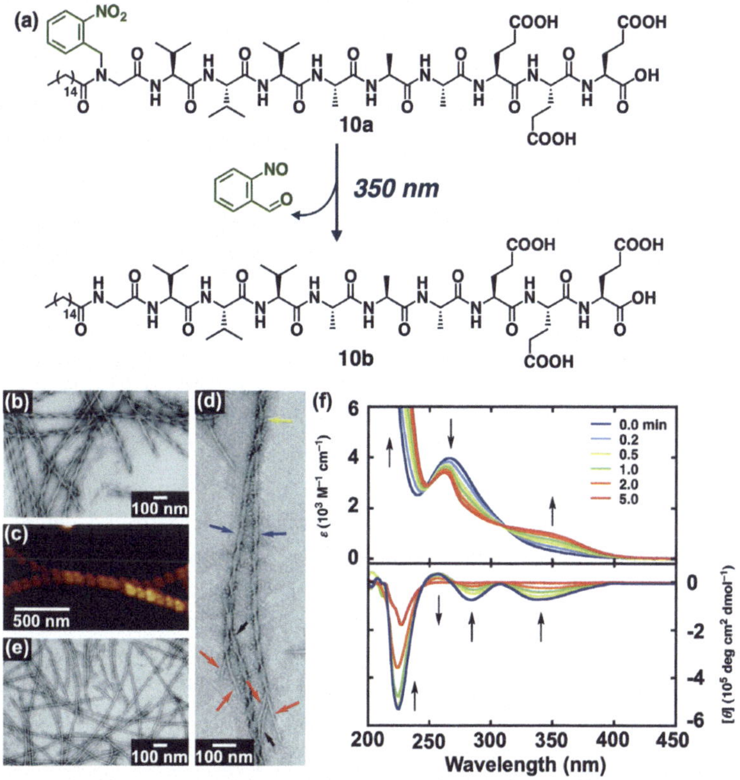

Fig. 9.13 **a** Schematic diagram of the photo-controlled caging of **10a** conversed to **10b** with 350 nm light irradiation. **b** TEM and **c** AFM images of **10a** in aqueous conditions (pH 11). **d** Quadruple nanofibers imaged by TEM. **e** **10a** after photoirradiation imaged by TEM. **f** Absorption spectra (top) and CD spectra (bottom) of **10a** upon 350 nm light irradiation. Adapted image with permission from Ref. [162]. Copyright (2008) (American Chemical Society)

strand non-helical nanofibers (Fig. 9.13e–f). The current study provided a strategy to produce functional photoresponsive helical supramolecular structures with prospects for sensing and actuation.

Later, Stupp and co-workers reported a new photoresponsive molecular amphiphile (**11a**) with a segment of a fibronectin epitope Arg-Gly-Asp-Ser (Fig. 9.14a) [163]. The structural modification of the β-sheet domains can alter the resulting assembled structures. Due to the weakened β-sheet forming segment of **11a** [164], a clear solution remained under the assembling condition of 4.0×10^{-4} M in 0.1 M $CaCl_2$ solution (Fig. 9.14b). In comparison to **10a,** the quadruple-helix was transformed into nanofibers and gels. Upon 350 nm irradiation of **11a**, the obtained **11b** provided a transparent gel to demonstrate a photoresponsive sol–gel transformation (Fig. 9.14c). Nanofibers with diameter of 11 nm were observed in **11a**, and nanospheres (12 nm in diameter) were observed in **11b** (Fig. 9.14d, e). Because of

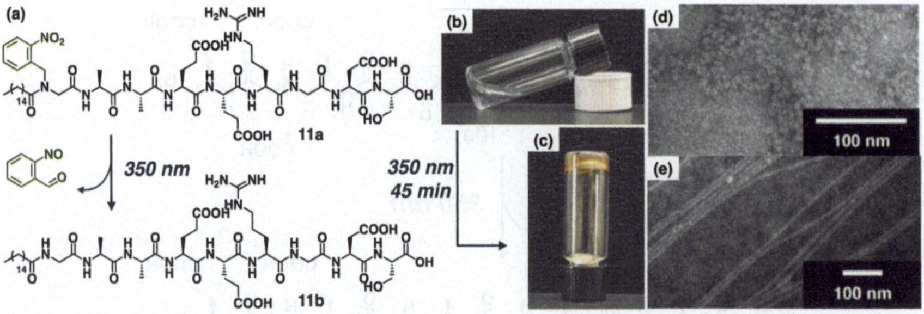

Fig. 9.14 Schematic diagram of **a** the photo-controlled caging of **11a** and the corresponding sol–gel transformations **b** before and **c** after 350 nm light irradiation. **11a d** before and **e** after 350 nm light irradiation imaged by TEM. Adapted image with permission from Ref. [163]. Copyright (2009) (John Wiley and Sons)

the limited cytotoxicity of both **11a** and **11b**, the current photoresponsive gelation process can potentially be applied for photoresponsive cell scaffold development.

In addition to multiple helical supramolecular structures, the design of highly dynamic and well-defined assembly structures with adaptive behavior to external stimulations would be considered one of the grand challenges for man-made supramolecular structures. In 2011, Feringa and co-workers reported a photoresponsive supramolecular assembly of vesicle-capped nanotubes of molecular amphiphile **12** in aqueous media (Fig. 9.15a) [165]. The phospholipid 1,2-dioleoyl-*sn*-glycero-3-phosphocholine (DOPC) was premixed to improve amphiphile solubility in aqueous media to provide unique vesicle-capped nanotubes, observed by cryogenic transmission electron microscopy (cryo-TEM). The vesicle-capped nanotube was chemically modified into nanotubes only by adding Triton X-100 and reformed by removing the Triton X-100 from the medium (Fig. 9.15c). Furthermore, the vesicle-capped nanotubes were selectively disassembled by photoirradiation (Fig. 9.15b). Due to the intrinsic fluorescent property of the overcrowded alkene core of **12**, the supramolecular transformation was monitored in-situ by fluorescence microscopy. The dynamic vesicle removal and photo-controlled disassembly process of **12** provided a novel strategy to control supramolecular assembly structure by both chemical and light stimulations. Later the vesicle in the vesicle-capped nanotubes showed sensitivity to change in osmotic pressure, providing assembly structure with vesicle-encapsulated nanotube (Fig. 9.15d) [166]. The photo-controlled disassembly of nanotubes allowed the release of encapsulated vesicles. This work demonstrated the multiple-responsive properties of these novel co-assembled structures.

Even though UV-light is commonly used in photoresponsive supramolecular systems, the low bio-tissue penetration and serious bio-tissue damage properties of UV-light limit the bio-related application of UV-responsive systems. Wang and co-workers reported a co-assembly system of a charged coumarin unit with common lipids was able to prepare near-infrared (NIR) responsive liposomes for controlled drug delivery [167], although no detailed assembly transformation process and origin of NIR responsiveness were mentioned (Fig. 9.16) [168]. The co-assembly

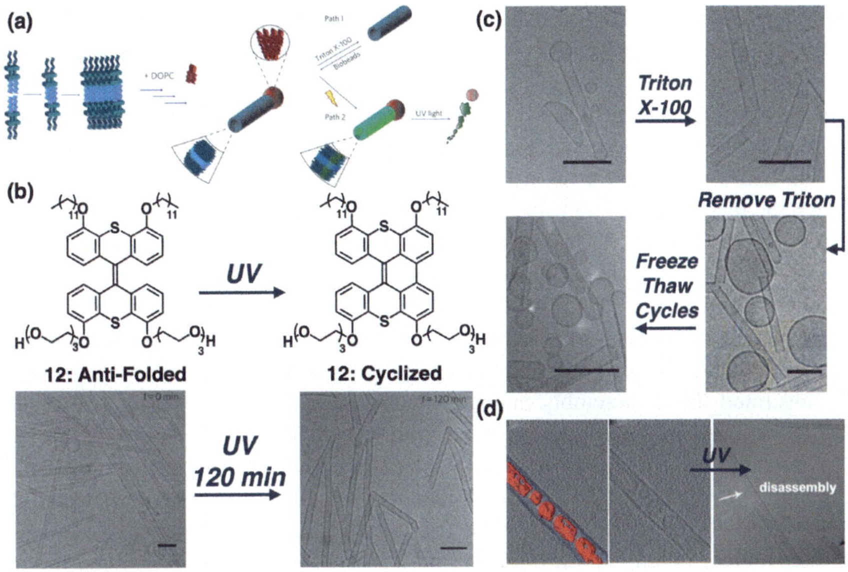

Fig. 9.15 **a** Schematic diagram of supramolecular transformations of co-assembly of **12** with DOPC. **b** Structure of **12** imaged by cryo-TEM after photoirradiation, images recorded at 0 min and 120 min. **c** Disassembly of vesicle imaged by cryo-TEM. **d** Osmosis caused encapsulation of vesicles imaged by cryo-TEM. Scale bars in **b**, **c** = 100 nm, in **d** = 50 nm. Adapted **a–c** with permission from Ref. [165]. Copyright (2011) (Springer Nature). Adapted **d** with permission from Ref. [166]. Copyright (2015) (John Wiley and Sons)

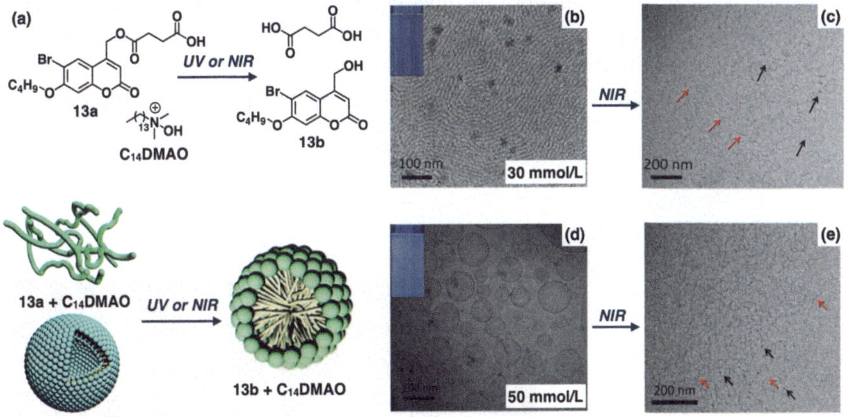

Fig. 9.16 **a** Schematic diagram of the photo-controlled caging process and assembly transformation of **13a** to **13b**. Solutions of **13a** (30 mM) with C_{14}DMAO **b** before and **c** after NIR irradiation imaged by cryo-TEM. Solutions of **13b** (50 mM) and C_{14}DMAO **d** before and **e** after NIR irradiation imaged by cryo-TEM. Adapted image with permission from Ref. [168]. Copyright (2017) (Royal Society of Chemistry)

of amphiphile **13a** and tetradecyldimethylamine oxide (C_{14}DMAO) assembled into worm-like micelles (in 30 mM of **13a**) and vesicles (in 50 mM of **13a**). Although the origin of the NIR responsiveness remained unclear, the worm-like micelles and vesicles were able to transform into spherical micelles with NIR irradiation through the photocleavage of **13a**.

The irreversible photoresponsive charged molecular amphiphilic systems have shown the feasibility of controllable supramolecular assembly structures in aqueous conditions ranging from 1D nanostructures transformations to isotropic entangled 3D sol–gel transformations. In contrast, reversible supramolecular assembly transformations can afford scientifically interesting opportunities toward developments of smart soft materials. Some pioneering works of supramolecular assemblies of photoresponsive molecular amphiphiles, reported by Engberts and co-workers, were demonstrated the co-assembly and molecular interactions of common surfactants and azobenzene amphiphiles to provide solid foundations for the later supramolecular assemblies of photoresponsive molecular amphiphiles developments [169–172]. Stupp and co-workers reported supramolecular helical nanofibers with a reversible controlled helical pitch, which were formed from a photoresponsive positively charged molecular amphiphile (azobenzene-based) **14** (Fig. 9.17a) [173]. The photoisomerization of nanofibers of *trans*-**14** to *cis*-**14** was driven by 360 nm light source. Due to the reduced planarity of *cis*-**14**, the helical pitch of nanofibers of *trans*-**14** could be reduced by the increased sterically induced torque. The nanofibers helical pitch was reduced from 78 to 56 nm upon photoirradiation, as confirmed by AFM (Fig. 9.17b, c). It is noted that the supramolecular assembly required the presence of organic solvents, which reduced the potential biocompatibility of the current system.

Supramolecular transformations in aqueous media of azobenzene-based photoresponsive molecular amphiphiles were widely studied, including cationic bola- and Gemini-molecular amphiphiles [174–183]. Some supramolecular assembly transformations from vesicles to other assemblies, such as nanofibers, can potentially be

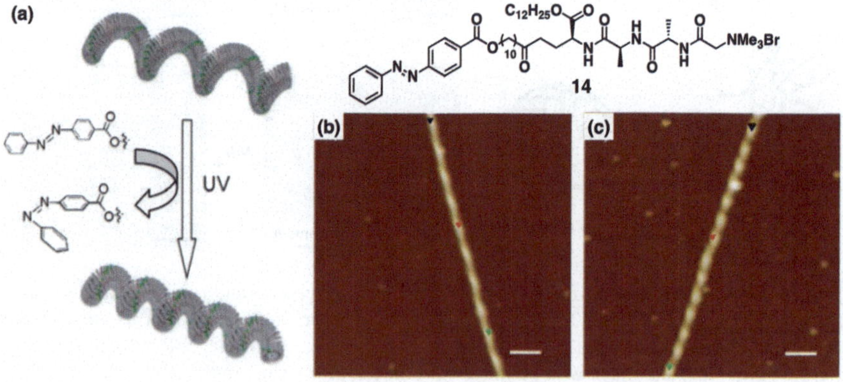

Fig. 9.17 **a** Molecular design of **14**. Helix pitch of **14** nanofibers **b** before and **c** after UV photoirradiation. Adapted image with permission from Ref. [173]. Copyright (2007) (John Wiley and Sons)

applied in drug encapsulation and delivery [55, 184–186]. However, the application potentiality of assemblies of azobenzene-based molecular amphiphiles might be limited by the two-states photoisomerization. The development of multiple-states assemblies using photoresponsive molecular amphiphiles in aqueous conditions remains highly challenging. A photoresponsive multiple-states nanostructure using co-assemblies of a molecular amphiphile **15** was reported, by Huang's group, with a cationic amphiphile cetyltrimethylammonium bromide (CTAB, Fig. 9.18) [187]. The co-assemblies transform reversibly between assemblies of worm-like micelles, vesicles, lamellar structures, and small micelles in aqueous media, which is controlled by the photoirradiation duration. The assembly transformations, controlled by degree of photoisomerization of **15**, resulted in significant change in macroscopic properties. The rheology behaviors of the co-assembly of **15** and CTAB, were studied by different photoirradiation duration, allowing classification of co-assembly structures into four classes, including (1) worm-like micelles (in the presence of 89% of *trans*-**15**), (2) bilayered vesicles with planar lamellae (68% of *trans*-**15**), (3) worm-like micelles (37% of *trans*-**15**), and (4) small micelles (17% of *trans*-**15**). But nanostructure studies with TEM in confirming those assembled states were not provided. Identical molecular amphiphile **15** was also applied in another co-assembly study with a surface-active ionic liquid to give worm-like micelles [188]. Upon 365 nm irradiation, the length of worm-like micelles was shortened and become entangled, affording increased viscosity. The supramolecular assemblies based on ionic liquids, due to the intrinsic properties, e.g., low vapor pressure, low melting point, thermal stability, have attracted significant attention lately.

Most of the discussed systems of photoresponsive charged molecular amphiphiles, their supramolecular transformations were observed at the equilibrium state before and after photoisomerization. *In-situ* supramolecular assembly transformations of photoresponsive glucose-based amphiphilic micelles were monitored by time-resolved small-angle neutron scattering (TR-SANS), reported by Wilkinson et al. [189]. The evolutions of the precise shape and aggregation number of micelles of

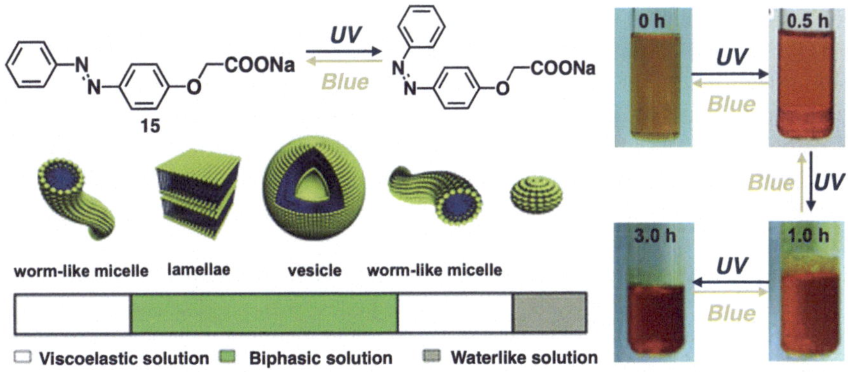

Fig. 9.18 Schematic diagram of photoresponsive co-assemblies. Adapted image with permission from Ref. [187]. Copyright (2010) (Royal Society of Chemistry)

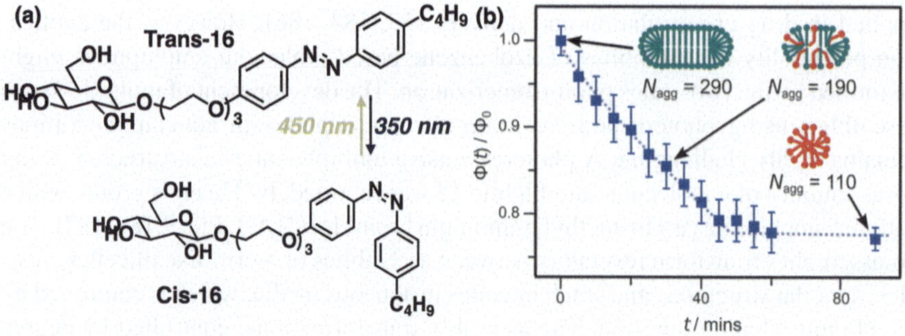

Fig. 9.19 **a** Photoisomerization of **16**. **b** Volume fraction of micelles of **16** in solution with photoisomerization progresses from the SANS data. Adapted image with permission from Ref. [189]. Copyright (2015) (Royal Society of Chemistry)

the *trans*- and *cis*-states of **16** as well as at discrete time intervals were recorded (Fig. 9.19). *Trans*-**16** aggregated into micelles with aggregation number ~290, and then transformed into aggregation number ~110 upon UV-light irradiation, revealing the size of micellar structures reduced [190]. The combined in-situ SANS with absorption spectroscopy can monitor the supramolecular transformations from worm-like micelles to fractal aggregates simultaneously with detailed photoisomerization processes upon photoirradiation. Time-resolved small-angle X-ray scattering (TR-SAXS) studies provided details information on the photo-controlled disassembling processes of an azobenzene-based amphiphile [191]. All these advanced measurements provide direct and *in-situ* studies of assembly transformations of photoresponsive molecular amphiphiles and deeper insight into the transformation mechanism.

Other supramolecular assembly examples of azobenzene-based amphiphiles in aqueous media, including a Gemini azobenzene-based amphiphile reported by Zhao [192] and a head–tail azobenzene-based amphiphile **17** reported by Jiang's group provided more detailed studies of photoresponsive macroscopic gel-sol transformations [193] The co-assembly of the cationic amphiphile **17** with sodium azophenol provided a dual stimuli-responsive 3D sol–gel transformations by light and gas purging (Fig. 9.20). The gel-sol transformation was controlled by the reversible structural transformations between **AzoONa** and **AzoOH** through alternative purging of carbon dioxide and nitrogen. Besides, the gel-sol transformation can also be controlled reversibly by the photoisomerization of amphiphile **17**. The dual stimuli-responsive gel-sol process can potentially be applied in microfluidics and tertiary oil recovery.

Other photoresponsive molecular amphiphiles with photoresponsive cores, e.g., molecular motors, spiropyrans, and diarylethenes will be discussed. A chiral diarylethene-based molecular amphiphile **18**, which featured with a hexa(ethylene glycol) side chains was reported by Irie and co-workers, to show light-dependent

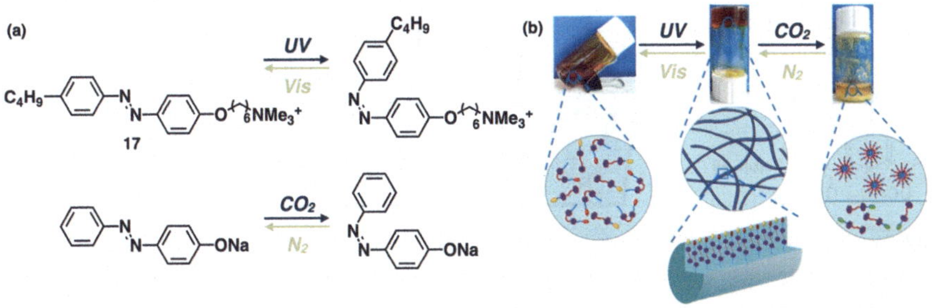

Fig. 9.20 **a** Photoisomerization of **17** with **AzoONa**. **b** A gas/photo-controlled 3D sol–gel process. Adapted image with permission from Ref. [193]. Copyright (2018) (Elsevier)

chirality transfer in aqueous media (Fig. 9.21) [194]. The supramolecular structures of open-**18** showed no supramolecular chirality induction but the photo-isomerized close-**18** did. The photo-controlled supramolecular chirality transformations might provide a novel strategy for photoswitching chiroptical properties in aqueous conditions.

In 2016, the design of motor amphiphile **19**, based on molecular design of amphiphile **12**, was reported by Feringa and co-workers, developing a multi-stage system for reversible supramolecular assembly in aqueous media (Fig. 9.22a) [195]. The photoisomerization and thermal helix inversion (THI) of motor amphiphile **19** were confirmed with a metastable-**19**/stable-**19** ratio of 95/5 at PSS, while a half-life of THI was determined as 270 h (at 20 °C) and 4.3 h (at 50 °C). The solution of motor amphiphile **19** with DOPC was co-assembled into nanotubes (Fig. 9.22b) and transformed into bilayered vesicles upon 365 nm irradiation for 15 min (Fig. 9.22c). The obtained solution after photoirradiation was subsequently heated at 50 °C for 16 h to give multi-lamellar vesicles (Fig. 9.22d). Nanotubes

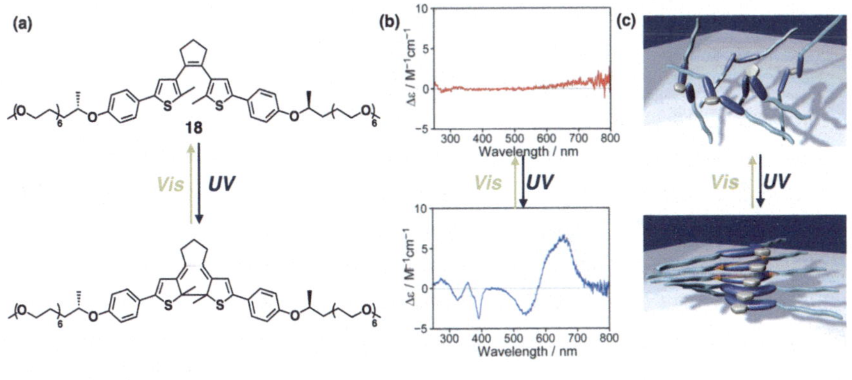

Fig. 9.21 **a** Photoisomerization of **18**. **b** CD spectra of **18** upon photoirradiation. **c** Schematic diagram of nanostructures of open-**18** (top) and closed-**18** (bottom). Adapted image with permission from Ref. [194]. Copyright (2006) (American Chemical Society)

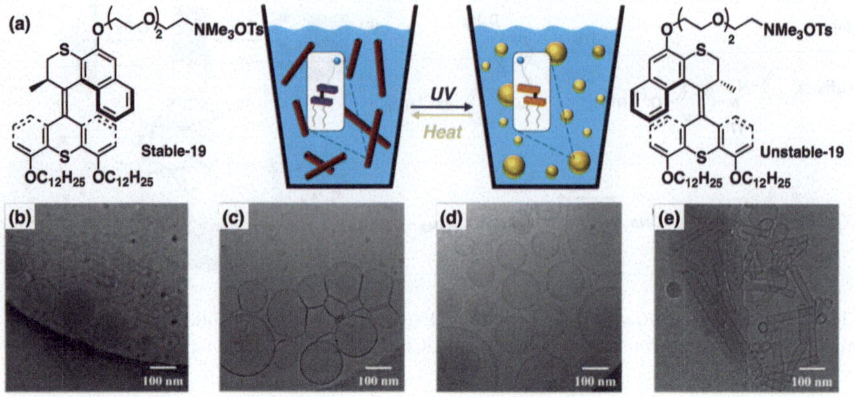

Fig. 9.22 **a** Schematic diagram of photoisomerization of **19** and its assembly transformation. Structures of stable-**19** with DOPC **b** before and **c** after photoirradiation, **d** subsequent heating, and **e** freeze-thawing imaged by cryo-TEM. Adapted image with permission from Ref. [195]. Copyright (2016) (American Chemical Society)

were regained only from multi-lamellar vesicles treated with freeze-thawing cycles (Fig. 9.22e), revealing a reversible supramolecular assembly process controlled by heat and light. This strategy might be paving a way for more sophisticated and dynamic artificial nano-systems in aqueous media.

To minimize the bio-damaging by UV-light in future bio-related applications, Kudernac and co-workers reported a novel visible-light responsive amphiphile **20** with two spiropyrans switching cores for demonstrating a reversible photo-controlled vesicle expansion (Fig. 9.23a) [196]. The two spiropyrans motifs were joined to a

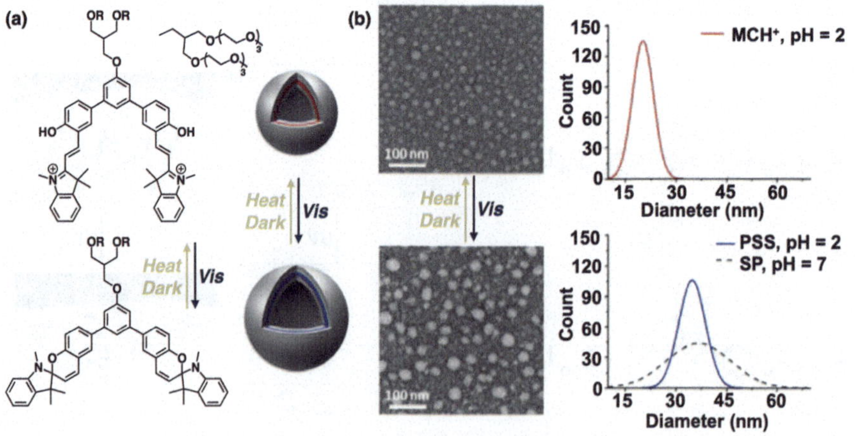

Fig. 9.23 Schematic diagram of **a** photoisomerization and assembly transformation of **20**. **b** Assembly transformation imaged by TEM. Adapted image with permission from Ref. [196]. Copyright (2018) (Royal Society of Chemistry)

bent aromatic unit with oligoether dendron as hydrophilic moiety giving the basic design of amphiphile **20**. At low pH, amphiphile **20** adopts the open protonated merocyanine form (MCH^+). This form can be converted to a ring-closed spiropyrans form (SP) transiently upon visible-light irradiation (Fig. 9.23a). Both SP and MCH^+ forms of **20** also assemble into vesicles with different average diameters (SP ~35 nm; MCH^+ ~20 nm), confirmed by TEM (Fig. 9.23b). Upon visible-light irradiation, the vesicle diameter of MCH^+ was increased from ~20 to ~35 nm, possibly due to the electrostatic interaction generated upon photoirradiation. The vesicle diameter of MCH^+ could be restored reversibly through thermal relaxation from SP to MCH^+. This strategy revealed a reversible vesicle expansion/contraction in aqueous media controlled by visible light.

To provide extra stimulative controls of the photoresponsive supramolecular assembling systems, recently, Leung and Feringa reported a multi-modal controlled assembly of a bola-amphiphile [197]. On the basis of the first-generation molecular motor core, the bola-amphiphile **21** was designed to connect the core with carboxylates group via alkyl-linkers, demonstrating UV-light, pH and counterion controlled assembly transformations in aqueous media (Fig. 9.24a). The supramolecular transformations from sheet-like structures to a mixture of sheet-like structures, vesicles, and micelles were observed upon 312 nm photoirradiation (Fig. 9.24b), possibly due to significant geometrical transformation from stable *trans*-**21** to unstable *cis*-**21**.

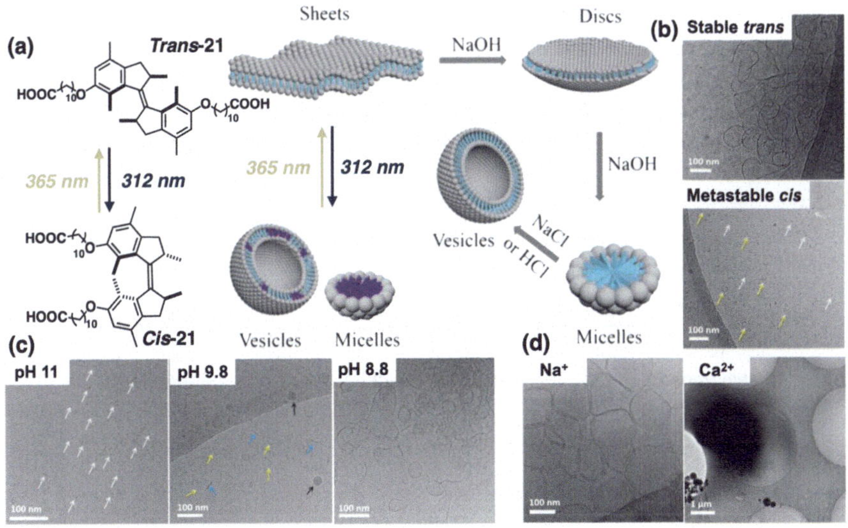

Fig. 9.24 **a** Schematic diagram of assembly transformations of **21**. **b** Photo-controlled, **c** pH-controlled, and **d** counterion controlled assembly transformations imaged by cryo-TEM. Adapted image with permission from Ref. [197]. Open access under a CC BY 4.0 license, https://creativecommons.org/licenses/by/4.0/

Turning pH from 11 to 8.8 of solution of stable *trans*-**21**, supramolecular transformations from micelles (pH 11) to disc-like structures (pH 9.8) and to sheet-like structures (pH 8.8) were observed (Fig. 9.24c), because of the packing parameter upon protonation. Due to charge screening effect, the vesicles were precipitated in the presence of calcium ions (Fig. 9.24d). This approach provides multi-modal controls of supramolecular assembly in aqueous media, but additional heating and cooling cycle was required for completing reversible photoresponsive assembly transformation. In 2020, Pujals and Albertazzi also reported an azobenzene-based charged amphiphile for multiple-responsive supramolecular transformations [198]. Through the controls of light, pH, ion, and temperature, the supramolecular helicity of the azobenzene-based charged amphiphile can be fine adjusted, as confirmed by TEM and circular dichroism (CD).

To provide instant effective photoresponsive supramolecular assembly transformations, Leung, Wang, and Feringa reported a new design of charged molecular motor amphiphile **22**, featuring with the first-generation molecular motor core functionalized with an alkyl-chain and a quaternary ammonium motif connected with triethylene glycol-linker (Fig. 9.25) [144]. Motor amphiphile **22** showed instant dynamic controls of supramolecular assemblies and consequently fine adjustment of macroscopic foam properties in aqueous media. Upon short duration of UV-irradiation (6 min), the supramolecular worm-like micelle structure of *trans*-**22** was transformed into a mixture of worm-like micelle and vesicles to reduce foaming ratio from ~13 to ~8, i.e., state 1 and state 2 (Fig. 9.25b). The 180° molecular rotation from stable *trans*-**22** to stable *cis*-**22** could be achieved by subsequent 254 nm photoirradiation and heating process, i.e., state 3. The detailed investigations of molecular isomerization, transformations of supramolecular assembly, in-situ surface tension, and macroscopic foam properties have revealed an amplification process from molecular motion to microscopic structural transformation, and to

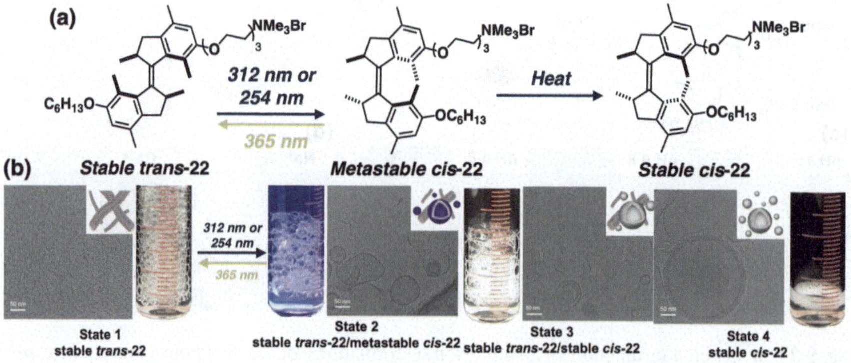

Fig. 9.25 **a** Photoisomerization and thermal helix inversion of **22**. **b** Dual light/heat-controlled assembly transformations imaged by cryo-TEM. Scale bars in cryo-TEM images are 50 nm. Adapted image with permission from Ref. [144]. Open access under a CC BY 4.0 license, https://creativecommons.org/licenses/by/4.0/

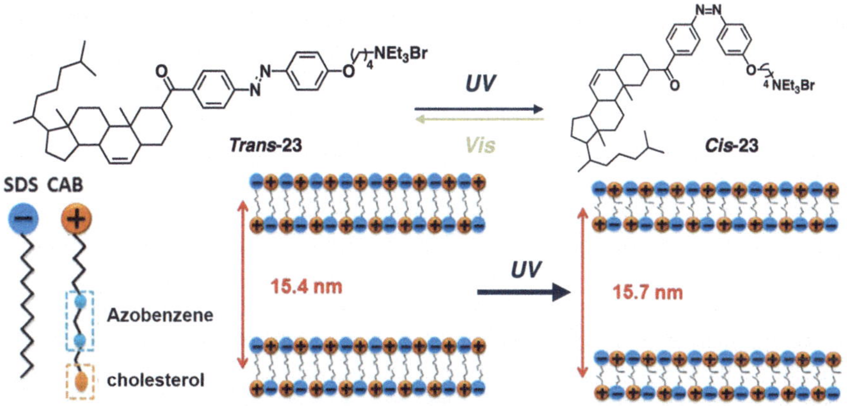

Fig. 9.26 Schematic diagram of photoisomerization of **23** and its vesicle increase of interlamellar spacing. Adapted image with permission from Ref. [199]. Open access under a CC BY 4.0 license, https://creativecommons.org/licenses/by/4.0/

macroscopic functions. The current system might provide unprecedented controls of multi-states supramolecular assembly transformations reversibly, without additional help from lipid or freeze–thaw cycles.

Through counterion co-assembly of an azobenzene-based amphiphile **23** with common amphiphile negative charged sodium dodecyl sulfate (SDS), it allowed forming multi-lamellar vesicles (Fig. 9.26), reported by Wang et al. [199]. The interlamellar spacing of the co-assembled vesicles was increased from 15.4 to 15.7 nm upon UV photoirradiation, confirmed by SAXS. The co-assembly is able to encapsulate rhodamine B (rhB), which was applied to rat retina for in-vivo drug delivery studies. A more significant release of rhB was observed upon UV photoirradiation. The current study illustrated the feasibility of photoresponsive supramolecular assembly in drug delivery applications.

The most representative examples in electrostatic interaction-based photoresponsive molecular amphiphiles (below 1.0 wt%) have been discussed to demonstrate how dynamic supramolecular assembled structural transformations in aqueous media could be achieved by photoirradiation. The discussed photoresponsive molecular amphiphiles show excellent potential as novel drug carriers for controlled drug release upon photoirradiation with high spatial–temporal precision.

9.4.2 Anisotropic Supramolecular Assembly of Photoresponsive Molecular Amphiphiles

Three-dimensional hierarchical supramolecular assembled structures are commonly found in biological systems, such as collagens and actin filaments [200–205]. The hierarchical collagen fibrils are composed of the assembly of the microfibrils of triple-stranded helices that are formed by the folding of three polypeptide chains. Actin filaments provide structural stability to cells and as major part of the contractile apparatus in muscle cells. Structurally, helical ribbon in actin filaments is constructed from two parallel strands of linear arrays of protein monomers, which the strands held tightly through multiple supramolecular interactions. With the inspiration from the natural hierarchical supramolecular assemblies, the novel designed and engineered synthetic molecular amphiphiles assembled precisely into anisotropic macroscopic structures have been recently developed. Anisotropic 3D hierarchical supramolecular structures, at macroscopic length scales, have provided exciting opportunities toward applications in regenerative biomedical materials, electronic/optoelectronic materials, anisotropic actuators, and soft robotics. Stupp and co-workers reported a peptide amphiphile assembled into nanofibers and, through a shear-flow method, the nanofibers were assembled hierarchically into a macroscopic string in the presence of calcium chloride [206]. The macroscopic string, composed of unidirectionally aligned nanofibers, was applied as artificial scaffolds for cell growth with directional preference, generating potential novel tissue regenerative materials [207, 208]. Later, the same group has shown a thermal responsive macroscopic actuation by co-assembled supramolecular nanofibers of peptide amphiphiles with thermal responsive polymers [209, 210].

With the well-established anisotropic 3D hierarchical supramolecular structures, a system with photo-responsiveness was reported by Liu and co-workers, featuring a cationic Gemini azobenzene amphiphile **24** (Fig. 9.27) [211]. Amphiphile **24** assembled from nanorods into crystalline helical twisted bundles (3 mm in length and 25 μm in diameter) using organic solvent evaporation for 2 days. The helical twisted

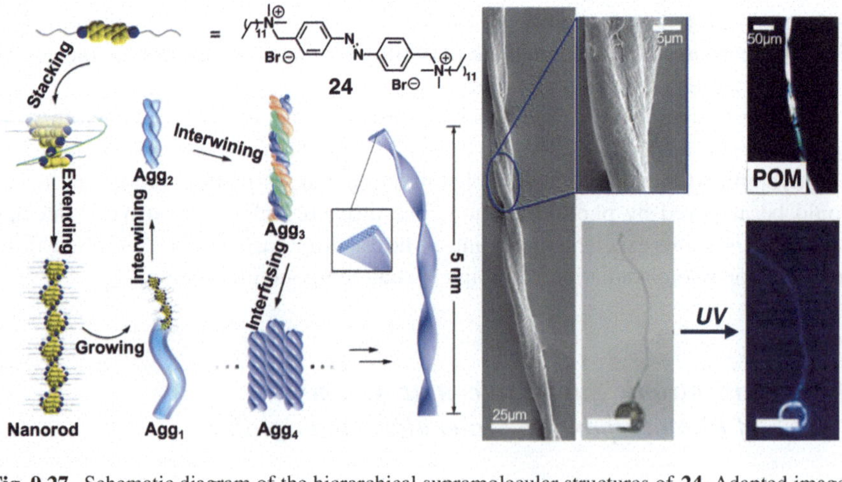

Fig. 9.27 Schematic diagram of the hierarchical supramolecular structures of **24**. Adapted image with permission from Ref. [211]. Copyright (2015) (Springer Nature)

bundles bent toward the incident light source (302 nm light) in 120 min, which actuated from 36° to a saturated flexion angle of 50° (actuation speed of 1.9×10^{-3} deg s^{-1}), affording a millimeter length-scale anisotropic actuation. However, the actuation mechanism of the hierarchical supramolecular assembly has remained unclear that is operated in organic medium.

In 2018, Feringa and co-workers reported the first anisotropic 3D hierarchical supramolecular structure of a photoresponsive molecular motor amphiphile in aqueous media, demonstrating macroscopic actuation in water and air (Fig. 9.28) [212]. The molecular motor amphiphile **25** is composed of two carboxyl groups linked with alkyl-linkers to lower half of motor core, while a dodecyl chain is linked to upper half of motor core (Fig. 9.28a, b). Due to the molecular phase separation of the motor amphiphile, well-structurally organized nanofibers were formed in aqueous media. The highly negatively charged surface of the nanofibers of **25** could form macroscopic supramolecular structure hierarchically by a shear-flow method in the presence of calcium chloride to afford unidirectionally aligned macroscopic string, confirmed by POM, and scanning electron microscopy (SEM). The resulting macroscopic string bent toward the light source to complete actuation from 0° to 90° saturated flexion angle within 60 s in aqueous media (Fig. 9.28c). Besides, the prepared macroscopic string could be pulled out from the aqueous media to perform actuation in air (Fig. 9.28d), and even could actuate a 0.4 mg piece of paper (Fig. 9.28e). To minimize the background scattering, *in-situ* SAXS measurements were performed with macroscopic string prepared in air. Upon photoisomerization of **25**, the increased free volume around molecular motor core increases the diameter of nanofibers from 5.45 to 5.80 nm, accompanied by decreased packing order of the lamellar structures. Due to constant total volume of the macroscopic string, the contraction of long axis of the string is expected. Meanwhile, by considering the degree of light penetration and the thickness of the string (~300 μm), the light intensity gradient allows the macroscopic bending of the string toward the light source. The current study has demonstrated the feasibility of effective energy conversion at nanoscale, acumination of strain within the hierarchical supramolecular structure, and amplification of molecular motion to macroscopic actuation of soft materials in aqueous media.

Feringa and co-workers have further investigated the electrostatic interaction between carboxylate groups of motor amphiphile **25** and counter cations, which allowed the stabilization and formation of nanofibers. The studies were focused on ion effect of the hierarchical supramolecular structure in the nanofiber formation, aggregation of nanofibers, structural orientation of macroscopic string, and its actuation speed (Fig. 9.29) [213]. A series of macroscopic strings were prepared from motor amphiphile **25** with different cations and measured with SAXS. The orientation order of resulting macroscopic is followed $Ca^{2+} > Mg^{2+} > Be^{2+} \approx Sr^{2+} \approx Sc^{3+} > Ba^{2+} \approx Li^{+} \approx Na^{+} \approx K^{+}$. The macroscopic strings prepared from solutions of $BaCl_2$, LiCl, NaCl, and KCl, revealed no alignment, and no actuation. In contrast, the actuation speed of macroscopic strings was shown: $Ca^{2+} > Mg^{2+} > Sr^{2+} \approx Sc^{3+} > Be^{2+}$, indicating structural orientation governed the actuation speed of resulting macroscopic string. Besides, the alkyl-linkers of the motor amphiphile was modified

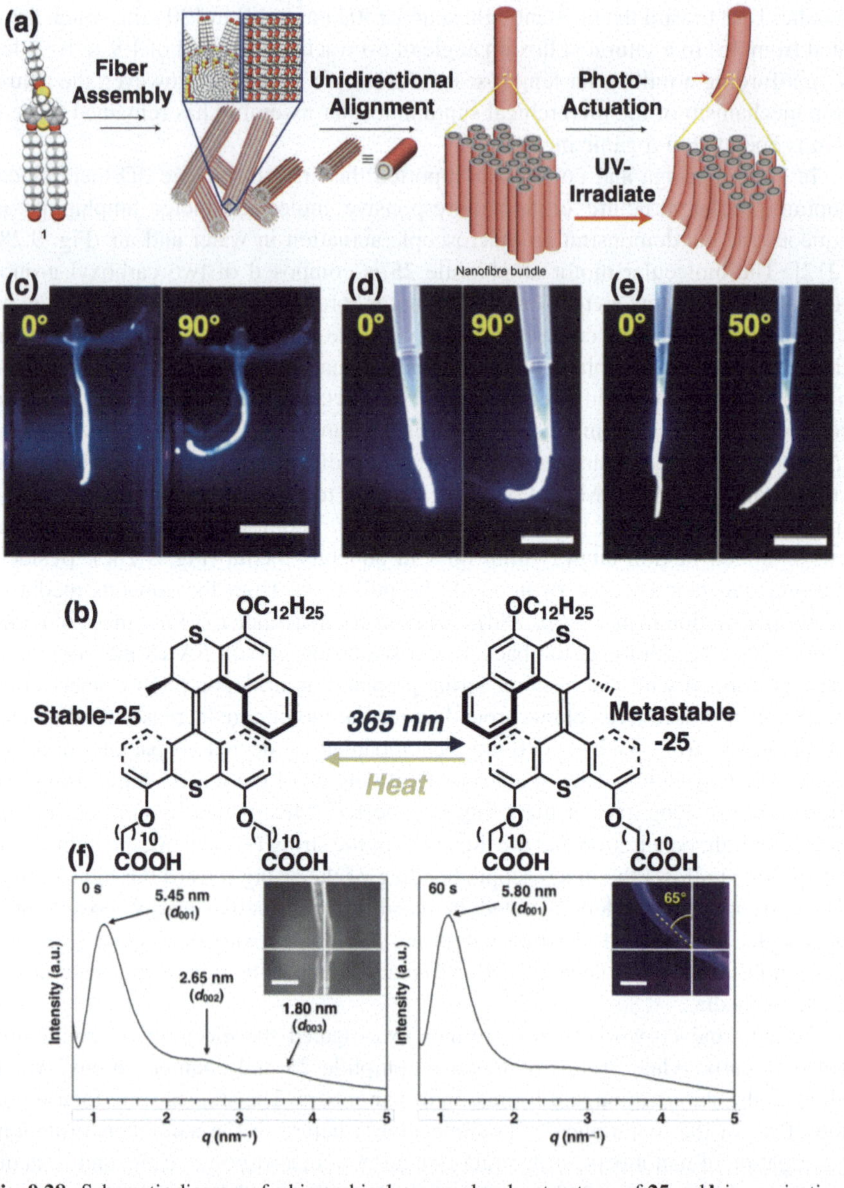

Fig. 9.28 Schematic diagram of **a** hierarchical supramolecular structures of **25** and **b** isomerization processes of **25**. A string of **25** bending upon UV-irradiation in **c** aqueous media, **d** air, and **e** with 0.4 mg paper, scale bar: 0.5 cm. **f** A **25** string before and after actuation analyzed by in-situ SAXS. Adapted image with permission from Ref. [212]. Copyright (2017) (Springer Nature)

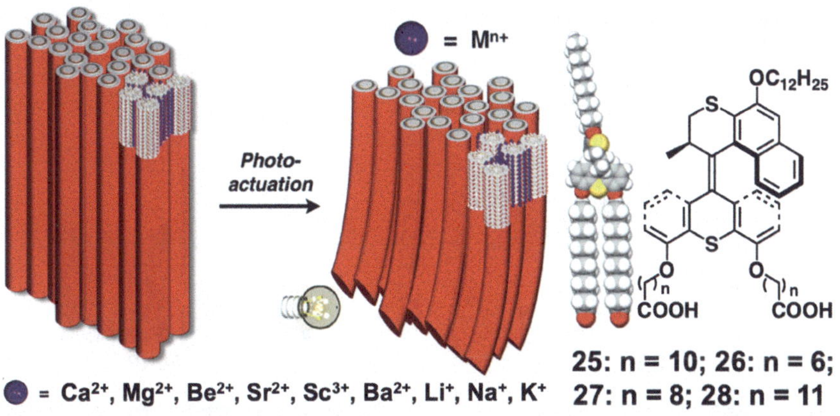

Fig. 9.29 Schematic diagram of s **25–28** with various alkyl-linkers lengths and the hierarchical organization the assembled structures. Adapted image with permission from Ref. [213]. Open access under a CC BY 4.0 license, https://creativecommons.org/licenses/by/4.0/

with different carbon number (n = 10, 6, 8, 11; **25–28**), providing modifications to supramolecular packing and resulting actuation functions (Fig. 9.29). The current study has clearly demonstrated that unidirectionally aligned hierarchical supramolecular structure and its actuation function could be finely adjusted simply by changing counterion for macroscopic string preparation, without covalent modification of the motor amphiphile.

Later, Leung and Feringa reported the first dual light/magnetic field controlled macroscopic actuation and cargo carrier, prepared from a supramolecular hierarchical assembled structure of motor amphiphiles. Motor amphiphile **29** is composed of two additional histidine motifs based on **25** to serve as the nucleation site for iron nanoparticles (FeNP) formation (Fig. 9.30a) [214]. To improve the structural orientation of macroscopic string of FeNP-**29**, nanofibers of **25** were added (FeNP-**29/25**), confirmed by POM, SEM, and SAXS measurements. Upon 365 nm photoirradiation, the macroscopic string of FeNP-**29/25** bent toward the light source from 0° to a saturation flexion angle of 90° in 25 s. Besides, the macroscopic string of FeNP-**29/25** moves toward a magnet with fast response (2 s, Fig. 9.30b). The dual-controlled macroscopic string was applied in cargo transport process (Fig. 9.30c–g), through sequential light/magnet stimulation, the string was able to carry a piece of paper away 2 cm from original position.

The discussed hierarchical supramolecular assembles of molecular motor amphiphiles in aqueous media with adjustable structural order and controllable actuation speed reveal unprecedented artificial muscle-like functions by using external stimulations, i.e., light and a magnetic field. The artificial muscle-like functions provide applications for cargo transport and weight lifting. Besides the first experimental demonstration of molecular energy conversion, accumulation of structural strain and energy amplification to macroscopic actuation was clearly demonstrated in

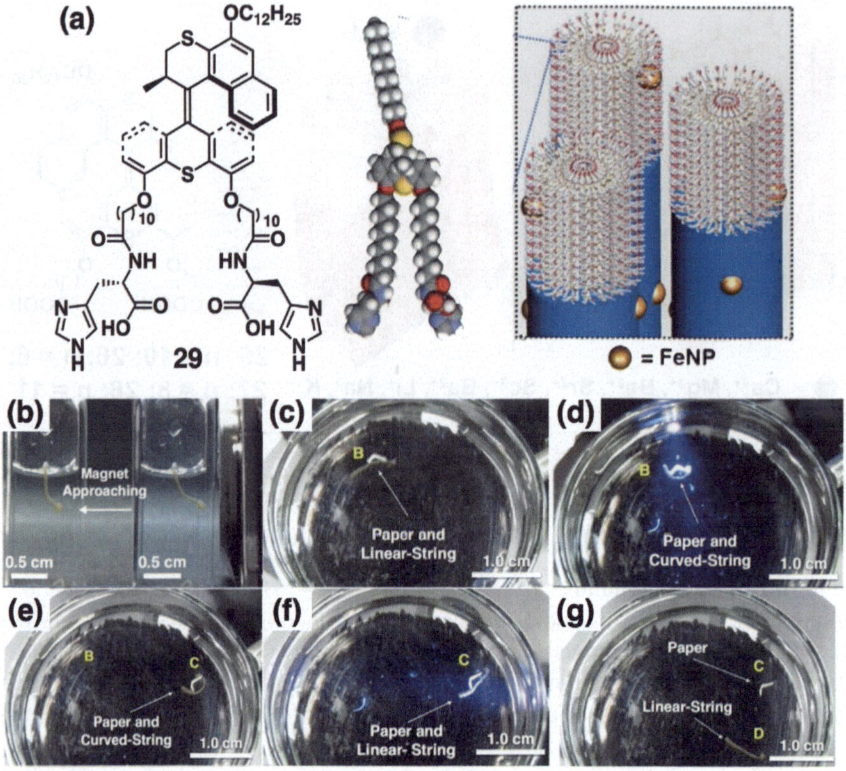

Fig. 9.30 Schematic diagram of **a** the assembly structure of FeNP-**29** and **b** a FeNP-**29/25** string moving towards a magnet. Images of **c** a FeNP-**29/25** string in position B, **d** changing to a curved shape, **e** carrying a piece of paper to position C, **f** changing to a linear shape, **g** unloading the paper. Adapted image with permission from Ref. [214]. Copyright (2019) (John Wiley and Sons)

using small molecule-based supramolecular systems. In addition, a related approach using polymeric gels driven by molecule motors was reported Guiseppone et al. [215–217]. The slow thermal helix inversion of molecular motor amphiphiles is limiting the concept of reversible macroscopic actuation under ambient conditions. The combination of photoresponsive units, i.e., photoswitches, with molecular amphiphilic structures in building hierarchical supramolecular assemblies might provide prospects to link up responsive molecules and macroscopic functions.

9.5 Conclusion

Supramolecular assembled structures of molecular amphiphiles are discovered in various living systems. As inspired by nature, different synthetic stimuli-responsive molecular amphiphiles in aqueous media provide well-defined supramolecular assembled structures with promising biomimetic functions. Light is applied as non-invasive and spatial controllable because of its unique advantages. The recent development of the photoresponsive molecular amphiphiles at various concentrations in different conditions such as Gibbs monolayers or aqueous media has been illustrated and summarized on the basis of their supramolecular assembled structures.

The morphological and physical/chemical parameters transformations in solution and at interfaces of air and water, i.e., Gibbs monolayers, are controlled with high precision by the hierarchical supramolecular assembled structures of photoresponsive molecular amphiphiles, which contain photoresponsive molecular machine units. At a molecular length-scale, dynamic adsorptions and desorptions processes of isomers produced by light at Gibbs monolayers result in photo-controlled Marangoni flows, to allow for centimeter length-scale photoresponsive dynamic functions, i.e., photo-controlled foams, crystal formations, liquid marble transportations, and particle depositions. The current discussed examples have shown that the photoresponsive amphiphiles prepared Gibbs monolayers at low concentration (<1.0 wt%) are able to control not only physical parameters but also smart responsive macroscopic functions, allowing for potential establishment of environmental more friendly industrial and production processing methodologies.

The photoresponsive amphiphiles with concentration less than 1.0 wt% show stimuli-responsive supramolecular assembled structures in aqueous solution, and its structural transformations supramolecularly enable macroscopic gel-sol transitions of 3D randomly organized structures. In advance, using a higher concentration of about 5.0 wt% of photoresponsive molecular amphiphiles in aqueous solutions, allows forming a large quantity of nanofibers. The nanofibers are unidirectionally aligned and oriented with a tight supramolecular packing into a macroscopic structure to show a lyotropic liquid crystal phase. The unidirectionally aligned hierarchical supramolecular structure of nanofibers with significant bundled structures allows an energy conversion from molecular rotations, acumination of tension strain within the highly packed structure, and amplification of molecular motions into macroscopic muscle-like motions upon light irradiation. Currently, simply in consideration of the amphiphilic molecular structure, concentration, and environments (at interfaces and in solution), it has created three distinct approaches to supramolecular chemistry research with promising and cutting-edge applications from the industrial processes to future soft robotics.

Acknowledgements This work was supported financially by the Hong Kong Research Grants Council, Early Career Scheme (ECS 25301320), the Croucher Foundation (Croucher Innovation Award), The Hong Kong Polytechnic University Start-up Fund (1-BE2H).

References

1. Vittorio Degiorgio P, Corti M (1985) Amphiphiles: micelles, vesicles and microemulsions. North-Holland Publishing Group, Amsterdam
2. Domb C, Lebowitz JL, Gompper G, Schick M (1994) Self-assembling amphiphilic systems, phase transitions and critical phenemena. Academic Press, London
3. Lombardo D, Kiselev MA, Magazù S, Calandra P (2015) Amphiphiles self-assembly: basic concepts and future perspectives of supramolecular approaches. Adv Condens Matter Phys 2015:1–22
4. Vittorio Degiorgio P (1985) Physics of amphiphiles, micelles and microemulsions. Europhys News 16:9–12
5. Sorrenti A, Illa O, Ortuño RM (2013) Amphiphiles in aqueous solution: well beyond a soap bubble. Chem Soc Rev 42:8200–8219
6. Song S, Dong R, Wang D et al (2013) Temperature regulated supramolecular structures via modifying the balance of multiple non-covalent interactions. Soft Matter 9:4209–4218
7. Li D, Yin P, Liu T (2012) Supramolecular architectures assembled from amphiphilic hybrid polyoxometalates. Dalt Trans 41:2853–2861
8. Dolbecq A, Dumas E, Mayer CR, Mialane P (2010) Hybrid organic-inorganic polyoxometalate compounds: from structural diversity to applications. Chem Rev 110:6009–6048
9. Qi W, Wu L (2009) Polyoxometalate/polymer hybrid materials: fabrication and properties. Polym Int 58:1217–1225
10. Song A, Hao J (2009) Self-assembly of metal-ligand coordinated charged vesicles. Curr Opin Colloid Interface Sci 14:94–102
11. Zhang X, Wang C (2011) Supramolecular amphiphiles. Chem Soc Rev 40:94–101
12. Chu Z, Dreiss CA, Feng Y (2013) Smart wormlike micelles. Chem Soc Rev 42:7174–7203
13. Chen YL, Chen S, Frank C, Israelachvili J (1992) Molecular mechanisms and kinetics during the self-assembly of surfactant layers. J Colloid Interface Sci 153:244–265
14. Israelachvili JN, Mitchell DJ, Ninham BW (1976) Theory of self-assembly of hydrocarbon amphiphiles into micelles and bilayers. J Chem Soc Faraday Trans 2(72):1525–1568
15. Ringsdorf H, Schlarb B, Venzmer J (1988) Molecular architecture and function of polymeric oriented systems: models for the study of organization, surface recognition, and dynamics of biomembranes. Angew Chem Int Ed Engl 27:113–158
16. Aida T, Meijer EW (2020) Supramolecular polymers—we've come full circle. Isr J Chem 60:33–47
17. Mendes AC, Baran ET, Reis RL, Azevedo HS (2013) Self-assembly in nature: using the principles of nature to create complex nanobiomaterials. Wiley Interdiscip Rev Nanomed Nanobiotechnol 5:582–612
18. Kim HJ, Kim T, Lee M (2011) Responsive nanostructures from aqueous assembly of rigid—flexible block molecules. Acc Chem Res 44:72–82
19. Krieg E, Rybtchinski B (2011) Noncovalent water-based materials: robust yet adaptive. Chem A Eur J 17:9016–9026
20. Lim YB, Moon KS, Lee M (2009) Recent advances in functional supramolecular nanostructures assembled from bioactive building blocks. Chem Soc Rev 38:925–934
21. Oshovsky GV, Reinhoudt DN, Verboom W (2007) Supramolecular chemistry in water. Angew Chem Int Ed 46:2366–2393
22. Kato T, Mizoshita N, Kishimoto K (2005) Functional liquid-crystalline assemblies: self-organized soft materials. Angew Chem Int Ed 45:38–68
23. Luk YY, Abbott NL (2002) Applications of functional surfactants. Curr Opin Colloid Interface Sci 7:267–275
24. Bong DT, Clark TD, Granja JR, Reza Ghadiri M (2001) Self-assembling organic nanotubes. Angew Chem Int Ed 40:988–1011
25. Moore JS, Kraft ML (2008) Synchronized self-assembly. Science 320:620–621
26. Whitesides GM, Grzybowski B (2002) Self-assembly at all scales. Science 295:2418–2421

27. Krieg E, Niazov-Elkan A, Cohen E et al (2019) Noncovalent aqua materials based on perylene diimides. Acc Chem Res 52:2634–2646
28. Hoffmann H (1994) Fascinating phenomena in surfactant chemistry. Adv Mater 6:116–129
29. Song S, Song A, Hao J (2014) Self-assembled structures of amphiphiles regulated via implanting external stimuli. RSC Adv 4:41864–41875
30. Wang C, Wang Z, Zhang X (2012) Amphiphilic building blocks for self-assembly: from amphiphiles to supra-amphiphiles. Acc Chem Res 45:608–618
31. Soc C, Sato K, Hendricks MP et al (2018) Peptide supramolecular materials for therapeutics. Chem Soc Rev 47:7539–7551
32. Goor OJGM, Hendrikse SIS, Dankers PYW, Meijer EW (2017) From supramolecular polymers to multi-component biomaterials. Chem Soc Rev 46:6621–6637
33. Krieg E, Bastings MMC, Besenius P, Rybtchinski B (2016) Supramolecular polymers in aqueous media. Chem Rev 116:2414–2477
34. Würthner F, Saha-Möller CR, Fimmel B et al (2016) Perylene bisimide dye assemblies as archetype functional supramolecular materials. Chem Rev 116:962–1052
35. Dong R, Zhou Y, Huang X et al (2015) Functional supramolecular polymers for biomedical applications. Adv Mater 27:498–526
36. Du X, Zhou J, Xu B (2014) Supramolecular hydrogels made of basic biological building blocks. Chem An Asian J 9:1446–1472
37. Ma X, Tian H (2014) Stimuli-responsive supramolecular polymers in aqueous solution. Acc Chem Res 47:1971–1981
38. Matile S, Jentzsch AV, Montenegro J, Fin A (2011) Recent synthetic transport systems. Chem Soc Rev 40:2453–2474
39. Mclntosh TJ, Simon SA (1994) Long- and short-range interactions between phospholipid/ganglioside GM1 bilayers. Biochemistry 33:10477–10486
40. Israelachvili JN, Marcelja S, Horn RG (1980) Physical principles of membrane organization. Q Rev Biophys 13:121–200
41. Wehner M, Würthner F (2020) Supramolecular polymerization through kinetic pathway control and living chain growth. Nat Rev Chem 4:38–53
42. Kazantsev RV, Dannenhoffer AJ, Weingarten AS et al (2017) Crystal-phase transitions and photocatalysis in supramolecular scaffolds. J Am Chem Soc 139:6120–6127
43. Fukui T, Kawai S, Fujinuma S et al (2017) Control over differentiation of a metastable supramolecular assembly in one and two dimensions. Nat Chem 9:493–499
44. Tantakitti F, Boekhoven J, Wang X et al (2016) Energy landscapes and functions of supramolecular systems. Nat Mater 15:469–476
45. Aliprandi A, Mauro M, De Cola L (2016) Controlling and imaging biomimetic self-assembly. Nat Chem 8:10–15
46. Ogi S, Sugiyasu K, Manna S et al (2014) Living supramolecular polymerization realized through a biomimetic approach. Nat Chem 6:188–195
47. Korevaar PA, De Greef TFA, Meijer EW (2014) Pathway complexity in π-conjugated materials. Chem Mater 26:576–586
48. Boekhoven J, Poolman JM, Maity C et al (2013) Catalytic control over supramolecular gel formation. Nat Chem 5:433–437
49. Korevaar PA, George SJ, Markvoort AJ et al (2012) Pathway complexity in supramolecular polymerization. Nature 481:492–496
50. Yan Q, Zhao Y (2013) CO_2-stimulated diversiform deformations of polymer assemblies. J Am Chem Soc 135:16300–16303
51. Eastoe J, Vesperinas A (2005) Self-assembly of light-sensitive surfactants. Soft Matter 1:338–347
52. Polarz S, Kunkel M, Donner A, Schlötter M (2018) Added-value surfactants. Chem A Eur J 24:18842–18856
53. Santer S (2018) Remote control of soft nano-objects by light using azobenzene containing surfactants. J Phys D Appl Phys 51:1–17

54. Zhu H, Shangguan L, Shi B et al (2018) Recent progress in macrocyclic amphiphiles and macrocyclic host-based supra-amphiphiles. Mater Chem Front 2:2152–2174
55. Basílio N, García-Río L (2017) Photoswitchable vesicles. Curr Opin Colloid Interface Sci 32:29–38
56. Liu Y, Jessop PG, Cunningham M et al (2006) Switchable surfactants. Science 313(80–):958–960
57. Wang A, Shi W, Huang J, Yan Y (2016) Adaptive soft molecular self-assemblies. Soft Matter 12:337–357
58. Brown P, Alan Hatton T, Eastoe J (2015) Magnetic surfactants. Curr Opin Colloid Interface Sci 20:140–150
59. Frisch H, Besenius P (2014) pH-switchable self-assembled materials. Macromol Rapid Commun 36:346–363
60. Singh J, Ranganathan R, Angayarkanny S et al (2013) pH-responsive aggregation states of chiral polymerizable amphiphiles from l-tyrosine and l-phenyl alanine in water. Langmuir 29:5734–5741
61. Seki T, Lin X, Yagai S (2013) Supramolecular engineering of perylene bisimide assemblies based on complementary multiple hydrogen bonding interactions. Asian J Org Chem 2:708–724
62. Song A, Dong S, Jia X et al (2005) An onion phase in salt-free zero-charged catanionic surfactant solutions. Angew Chem Int Ed 44:4018–4021
63. Brown P, Butts CP, Eastoe J (2013) Stimuli-responsive surfactants. Soft Matter 9:2365–2374
64. Lloyd GO, Steed JW (2009) Anion-tuning of supramolecular gel properties. Nat Chem 1:437–442
65. Liu X, Abbott NL (2009) Spatial and temporal control of surfactant systems. J Colloid Interface Sci 339:1–18
66. George M, Weiss RG (2001) Chemically reversible organogels via "latent" gelators. Aliphatic amines with carbon dioxide. J Am Chem Soc 123:10393–10394
67. Hirst AR, Escuder B, Miravet JF, Smith DK (2008) High-tech applications of self-assembling supramolecular nanostructured gel-phase materials: from regenerative medicine to electronic devices. Angew Chem Int Ed 47:8002–8018
68. Vigier-Carrière C, Boulmedais F, Schaaf P, Jierry L (2018) Surface-assisted self-assembly strategies leading to supramolecular hydrogels. Angew Chem Int Ed 57:1448–1456
69. Tabor RF, McCoy TM, Hu Y, Wilkinson BL (2018) Physicochemical and biological characterisation of azobenzene-containing photoswitchable surfactants. Bull Chem Soc Jpn 91:932–939
70. Balzani V, Credi A, Venturi M (2009) Light powered molecular machines. Chem Soc Rev 38:1542–1550
71. Wang L, Li Q (2018) Photochromism into nanosystems: towards lighting up the future nanoworld. Chem Soc Rev 47:1044–1097
72. Wang C, Chen Q, Xu H et al (2010) Photoresponsive supramolecular amphiphiles for controlled self-assembly of nanofibers and vesicles. Adv Mater 22:2553–2555
73. Carswell ADW, O'Rear EA, Grady BP (2003) Adsorbed surfactants as templates for the synthesis of morphologically controlled polyaniline and polypyrrole nanostructures on flat surfaces: from spheres to wires to flat films. J Am Chem Soc 125:14793–14800
74. Yagai S, Kitamura A (2008) Recent advances in photoresponsive supramolecular self-assemblies. Chem Soc Rev 37:1520–1529
75. Ball P (2008) Water as an active constituent in cell biology. Chem Rev 108:74–108
76. Lsraelachvili J, Wennerstrom H (1996) Role of hydration and water structure in biological and colloidal interactions. Nature 379:219–225
77. Myers D (2006) Surfactant science and technology. Wiley, Hoboken
78. Milton R (1989) Surfactants and interfacial phenomena. Wiley, Hoboken
79. Kunitake T (1992) Synthetic bilayer membranes: molecular design, self-organization, and application. Angew Chem Int Ed Engl 31:709–726

80. Monger FM, Littau CA (1991) Gemini surfactants; synthesis and properties. J Am Chem Soc 113:1451–1452
81. Fuhrhop J-H, Wang T (2004) Bolaamphiphiles. Chem Rev 104:2901–2937
82. Meister A, Bastrop M, Koschoreck S et al (2007) Structure-property relationship in stimulus-responsive bolaamphiphile hydrogels. Langmuir 23:7715–7723
83. Goulet-Hanssens A, Eisenreich F, Hecht S (2020) Enlightening materials with photoswitches. Adv Mater 32:1905966
84. Yagai S, Karatsu T, Kitamura A (2005) Photocontrollable self-assembly. Chem A Eur J 11:4054–4063
85. Velázquez MM, Alejo T, López-Díaz D et al (2016) Langmuir–Blodgett methodology: a versatile technique to build 2D material films. In: Two-dimensional materials: synthesis, characterization and potential applications. InTech, pp 21–42
86. Rogalska E, Bilewicz R, Brigaud T et al (2000) Formation and properties of Langmuir and Gibbs monolayers: a comparative study using hydrogenated and partially fluorinated amphiphilic derivatives of mannitol. Chem Phys Lipids 105:71–91
87. Ariga K, Yamauchi Y, Mori T, Hill JP (2013) 25th anniversary article: what can be done with the Langmuir-Blodgett method? Recent developments and its critical role in materials science. Adv Mater 25:6477–6512
88. Holden DA, Ringsdorf H, Deblauwe V, Smets G (1984) Photosensitive monolayers. Studies of surface-active spiropyrans at the air-water interface. J Phys Chem 88:716–720
89. Rossos AK, Katsiaflaka M, Cai J et al (2018) Photochromism of amphiphilic dithienylethenes as Langmuir-Schaefer films. Langmuir 34:10905–10912
90. Gong HF, Tang JA, Wang CM et al (2003) In situ observation of the photochromism in the Langmuir monolayer of a non-typical amphiphilic spiropyran derivative at the air/water interface. Chinese J Chem 21:387–391
91. Yamaguchi T, Kajikawa K, Takezoe H, Fukuda A (1992) Observation of photochromic reactions in spiropyran monolayers by surface potential measurement. Jpn J Appl Phys 31:1160–1163
92. Miyata A, Unuma Y, Higashigaki Y (1993) Optical properties and molecular orientation of aggregates in Langmuir-Blodgett films of A long-chain spiropyran. Bull Chem Soc Jpn 66:993–998
93. Tachibana H, Yamanaka Y, Matsumoto M (2002) Surface and photochemical properties of Langmuir monolayer and Langmuir-Blodgett films of a spiropyran derivative. J Mater Chem 12:938–942
94. Bubeck C (1988) Reactions in monolayers and Langmuir-Blodgett films. Elsevier Sequoia, The Netherlands
95. Ando E, Moriyama K, Arita K, Morimoto K (1990) Photochromic behaviors of long alkyl chain spiropyrans at the air-water interface and in LB films. Langmuir 6:1451–1454
96. Miyata A, Unuma Y, Higashigali Y (1991) Aggregates in Langmuir-Bladgett films of spiropyrans having hydroxyl or hydroxymethyl group. Bull Chem Soc Jpn 64:1719–1725
97. Whitten DG (1993) Photochemistry and photophysics of trans-stilbene and related alkenes in surfactant assemblies. Acc Chem Res 26:502–509
98. Cheng J, Štacko P, Rudolf P et al (2017) Bidirectional photomodulation of surface tension in Langmuir films. Angew Chem Int Ed 56:291–296
99. Backus EHG, Kuiper JM, Engberts JBFN et al (2011) Reversible optical control of monolayers on water through photoswitchable lipids. J Phys Chem B 115:2294–2302
100. Yamamoto T, Umemura Y, Sato O, Einaga Y (2004) Photoswitchable magnetic films: Prussian blue intercalated in Langmuir-Blodgett films consisting of an amphiphilic azobenzene and a clay mineral. Chem Mater 16:1195–1201
101. Nakazawa T, Azumi R, Sakai H et al (2004) Brewster angle microscopic observations of the Langmuir films of amphiphilic spiropyran during compression and under UV illumination. Langmuir 20:5439–5444
102. Kim I, Rabolt JF, Stroeve P (2000) Dynamic monolayer behavior of a photo-responsive azobenzene surfactant. Colloids Surf A Physicochem Eng Asp 171:167–174

103. Karthaus O, Shimomura M, Hioki M et al (1996) Reversible photomorphism in surface monolayers. J Am Chem Soc 118:9174–9175
104. Song B, Zhao J (2010) Orientation of the azobenzene spacer of carboxylic methyl ester gemini surfactants in Langmuir monolayer. Chinese J Chem 28:189–192
105. Kharlamov AA, Lyubimov AV, Vinogradov AM (1994) The photoinduced surface pressure relaxation processes in amphiphilic spiropyrane and spiroindolinonaphthooxazine. Thin Solid Films 244:962–965
106. Sakai K, Imaizumi Y, Oguchi T et al (2010) Adsorption characteristics of spiropyran-modified cationic surfactants at the silica/aqueous solution interface. Langmuir 26:9283–9288
107. Eastoe J, Dominguez MS, Wyatt P et al (2002) Properties of a stilbene-containing gemini photosurfactant: Light-triggered changes in surface tension and aggregation. Langmuir 18:7837–7844
108. Kang HC, Lee BM, Yoon J, Yoon M (2000) Synthesis and surface-active properties of new photosensitive surfactants containing the azobenzene group. J Colloid Interface Sci 231:255–264
109. Dunkin IR, Gittinger A, Sherrington DC, Whittaker P (1996) Synthesis, characterization and applications of azo-containing photodestructible surfactants. J Chem Soc Perkin Trans 2:1837–1842
110. Yang L, Takisawa N, Hayashita T, Shirahama K (1995) Colloid chemical characterization of the photosurfactant 4-ethylazobenzene 4′-(oxyethyl)trimethylammonium bromide. J Phys Chem 99:8799–8803
111. Hayashita T, Kurosawa T, Miyata T et al (1994) Effect of structural variation within cationic azo-surfactant upon photoresponsive function in aqueous solution. Colloid Polym Sci 272:1611–1619
112. Dunkin IR, Gittinger A, Sherrington DC, Whittaker P (1994) A photodestructible surfactant. J Chem Soc Chem Commun 2245–2246
113. Drummond CJ, Albers S, Furlong DN, Wells D (1991) Photocontrol of surface activity and self-assembly with a spirobenzopyran surfactant. Langmuir 7:2409–2411
114. Shinkai S, Matsuo K, Harada A, Manabe O (1982) Photocontrol of micellar catalyses. J Chem Soc Perkin Trans 1:1261–1265
115. Shin JY, Abbott NL (1999) Using light to control dynamic surface tensions of aqueous solutions of water soluble surfactants. Langmuir 15:4404–4410
116. Cicciarelli BA, Hatton TA, Smith KA (2007) Dynamic surface tension behavior in a photoresponsive surfactant system. Langmuir 23:4753–4764
117. Chevallier E, Mamane A, Stone HA et al (2011) Pumping-out photo-surfactants from an air-water interface using light. Soft Matter 7:7866–7874
118. Chevallier E, Monteux C, Lequeux F, Tribet C (2012) Photofoams: remote control of foam destabilization by exposure to light using an azobenzene surfactant. Langmuir 28:2308–2312
119. Chevallier E, Saint-Jalmes A, Cantat I et al (2013) Light induced flows opposing drainage in foams and thin-films using photosurfactants. Soft Matter 9:7054–7060
120. Mamane A, Chevallier E, Olanier L et al (2017) Optical control of surface forces and instabilities in foam films using photosurfactants. Soft Matter 13:1299–1305
121. Jiang J, Ma Y, Cui Z (2017) Smart foams based on dual stimuli-responsive surfactant. Colloids Surf A Physicochem Eng Asp 513:287–291
122. Lei L, Xie D, Song B et al (2017) Photoresponsive foams generated by a rigid surfactant derived from dehydroabietic acid. Langmuir 33:7908–7916
123. Chen S, Wang C, Yin Y, Chen K (2016) Synthesis of photo-responsive azobenzene molecules with different hydrophobic chain length for controlling foam stability. RSC Adv 6:60138–60144
124. Chen S, Zhang W, Wang C, Sun S (2016) A recycled foam coloring approach based on the reversible photo-isomerization of an azobenzene cationic surfactant. Green Chem 18:3972–3980
125. Chen S, Zhang Y, Chen K et al (2017) Insight into a fast-phototuning azobenzene switch for sustainably tailoring the foam stability. ASC Appl Mater Interfaces 9:13778–13784

126. Fei L, Ge F, Yin Y, Wang C (2019) Photo-responsive foam control base on nonionic azobenzene surfactant as stabilizer. Colloids Surf A Physicochem Eng Asp 560:366–375
127. Chen S, Fei L, Ge F, Wang C (2019) Photoresponsive aqueous foams with controllable stability from nonionic azobenzene surfactants in multiple-component systems. Soft Matter 15:8313–8319
128. Chen S, Fei L, Ge F et al (2020) A versatile and recycled pigment foam coloring approach for natural and synthetic fibers with nearly-zero pollutant discharge. J Clean Prod 243:118504
129. Jiang X, Guo Q, Li H et al (2017) Photofoams and flotation mechanism of an azobenzene-based surfactant on quartz. Colloids Surf A Physicochem Eng Asp 535:201–205
130. Jiang X, Guo Q, He Y et al (2018) Using light to control the floatability of solid particles in aqueous solution of a Gemini surfactant. Colloids Surf A Physicochem Eng Asp 553:218–224
131. Varanakkottu SN, Anyfantakis M, Morel M et al (2016) Light-directed particle patterning by evaporative optical Marangoni assembly. Nano Lett 16:644–650
132. Lv C, Varanakkottu SN, Baier T, Hardt S (2018) Controlling the trajectories of nano/micro particles using light-actuated marangoni flow. Nano Lett 18:6924–6930
133. Diguet A, Guillermic RM, Magome N et al (2009) Photomanipulation of a droplet by the chromocapillary effect. Angew Chem Int Ed 48:9281–9284
134. Baigl D (2012) Photo-actuation of liquids for light-driven microfluidics: state of the art and perspectives. Lab Chip 12:3637–3653
135. Kavokine N, Anyfantakis M, Morel M et al (2016) Light-driven transport of a liquid marble with and against surface flows. Angew Chem Int Ed 55:11183–11187
136. Vialetto J, Anyfantakis M, Rudiuk S et al (2019) Photoswitchable dissipative two-dimensional colloidal crystals. Angew Chem Int Ed 58:9145–9149
137. Schnurbus M, Stricker L, Ravoo BJ, Braunschweig B (2018) Smart air-water interfaces with arylazopyrazole surfactants and their role in photoresponsive aqueous foam. Langmuir 34:6028–6035
138. Honnigfort C, Campbell RA, Droste J et al (2020) Unexpected monolayer-to-bilayer transition of arylazopyrazole surfactants facilitates superior photo-control of fluid interfaces and colloids. Chem Sci 11:2085–2092
139. Sakai H, Ebana H, Sakai K et al (2007) Photo-isomerization of spiropyran-modified cationic surfactants. J Colloid Interface Sci 316:1027–1030
140. Moo JGS, Presolski S, Pumera M (2016) Photochromic spatiotemporal control of bubble-propelled micromotors by a spiropyran molecular switch. ACS Nano 10:3543–3552
141. Schnurbus M, Kabat M, Jarek E et al (2020) Spiropyran sulfonates for photo- and pH-responsive air-water interfaces and aqueous foam. Langmuir 36:6871–6879
142. Zhmud BV, Tiberg F, Kizling J (2000) Dynamic surface tension in concentrated solutions of CnEm surfactants: a comparison between the theory and experiment. Langmuir 16:2557–2565
143. Beneventi D, Carre B, Gandini A (2001) Role of surfactant structure on surface and foaming properties. Colloids Surf A Physicochem Eng Asp 189:65–73
144. Chen S, Chen S, Leung FKC et al (2020) Dynamic assemblies of molecular motor amphiphiles control macroscopic foam properties. J Am Chem Soc 142:10163–10172
145. Nagarajan R (2002) The neglected role of the surfactant tail self-assembly: the neglected role of the surfactant tail. Langmuir 18:31–38
146. Li J, Zhao M, Zhou H et al (2012) Photo-induced transformation of wormlike micelles to spherical micelles in aqueous solution. Soft Matter 8:7858–7864
147. Wang D, Wei G, Dong R, Hao J (2013) Multiresponsive viscoelastic vesicle gels of nonionic $C_{12}EO_4$ and anionic AzoNa. Chem A Eur J 19:8253–8260
148. Tabor RF, Tan DD, Han SS et al (2014) Reversible pH- and photocontrollable carbohydrate-based surfactants. Chem A Eur J 20:13881–13884
149. Tu Y, Chen Q, Shang Y et al (2019) Photoresponsive behavior of wormlike micelles constructed by Gemini surfactant 12-3-12·2Br– and different cinnamate derivatives. Langmuir 35:4634–4645
150. Fameau AL, Arnould A, Lehmann M, Von Klitzing R (2015) Photoresponsive self-assemblies based on fatty acids. Chem Commun 51:2907–2910

151. Jia K, Hu J, Dong J, Li X (2016) Light-responsive multillamellar vesicles in coumaric acid/alkyldimethylamine oxide binary systems: effects of surfactant and hydrotrope structures. J Colloid Interface Sci 477:156–165
152. Blayo C, Houston JE, King SM, Evans RC (2018) Unlocking structure-self-assembly relationships in cationic azobenzene photosurfactants. Langmuir 34:10123–10134
153. Shimizu T, Masuda M, Minamikawa H (2005) Supramolecular nanotube architectures based on amphiphilic molecules. Chem Rev 105:1401–1443
154. Chen S, Costil R, Leung FKC, Feringa BL (2021) Self-assembly of photoresponsive molecular amphiphiles in aqueous media. Angew Chem Int Ed 60:11604–11627
155. De JJJD, Lucas LN, Kellogg RM (2004) Supramolecular chirality into molecular chirality. Science 304:278–281
156. Eelkema R, Feringa BL (2006) Amplification of chirality in liquid crystals. Org Biomol Chem 4:3729–3745
157. de Jong JJD, van Rijn P, Tiemersma-Wegeman TD et al (2008) Dynamic chirality, chirality transfer and aggregation behaviour of dithienylethene switches. Tetrahedron 64:8324–8335
158. Katsonis N, Lacaze E, Feringa BL (2008) Molecular chirality at fluid/solid interfaces: expression of asymmetry in self-organised monolayers. J Mater Chem 18:2065–2073
159. Barclay TG, Constantopoulos K, Matisons J (2014) Nanotubes self-assembled from amphiphilic molecules via helical intermediates. Chem Rev 114:10217–10291
160. Vandijken DJ, Beierle JM, Stuart MCA et al (2014) Autoamplification of molecular chirality through the induction of supramolecular chirality. Angew Chem Int Ed 53:5073–5077
161. Liu M, Zhang L, Wang T (2015) Supramolecular chirality in self-assembled systems. Chem Rev 115:7304–7397
162. Muraoka T, Cui H, Stupp SI (2008) Quadruple helix formation of a photoresponsive peptide amphiphile and its light-triggered dissociation into single fibers. J Am Chem Soc 130:2946–2947
163. Muraoka T, Koh CY, Cui H, Stupp SI (2009) Light-triggered bioactivity in three dimensions. Angew Chem Int Ed 48:5946–5949
164. Paramonov SE, Jun HW, Hartgerink JD (2006) Self-assembly of peptide-amphiphile nanofibers: the roles of hydrogen bonding and amphiphilic packing. J Am Chem Soc 128:7291–7298
165. Caroli G, Coleman AC, Beierle JM et al (2011) Light-induced disassembly of self-assembled vesicle-capped nanotubes observed in real time. Nat Nanotechnol 6:547–552
166. Erne PM, Van Bezouwen LS, Štacko P et al (2015) Loading of vesicles into soft amphiphilic nanotubes using osmosis. Angew Chem Int Ed 54:15122–15127
167. Sun Y, Ji Y, Yu H et al (2016) Near-infrared light-sensitive liposomes for controlled release. RSC Adv 6:81245–81249
168. Wang D, Hou X, Ma B et al (2017) UV and NIR dual-responsive self-assembly systems based on a novel coumarin derivative surfactant. Soft Matter 13:6700–6708
169. Wang G, Engberts JBFN (1994) Induction of aggregate formation of cationic polysoaps and surfactants by low concentrations of additives in aqueous solution. Langmuir 10:2583–2587
170. Buwalda RT, Jonker JM, Engberts JBFN (1999) Aggregation of Azo dyes with cationic amphiphiles at low concentrations in aqueous solution. Langmuir 15:1083–1089
171. Buwalda RT, Engberts JBFN (2001) Aggregation of dicationic surfactants with methyl orange in aqueous solution. Langmuir 17:1054–1059
172. Buwalda RT, Stuart MCA, Engberts JBFN (2002) Interactions of an azobenzene-functionalized anionic amphiphile with cationic amphiphiles in aqueous solution. Langmuir 18:6507–6512
173. Li LS, Jiang H, Messmore BW et al (2007) A torsional strain mechanism to tune pitch in supramolecular helices. Angew Chem Int Ed 46:5873–5876
174. Song X, Perlstein J, Whitten DG (1997) Supramolecular aggregates of azobenzene phospholipids and related compounds in bilayer assemblies and other microheterogeneous media: structure, properties, and photoreactivity. J Am Chem Soc 119:9144–9159

175. Sakai H, Matsumura A, Yokoyama S et al (1999) Photochemical switching of vesicle formation using an azobenzene-modified surfactant. J Phys Chem B 103:10737–10740
176. Khairutdinov RF, Hurst JK (2004) Light-driven transmembrane ion transport by spiropyran-crown ether supramolecular assemblies. Langmuir 20:1781–1785
177. Lee CT, Smith KA, Hatton TA (2004) Photoreversible viscosity changes and gelation in mixtures of hydrophobically modified polyelectrolytes and photosensitive surfactants. Macromolecules 37:5397–5405
178. Bonini M, Berti D, Di Meglio JM et al (2005) Surfactant aggregates hosting a photoresponsive amphiphile: Structure and photoinduced conformational changes. Soft Matter 1:444–454
179. Faure D, Gravier J, Labrot T et al (2005) Photoinduced morphism of gemini surfactant aggregates. Chem Commun 16:1167–1169
180. Hubbard FP, Santonicola G, Kaler EW, Abbott NL (2005) Small-angle neutron scattering from mixtures of sodium dodecyl sulfate and a cationic, bolaform surfactant containing azobenzene. Langmuir 21:6131–6136
181. Shang T, Smith KA, Hatton TA (2006) Self-assembly of a nonionic photoresponsive surfactant under varying irradiation conditions: a small-angle neutron scattering and cryo-TEM study. Langmuir 22:1436–1442
182. Hubbard FP, Abbott NL (2007) Effect of light on self-assembly of aqueous mixtures of sodium dodecyl sulfate and a cationic, bolaform surfactant containing azobenzene. Langmuir 23:4819–4829
183. Sakai H, Orihara Y, Kodashima H et al (2005) Photoinduced reversible change of fluid viscosity. J Am Chem Soc 127:13454–13455
184. Alvarez-Lorenzo C, Bromberg L, Concheiro A (2009) Light-sensitive intelligent drug delivery systems. Photochem Photobiol 85:848–860
185. Fomina N, Sankaranarayanan J, Almutairi A (2012) Photochemical mechanisms of light-triggered release from nanocarriers. Adv Drug Deliv Rev 64:1005–1020
186. Liu X, Yang B, Wang Y et al (2005) New nanoscale pulsatile drug delivery system shanghai institute of organic chemistry, Chinese academy applications in drug delivery and controlled release. This is because liposomes with diameters approximately of 100 nm can be delivered to tumor tissue. Chem Mater 17:2792–2795
187. Lin Y, Cheng X, Qiao Y et al (2010) Creation of photo-modulated multi-state and multi-scale molecular assemblies via binary-state molecular switch. Soft Matter 6:902–908
188. Bi Y, Wei H, Hu Q et al (2015) Wormlike micelles with photoresponsive viscoelastic behavior formed by surface active ionic liquid/azobenzene derivative mixed solution. Langmuir 31:3789–3798
189. Tabor RF, Pottage MJ, Garvey CJ, Wilkinson BL (2015) Light-induced structural evolution of photoswitchable carbohydrate-based surfactant micelles. Chem Commun 51:5509–5512
190. Kelly EA, Houston JE, Evans RC (2019) Probing the dynamic self-assembly behaviour of photoswitchable wormlike micelles in real-time. Soft Matter 15:1253–1259
191. Lund R, Brun G, Chevallier E et al (2016) Kinetics of photocontrollable micelles: light-induced self-assembly and disassembly of azobenzene-based surfactants revealed by TR-SAXS. Langmuir 32:2539–2548
192. Song B, Hu Y, Zhao J (2009) A single-component photo-responsive fluid based on a gemini surfactant with an azobenzene spacer. J Colloid Interface Sci 333:820–822
193. Zhang D, Lu X, Li Y et al (2018) Dual stimuli-responsive wormlike micelles base on cationic azobenzene surfactant and sodium azophenol. Colloids Surf A Physicochem Eng Asp 543:155–162
194. Hirose T, Matsuda K, Irie M (2006) Self-assembly of photochromic diarylethenes with amphiphilic side chains: reversible thermal and photochemical control. J Org Chem 71:7499–7508
195. Van Dijken DJ, Chen J, Stuart MCA et al (2016) Amphiphilic molecular motors for responsive aggregation in water. J Am Chem Soc 138:660–669
196. Kwangmettatam S, Kudernac T (2018) Light-fuelled reversible expansion of spiropyran-based vesicles in water. Chem Commun 54:5311–5314

197. Xu F, Pfeifer L, Stuart MCA et al (2020) Multi-modal control over the assembly of a molecular motor bola-amphiphile in water. Chem Commun 56:7451–7454
198. Fuentes E, Gerth M, Berrocal JA et al (2020) An azobenzene-based single-component supramolecular polymer responsive to multiple stimuli in water. J Am Chem Soc 142:10069–10078
199. Geng S, Wang Y, Wang L et al (2017) A light-responsive self-assembly formed by a cationic Azobenzene derivative and SDS as a drug delivery system. Sci Rep 7:1–13
200. Simmons NS, Blout ER (1960) The structure of tobacco mosaic virus and its components: ultraviolet optical rotatory dispersion. Biophys J 1:55–62
201. Iino T (1974) Assembly of Salmonella Flagellin in vitro and vivo. J Supramol Struct 2:372–384
202. Prockop DJ, Fertala A (1998) The collagen fibril: the almost crystalline structure. J Struct Biol 122:111–118
203. Tsai CJ, Ma B, Kumar S et al (2001) Protein folding: Binding of conformationally fluctuating building blocks via population selection. Crit Rev Biochem Mol Biol 36:399–433
204. Fletcher DA, Mullins RD (2010) Cell mechanics and the cytoskeleton. Nature 463:485–492
205. Huber F, Schnauß J, Rönicke S et al (2013) Emergent complexity of the cytoskeleton: from single filaments to tissue. Adv Phys 62:1–112
206. Zhang S, Greenfield MA, Mata A et al (2010) A self-assembly pathway to aligned monodomain gels. Nat Mater 9:594–601
207. Angeloni NL, Bond CW, Tang Y et al (2011) Regeneration of the cavernous nerve by Sonic hedgehog using aligned peptide amphiphile nanofibers. Biomaterials 32:1091–1101
208. McClendon MT, Stupp SI (2012) Tubular hydrogels of circumferentially aligned nanofibers to encapsulate and orient vascular cells. Biomaterials 33:5713–5722
209. Chin SM, Synatschke CV, Liu S et al (2018) Covalent-supramolecular hybrid polymers as muscle-inspired anisotropic actuators. Nat Commun 9:2395
210. Li C, Iscen A, Sai H et al (2020) Supramolecular–covalent hybrid polymers for light-activated mechanical actuation. Nat Mater 19:900–909
211. Sheng Y, Chen Q, Yao J et al (2015) Hierarchical assembly of a dual-responsive macroscopic insulated molecular wire bundle in a gradient system. Sci Rep 5:1–6
212. Chen J, Leung FKC, Stuart MCA et al (2018) Artificial muscle-like function from hierarchical supramolecular assembly of photoresponsive molecular motors. Nat Chem 10:132–138
213. Leung FKC, van den Enk T, Kajitani T et al (2018) Supramolecular packing and macroscopic alignment controls actuation speed in macroscopic strings of molecular motor amphiphiles. J Am Chem Soc 140:17724–17733
214. Leung FKC, Kajitani T, Stuart MCA et al (2019) Dual-controlled macroscopic motions in a supramolecular hierarchical assembly of motor amphiphiles. Angew Chem Int Ed 58:10985–10989
215. Li Q, Fuks G, Moulin E et al (2015) Macroscopic contraction of a gel induced by the integrated motion of light-driven molecular motors. Nat Nanotechnol 10:161–165
216. Foy JT, Li Q, Goujon A et al (2017) Dual-light control of nanomachines that integrate motor and modulator subunits. Nat Nanotechnol 12:540–545
217. Goujon A, Mariani G, Lang T et al (2017) Controlled sol−gel transitions by actuating molecular machine based supramolecular polymers. J Am Chem Soc 139:4923–4928

Chapter 10
Organic Salts as Tectons for Self-assembly Processes in Solution

Salvatore Marullo, Carla Rizzo, and Francesca D'Anna

Abstract The literature, covering the last decade, about the self-assembly processes of organic salts in conventional solvents was analyzed. In particular, data reported about imidazolium and ammonium salts have been considered. The analysis shows that these processes are highly determined by structural features of the salts. Indeed, besides the nature of the cationic head, features of the alkyl chain borne on the cation structure and its possible functionalization, as well as the nature of the anion play a pivotal role. These factors determine not only the nature of the solvent in which the process occurs but also the nature of the self-assembly mechanism. Consequently, the structure tunability of the salts affects the characteristics of the aggregates and, in turn, their possible applications. To this aim, some interesting applications in the biomedical field are reviewed and discussed.

10.1 Introduction

Self-assembly of organic species is a topic that has raised a surge of interest in the last decades. This is mainly due to the significant changes both in properties and functions that can be detected moving from the simple tecton to the self-assembled system. Indeed, the organization of small organic molecules in bi- or tridimensional supramolecular systems frequently allows the obtaining of functional materials with specific applications.

S. Marullo · C. Rizzo · F. D'Anna (✉)
Dipartimento di Scienze e Tecnologie Biologiche, Chimiche, Farmaceutiche (STEBICEF), Università degli Studi di Palermo, Viale delle Scienze, Ed. 17, 90128 Palermo, Italy
e-mail: francesca.danna@unipa.it

S. Marullo
e-mail: salvatore.marullo@unipa.it

C. Rizzo
e-mail: carla.rizzo03@unipa.it

M. A. Aboudzadeh and A. Frontera (eds.), *Supramolecular Assemblies Based on Electrostatic Interactions*,
https://doi.org/10.1007/978-3-031-00657-9_10

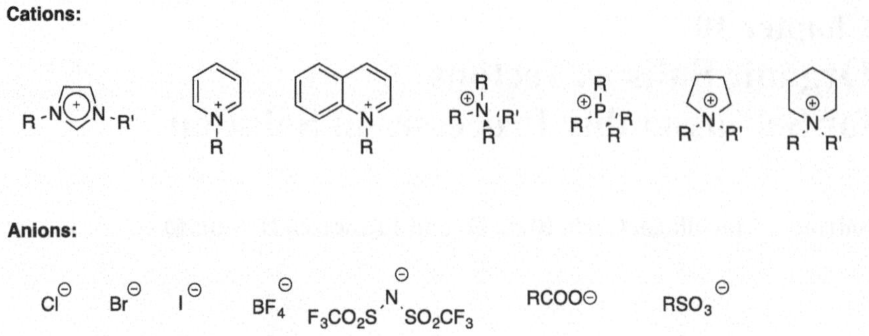

Fig. 10.1 Structures of some of the most common cations and anions in organic salts

Among organic species that can give rise to self-assembled processes, organic salts play a pivotal role. One of the first reports about organic salts dates back to 1839 [1] and it describes first evidence of the obtainment of salts arising from organic bodies. Since then, they have been widely investigated and used. Organic salts are usually defined as ionic species mainly composed of organic cations and organic or inorganic anions. Among different classes, imidazolium, pyridinium, ammonium, phosphonium, quinolinium, piperidinium, and pyrrolidinium ones hold particular interest (Fig. 10.1).

Their main feature stands in their structure tunability. Indeed, the skeleton of organic salts clearly shows three possible points of modification: (i) the cationic head; (ii) the alkyl tail and (iii) the anionic counterpart.

A raise in possibilities can be obviously detected as far as dicationic organic salts are considered as, in this case, also the nature of the spacer between the cationic heads and the symmetric or asymmetric substitution of the spacer plays a role. The most relevant consequence of the action of the above structural features are the significant changes in the salt properties and consequently, in their possible applications.

Organic salts may give self-assembly processes both in bulk and in solution. As far as self-assembly in bulk is considered, they can give rise to the formation of thermotropic liquid-crystalline organization and some interesting examples have been already reported [2], together with valuable review articles [3].

On the other hand, if the use of organic salts as tectons for self-assembly processes in solution is taken in consideration, all the above statements are still valid and, in the last two decades, a gradual increase in the number of publications concerning this topic has been detected (Fig. 10.2).

Consequently, amphiphilic cationic compounds have been used as self-assembled systems able to act as micellar catalysts [4], nanocontainers for drug delivery [5], non-viral vectors [6], and so on. However, they also demonstrated the ability to inhibit the growth of various pathogens [7, 8] and have been applied as basis of gels [8, 9] and emulsions [10] suitable for biomedical applications.

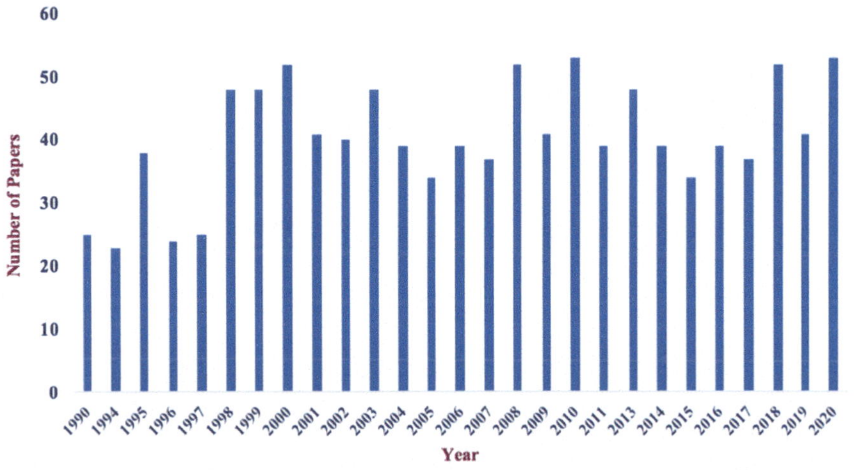

Fig. 10.2 Number of papers published in the last three decades, using self-assembly and organic salts as keywords, on the Scopus database

In the light of the above considerations, this chapter reviews the literature published in the last decade, about self-assembly processes occurring in solution and involving organic salts. In particular, special attention has been devoted to systems based on imidazolium and ammonium salts. Data have been discussed as a function of structural features of tectons, to better highlight their implications on the different applications.

10.2 Imidazolium Salts

The nucleus of the imidazolium ion has been considered a valuable tecton for the obtainment of self-assembled systems. This mainly derives from some key structural features, like the presence of an aromatic structure that allows the establishment of π-π interactions, the presence of relatively acidic hydrogens favoring the occurrence of hydrogen bonds and last but not least, the presence of alkyl chain responsible for the occurrence of van der Waals and hydrophobic interactions. All these interactions are able to drive the formation of organized supramolecular systems in which the strength and the nature of interactions are also modulated by the nature of the anion.

The above consideration clearly explains why imidazolium salts, in the liquid state, giving rise to the class of ionic liquids, have been claimed as organized supramolecular fluids [11], and the effect of their structural organization proved to be pivotal in determining reactivity in such kind of solvents [12, 13]. When used as solutes, imidazolium salts have allowed the obtainment of assembled systems showing peculiar features and giving a plethora of different applications. To this aim, self-aggregation behavior of amino acid-derived salts, **[C_1C_{14}mim][AA]**, has

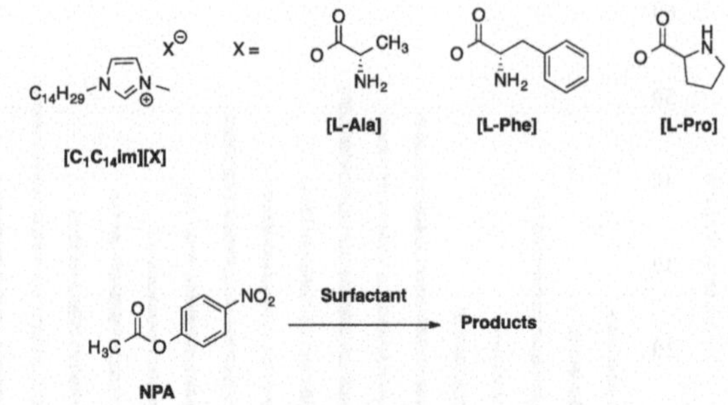

Fig. 10.3 Structure of $[C_1C_{14}im][AA]$ and representation of the probe reaction studied in the micellar aggregates

been successfully studied in water solution. In particular, salts different in the nature of the amino acid anion, like L-Phe, L-Pro, and L-Ala were considered, together with the bromide salt (Fig. 10.3) [14].

The combined approach of different techniques, like surface tension measurements, steady-state fluorescence, and isothermal titration calorimetry (ITC) allowed evidencing the formation of micellar aggregates. Critical micellar concentration (cmc) heavily depended on the anion nature, increasing in the order: $\mathbf{[C_1C_{14}im][Phe]} < \mathbf{[C_1C_{14}im][Ala]} < \mathbf{[C_1C_{14}im][Pro]} < \mathbf{[C_1C_{14}im][Br]}$, indicating that self-aggregation was driven by the anion hydrophobicity. Gibbs free energy ($\Delta G^{\circ}_{\text{mic}}$) obtained from ITC investigation shed light on the spontaneous nature of the self-aggregation process, with no significant differences detected as function of the different nature of the amino acid.

Micellar aggregates were used as organized reaction media for the hydrolysis of *p*-nitrophenylacetate (NPA). Analysis of reaction products allowed to identify the occurrence of two parallel reaction pathways, as accounted for by the presence of acetate anion and acetyl derivatives of amino acids. This latter is derived from the aminolysis reaction of NPA. The target reaction occurred faster in the aggregates systems than in water solution. However, the catalytic effect of micellar aggregates was observed until a limit value of organic salts concentration was used (~0.02 M). Above this value, the increased amount of micellar aggregates induced a separation of the reagents, decreasing, on the whole, the reaction rate. Kinetic constants determined by UV–vis measurements, changed along the trend: $\mathbf{[C_1C_{14}im][Br]} < \mathbf{[C_1C_{14}im][Phe]} < \mathbf{[C_1C_{14}im][Ala]} < \mathbf{[C_1C_{14}im][Pro]}$. The highest reactivity detected in the presence of [Pro]-based imidazolium aggregates perfectly accounted for the lowest basicity and highest nucleophilic character of the amino acid.

The occurrence of molecular assembly for imidazolium-based salts has been detected for simple protic cations such as 1-methylimidazolium triflate ($\mathbf{[HC_1im][CF_3SO_3]}$) and 1H-imidazolium triflate ($\mathbf{[H_2im][CF_3SO_3]}$, Fig. 10.4) [15].

The combined use of ^{1}H and ^{13}C-NMR and PFG-NMR experiments evidenced, at temperature lower than 60 °C and in D_2O solution, the unusual magnetic equivalence of hydrogens and carbons on the C2 and C3 of imidazolium ring of $[HC_1im^+]$ cation. Furthermore, with respect to $[H_2im^+]$ cation, PFG-NMR experiments, performed at

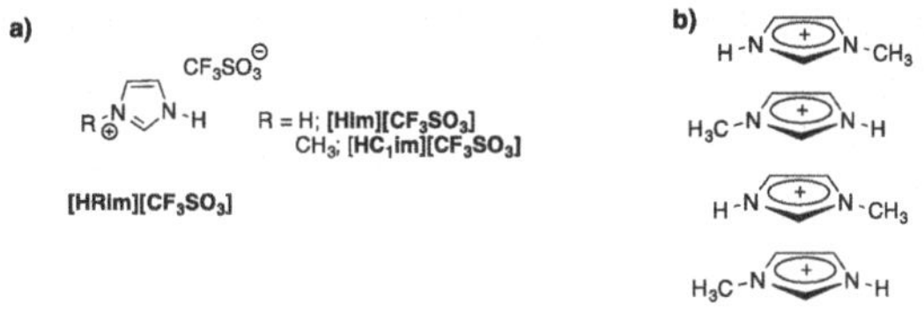

Fig. 10.4 **a** Structure of protic imidazolium salts and **b** representation of the column-like structures of the aggregates

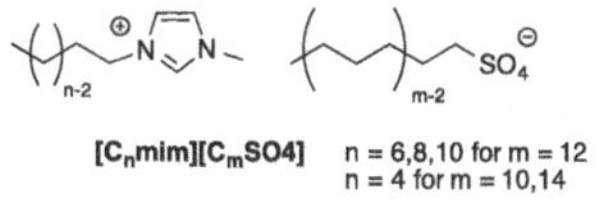

Fig. 10.5 Structure of surface-active ionic liquids

$T < 60$ °C, revealed a higher diffusion for N–H/D_2O protons and a lower diffusion for the cation. The above results were ascribed to the incomplete solvation of [HC_1im^+] species, as a consequence of its hydrophobicity, which thanks to the occurrence of π-π interactions, gave rise to the formation of dimers or ring clusters with a column-like structure. In the above self-assembled structures, the imidazolium rings bear the methyl groups on opposite sides, minimizing steric hindrance.

This arrangement created an axial symmetric electronic shielding for H2/H3. The increase in temperature-induced the melting of the aggregates, causing the split of H2/H3 peak and the drop in the diffusion coefficient of N–H/D_2O.

The aggregation behavior of imidazolium salts has been also deeply investigated considering salt-free catanionic surface-active ionic liquids (SAILs). In particular, organic alkyl sulfates bearing alkyl chains of different lengths on both cation and anion structures were used (Fig. 10.5) [16].

The self-assembly behavior was investigated in water solution, by using surface activity, fluorescence, and electrical conductivity measurements. Collected results demonstrated a higher tendency of the used organic salts to form micelles, with respect to corresponding cationic or anionic surfactants, as accounted for by the lower cmc values. Plots of cmc values as a function of the sum of carbon atom numbers in the hydrophobic chain, showed that cmc follows the empirical Stauff-Klevens rule [17]:

$$\log \text{cmc} = A - BN_C$$

where A is the parameter that accounts for the contribution of the polar head group to micelles formation while B gives the effect of additional methylene unit to cmc. The SAILs investigated gave different linear correlations in dependence of the length of the alkyl chain on the cation or anion. In particular, as far as B values are concerned, both changing the alkyl tail on the anion or on the cation gave values out of ranges traditionally obtained for surfactants. The above result indicated that these SAILs behaved differently from the mixtures of cationic-anionic surfactants and, in the case of the elongation of the tail on the anion, testified that longer alkyl chain induced a higher propensity of the anion to participate in the formation of the aggregates

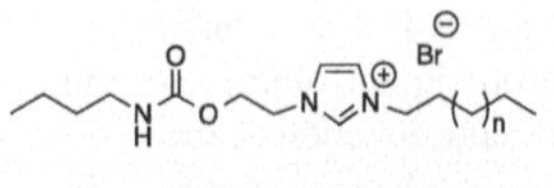

Fig. 10.6 Structure of amphiphilic urethane-based imidazolium salts

structure. Electrical conductivity measurements as a function of the temperature indicated that ΔG_m became more negative as a consequence of the elongation of the alkyl chain. Furthermore, negative values for ΔH_m and $-T\Delta S_m$ indicated that the micellization process is an exothermic and entropy-driven process.

Self-assembly behavior of amphiphilic imidazolium salts bearing a urethane moiety has been investigated also in the light of the antimicrobial activity that such kinds of salts are able to exert. With respect to conventional neutral antimicrobial agents, this kind of substrate induces the disintegration of microbial membranes through both electrostatic and hydrophobic interactions. Consequently, structural features of amphiphilic compounds play a pivotal role in determining their activity. However, as the same structural features also determine the presence of the amphiphile as monomer or aggregate species, the study of parameters operating on the self-assembly process becomes relevant. On this subject, aggregation behavior of imidazolium amphiphilic compounds, bearing urethane fragments with butyl substituent and various hydrophobic tails, has been studied in aqueous solution, using tensiometry, conductivity, dynamic and electrophoretic light scattering measurements (Fig. 10.6) [18].

Data reported demonstrated that the elongation of the alkyl tail induces a systematic decrease in the cmc values, as well as the elongation of the urethane radical moiety from the ethyl to the butyl, improving the aggregation capacity by 1.5–3 fold. The elongation of the alkyl tail also caused a decrease in ΔG_m due to the increase in density of packing in the absorption layer. Calculation of packing parameters gave in all cases values lower than 0.33 and shed light on the formation of spherical micellar aggregates.

Interestingly, fluorescence of pyrene was used to calculate the aggregation number. In the case of **[C$_4$UC$_{14}$im]Br**, the above parameter increased with the concentration. Differently, for **[C$_4$UC$_{16}$im]Br**, it firstly increased and then decreased probably as a consequence of a morphological rearrangement of the aggregates. The above hypothesis was also supported by DLS investigation that detected a decrease in the hydrodynamic radius from $D_H = 6$ nm down to $D_H = 2$ nm, in the cmc region. The reorganization was ascribed to a different balance of π-π stacking interactions and hydrophobic effect induced by changes in the amphiphile concentration. Amphiphiles were tested as antimicrobial agents and the highest activity was detected for **[C$_4$UC$_{14}$im]Br**. The activity was mainly related to the ability to mediate the amphiphile integration with lipid bilayer and this, in turn, depends on the free volume that the alkyl tail is able to induce in the membrane bilayer.

Homologous imidazolium-based organic salts, with alkyl tails borne on the urethane moiety varying from 14 up to 18 carbon atoms, were further investigated for their self-assembly behavior [19]. Besides the significant decrease in cmc values

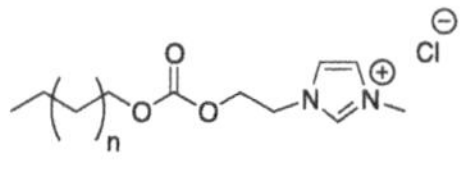

Fig. 10.7 Structures of 1-alkyloxycarbonyloxyethyl-3-methylimidazolium chlorides

that dropped from 1.3 down to 0.1 mM with the alkyl chain elongation, the formation of different morphological features, like small micelles (C_{14}), rod-like or elongated structures (C_{18}) were detected by combining results deriving from DLS and TEM investigation. Formation of such kinds of structures was considered as the result of both hydrophobic and π-π stacking interactions. Furthermore, hydrogen bonds among urethane units also played a role. The aggregates obtained proved able to act as nanocontainers for hydrophobic compounds, with solubilizing ability increasing with the elongation of the alkyl tail. They also showed a good capability as vehicles for delivery of genetic material, as well as antimicrobial agents. In this latter case, the tetradecyl derivative was the most efficient one, probably as a consequence of its pronounced membranotropic properties.

A further structural parameter that has been considered one of the most important in affecting the self-assembly process of imidazolium salts, is the one related to the insertion of a functionalized ester group in the side chain. To this aim, 1-alkyloxycarbonyloxyethyl-3-methylimidazoliun chlorides **[$C_n$$CC_1$im]Cl**, with alkyl chains ranging from 10 up to 14 carbon atoms were considered (Fig. 10.7) [20].

Cmc values, calculated by means of conductivity and surface tension measurements, gradually decreased with the alkyl chain elongation. However, they proved four times lower than those corresponding to their non-functionalized counterparts, probably as a consequence of the establishment of hydrogen bonds in the headgroup region. Values for counterion binding (β) were lower than the ones measured for the single chain functionalized imidazolium salts. However, also, in this case, the decrease in ΔG_m detected for longer alkyl chain was considered as a result of the relevance of the hydrophobic effect in determining the outcome of the micellization process. Interestingly, determination of the hydrodynamic radius, through DLS measurements, allowed detecting the presence of aggregates of different sizes, as accounted for by D_H values (D_H = 2.5–3.3 nm and 110–130 nm). The aggregates of smaller size proved to be more than two times more compacted with respect to the ones formed by the corresponding nonfunctionalized imidazolium salts. The **[$C_n$$CC_1$im]Cl** salts were tested for their antimicrobial and antifungal activity. In particular, the antimicrobial activity increased with the elongation of the alkyl chain. As the same factor induced a decrease in cmc, it was supposed that with the elongation of the alkyl tail, the amount of monomeric salt at the cell membrane became low and the high activity was ascribed to the high adsorption efficiency (pC_{20}) of the carbonate-functionalized salts, which would promote the interaction with cell membrane.

The self-assembly behavior of 1-alkyl-3-methylimidazolium bromides, in aqueous solution, has been investigated using alkyl chains ranging from tetra- to octadecyl (Fig. 10.8) [21].

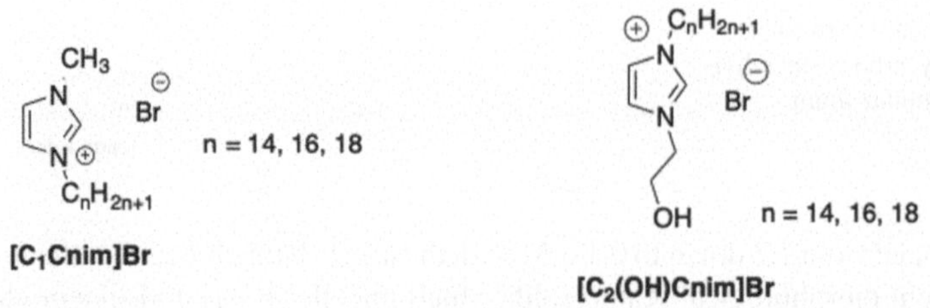

Fig. 10.8 Structures of 1-alkyl-3-methylimidazolium bromides and of 1-alkyl-3-hydroxyethylimidazolium bromides

Results collected on the grounds of tensiometry and conductometry investigation demonstrated that the cmc decreased from 2.5 mM for C_{14}-derivative down to 0.2 mM for C_{18}-derivative. DLS investigation shed light on the occurrence of a reorganization process driven by the salt concentration. Indeed, below the cmc, the presence of aggregates of higher hydrodynamic radius (D_H = 6–10 nm) was detected. These, beyond cmc, are reorganized to give smaller aggregates (D_H = 2–4 nm). The detected process was ascribed to the occurrence of different interactions operating on the self-assembly process: π-π stacking in the premicellar region and hydrophobic effect in the micellar ones. The obtained aggregates proved able to interact with oligonucleotides. However, differently from what was previously observed for ammonium salts, the higher charge delocalization on the imidazolium ion did not induce differences in the electrostatic interactions established between oligonucleotides and surfactant molecules or aggregates. A very interesting behavior was detected for membranotropic properties. Indeed, in the presence of dipalmitoylphosphocholine (DPPC), as model liposomes, a decrease in the gel/liquid crystal main transition was observed for the C_{14}-derivative. Differently, the increase in the alkyl tail induced an opposite effect, indicating a stabilizing effect on the DPPC lipid bilayer and an improvement of the packing efficiency of amphiphiles hydrophobic tails in a matrix of phospholipid alkyl moiety.

The study of self-assembly behavior of imidazolium-based amphiphiles has been extended to functional derivatives. In particular, 1-alkyl-2-(hydroxyethyl)imidazolium bromides have been considered, with alkyl tails ranging from 14 up to 18 carbon atoms (Fig. 10.8) [22].

Determination of cmc values by conductometric investigation and comparison with corresponding values obtained for 1-alkyl-2-ethylimidazolium bromides allowed stating that the presence of hydroxyethyl fragment induces a slight increase in the self-assembly tendency, probably as a consequence of the intermolecular hydrogen bond formation. On the other hand, the elongation of the alkyl tail by two methylene units caused up to a threefold decrease of cmc. A combined investigation performed by means of dynamic and electrophoretic light scattering allowed establishing that C_{14} derivative formed micellar aggregates with D_H ranging from 2 up to 4 nm. On the other hand, for higher homologous, C_{16} and C_{18}, the presence of rod-like

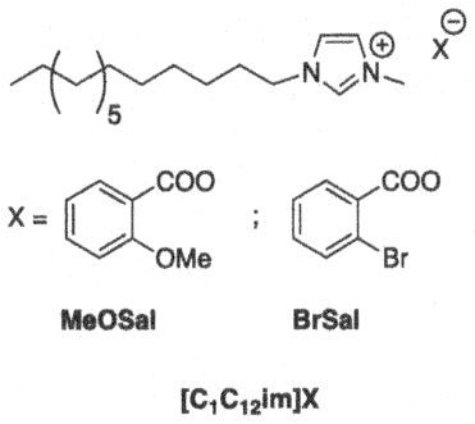

Fig. 10.9 Structure of 1-dodecyl-3-methylimidazolium salicylate organic salts

aggregates was hypothesized, as accounted for by D_H values (>100 nm). This hypothesis was supported by the increase in the zeta potential, which ranged from 60 mV (C_{14}) up to 90 mV (C_{18}), by TEM micrographs obtained from aqueous solutions and from aggregation number obtained through fluorescence measurements.

In the case of larger aggregates, besides hydrogen bond and hydrophobic interactions, also π-π stacking became responsible for their formation.

The **[C$_2$(OH)C$_n$im]Br** aggregates proved able to act as nanocontainers toward hydrophobic compounds like dye Orange OT. In particular, the elongation of the alkyl tail induced a decrease in the surfactant concentration corresponding to the initial solubilization activity. Furthermore, they exhibited antimicrobial properties, with C_{14} proving more active than norfloxacin against *Staphylococcus aureus* bacteria, whereas C_{18} exhibited activity comparable to ketoconazole toward *Candida albicans* fungi.

The aggregation of surface-active ionic liquids has been investigated also as a function of the different nature of the anion. To this aim, the behavior of aqueous solution of **[C$_1$C$_{12}$im][MeOSal]** and **[C$_1$C$_{12}$im][BrSal]** (Fig. 10.9) has been investigated by using surface tension, conductivity, rheology, SAXS, and POM measurements [23].

Determination of cmc values demonstrated that aggregation occurred at lower SAILs concentration in the presence of the aromatic anions with respect to the corresponding bromide salt. In general, the presence of aromatic anion induced an increase in hydrophobic interactions, favoring the micelles formation. The aggregation was driven by both hydrophobic and electrostatic interactions and lower cmc determined for **[C$_1$C$_{12}$im][BrSal]** allowed stating that the presence of bromide substituent on the aromatic anion significantly reduced electrostatic repulsions among the imidazolium heads. Electrical conductivity measurements performed as a function of temperature shed light on the spontaneous nature of the micellization process, which was exotermic, driven by the entropic contribution, and was favored in the presence of bromide, as accounted for by ΔG_m values. POM and SAXS investigations were performed as a function of the concentration. Up to 40 wt% both salts exhibited a lyotropic liquid crystal phase, presenting fan-like or pleated-ribbon texture.

However, when the concentration raised up to 80 wt%, transparent and clear phases formed. The above observations were supported by SAXS data. In particular, the comparison among lattice parameters (a_0) corresponding to **[C$_1$C$_{12}$im][Sal]**, **[C$_1$C$_{12}$im][MeOSal]**, and **[C$_1$C$_{12}$im][BrSal]** demonstrated that the densest arrangement was obtained in the first case, as the presence of substituents on the aromatic anions induced a significant increase in a_0. Interestingly, rheology investigation

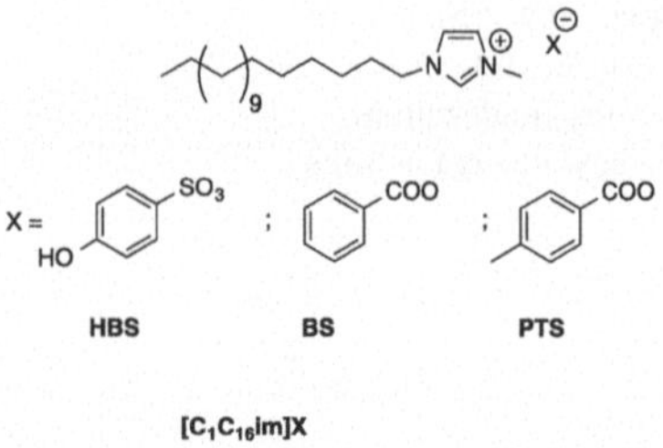

Fig. 10.10 Structures of 1-hexadecyl-3-methylimidazolium based organic salts bearing different benzenesulfonate anions

performed at 60 wt% indicated that in such systems viscosity is predominant over the elastic property.

The role played by the aromatic anions in the self-assembly process of imidazolium salts has been deeply investigated also taking into consideration the $[C_1C_{16}im]^+$ cation in combination with aromatic anions, like the 4-hydroxybenzenesulfonate (HBS), the benzenesulfonate (BS) and *p*-toluenesulfonate (PTS, Fig. 10.10) [24].

A combined experimental approach, based on surface tension and conductivity measurements, allowed demonstrating that cmc changed along with the order: $[HBS^-] > [BS^-] > [PTS^-]$, according to hydrophobicity or bulkiness of the substituent. Cation being the same, the presence of the aromatic counterion induced a 2–3 folds decrease in cmc with respect to analogous salts bearing different counterions. Conductivity investigation allowed determining the degree of counterion binding (β), which increased with the anion hydrophobicity, indicating a strong cation–anion interaction also in the micellar phase.

From a thermodynamic point of view, negative ΔG_m values, decreasing in parallel with anion hydrophobicity, indicated that the aggregation process was hydrophobically driven. Thanks to the use of fluorescence measurements, based on pyrene emission, important information about the cybotactic region as well as the nature and the composition of both Stern and Palisade layers was obtained. The cybotactic region represents the region around the solute molecule where ordering of the solvent is modified by the presence of solute. The Stern layer of micelle is considered as the layer around the core of an ionic micelle formed by ionic head groups and tightly bounded counterions. Finally, the Palisade layer of micelle is the area between oil-like hydrophobic core and Stern layer comprised of few carbon atoms. On the grounds of fluorescence measurements, it was demonstrated that the polarity of the cybotactic region decreased along the order: **$[C_1C_{16}im][HBS]$** > **$[C_1C_{16}im][BS]$** > **$[C_1C_{16}im][PTS]$**. Analysis of emission intensity (I_{em}) as a function of the salt concentration allowed hypothesizing the insertion of $[HBS^-]$ between the imidazolium groups in the Stern layer of the micelles, whereas $[BS^-]$ and $[PTS^-]$ anions were located in Palisade layer of the micelle and this hypothesis was corroborated by ^1H NMR investigation (Fig. 10.11).

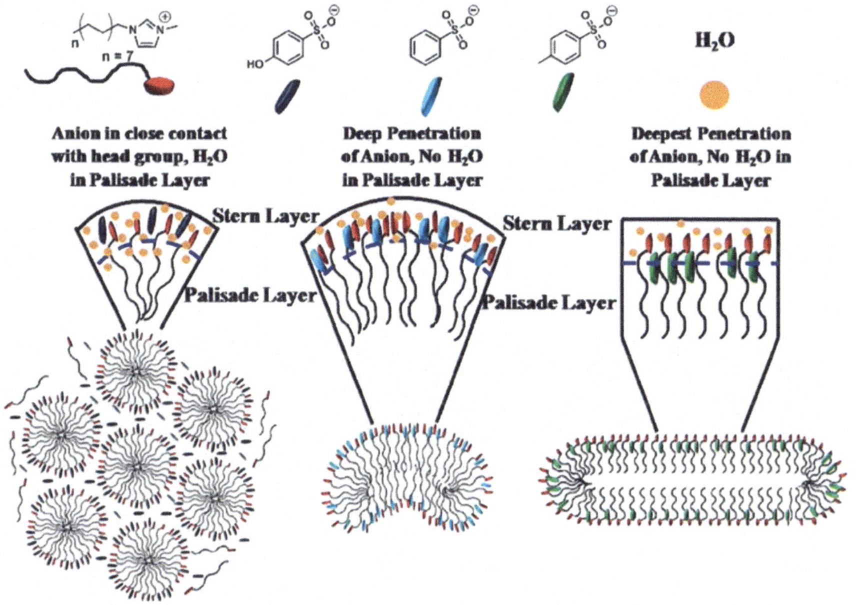

Fig. 10.11 Schematic representation showing the relative position of aromatic anions and water molecules with respect to $[C_1C_{16}im]$ cation in micelles of different ILs with varying shapes and sizes in aqueous medium. Adapted with permission from Ref. [24], Copyright 2016 American Chemical Society

Furthermore, the nature of the counterion also determined the shape of the aggregates that, on the grounds of DLS and TEM measurements, proved to be spherical micelles, in the case of $[HBS^-]$ anion and rod-like micelles in the case of $[BS^-]$ and $[PTS^-]$.

In a different investigation, besides the nature of the anion and the length of the alkyl chain, also the nature of the cationic head was taken into consideration, studying the self-assembly behavior of surface-active zwitterionic organic salts (Fig. 10.12) [25].

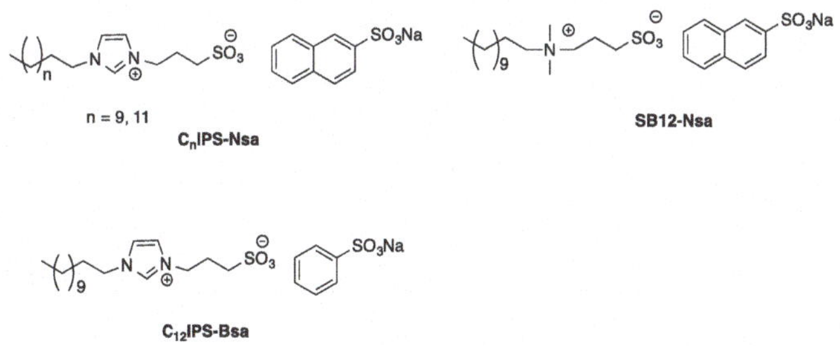

Fig. 10.12 Structures of zwitterionic organic salts

In particular, ammonium and imidazolium butane sulfonate salts, bearing alkyl chains ranging from C_{12} up to C_{14} and paired with naphthalene or benzene sodium salts were considered.

Analysis of the results collected by means of surface tension measurements revealed that cation being the same (imidazolium), cmc decreased on going from **C_{12}IPS-Nsa** to **C_{14}IPS-Nsa**. Furthermore, it was affected by the anion nature, going from [Bsa$^-$] to [Nsa$^-$], according to the anion hydrophobicity.

On the other hand, cation hydrophobicity and charge delocalization were considered the main factors accounting for the lower cmc detected for **C_{12}IPS-Nsa** with respect to **SB_{12}-Nsa**.

In the case of imidazolium salts, it was hypothesized that the formation of spherical micelles was supported by π-π interactions among imidazolium and aromatic anions. The above interactions proved to be stronger in the presence of [Nsa$^-$] anions, as accounted for by the more significant upfield shift detected for the alkyl chain protons, in the case of **C_{14}IPS-Nsa, C_{12}IPS-Nsa,** and **SB_{12}-Nsa**. This was a consequence of the penetration of the aromatic anion in the hydrophobic region.

The above hypothesis was also supported by calculated interaction energies of zwitterionic salt/water (1:1) complexes, by DFT calculations, that accounted for more negative energies in the case of **C_{12}IPS-Bsa** and **SB_{12}-Nsa/H_2O**, according to their difficulty to form micelles.

The study of the self-assembly behavior of 1-dodecyl-3-methylimidazolium based organic salts bearing aromatic anions like salicylate and 3-hydroxy-2-naphthoate was also performed in water solution, as a function of salt concentration (Fig. 10.13) [26].

The use of surface tension and conductivity measurements allowed us to demonstrate that cmc values gradually increased on going from **[C_1C_{12}im][HNC]** to **[C_1C_{12}im][Sal]**. The above trend was ascribed to the higher hydrophobicity of the [HNC$^-$] anion that favors micelles formation. Conductometric measurements carried out as a function of the temperature, allowed to obtain ΔG_m values that resulted more negative in the case of **[C_1C_{16}im][HNC]**, accounting for a spontaneous aggregation process driven by the hydrophobicity.

Analysis of the phase diagrams of the imidazolium salts as a function of the concentration shed light on the occurrence of three different systems: (i) the isotropic solution phase; (ii) the anisotropic hexagonal liquid-crystalline phase (H_1) and (iii) the isotropic cubic liquid-crystalline phase (V_1). Obviously, the concentration range

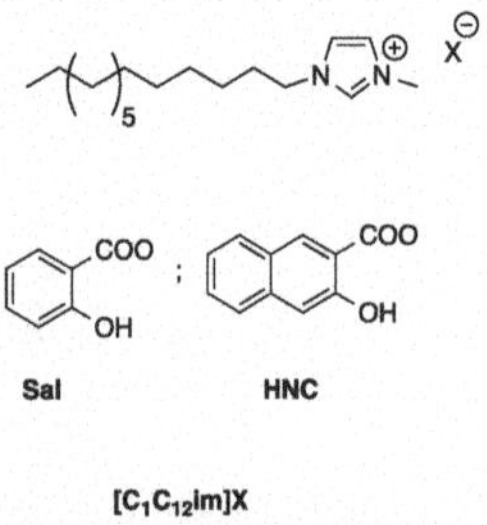

Fig. 10.13 Structures of 1-dodecyl-3-methylimidazolium salts bearing aromatic anions

corresponding to different phases was affected by the anion nature. SAXS patterns allowed demonstrating that H_I phase was featured by the presence of cylinder like aggregates, packed in a hexagonal array, and separated by a continuous solvent region. In the case of **[C$_1$C$_{12}$im][Sal]**, increasing the concentration, a denser arrangement of surface-active ILs molecules was observed, as accounted for by the decrease in the radius of the cylindrical micelles. The phase behavior was also investigated as a function of the temperature. In both cases, the increase in temperature induced a decrease in the radius of cylindrical aggregates and in the thickness of the solvent layer between cylinders. For **[C$_1$C$_{12}$im][Sal]**, this was ascribed to the higher softness of the molecules that extruded each other to a higher degree. Differently, in the case of **[C$_1$C$_{12}$im][HNC]**, the observed changes were ascribed to the occurrence and more interdigitate systems. DFT calculations demonstrated that, in the aggregates, the [Sal$^-$] anion interacts with the cationic headgroup, whereas [HNC$^-$], penetrates in the hydrophobic region. The above result was ascribed to the lower electronegativity of [HNC$^-$] that proved less able to screen the electrostatic repulsions between the imidazolium ions. However, its higher hydrophobicity better-promoted aggregates formation. Further differences in the aggregation behavior were also elucidated on the grounds of the determination of the critical packing parameter, which was equal to 0.46 and 0.30 in the case of **[C$_1$C$_{12}$im][Sal]** and **[C$_1$C$_{12}$im][HNC]**, accounting for the inclination to pack in a hexagonal array and to form spherical micelles.

Imidazolium salts have been engaged in self-assembled systems also in the form of complexes with lanthanides. In particular, these organic cations have been used to prevent the insertion of active hydroxyls of solvent in the lanthanide coordination sphere. This is frequently responsible for the fluorescence quenching. To this aim, complexes of imidazolium organic salts, bearing naphthyl and branched alkyl chain ([C$_m$naphim$^+$]) with the anion [(thenoyltrifluoroacetone)$_4$Eu$^-$] ([Eu(tta)$_4$$^-$]) were prepared (Fig. 10.14) [27].

Ethanol solutions of **[C$_m$naphim][Eu(tta)$_4$]** were dropped in stirred water, observing a significant increase in the solution turbidity as V_{H_2O} exceeded 50%. The rise in turbidity was ascribed to aggregates formation. The above hypothesis was supported by both UV–vis and fluorescence investigation. In particular, in the

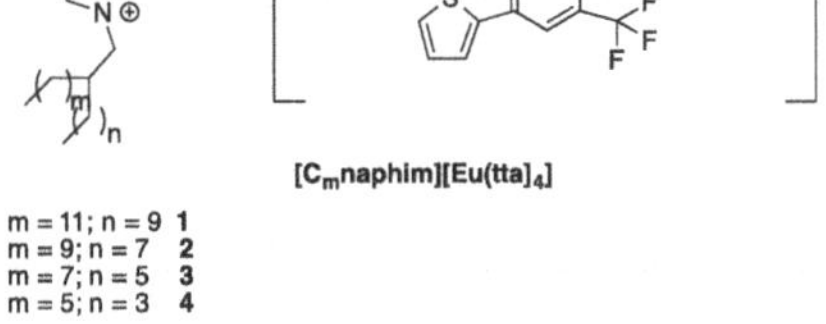

Fig. 10.14 Structure of [(thenoyltrifluoroacetone)$_4$Eu$^-$] based imidazolium salts

case of UV–vis investigation, two main absorption bands were detected. The first one, occurring at 275 nm (A_1), was assigned to the naphthyl group, whereas the second one located at 338 nm (A_2) was attributed to the absorption of [Eu(tta)$_4^-$]. These bands showed a different behavior as a function of the water content. Indeed, at $V_{H_2O} < 50\%$, the A_1 slightly increased with water content, whereas the A_2 band continuously decreased, At $V_{H_2O} > 50\%$, they became similar and weakened. In a similar way, in the emission spectra, two main bands were detected, at 377 and 612 nm, corresponding to the emission of naphthyl group and [Eu(tta)$_4^-$], respectively.

As far as the [Eu(tta)$_4^-$] band was concerned, it was observed a gradual increase in the emission for $V_{H_2O} > 50\%$, which reached the maximum value at $V_{H_2O} = 70\%$. This was ascribed to the aggregates formation and the enhanced emission was considered a consequence of the aggregates ability to shield the Eu(III) from the active hydroxyl groups of the solvent.

Features of the aggregates were investigated using a Malavern size particle analyzer, DLS measurements, TEM, and SEM. Interestingly, the size of the aggregates, as well as fluorescence emission, decreased with the water amount (for $V_{H_2O} > 70\%$) and complex concentration. In both cases, this was ascribed to the dissociation of **[C$_m$naphim][(Eu(tta)$_4$]** complex, which induced the diffusion of [Eu(tta)$_4^-$] in the bulk solution with [C$_m$naphim$^+$] acting as counterion.

On the other hand, features of the aggregates depended on the length of the alkyl chain, as accounted for by the decrease in fluorescence emission with the increase in the above parameter.

The architecture of the self-assembled systems formed by organic salts also depends on the molecular design of the individual constituents. Under this light, a raise of interest has been detected toward the obtainment of self-assembled systems based on host–guest interactions involving organic salts. In this context, the use of differently functionalized imidazolium salts and their interaction with cucurbit[8]uril has been considered (Fig. 10.15) [28].

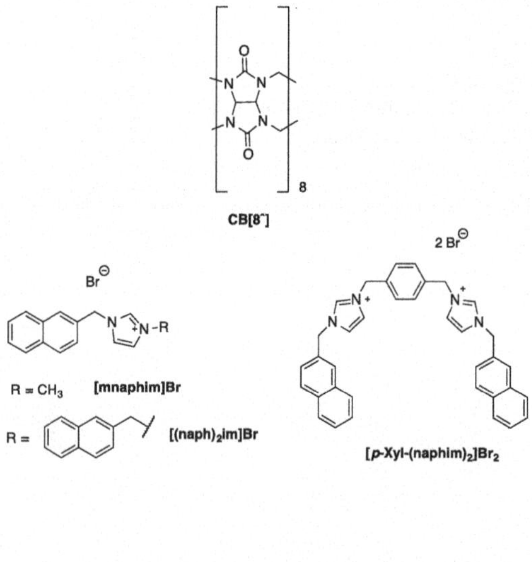

Fig. 10.15 Structures of cucurbit[8]uril and imidazolium-based organic salts

In particular, using equimolar mixtures of **[mnaphim]Br** and CB[8] or **[(naph)$_2$im]Br** and CB[8], nanoscopic crystalline structures were obtained. ^{1}H NMR investigation allowed to detected a set of broad and ill-defined signals that could not unambiguously be assigned to the presence of polymeric species. To clarify and support the above hypothesis, a combination of small-angle neutron scattering, AFM, and TEM measurements were performed that allowed identifying the presence of elongated plate-like aggregates, in the case of **[mnaphim]Br/CB[8]**, and the presence of large nanoscale assemblies in the case of **[(naph)$_2$im]Br/CB[8]**. In particular, in the latter case, the presence of fibers was detected and the mechanism of the self-assembly process favoring the fibers formation was elucidated (Fig. 10.16).

In the packing model, the formation of zigzagged strands formed by single layer of [(naph)$_2$im]Br/CB[8] host–guest complexes was hypothesized, in which the mainly stabilizing interactions were represented by C-H••••O=C hydrogen bonds, between the imidazolium groups and the carbonyl portals, and ion–dipole interactions.

Fibers formation was also investigated using rheology measurements that gave trends reminiscent of Maxwellian behavior found in many viscoelastic worm-like micellar solutions. Analysis of kinetic data obtained, performing rheology measurements as a function of time, and using Avrami theory, supports the hypothesis of a one-dimensional and interfacial-controlled growth process.

In a further example, complexes formed by dimethoxypillar[5]arene and anthracenyl-based imidazolium salts, differing in the alkyl chain length and the anion nature, were investigated both in solution and in the solid-state (Fig. 10.17) [29].

Fluorescence investigation performed in the solid-state evidenced how the addition of DMP[5]A to the imidazolium salts gave rise to the occurrence of aggregation-induced emission (AIE) processes, as accounted for by the significant increase in the fluorescence emission.

Self-assembly of diimidazolium salts bearing rigid aromatic spacers and alkyl chains of different lengths has been also investigated. The interest for such kinds of tectons is derived from their structural flexibility and also from the possibility that they, like cationic surfactants, could be used for drug complexation and delivery, above all in the case of drugs bearing carboxylate groups.

To this aim, the self-assembly of diimidazolium salts bearing alkyl chain varying from five up to nine carbon atoms and bearing a *meta*-xylyl spacer was investigated by tensiometry, fluorescence spectroscopy, and pulse-gradient spin-echo measurements (Fig. 10.18) [30].

The results reported evidence of the ability of organic salts to form micellar aggregates and clearly shed light on the dependence of cmc values on the length of the alkyl chain. Furthermore, PGSE measurements indicate a reduction of the self-diffusion coefficient with the increase of fraction of micellized surfactants.

In the case of $n = 7$, such kind of investigation allowed us to calculate a hydrodynamic radius equal to 1.8 nm. On the other hand, SANS investigation gave a model for the aggregates as formed by a hydrophobic core, surrounded by a polar shell that included the surfactant headgroup and hydrating water. The only case in which the SANS model did not give the expected results was the case of $n = 5$. The self-assembly behavior of diimidazolium salts was further analyzed in the presence

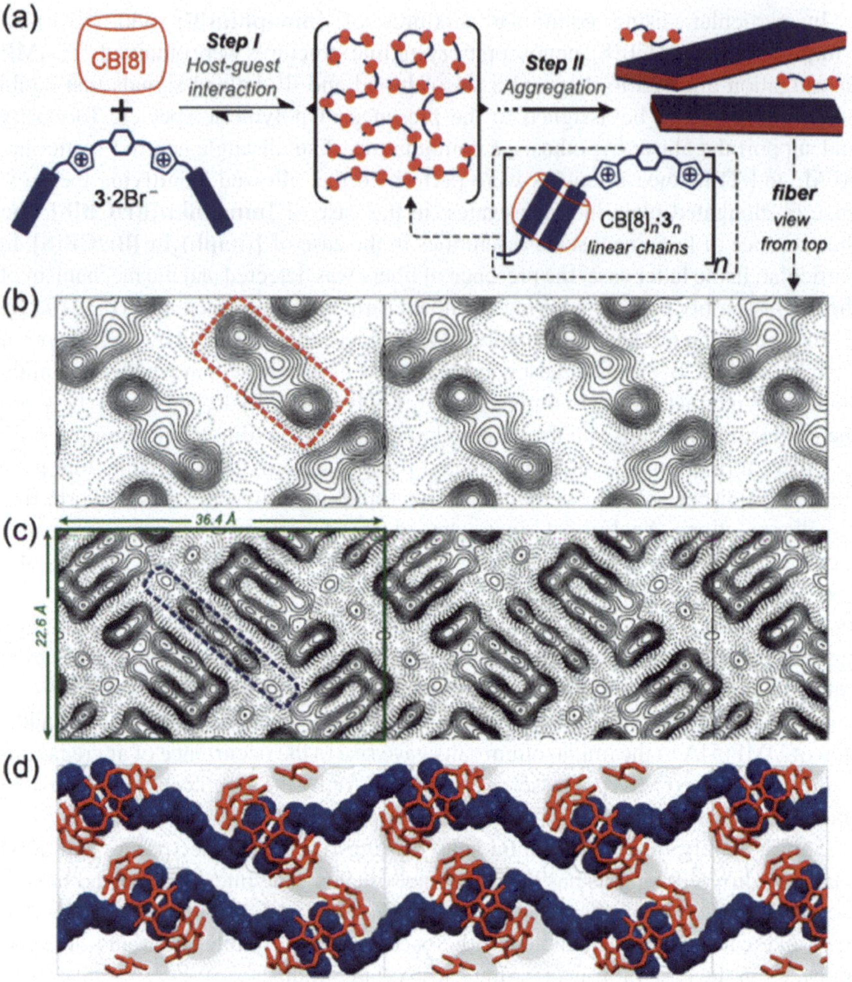

Fig. 10.16 **a** Self-assembly scheme of CB[8] and [p-Xyl-(naphim)$_2$]Br; 2cryo-TEM maps at a resolution of 5 (**b**) and 4 Å (**c**), and packing model for the CB[8]+3 fibers (**d**) superimposed to panel **b**. Unit cell parameters (in green) are shown in panel **c**. Adapted with permission from Ref. [28], Copyright 2019 American Chemical Society

Fig. 10.17 Structures of DMP[5]A and anthracenyl-based imidazolium salts

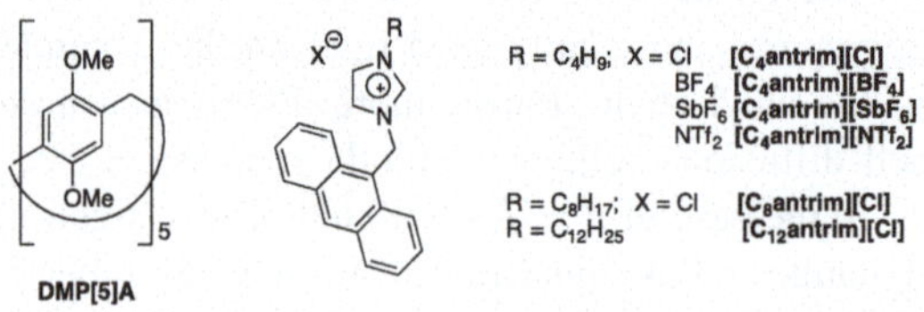

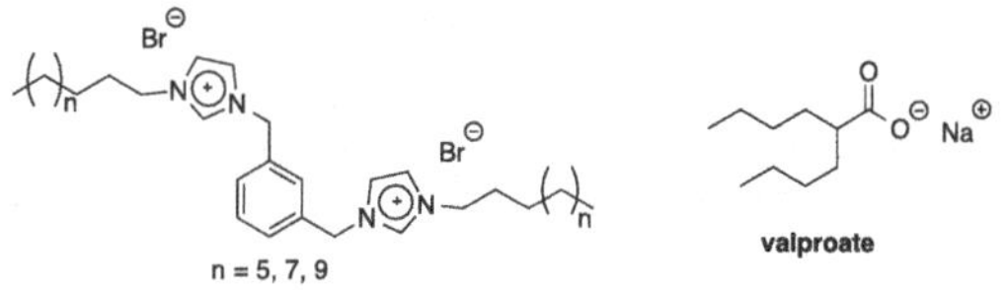

[m-Xyl-($C_n$im)$_2$][Br]$_2$

Fig. 10.18 Structures of diimidazolium based organic salts

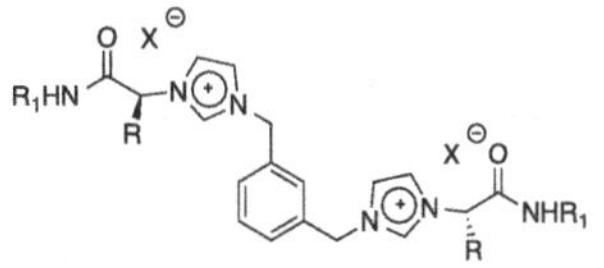

R = CH_2Ph; R1 = Bn
R = $CH(CH_3)_2$; R1 = Bn
R = $CH(CH_3)_2$; R1 = $C_{12}H_{25}$
X = Br; NTf_2

Fig. 10.19 Structures of diimidazolium based organic salts bearing amino acids residues in the alkyl tail

of increasing concentration of the carboxylate drug valproate. The obtained self-diffusion coefficients supported the hypothesis of the interaction between the drug and micellar aggregates, as accounted for by both the decrease and the increase in diffusion coefficients of aggregates and drug, respectively.

In the attempt to obtain chiral selectors for the enantiodiscrimination of dicarboxylate salts of biological relevance, some chiral diimidazolium salts featured by the presence of amino acids in the side chain have been synthesized (Fig. 10.19) [31].

The conformation of these ditopic receptors was firstly analyzed using, ATR-FTIR, Monte Carlo simulations ^{1}H-NMR, and ^{1}D-NOESY experiments. Data collected from this combined approach allowed us to state that conformation of all tested organic salts heavily depended on the nature of the anion. Indeed, for Br^--based salts a cleft folded structure, in which the aromatic rings of the N-benzyl groups occupied the final section of the cleft, was suggested. Differently, for [NTf_2^-]-based salts an open conformation was proposed, in which the bigger anion was located in the other section and interacted only with one amide fragment.

The possibility for such structures to form supramolecular aggregates was investigated by using ^{1}H NMR spectroscopy in $CDCl_3$ solution. Once again, results testified to the dependence of the salts' behavior on the anion nature. Indeed, spectra of [NTf_2^-]-based salts exhibited a negligible dependence on the salt concentration. Differently, for Br^--based salts, increasing the concentration from 2 up to 80 mM, gave rise to a significant upfield shift for H_A and H_G protons, whereas downfield shift was observed for H_B (Fig. 10.20).

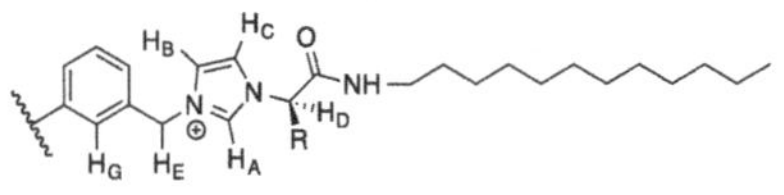

Fig. 10.20 Structures of diimidazolium based organic salts bearing amino acids residues in the alkyl tail with labeled protons

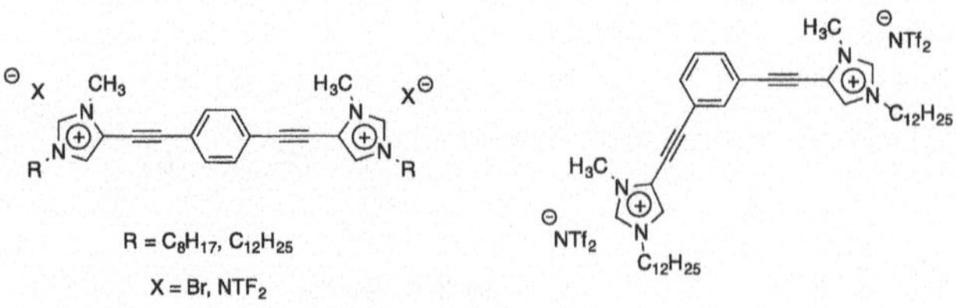

Fig. 10.21 Self-assembling phenylene ethynylene-based diimidazolium salts

On the whole, the above trend indicated a major involvement of H_A in intramolecular interactions that became more relevant at higher concentrations. In the case of [Br^-] salts, morphology of the aggregates obtained from water/MeOH mixtures (5%) were affected by the nature of the side chain on the imidazolium cation with well-defined fractal structures observed for phenylalanine derivatives and spherical aggregates detected for leucine derivatives, as accounted for by SEM images.

Imidazolium salts can give a diverse range of aggregated structures, upon self-assembly when coupled with a π-conjugated spacer. On this topic, it has been demonstrated that aggregation of diimidazolium salts is affected by the nature of the conjugated spacer interposed between the cationic heads. This is the case of diimidazolium salts bearing the *m*- and *p*-phenylene ethynylene spacers (Fig. 10.21) [32].

The salts considered also differed for the alkyl chain length borne on the imidazolium rings as well as for the anion. Spectroscopic investigation revealed that the different substitution or alkyl chain lengths affect the solvents in which aggregation occurs, such as dipolar aprotic solvents like acetonitrile and tetrahydrofuran or short-chain alcohols. Notably, the aggregates retained fluorescence in the solid-state, and the higher tendency to self-assemble observed for the *p*-substituted salts could be ascribed to a different geometry of the aggregates, as suggested by DFT-level calculations.

Changing the spacer, the same group investigated the aggregation of diimidazolium salts with a perylene bisimide-based spacer (Fig. 10.22), differing for the alkyl chain length [32], or the anion [33].

UV–vis and fluorescence spectroscopy investigations pointed out that the salts bearing iodide anions and different alkyl chains on the cationic moiety, aggregated via an isodesmic pathway, and that the alkyl chain length mainly affected the stability of the aggregates, as estimated by the equilibrium formation constants [32]. Furthermore, SEM images showed that varying the alkyl chain length barely affected the morphology of the aggregates, which in all cases appeared as rod-like objects. A remarkably different result was observed when considering salts sharing the same

Fig. 10.22 Structure of perylene bisimide functionalized diimidazolium salts

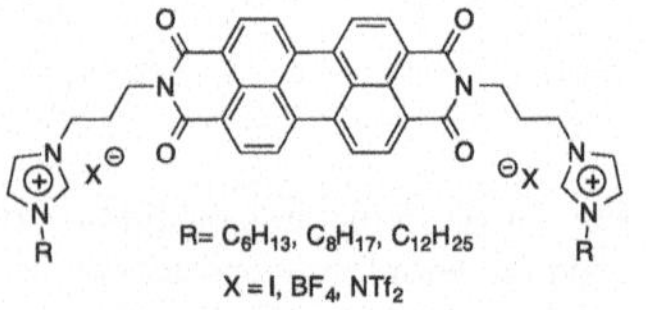

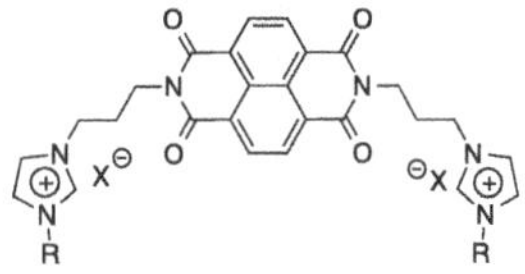

Fig. 10.23 Structure of NDI-functionalized diimidazolium salts

alkyl chains on the imidazolium ring, but differing in the counteranions [33]. In this case, this structural variation not only affected the aggregates stability, but also their morphology. In particular, when [NTf_2^-] was the anion, morphology changed from rod-like to irregularly shaped objects.

A similar approach was used to investigate the self-assembly of diimidazolium salts tethered to a naphthalene diimide (NDI) spacer, reported in Fig. 10.23, once again differing either for the alkyl chain length on the imidazolium ring [34], or the anion [35].

In the first case, salts with both alkyl and fluoroalkyl chains were considered. UV–vis and spectroscopic investigations evidenced that all the salts aggregate in solvents like DMF or THF forming *H*-aggregates [34]. Notably, an odd–even effect was observed, given that the concentration corresponding to the onset of aggregation follows different trends as a function of the alkyl chain length, depending on whether chains with an odd- or even number of carbon atoms are considered. Furthermore, the same odd–even effect was observed for the variation of the melting points of the salts, as a function of the alkyl chain length. In general, introduction of fluorinated chains resulted in less stable aggregates, and SEM investigation showed that these salts aggregated into spherical or disc-like structures, depending on the alkyl chain length as well as the solvent.

On the other hand, changing the anion significantly altered the properties of the aggregates [35]. In particular, for NDI-salts bearing octyl chains, changing the anion from I^- to [BF_4^-] or [NTf_2^-] induced the shift from an aggregation caused quenching (ACQ) in the first case, to aggregation enhanced emission (AEE) in the other ones. However, the same structural change also induced a reduction of the formation constant of the aggregates. Finally, changing the anion resulted in different aggregate morphology, as revealed by SEM images.

Markedly different properties were observed when salts with a much more rigid and extensively conjugated spacer were considered [36]. In this case, the spacers interpose between an imidazolium cation and a neutral imidazole unit, so that a potentially push–pull system can be realized. The two salts differ for the π-surface extension of the aromatic portion of the spacer, comprising a phenyl or a pyrene unit (Fig. 10.24).

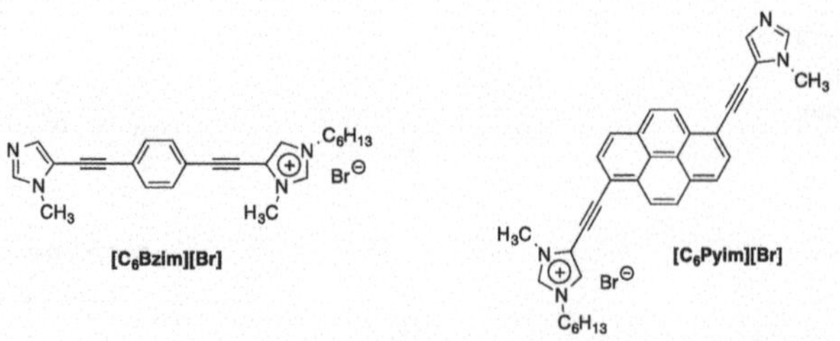

Fig. 10.24 Structures 1,4-phenyldiethynyl- and 1,6-pyrenediethynyl-functionalized imidazolium salts

Differently from the previous cases, these salts were able to self-assemble not only in conventional solvents but also in ionic liquids. The presence of an electron-donating group (imidazole) linked through extensive conjugation with an electron-accepting group (imidazolium cation), led to significant AIE in solution. Notably, aggregation occurred following different pathways, depending on the nature of the spacer; in particular isodesmic aggregation was observed for **[C$_6$Pyim][Br]**, whereas **[C$_6$Bzim][Br]** aggregated through a cooperative mechanism.

Solid-state fluorescence investigation also showed that **[C$_6$Bzim][Br]** forms highly emissive *J*-aggregates in the solid-state. In contrast, **[C$_6$Pyim][Br]** forms *H*-aggregates in the solid-state, and emission quenching is observed.

The versatility of imidazolium salts tethered to emissive aromatic assembling motifs is demonstrated by the application of self-assembling 1,8-naphthalimide imidazolium salts to obtain bioimaging and theranostic soft materials [37]. In particular, salts with different lengths of the alkyl chain borne on the imidazolium cation were considered, and also differing for the anion, **[C$_n$NIm][X]** (Fig. 10.25).

Solvent-dependent spectroscopic UV–vis and fluorescence investigation showed that these salts are scarcely fluorescent in DMSO, but intense emission is switched on by adding increasing amounts of water. This is due to occurrence of *J*-aggregation and AIE. Notably, fluorescence in solution is persistent over an extended period of time, such as 8 months (Fig. 10.25b). At higher concentrations, these salts formed conductive, fluorescent supramolecular gels in H_2O/DMSO as well as aqueous buffers/DMSO mixtures. Fluorescence microscopy showed that these salts, when dissolved at concentration lower than IC_{50} were efficiently internalized in cancer cells of different strains, such as HeLa, MCF7 or HCT117, exhibiting intense blue emission localized in the organelles. At higher concentration, they exert cytotoxic activity on these cells and the **[C$_{14}$NIm][glu]** salt show a higher selectivity index toward cancer cells over non-cancer strain.

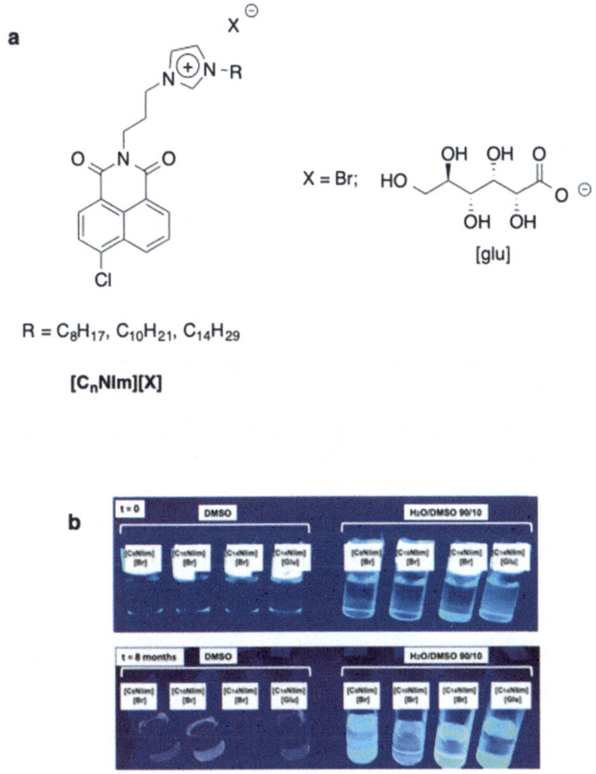

Fig. 10.25 **a** 1,8-Naphthalimide-based imidazolium salts, **b** pictures of salts solution under UV light. Adapted with permission from Ref. [37], Copyright 2020 American Chemical Society

10.3 Ammonium Salts

Ammonium salts are among the most frequently used organic salts with self-assembling ability. This is likely due to the easy introduction of an ammonium functionality in a self-assembling molecule, often carried out to allow or enhance solubility in water.

The self-assembly of surfactants can be modulated by adding external cations. In this regard, it has been described how the addition of Ca^{2+} or Gd^{3+} ions modulate the morphology of the nanostructures formed by the saturated phospholipid 1,2-dimyristoyl-*sn-glycero*-3-phosphocholine (**DMPC**, Fig. 10.26), in organic solvents [38].

In particular, small-angle X-ray scattering (SAXS), dynamic light scattering (DLS) investigations as well as transmission electron microscopy (TEM) imaging revealed that at low concentrations, the added cations led to the formation of cylindrical micelles, which resulted in gel-like materials. At higher concentrations of

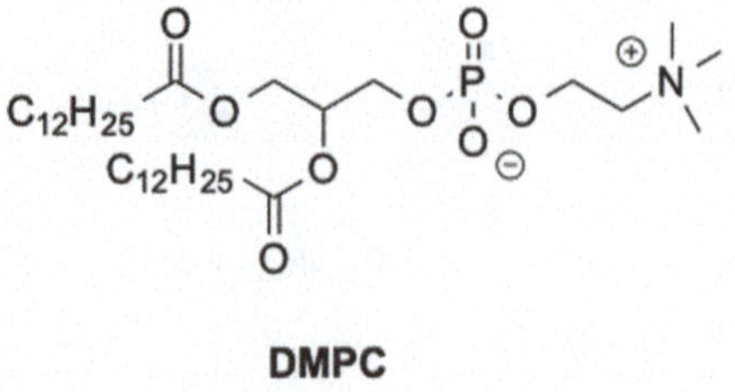

Fig. 10.26 Structure of DMPC surfactant

cations, the surfactant assembles into reverse onions which revert to lamellar stacks at even higher cation concentrations.

In a recent example, an ammonium-appended tetraphenylene-salt, **TPE-Gir**, gave rise to self-assembled structures able to recognize nucleic acids (Fig. 10.27) [39].

The tetraphenylene core is a typical motif for obtaining aggregation-induced emission (AIE), which results in enhanced fluorescence emission upon aggregation. Moreover, such a π-extended molecule could give rise to supramolecular polymers underpinned by π-stacking interactions, thus leaving the positively charged ammonium groups available to interact with anionic DNA strands. In particular, detailed spectroscopic investigation revealed that the self-assembly of **TPE-Gir** in THF/water solutions leads to fluorescent nanoparticles with diameters in the range of 100–120 nm. Further spectroscopic evidence showed that these nanoparticles had a strong tendency to intercalate with single-stranded DNA and bound to the minor

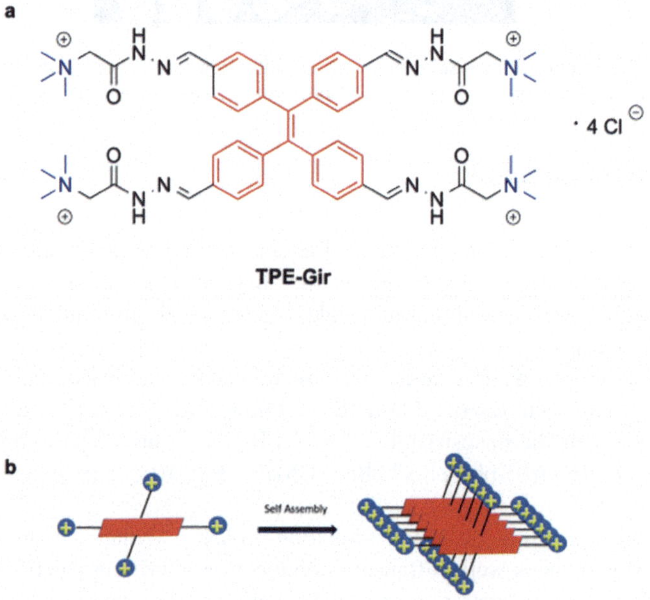

Fig. 10.27 **a** Structure of **TPE-Gir** and **b** pictorial representation of self-assembled structures

groove of double-stranded DNA. Thermodynamic investigations, by ITC, showed also selective binding to G-quadruplex DNA.

The versatility of ammonium salts in self-assembly is also demonstrated by their ability to form supramolecular network polymers in the presence of suitable macrocycles. In this regard, it has been reported the obtainment of a hexagonal supramolecular polymer formed by the planar tritopic salt **TAPB** (Fig. 10.28) in the presence of the cucurbit[5]uril-based host **Me$_{10}$TD[5]** [40].

NMR and UV–vis investigations proved that the formation of the supramolecular polymer was driven by the H-bond interaction involving each ammonium moiety in **TAPB** with the carbonyl groups of the cucurbit[5]uril-based host. The structure of the polymer was further confirmed by DLS measurements and TEM images.

A convenient characteristic of self-assembly by organic salts is the possibility to radically change the morphology or the spectral features of the aggregates in a straightforward way by changing the ions. In this regard, it has been reported that the morphology and emission of the cyanine-based salts **[PIC][NTf$_2$]** and **[PIC][BETI]**, Fig. 10.29, change drastically as a function of the anion [41].

TEM and SEM images revealed that while **[PIC][NTf$_2$]** forms diamond-like structures, **[PIC][BETI]** assembled forming rod-shaped objects. This difference in

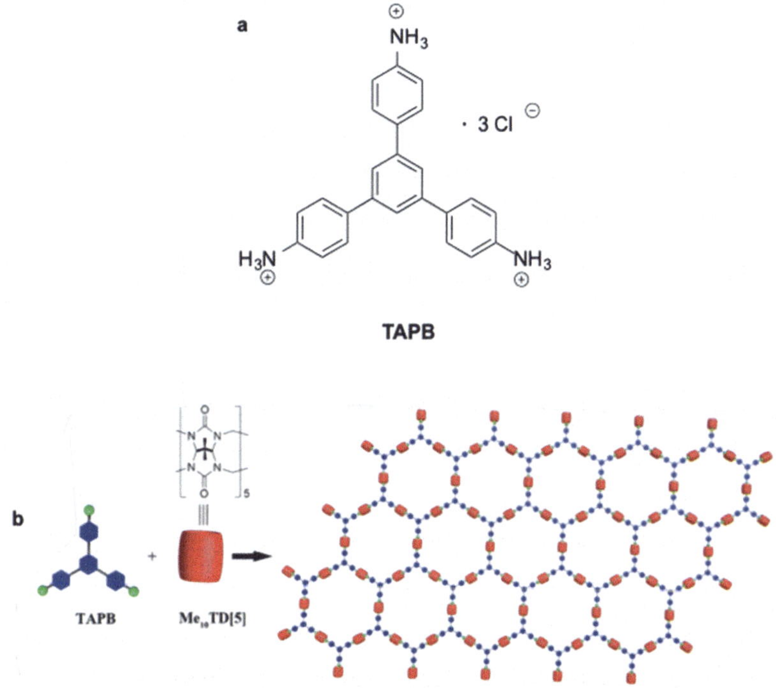

Fig. 10.28 **a** Structure of **TAPB** and **b** pictorial representation of supramolecular network polymer. Reproduced with permission from Ref. [40], Copyright 2018 Springer Nature

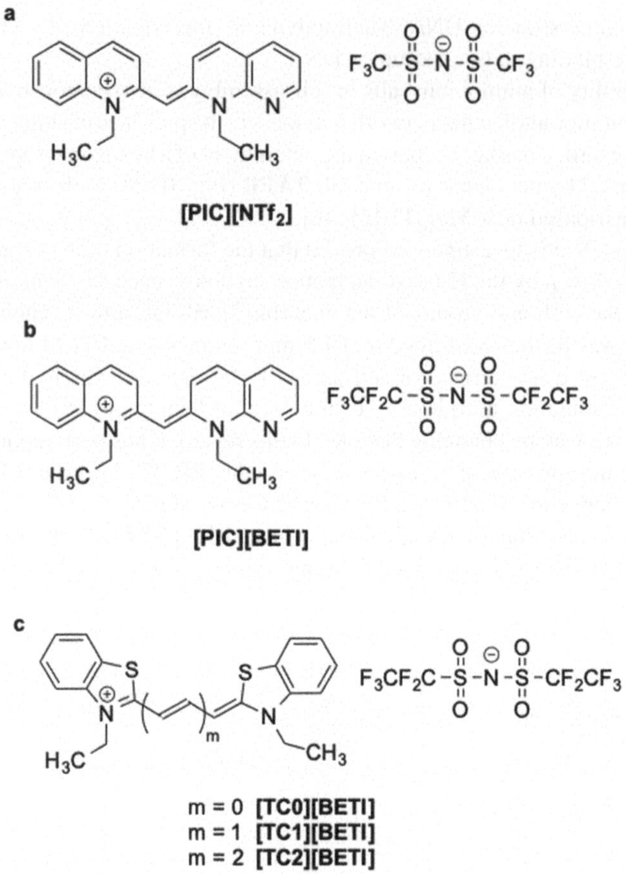

Fig. 10.29 **a** [NTf_2]- and **b** [BETI]-self-assembling cyanine-based salts and **c** thiacarbacyanine-based salts

morphology resulted from a different mode of aggregation, as revealed by UV–vis and fluorescence investigations. In particular, **[PIC][NTf_2]** formed *J*-aggregates and showed aggregation-induced emission (AIE) whereas the other salts form *H*-aggregates.

Notably, a similar effect has been reported by the same group, by variation of the cation, in thiacarbacyanine salts [41] differing for the length of the cation aliphatic spacer.

In this case, aggregation of all these salts yielded rod-like objects, and the length of the spacer mainly affected the size and aspect ratios of the objects, as revealed by SEM and TEM images. In addition, structural variations in the cation also induced different aggregation modes, as exemplified by **[TC1][BETI]** forming *J*-aggregates, whereas *H*-aggregates were observed in the case of **[TC2][BETI]**. Finally, further variation of morphology and emission maxima could be achieved by mixing these

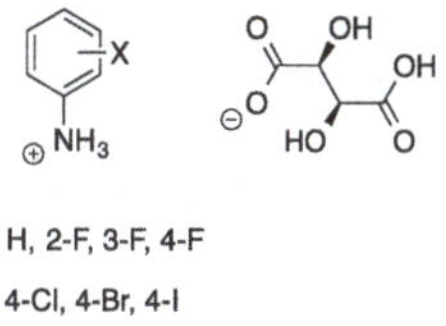

Fig. 10.30 Structures of anilinium tartrate self-assembling salts

salts, evidencing a concentration-dependent behavior. Self-assembly of mixture of salts also exhibited fluorescence resonance energy transfer (FRET).

It is worth noting how subtle structural change in the ammonium cation can significantly alter the self-assembly ability of the relevant salts. This is exemplified by looking at the self-assembling ability of differently haloanilinium tartrate salts (Fig. 10.30) [42].

Among the salts considered, only the 4-F and 4-I substituted salts were able to assemble forming organogels and helical fibers, whereas the other ones only formed crystalline flakes or needle-like structures. Interestingly, SEM and circular dichroism measurements revealed that the helicity of the fibers could be switched upon changing the gelation solvents, from methanol to tetrahydrofuran.

Simple organic salts can be useful in allowing solubilization of ionic surfactants in apolar solvents, while at the same time modulating their aggregation. Along this line, salts like sodium salicylate or benzoate allowed solubilization and aggregation of the quaternary ammonium gemini surfactants 12-s-12, differing for the spacer length (Fig. 10.31) [43].

The spacer length had a clear effect on the aggregation of the surfactants, since, in the presence of sodium salicylate, **12-2-12** and **12-6-12** formed a gel in cyclohexane, whereas this was not the case for **12-10-12**. The structure of the aggregates was probed by SAXS and POM investigations, showing that the **12-2-12** surfactants

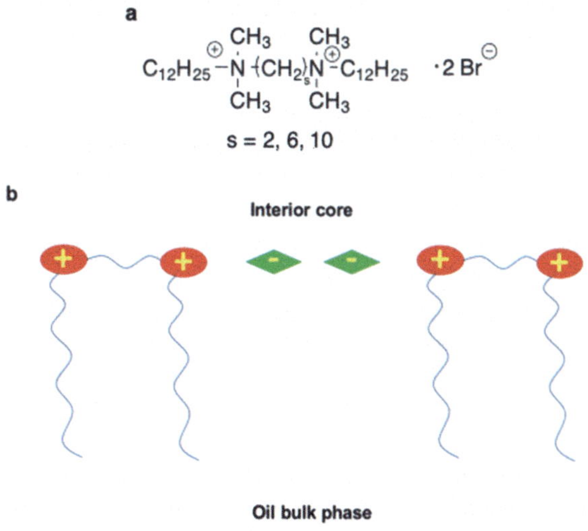

Fig. 10.31 **a** Structure of the **12-s-12** surfactants, **b** pictorial depiction of organic salts effect on the aggregation of 12-s-12 in cyclohexane

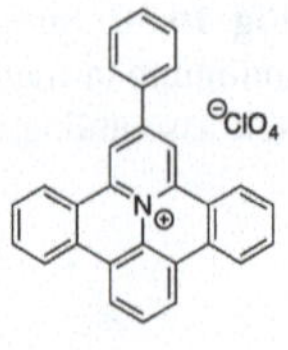

Fig. 10.32 Structure of $\mathbf{PQPClO_4}$

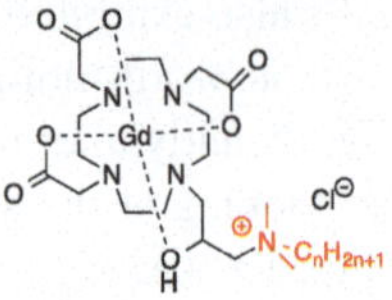

Fig. 10.33 Structure of ammonium-appended Gd(III) complex

formed hexagonal liquid-crystalline structures, whereas a coexistence of different liquid-crystalline textures was found for **12-6-12**. Differently, the longest spacer in **12-10-12** gave rise to reverse vesicles. Finally, the effect of the organic salts could be rationalized by hypothesizing that the organic salts interpose within the cationic heads of the surfactants, within the interior core of the aggregates.

An intriguing property of self-assembling organic salts is that, in principle, they can be strongly responsive to electrochemical stimuli. To this aim, it has been described the possibility of electrochemical control of the aggregation of a polycyclic aromatic salts, $\mathbf{PQPClO_4}$ (Fig. 10.32) [44].

Self-assembly of $\mathbf{PQPClO_4}$ was investigated at the liquid–solid interface between a gold surface and a perchloric acid solution. Varying the potential applied induced a switch from monolayer to bilayer assembly, as evidenced by STM-high resolution images.

The presence of an ammonium tag on a Gd(III) complex with a cyclen-based ligand favored not only solubilization in water but also self-assembly into micelles (Fig. 10.33) [45].

In particular, it has been demonstrated that in water these complexes self-assembled into micelles when the ammonium moiety bears the two longest alkyl chains. The ammonium cationic head also enabled interaction of the micelles with a polyelectrolyte such as sodium polyacrylate (PAA). The resulting hybrid material displayed convenient properties as MRI contrast agent and was applied successfully to in vivo imaging of bladder and liver regions.

Another example of self-assembly originated from a supramolecular complex, the fabrication of a supramolecular amphiphile constituted by an inclusion complex of the ammonium salt bearing a *trans*-diazobenzene (**trans-G2**, Fig. 10.34) group in a pillar[5]arene-based host was described [46].

As evidenced by SEM and TEM images, the complex in water self-organized into nanosheet-like structures, which were broken down by raising the temperature to 55 °C. Interestingly, the nanosheet could be reversibly restored on cooling to 25 °C.

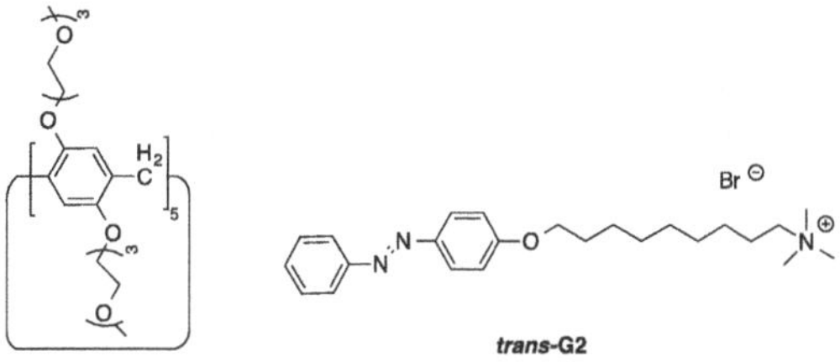

Host

trans-G2

Fig. 10.34 Pillar[5]arene-based host and quaternary ammonium guest

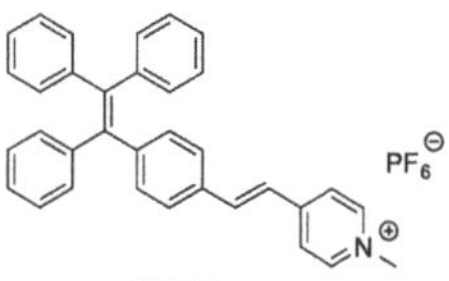

TPE-Py

Fig. 10.35 Structure of tetraphenylene-substituted pyridinium salt

The complex could be destroyed by UV irradiation, due to isomerization of the diazo-moiety into the *cis*-form. Furthermore, adding β-cyclodextrin to the system induced the formation of a ternary complex which self-assembled into vesicles. Notably, these vesicles could be transformed back to the nanosheets upon UV irradiation, which makes this a triply responsive supramolecular system.

Moving to the self-assembly of pyridinium salts, the aggregation of a tetraphenylene-substituted pyridinium salt, **TPE-Py** (Fig. 10.35), was described [47].

This salt had low emission when molecularly dissolved in THF but becomes strongly emissive upon adding increasing amounts of water, showing AIE. In the solid-state, the aggregates were still emissive and existed in two forms, amorphous and crystalline. These forms emitted at different wavelengths and could be reversibly switched by fuming-heating or grinding-heating. TEM and SEM images evidenced that the crystalline form, which displayed a microrod-morphology, could be successfully used as optical wavelength guide, with very low optical loss. Finally, the emissive aggregates of **TPE-Py** enabled specific staining of mitochondria in living cells.

Anionic lipid amphiphiles are well-known aggregation motifs, which can confer self-assembled ability to metal complexes when used as counterions. In this regard, the self-assembly of Ru(II) bipyridine complexes $[Ru(bpy)_3]^{2+}$ bearing the anionic lipids (Fig. 10.36) as counterions has been reported [48].

Depending on the anion used, the complexes obtained self-assembled in ethanol forming ribbons, helical ribbons as well as sheet-like structures, as evidenced by TEM

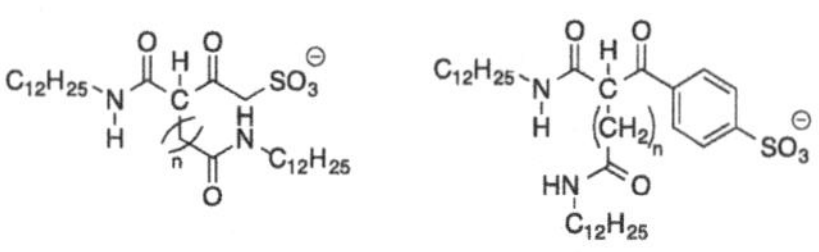

Fig. 10.36 Structure of self-assembly anionic lipids used as counterions of Ru(II) complexes

images and circular dichroism measurements. Spectroscopic investigations revealed also that all the complexes were phosphorescent.

It is worth noting that the aggregation of amphiphiles and surfactants can also be affected by the presence of added salts, either in water or in mixtures with organic solvent, as exemplified by the drop in cmc of sodium dodecyl sulfate (SDS) upon adding alkali metal halides like NaCl, KBr or KCl [49].

Aggregation of IL-based surfactants can also lead to ionic liquid crystals. For instance, self-assembly of *N*-methyl-*N*-hexadecyl pyrrolidinium bromide in water led to formation of lyotropic ionic liquid crystals, when the amount of IL was comprised between 35 and 70% [50]. Due to the absence of aromatic moieties and, consequently, of π-π interactions, this surfactant assembles less tightly than the related imidazolium-based counterparts.

10.4 Conclusions

Analysis of the literature, published in the last decade concerning the self-assembly processes of organic salts in conventional solvents, clearly shows the wide potential of such kinds of systems. The structural tunability featuring organic salts induces a high tunability of the properties of the aggregates systems that, consequently, find application in very different fields.

This represents the main advantage with respect to non-ionic amphiphilic systems, above all for the very simple synthetic procedures that lead to them and for the high level of modularity for the wide range of cations and anions available. Furthermore, with respect to non-ionic species, the self-assembly of organic salts can afford conductive materials and, in some cases, ionic materials show significantly higher thermal stability with respect to the ones formed by neutral amphiphilic species.

In the case of organic salts, data analyzed how the elongation of the alkyl chain, borne on the charged nucleus, can induce significant changes in the aggregates organization and in the nature of the application.

Analogously, the nature of the aggregated systems is highly modulated by the tecton concentration, which affects the nature of the interaction occurring in the supramolecular systems. On the other hand, cation being the same, the different anion nature significantly determines the structure of the aggregates, which may change also in dependence on the counterion location.

A further structural element that can drive the self-assembly process of charged species is the nature of the spacer bearing the cationic heads. This, together with alkyl chain length and anion nature, can induce significant variations in the nature of the solvent in which the aggregation occurs, as well as the mechanism of the self-assembly process.

Interestingly, given the charged nature of the tectons, they easily interact with aromatic supramolecular hosts. In this case, the formation of self-assembled systems is driven by the features of host–guest complexes.

As for the applications, the analyzed literature demonstrated that these aggregated systems are widely used in biomedical field, spanning from drug delivery to antimicrobial and bioimaging agents.

Obviously, the reported analysis is not exhaustive but represents a first approach to demonstrate the strict relationship operating between the properties of self-assembled systems based on organic salts and the structural features of the ionic tectons.

Acknowledgements We thank the University of Palermo for financial support.

References

1. Regnault MV (1839) J Franklin Inst 27:2
2. Salikolimi K, Sudhakar AA, Ishida Y (2020) Functional ionic liquid crystals. Langmuir 36(40):11702–11731
3. Goossens K, Lava K, Bielawski CW, Binnemans K (2016) Ionic liquid crystals: versatile materials. Chem Rev (Washington, DC, US) 116(8):4643–4807
4. Samarkina DA, Gabdrakhmanov DR, Semenov VE, Valeeva FG, Gubaidullina LM, Zakharova LY, Reznik VS, Konovalov AI (2016) Self-assembling catalytic systems based on new amphiphile containing purine fragment, exhibiting substrate specificity in hydrolysis of phosphorus acids esters. Russ J Gen Chem 86(3):656–660
5. Mirgorodskaya AB, Kushnazarova RA, Lukashenko SS, Zakharova LY (2019) Self-assembly of mixed systems based on nonionic and carbamate-bearing cationic surfactants as a tool for fabrication of biocompatible nanocontainers. J Mol Liq 292:111407
6. Hao J, Qin T, Zhang Y, Li Y, Zhang Y (2019) Synthesis, surface properties and antimicrobial performance of novel gemini pyridinium surfactants. Colloids Surf B 181:814–821
7. Cancemi P, Buttacavoli M, D'Anna F, Feo S, Fontana RM, Noto R, Sutera A, Vitale P, Gallo G (2017) The effects of structural changes on the anti-microbial and anti-proliferative activities of diimidazolium salts. New J Chem 41(9):3574–3585
8. Garcia MT, Ribosa I, Perez L, Manresa A, Comelles F (2013) Aggregation behavior and antimicrobial activity of ester-functionalized imidazolium- and pyridinium-based ionic liquids in aqueous solution. Langmuir 29(8):2536–2545
9. Rizzo C, Arrigo R, Dintcheva NT, Gallo G, Giannici F, Noto R, Sutera A, Vitale P, D'Anna F (2017) Supramolecular hydro- and ionogels: a study of their properties and antibacterial activity. Chem Eur J 23(64):16297–16311
10. Lei J, Gao Y, Ma Y, Zhao K, Du F (2019) Improving the emulsion stability by regulation of dilational rheology properties. Colloids Surf A: Physicochem Eng Asp 583:123906
11. Consorti CS, Suarez PAZ, de Souza RF, Burrow RA, Farrar DH, Lough AJ, Loh W, da Silva LHM, Dupont J (2005) Identification of 1,3-dialkylimidazolium salt supramolecular aggregates in solution. J Phys Chem B 109(10):4341–4349
12. D'Anna F, Marullo S, Vitale P, Noto R (2011) The effect of the cation π-surface area on the 3D organization and catalytic ability of imidazolium-based ionic liquids. Eur J Org Chem 28:5681–5689
13. Rizzo C, D'Anna F, Marullo S, Noto R (2014) Task specific dicationic ionic liquids: recyclable reaction media for the mononuclear rearrangement of heterocycles. J Org Chem 79(18):8678–8683
14. Rojas M, Santos JG, Orth E, Figueroa R, Pavez P (2020) The effect of imidazolium salts with amino acids as counterions on the reactivity of 4-nitrophenyl acetate: a kinetic study. J Mol Liq 310: 113206

15. Zhu H, Vijayaraghavan R, MacFarlane DR, Forsyth M (2019) Self-assembled structure and dynamics of imidazolium-based protic salts in water solution. Phys Chem Chem Phys 21(5):2691–2696
16. Jiao J, Han B, Lin M, Cheng N, Yu L, Liu M (2013) Salt-free catanionic surface active ionic liquids 1-alkyl-3-methylimidazolium alkylsulfate: aggregation behavior in aqueous solution. J Colloid Interface Sci 412:24–30
17. Bai G, Wang J, Wang Y, Yan H, Thomas RK (2002) Thermodynamics of hydrophobic interaction of dissymmetric gemini surfactants in aqueous solutions. J Phys Chem B 106(26):6614–6616
18. Kuznetsova DA, Gabdrakhmanov DR, Kuznetsov DM, Lukashenko SS, Sapunova AS, Voloshina AD, Nizameev IR, Kadirov MK, Zakharova LY (2020) Biocompatible supramolecular systems based on novel cationic imidazolium- and urethane-containing amphiphiles: self-assembly and antimicrobial properties. J Mol Liq 319:114094
19. Kuznetsova DA, Gabdrakhmanov DR, Ahtamyanova LR, Lukashenko SS, Kusova AM, Zuev YF, Voloshina AD, Sapunova AS, Kulik NV, Kuznetsov DM, Nizameev IR, Kadirov MK, Zakharova LY (2020) Novel self-assembling systems based on imidazolium amphiphiles with cleavable urethane fragment for construction of soft nanocontainers for biomedicine application. J Mol Liq 298:111961
20. Garcia MT, Ribosa I, Perez L, Manresa A, Comelles F (2017) Micellization and antimicrobial properties of surface-active ionic liquids containing cleavable carbonate linkages. Langmuir 33(26):6511–6520
21. Samarkina DA, Gabdrakhmanov DR, Lukashenko SS, Khamatgalimov AR, Kovalenko VI, Zakharova LY (2017) Cationic amphiphiles bearing imidazole fragment: from aggregation properties to potential in biotechnologies. Colloids Surf A 529:990–997
22. Kuznetsova DA, Gabdrakhmanov DR, Lukashenko SS, Voloshina AD, Sapunova AS, Kulik NV, Nizameev IR, Kadirov MK, Kashapov RR, Zakharova LY (2019) Supramolecular systems based on cationic imidazole-containing amphiphiles bearing hydroxyethyl fragment: aggregation properties and functional activity. J Mol Liq 289:111058
23. Ma H, Ke H, Wang T, Xiao J, Du N, Yu L (2017) Self-assembly of imidazolium-based surface active ionic liquids in aqueous solution: the role of different substituent group on aromatic counterions. J Mol Liq 240:556–563
24. Singh G, Singh G, Kang TS (2016) Micellization behavior of surface active ionic liquids having aromatic counterions in aqueous media. J Phys Chem B 120(6):1092–1105
25. Sun P, Shi L, Lu F, Zheng L (2016) Aggregation behavior of zwitterionic surface active ionic liquids with different counterions, cations, and alkyl chains. RSC Adv 6(33):27370–27377
26. Xu W, Wang T, Cheng N, Hu Q, Bi Y, Gong Y, Yu L (2015) Experimental and DFT studies on the aggregation behavior of imidazolium-based surface-active ionic liquids with aromatic counterions in aqueous solution. Langmuir 31(4):1272–1282
27. Zhang G, Zhu H, Chen M, Pietraszkiewicz M, Pietraszkiewicz O, Li H, Hao J (2018) Aggregation-induced emission of Eu^{III} complexes balanced with bulky and amphiphilic imidazolium cations in ethanol/water binary mixtures. Chem Eur J 24(59):15912–15920
28. Barrio JD, Liu J, Brady RA, Tan CS, Chiodini S, Ricci M, Fernández-Leiro R, Tsai CJ, Vasileiadi P, Di Michele L, Lairez D, Toprakcioglu C, Scherman OA (2019) Emerging two-dimensional crystallization of cucurbit[8]uril complexes: from supramolecular polymers to nanofibers. J Am Chem Soc 141(36):14021–14025
29. D'Anna F, Rizzo C, Vitale P, Marullo S, Ferrante F (2017) Supramolecular complexes formed by dimethoxypillar[5]arenes and imidazolium salts: a joint experimental and computational investigation. New J Chem 41(21):12490–12505
30. Casal-Dujat L, Griffiths PC, Rodríguez-Abreu C, Solans C, Rogers S, Pérez-García L (2013) Nanocarriers from dicationic bis-imidazolium amphiphiles and their interaction with anionic drugs. J Mater Chem B 1(38):4963–4971
31. González-Mendoza L, Escorihuela J, Altava B, Burguete MI, Luis SV (2015) Application of optically active chiral bis(imidazolium) salts as potential receptors of chiral dicarboxylate salts of biological relevance. Org Biomol Chem 13(19):5450–5459

32. D'Anna F, Marullo S, Lazzara G, Vitale P, Noto R (2015) Aggregation processes of perylene bisimide diimidazolium salts. Chem Eur J 21(42):14780–14790
33. Marullo S, Feroci M, Noto R, D'Anna F (2017) Insights into the anion effect on the self assembly of perylene bisimide diimidazolium salts. Dyes Pigments 146:54–65
34. Billeci F, D'Anna F, Marullo S, Noto R (2016) Self-assembly of fluorescent diimidazolium salts: tailor properties of the aggregates changing alkyl chain features. RSC Adv 6(64):59502–59512
35. Billeci F, D'Anna F, Chiarotto I, Feroci M, Marullo S (2017) The anion impact on the self-assembly of naphthalene diimide diimidazolium salts. New J Chem 41(22):13889–13901
36. Rizzo C, Marullo S, Feroci M, Accurso V, D'Anna F (2021) Insights into the effect of the spacer on the properties of imidazolium based AIE luminogens. Dyes Pigments 186:109035
37. Rizzo C, Cancemi P, Mattiello L, Marullo S, D'Anna F (2020) Naphthalimide imidazolium-based supramolecular hydrogels as bioimaging and theranostic soft materials. ACS Appl Mater Interfaces 12(43):48442–48457
38. Lee H-Y, Hashizaki K, Diehn K, Raghavan SR (2013) Reverse self-assembly of lipid onions induced by gadolinium and calcium ions. Soft Matter 9(1):200–207
39. Coste M, Kotras C, Bessin Y, Gervais V, Dellemme D, Leclercq M, Fossépré M, Richeter S, Clément S, Surin M, Ulrich S (2021) 2021, Synthesis, self-assembly, and nucleic acid recognition of an acylhydrazone-conjugated cationic tetraphenylethene ligand. Eur J Org Chem 7:1123–1135
40. Qian Z, Yuan T, Wang Q (2018) Supramolecular hexagonal network based on a tritopic amine hydrochloride and a cucurbit[5]uril analogue. Res Chem Intermed 44(10):6445–6451
41. de Rooy SL, Das S, Li M, El-Zahab B, Jordan A, Lodes R, Weber A, Chandler L, Baker GA, Warner IM (2012) Ionically self-assembled, multi-luminophore one-dimensional micro- and nanoscale aggregates of thiacarbocyanine GUMBOS. J Phys Chem C 116(14):8251–8260
42. Yoshii Y, Hoshino N, Takeda T, Moritomo H, Kawamata J, Nakamura T, Akutagawa T (2014) The formation of organogels and helical nanofibers from simple organic salts. Chem Eur J 20(49):16279–16285
43. Deng S-L, Zhao J-X, Wen Z-X (2018) Self-assembly of quaternary ammonium gemini surfactants in cyclohexane upon reinforcement by simple counterions. RSC Adv 8(34):18880–18888
44. Cui K, Mali KS, Ivasenko O, Wu D, Feng X, Walter M, Müllen K, De Feyter S, Mertens SFL (2014) Squeezing, then stacking: from breathing pores to three-dimensional ionic self-assembly under electrochemical control. Angew Chem Int Ed 53(47):12951–12954
45. Zhu Q, Yang H, Li Y, Tian Y, Wang W, Tang W, Yuan Y, Hu A (2016) HP-DO3A-based amphiphilic MRI contrast agents and relaxation enhancement through their assembly with polyelectrolytes. J Mater Chem B 4(45):7241–7248
46. Zhou Y, Jie K (2017) A triply-responsive supramolecular amphiphile fabricated by a thermal-responsive pillar[5]arene-based host–guest recognition motif. Tetrahedron Lett 58(23):2217–2222
47. Zhao N, Li M, Yan Y, Lam JWY, Zhang YL, Zhao YS, Wong KS, Tang BZ (2013) A tetraphenylethene-substituted pyridinium salt with multiple functionalities: synthesis, stimuli-responsive emission, optical waveguide and specific mitochondrion imaging. J Mater Chem C 1(31):4640–4646
48. Hatakeda M, Toohara S, Nakashima T, Sakurai S, Kuroiwa K (2019) Helical-ribbon and tape formation of lipid packaged $[Ru(bpy)_3]^{2+}$ complexes in organic media. Int J Mol Sci 20(13):3298
49. Niraula TP, Chatterjee SK, Bhattarai A (2018) Micellization of sodium dodecyl sulphate in presence and absence of alkali metal halides at different temperatures in water and methanol-water mixtures. J Mol Liq 250:287–294
50. He JH, Li CH, Qian LA, Liu JH, Zhang QL (2012) Self-assembly behavior of N-methly-N-hexadecyl-pyrrolidinium bromide in water. Adv Mater Res 557–559:611–614

Chapter 11
Computational Modelling of Supramolecular Polymers

Azahara Doncel-Giménez, Joaquín Calbo, Enrique Ortí, and Juan Aragó

Abstract Supramolecular polymers are a type of polymers where the monomeric units are held together via non-covalent interactions and, thus, exhibit a dynamic character similar to that found in biological systems (e.g., proteins). Over the last years, supramolecular polymers have received a deal of attention due to their potential as functional materials. Nevertheless, the characterization of the final supramolecular polymeric architectures at atomistic resolution (i.e., dominant interactions, polymerization mechanism, transfer and amplification of chirality) is quite challenging for the current experimental techniques but accessible by computational modelling. In this chapter, we highlight, by a selection of research examples, that computational chemistry methods (at quantum and classical level) can be seen as powerful techniques to gain insights not only into the structure of the final supramolecular assemblies but also into the nature of the key intermolecular interactions and the mechanism that control the supramolecular polymerization growth as well as the chiral response. The computational tools discussed in this chapter can be also applied to the theoretical characterization of other supramolecular systems; for instance, those mainly governed by intermolecular electrostatic interactions.

11.1 Introduction

Supramolecular polymers (SPs) are dynamic self-assembled structures formed by monomeric units interacting via non-covalent interactions (electrostatic interactions, hydrogen-bonding, π-π stacking, hydrophobic interactions, metal coordination, etc.; see Fig. 11.1) [1, 2]. The main difference with respect to covalent polymers lies in the nature of the interactions between the monomeric units. In supramolecular polymers, non-covalent interactions are much weaker and reversible, and, therefore, the final polymeric structure exhibits a dynamic character. In fact, it is this dynamic character

A. Doncel-Giménez · J. Calbo · E. Ortí · J. Aragó (✉)
Instituto de Ciencia Molecular (ICMol), Universidad de Valencia, c/ Catedrático José Beltrán, 2, 46980 Paterna, Spain
e-mail: juan.arago@uv.es

M. A. Aboudzadeh and A. Frontera (eds.), *Supramolecular Assemblies Based on Electrostatic Interactions*,
https://doi.org/10.1007/978-3-031-00657-9_11

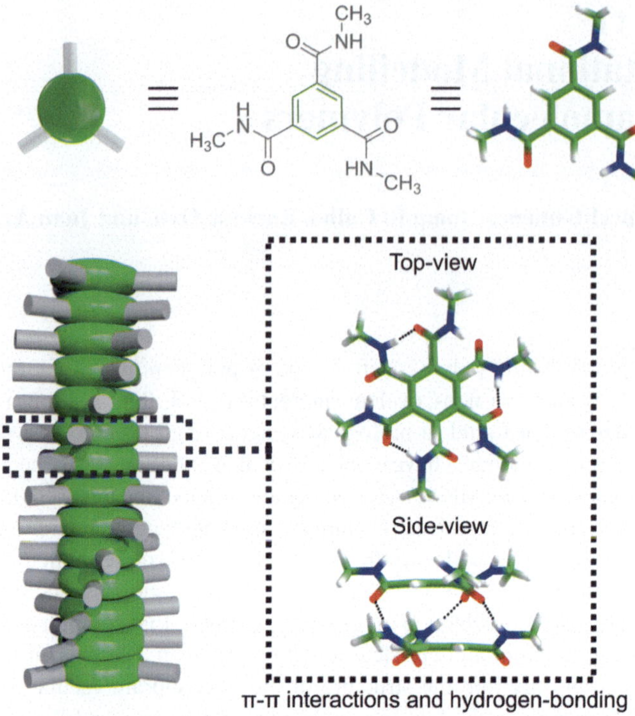

Fig. 11.1 Schematic representation of a helical columnar supramolecular polymer based on a benzene-1,3,5-tricarboxamide (BTA) derivative. The BTA core is shown in green whereas the three arms in grey sticks. In the top and side views of the central BTA dimer, carbon atoms of the BTA are in green, hydrogen in white, nitrogen in blue, and oxygen in red

that endows SPs with interesting bio-inspired properties such as self-healing, shape memory, stimuli-responsiveness, and adaptability. SPs have emerged as promising candidates to build novel functional materials [3, 4]. Furthermore, the electronic and optical properties of SPs can be easily tailored by employing different building monomeric units or by synthetic modification of the peripheral groups attached to a π-conjugated scaffold.

Over the past decade, SPs have been explored for applications ranging from biomaterials [3] to electronics [5]. An interesting example in the context of biomaterials is the ureido-pyrimidone (ur-Py), which is able to assemble into one-dimensional (1D) fibres. Ur-Py moieties dimerize through four hydrogen bonds, and the addition of urea groups allows lateral stacking in fibres joined by a combination of hydrogen bonding, hydrophobic interactions, and π-π stacking. In dilute forms, ur-Py fibres together with cationic residues allow intracellular delivery of siRNA [6]. In the area of molecular electronics, Headler et al. have obtained highly-ordered helical SPs based on heterotriangulenes (Fig. 11.2) capable of transporting energy very efficiently at

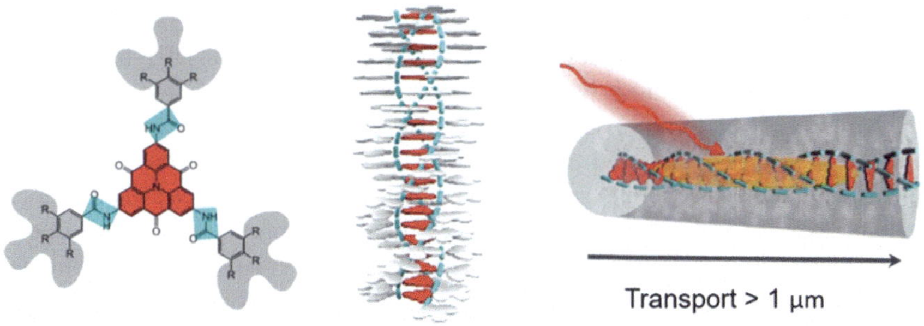

Fig. 11.2 An *N*-heterotriangulene-based supramolecular polymer capable of transporting energy efficiently. Image adapted with permission from [8]. Copyright 2020 American Chemical Society

room temperature, which is of great interest for nanophotonic devices and quantum information technology [7, 8].

The field of SPs began with Prof. Jean-Marie Lehn, one of the pioneers in Supramolecular Chemistry. In 1990, Lehn et al. reported the first supramolecular polymer [9], which was constituted by two different monomers, one composed by two diaminopyridines connected by a tartrate bridge and the other by two uracil derivates. The supramolecular aggregation process occurred through a triple H-bond array between the diaminopyridene and uracil moieties, resulting in a helical supramolecular polymer with liquid crystallinity. Since 1990, impressive experimental efforts have been devoted to the synthesis and characterization of a wide variety of SPs, with special attention over the control of the supramolecular growth [10]. Despite the significant progress in the field, there are still fundamental questions about SPs that are difficult to address experimentally [11]. In particular, it is difficult to unveil, at atomistic resolution, the physical factors that control monomer–monomer and monomer–solvent interactions, and, in turn, the final supramolecular polymeric structure. In this context, Computational Chemistry emerges as a powerful tool to study in-depth and systematically the factors governing the self-assembly process at molecular scale (electrostatic interactions, hydrogen-bonding, π-π stacking, and hydrophobic interactions) as well as the electronic and optical properties of the resulting supramolecular polymers [2, 12–15]. Currently, theoretical calculations are enormously helpful to offer atomistic models for the supramolecular polymeric architectures and interpret the experimental results, which entail great complexity [16–19].

Broadly speaking, the computational chemistry techniques mostly employed for the characterization of SPs can be classified according to their physical roots (quantum or classical). Amongst the quantum mechanical approximations, Density Functional Theory (DFT) is widely used to study the molecular and electronic structure of supramolecular polymer models [12, 20–22]. At this DFT level of theory, simplifications in the structure of the supramolecular systems and the treatment of implicit solvent approaches are usually necessary owing to the high computational cost. Semi-empirical methods, which are able to capture quantum effects and are computationally efficient, represent a feasible alternative to model SPs of large size at a reasonable level of accuracy [14, 23, 24]. A good choice of DFT-based semi-empirical methods are those based on a tight-binding approximation such as the DFTB3-D3 [25] and GFN2-xTB [26] models since they provide accurate results in small aggregates compared to more sophisticated methods. Although quantum mechanical methods generally guarantee great precision for structure and energetics, dynamic aspects of the self-assembly process cannot be properly modelled due to the extremely huge computational cost [27]. Classical computational techniques based on all-atom (AA) or coarse-grained (CG) molecular dynamics (MD) simulations have been employed to successfully investigate the dynamic aspects during the supramolecular polymerization. The classical nature of the MD technique allows to deal with large systems including explicit treatment of the solvent [2, 4, 28, 29].

In this chapter, we offer a personal perspective to highlight how computational chemistry techniques are particularly helpful for a comprehension at molecular scale of the complex supramolecular polymerization process as well as the relevant properties that the final self-assembling architectures can acquire (e.g., chirality). To illustrate this objective, we showcase a variety of research examples in which computational modelling gave important clues to understand the self-assembling process in some SPs.

11.2 Methodology

Over the last years, computational modelling is becoming a transversal discipline in all fields of the natural science. The progress in theoretical and computational principles and algorithms, together with the development of increasingly faster computers, has enabled a huge leap in the precision and speed of current computational methods. In chemistry, the computational techniques, based on both quantum and classical mechanics, have reached such a fantastic ratio between precision and computational time that some theoretical applications inconceivable several years ago can be routinely performed nowadays. Depending on the chosen methodology, small-to-large molecular systems can be theoretically modelled at different levels of accuracy (Fig. 11.3). In this section, we briefly summarize different computational chemistry techniques commonly used to study supramolecular polymers at different scale and precision levels according to the problem pursued.

11.2.1 Density Functional Theory

Density functional theory (DFT) is by far the most used quantum–mechanical theory to study the electronic structure of many-body systems. The success of DFT lies mainly in its favourable trade-off between accuracy and computational cost to

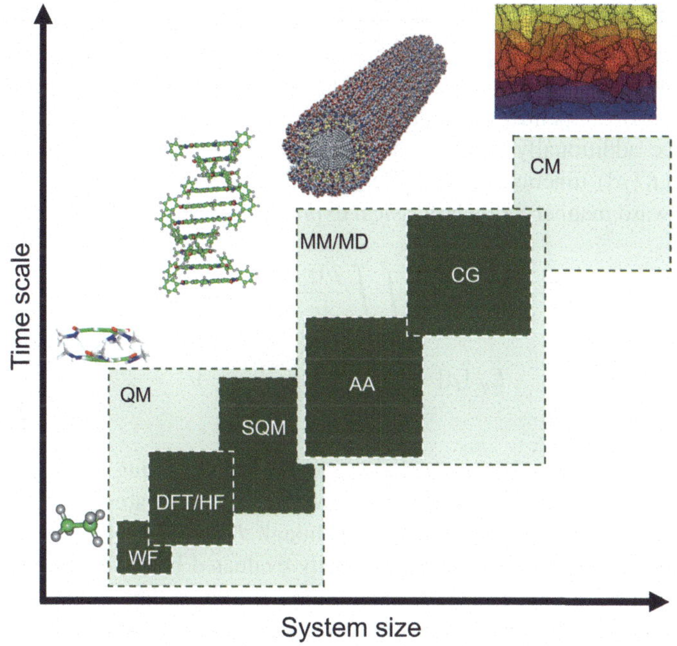

Fig. 11.3 Schematic representation of the available computational approaches mostly used in the calculation of chemical systems according to molecular size and time scale. The computational methods are divided in three groups based on quantum mechanics (QM), classical molecular mechanics and molecular dynamics (MM/MD), and continuum models (CM). Within the QM methods, the wavefunction (WF), density functional theory (DFT), Hartree–Fock (HF), and semi-empirical quantum mechanical (SQM) methods are included. MM/MD approaches encompass all-atom (AA) and coarse-grained (CG) models

predict general properties such as thermochemistry, kinetics parameters, spectroscopic constants, and non-covalent interactions. For the latter, additional corrections are required as will be discussed below.

DFT is based on the Hohenberg and Kohn theorem, which states that the ground-state electronic energy of a system is determined by its electron density ρ; [30] i.e., the ground-state energy is now a functional of the electron density. Conceptually, this theorem possesses an important simplification for the theoretical description of N-electron systems since the ground-state energy depends on the electron density and this density only depends on the spatial coordinates x, y, and z (i.e., only 3 variables). This contrasts significantly to traditional wavefunction methods where the energy expressions are functionals of complex wavefunctions, which depend on the coordinates of all electrons (3 N variables). Unfortunately, the exact mathematical expression for the functional that connects the electron density of a system and its energy is unknown. In the last three decades, the main developments in the DFT field have been focused on the design of density functionals with different degree of accuracy and computational cost.

In practical DFT derivations, the ground-state energy functional can be divided into three components: kinetic energy ($T[\rho]$), electron-nuclei attraction ($E_{\text{ne}}[\rho]$), and electron–electron repulsion ($E_{\text{ee}}[\rho]$). The nuclear–nuclear repulsion is a constant within the Born–Oppenheimer approximation [31]. The electron–electron repulsion term can be additionally divided into two components, the Coulomb ($J[\rho]$) and exchange ($K[\rho]$) functionals. The $J[\rho]$ and $E_{\text{ne}}[\rho]$ terms can be calculated in a straightforward manner by their classical expressions [32],

$$J[\rho] = \frac{1}{2}\int\int \frac{\rho(r)\rho(r')}{|r - r'|} \mathrm{d}r\mathrm{d}r' \tag{11.1}$$

$$E_{ne}[\rho] = -\sum \int \frac{Z_A \rho(r)}{|R_A - r|} \mathrm{d}r \tag{11.2}$$

where the factor of ½ in $J[\rho]$ is needed to avoid double counting by integrating over all space for both electronic coordinates r and r'. Z_{A} and R_{A} are the atomic number and the position of atom A, respectively. Although $J[\rho]$ and $E_{\text{ne}}[\rho]$ depend directly on the density, these terms cannot be directly evaluated because the exact density matrix is not known. To overcome this difficulty, the Kohn–Sham (KS) formulation of DFT is generally employed where the unknown electron density for the N-electron system of interest is assumed to be similar to that obtained for an ideal non-interacting system with the same number of electrons. Then, the approximate density $\rho(r)$ can be written in terms of a set of auxiliary one-electron functions, the KS orbitals $\phi_i(r)$, by the following expression [32]:

$$\rho(r) = \sum_{i=1}^{N} |\phi_i(r)|^2 \tag{11.3}$$

In the KS DFT formulation, the kinetic energy functional $T[\rho]$ is split into two parts. One part can be calculated exactly by applying the classical kinetic operator on KS molecular orbitals, and the other corresponds to the kinetic correlation energy part. The classical kinetic energy functional is thus written as

$$T_S = \sum_{i=1}^{N} \langle\phi_i| - \frac{1}{2}\nabla|\phi_i\rangle \tag{11.4}$$

where the subscript S implies that the kinetic energy is calculated from a Slater determinant. The remaining part of the kinetic energy not contemplated in the previous functional is included in the exchange–correlation functional $E_{\text{xc}}[\rho]$. Therefore, a general DFT energy functional can be expressed as

$$E_{\text{DFT}}[\rho] = T_S + E_{\text{ne}}[\rho] + J[\rho] + E_{\text{xc}}[\rho] \tag{11.5}$$

The main difficulty in density functional theory lies in finding suitable expressions for $E_{xc}[\rho]$, since its exact expression is completely unknown. If $E_{xc}[\rho]$ was exactly derived, the exact ground-state energy for a particular N-electron system could be calculated including electronic correlation effects. Nevertheless, if we assume that the $E_{xc}[\rho]$ expression is available, the problem is similar to that found for the fundamental Hartree–Fock iterative method but using the KS equations; i.e., we need to determine a set of orthogonal KS orbitals that minimize the ground-state electronic energy.

Despite the unknown exact exchange–correlation energy functional, different affordable strategies to provide approximate mathematical expressions for the $E_{xc}[\rho]$ functional have been successfully developed and can be classified in a hierarchical manner regarding their accuracy (DFT rungs, see Fig. 11.4). The simplest method to approximate the exchange–correlation functional is the local-density approximation (LDA) [32]. In the LDA approach, the uniform electron gas model is invoked and the energy functional only depends on the local density at each point. Rising in the Perdew Jacob's ladder of density functional approximations, the generalized gradient approximation (GGA) is found [33]. In GGA density functionals, the exchange–correlation energy not only depends on the value of the electron density at each point but also on its gradient to account for the non-homogeneity of the true electron density. The corrections introduced by the GGA approach improve the prediction of equilibrium geometries, harmonic frequencies, and charge densities compared to the LDA

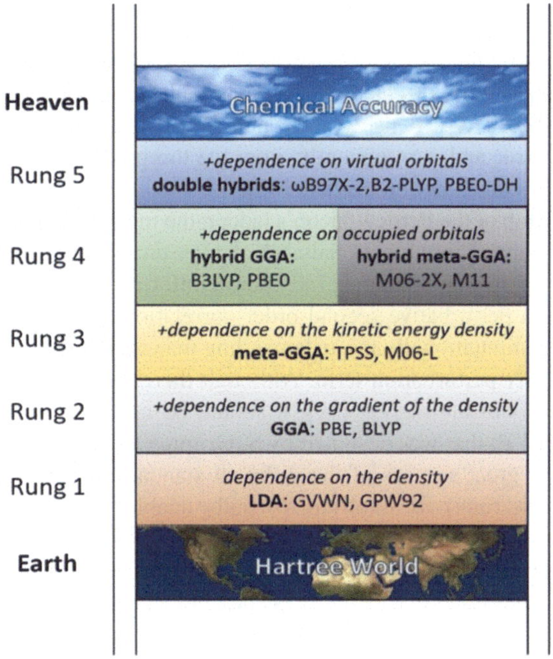

Fig. 11.4 Schematic diagram of Jacob's ladder of density functional approximations suggested by Perdew. Image adapted with permission from [35]. Copyright 2015 Elsevier

approximation. A widely-used GGA functional for multiple applications in chemistry and condensed matter physics is the popular Perdew–Burke–Ernzerhof PBE functional [33]. In a further step, dependencies with the second derivate with respect to the density (kinetic energy density) have also been included in the exchange–correlation functional expression. These functionals are called meta-GGA and a representative example of this type of functional is the TPSS [34].

Another level of significant improvement was achieved by hybrid functionals, which incorporate a part of the exact Hartree–Fock exchange in its mathematical formulation. Hybrid functionals have provided excellent results for the thermochemistry tests and high performances in the calculation of equilibrium geometries and harmonic frequencies and in many difficult chemical applications. These are the reasons why hybrid functionals have become so popular in the quantum chemistry community since early 90s. Amongst hybrid functionals, the B3LYP functional is likely to be the most widely used density functional according to the number of citations.

Even though standard hybrid functionals provide reasonable accuracy for many chemically relevant properties, they still present problems. One of these drawbacks is the "self-interaction" error, which arises because electrons can unphysically interact with themselves [32]. A popular approach to mitigate this flaw is to use the long-range corrected approximation, which partitions the electron–electron interaction ($1/r_{ij}$) into long-range and short-range regions controlled by a fitted parameter (ω). In most of the long-range corrected hybrid density functionals, the short-range region is treated with a semi-local GGA exchange–correlation functional whereas for the long-range region the exact Hartree–Fock exchange functional is employed. Some popular long-range corrected hybrid functionals are the LC-ωPBE [36], CAM-B3LYP [37], and ωB97X-D density functionals [38]. These functionals minimize the self-interaction error and, consequently, improve the prediction of charge-transfer excitation energies and reaction barriers [32].

Finally, at the top of the Perdew Jacob's ladder, the modern double-hybrid density functionals are found. These functionals, which have turned out to be highly accurate for numerous chemical applications, incorporate information about the unoccupied orbitals through a perturbative second-order correlation term. Nevertheless, these functionals are computationally more demanding than their hybrid counterparts due to the second-order perturbation term. Although there are different double-hybrid density functionals in the market [39], it should be noted the first double-hybrid functional (B2PLYP) that was designed by S. Grimme [40].

Despite of the great success of the different standard density functionals, they (irrespective of the rung) are totally or partially unable to capture the long-range electron correlation phenomenon responsible for non-covalent interactions. That flaw compromises the use of standard density functionals in the field of supramolecular chemistry. Nonetheless, a great effort by the theoretical community to develop DFT-based approximations able to accurately describe weakly-interacting supramolecular systems has been carried out in the last fifteen years. Amongst the different approximations to include non-covalent interactions in DFT calculations, it is necessary to stress the DFT symmetry-adapted intermolecular perturbation theory (DFT-SAPT) [41], the Tkatchenko–Scheffler (TS) approximation [42], the Becke–Johnson exchange–hole dipole model (XDM) [43], the elegant and seamless non-local van der Waals correction [44–46], and the Grimme's dispersion correction [47–51]. The latter has become by far the most popular approach to deal with dispersion interactions within the DFT framework due to its excellent computational cost.

The basic idea behind the Grimme's correction is to add an atom-pairwise correction to the conventional KS density functional to compute the total energy as:

$$E_{\text{DFT-D}} = E_{\text{DFT}} + E_{\text{disp}}, \tag{11.6}$$

where E_{DFT} corresponds to the self-consistent KS energy as obtained from the selected density functional, and E_{disp} is the dispersion correction term. Within the Grimme's correction, the E_{disp} term is derived on the basis of the classical definition of long distance forces and can be calculated according to the following pairwise energy expression:

$$E_{\text{disp}} = -\sum_{n=6,8} s_n \sum_{i}^{N_{\text{at}}-1} \sum_{j=i+1}^{N_{\text{at}}} \frac{C_{n,\text{AB}}}{(r_{\text{AB}})^n} f_{\text{damp}}(r_{ij}), \tag{11.7}$$

where s_n are specific parameters defined for each density functional, $C_{n,\text{AB}}$ denotes the nth-order dispersion coefficient for atom pair AB, and $(r_{\text{AB}})^n$ is the intermolecular AB distance (n value can be 6, 8, 10, …) [47]. The $C_{n,\text{AB}}$ dispersion coefficients are geometry-dependent in Grimme's D3 version since they are adjusted on the basis of local geometry (coordination number) around atoms A and B, whereas f_{damp} is a damping function that depends on the distance r_{AB} and can adopt several forms to attenuate the dispersion correction term in the short range [52]. Indexes i and j run over the total number of atoms N_{at}.

The Grimme's dispersion scheme can be further corrected to take into account three-body interactions by using the following expression:

$$E^{\text{ABC}} = \sum_{\text{A}<\text{B}<\text{C}}^{N_{at}} \frac{C_{9,\text{ABC}}(3\cos(\sigma_a)\cos(\sigma_b)\cos(\sigma_c)+1)}{(r_{\text{AB}}r_{\text{BC}}r_{\text{AC}})^3} f_n(r_{\text{AB}}r_{\text{BC}}r_{\text{AC}}) \tag{11.8}$$

$$C_{9,\text{ABC}} \approx -\sqrt{C_{9,\text{AB}}C_{9,\text{BC}}C_{9,\text{AC}}} \tag{11.9}$$

where σ_i are the internal angles of the triangle formed by the interatomic distances between atoms A, B, and C $(r_{\text{AB}}r_{\text{BC}}r_{\text{AC}})$, f_n is a damping function, and $C_{9,\text{ABC}}$ is the coefficient approximated calculated using Eq. 11.9.

The current version of the Grimme's dispersion correction is known as D4 [53], which incorporates an efficient way to compute the dipole–dipole molecular dispersion coefficients. Unlike the D3 version, the $C_{n,\mathrm{AB}}$ coefficients are now obtained on the fly by numerical integration of the atomic polarizabilities. Electronic density information is therefore incorporated via atomic partial charges computed at a semi-empirical quantum mechanical tight-binding level. Although the Grimme's D4 version has proven to be slightly more accurate than its D3 predecessor, the D3 variant is a general and reliable approximation. In fact, D3 is likely to be the most employed dispersion-corrected approach in the DFT framework as it is the approach implemented in general quantum-chemistry software packages.

11.2.2 Semi-empirical Methods

Many problems in chemistry require methods to handle hundreds-to-thousands atoms and, therefore, are beyond the scope of DFT methods. Semi-empirical quantum mechanical (SQM) methods are approaches to bridge the gap between first principles and molecular mechanics/dynamics methods. SQM methods, derived from either Hartree-Fock or Kohn-Sham DFT theory, typically consist of a valence-only, minimal basis, self-consistent field method, where several electronic integrals are substituted by some empirical parameters. This entails that they are faster by at least 2 orders of magnitude compared to ab initio QM methods. The parameters are adjusted to improve the agreement with either experimental data or ab initio calculations [54]. These, however, cause a significant reduction of accuracy and robustness. Nevertheless, if the parameters are carefully adjusted, SQM methods can be well-suited to calculate different properties.

In recent years, a family of extended tight-binding (xTB) SQM methods, derived from DFT, have received significant attention. The xTB methods have been specially developed for the accurate calculation of Geometries, vibrational Frequencies, and Non-covalent interactions (GFN). A great advantage of the GFN-xTB methods is to be parameterized for all elements up to radon [55]. Within the family of the GFN-xTB methods, the GFN2-xTB model (the second version in this family) can successfully identify the relevant structures for non-polar, polar, and strongly hydrogen-bonded systems compared to its predecessors GFN1-xTB and GFN0-xTB. Furthermore, GFN2-xTB does not employ classical FF-type corrections to describe hydrogen bonds, unlike others GFN-xTB methods. The GFN2-xTB basis set consists of a minimal valence basis set incorporating Slater functions (STO-mG) [56]. The energy expression of GFN2 is given by [57]

$$E_{\mathrm{GFN2\text{-}xTB}} = E_{\mathrm{rep}} + E_{\mathrm{disp}} + E_{\mathrm{EHT}} + E_{\mathrm{IES+IXC}} + E_{\mathrm{AES}} + E_{\mathrm{AXC}} + G_{\mathrm{Fermi}} \tag{11.10}$$

Here, the repulsion energy (E_{rep}) is calculated using an atom pairwise potential [58]. The dispersion energy (E_{disp}) is described by a modified D4 dispersion model. The E_{EHT} is the extended-Hückel energy. By invoking the well-known extended Hückel theory (EHT) [59], the description of covalent bonds becomes possible. $E_{\text{IES + IXC}}$ corresponds to energy that comes from isotropic electrostatic (IES) and isotropic exchange–correlation (IXC) contributions. E_{AES} and E_{AXC} are anisotropic electrostatic (AES) and anisotropic XC (AXC) energies, respectively. Due to the anisotropic electrostatic and XC terms, GFN2-xTB does not require any additional hydrogen- or halogen-bond corrections. The description of electrostatic interactions is already improved by E_{AES}. Finally, G_{Fermi} is the entropic contribution of an electronic free energy at finite electronic temperature. This term allows a finite electronic temperature treatment typically at low temperatures, to enable covalent bond dissociations [26].

In addition to the GFN-xTB methods, other SQM methods are widely used. For example, the methods based on the Neglect of Differential Diatomic Overlap (NDDO) integral approximation (MNDO [60], AM1 [61], PM3 [62], …). Other older methods use simpler integral schemes such as Complete Neglect of Differential Overlap (CNDO) and Intermediate Neglect of Differential Overlap (INDO) [63]. The three methods, NDDO, CNDO, and INDO, are included in the class of Zero Differential Overlap (ZDO) methods [64]. Amongst the ZDO methods, the PM7 method should be highlighted [65]. PM7 does not only remove some errors of the NDDO approximation that affect large systems, but also provides a reasonable description of large organic and solid systems. The most important improvement of PM7 is that it is partially able to capture weak non-covalent interactions, allowing its use in supramolecular complexes of increasing size. However, its performance is inferior to that obtained for the GFN2-xTB method [55].

11.2.3 Molecular Mechanics and Molecular Dynamics Simulations

Molecular Mechanics (MM) is a computational technique based on classical type models (known as force fields, FFs) to predict the energy of molecular systems. The main advantage of MM is its simplicity and efficiency; i.e., it is less time-consuming than quantum mechanical methods. Therefore, MM can be used to study large and complex systems as biological molecules or materials composed of thousands of atoms [66]. MM computes the potential energy surface for an arrangement of thousands of atoms using potential functions (i.e., FFs) that are derived for a particular set of systems using classical physics. The potential energy for a particular FF and at a given structure can be written as a sum of terms [67],

$$E_{\text{FF}} = E_{\text{covalent}} + E_{\text{noncovalent}} = (E_{\text{str}} + E_{\text{bend}} + E_{\text{tors}}) + (E_{\text{vdw}} + E_{\text{el}}) \quad (11.11)$$

where E_{str} is the energy function for the stretching of a bond between two atom types A and B, E_{bend} is the energy required for bending an angle formed by three atoms A - B - C, and E_{tors} is the torsional energy for a rotation around a B - C bond in four-atom sequence A - B - C - D. E_{vdw} (van der Waals energy describing the repulsion or attraction between atoms that are not directly bonded) and E_{el} (energy associate to the internal distribution of the electrons, creating positive and negative parts of the molecule) describe the non-bonded atom–atom interactions [66].

The intramolecular terms (E_{str}, E_{bend} and E_{tors}) are given by:

$$E_{\mathrm{covalent}} = \sum_{\mathrm{str}} K_r(r - r_{\mathrm{eq}})^2 + \sum_{\mathrm{bend}} K_\varphi(\varphi - \varphi_{\mathrm{eq}})^2 + \sum_{\mathrm{tors}} K_\phi[1 - \cos(n\phi - \delta)] \tag{11.12}$$

where, K_r K_φ, and K_ϕ are the stretching, bending, and torsion force constants, respectively. $r - r_{\mathrm{eq}}$ and $\varphi - \varphi_{\mathrm{eq}}$ correspond to the distance and angle variation with respect to the equilibrium position. In the last term, n is the multiplicity of the function, ϕ is the dihedral angle between $A - B - C - D$ atoms, and δ is the phase shift. There are cases where K_ϕ, that perfectly reproduces the energy of in-plane angular bending, is not high enough to reproduce the energy of out-of-plane motions. Therefore, an additional out-of-plane term must be included, usually in the form of an improper dihedral term, where the potential energy is harmonic as a function of the out-of-plane angle ϕ. If necessary, an extra term to take into account two (or more) coupled bonded terms (such as angles and bond lengths) may be added.

The non-bonded terms are more computationally intensive. A common choice is to limit interactions to atomic pairwise energies, which are calculated for those atom pairs belonging to different molecules or separated by at least 3 bonds. Non-covalent interactions are usually captured by a Lennard–Jones (LJ) potential and a Coulomb electrostatic term. The non-covalent energy term is generally given by

$$E_{\mathrm{noncovalent}} = E_{\mathrm{vdw}} + E_{\mathrm{el}} = \sum \left\{ \varepsilon_{ij} \left[\left(\frac{r_{\mathrm{min}}}{r_{ij}} \right)^{12} - 2 \left(\frac{r_{\mathrm{min}}}{r_{ij}} \right)^{6} \right] + \frac{q_i q_j}{4\pi \varepsilon_0 r_{ij}} \right\} \tag{11.13}$$

where r_{min} is the distance at which the potential reaches its minimum for atoms i and j, r_{ij} is the distance between both atoms, ε_{ij} is the potential depth and a measure of how strongly the two particle (i and j) attract each other, ε_0 is the dielectric constant, and q is the atomic charge. Both non-bonded terms can be buffered or scaled by a constant factor to account for electronic polarizability and provide better agreement with experimental observations.

Different FFs have been recently used to theoretically explore and understand the supramolecular polymerization process at atomistic scale in C_3 symmetric BTA-based compounds [2], bis-urea derivatives [68], oligophenylenevinylene (MOPV) systems, and rigid anionic azo dyes [69], just to mention a few.

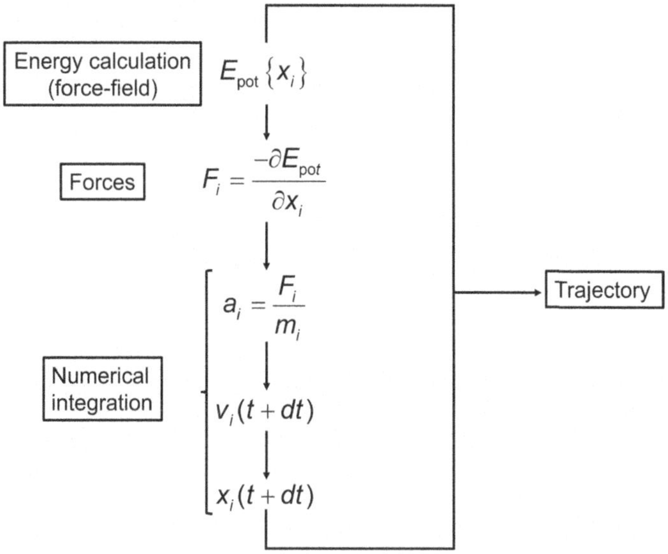

Fig. 11.5 Basic algorithm for molecular dynamics simulations

Apart from the potential energy of a particular system at MM level, the time evolution of a targeted system can be studied by the Molecular Dynamics (MD) technique. This technique generates configurations of the system as a time sequence (trajectory) by numerically solving the Newton's equations of motion, for a system of interacting particles, where forces between the particles and their potential energies are calculated using molecular mechanics force fields [70]. A standard MD algorithm is briefly schematized in Fig. 11.5. Initially, the coordinates, velocities, and the necessary parameters of the force field must be defined. Then, energies and forces that act on every atom are calculated. Note that the simplicity of general FFs assures that energy and force computations are extremely fast even for large systems. Once the forces are obtained, the classical Newton's laws are used to calculate accelerations and update the velocities and atom positions at time $t + dt$. The algorithm is repeated until the selected time is reached [66, 70]. MD simulations is an attractive computational technique since it allows the modelling of systems constituted by thousands of atoms with a reasonable computational cost.

The performance of an MD simulation is highly dependent on the choice of FF, its parametrization, and the accuracy thereof when describing the molecular system of interest.

Early investigations of supramolecular polymerization processes by MD simulations involved clusters of relatively small molecular size. Currently, as computational power increased, several softwares specialized in MD simulations have been developed to take the advantage of the acceleration provided by the use of GPUs. Therefore, all-atom MD simulations of relatively large supramolecular aggregates have become accessible from nano to microsecond time scales. Nevertheless, this time and length scales are not enough to simulate and reveal a complete picture of the self-assembly process. Recently, interesting coarse-grained force fields models have been successfully proposed for a variety of supramolecular polymers [2, 71], allowing MD simulations of larger time and size scales intractable with fully atomistic models [66, 72].

11.2.4 Coarse-Grained Molecular Dynamics

In recent years, there has been a growing interest in the development of coarse-grained (CG) force fields for a variety of supramolecular polymers [2, 66, 72, 73]. CG models simplify the complexity of the systems by grouping atoms together into single particles (beads), thereby reducing the number of interactions to deal with and, consequently, the computational cost (Fig. 11.6). For these models, a balanced dimensionality reduction is crucial; i.e., it is necessary to simplify the system as much as possible for a significant computational speed-up but, at the same time, the CG model must incorporate the sufficient details to preserve the essence of the targeted system. Despite the dimensionality reduction, CG models can be interestingly used to model thermodynamic properties since they can satisfactorily reproduce free energy differences compared to fully atomistic simulations. This arises from an error compensation since individual enthalpy and entropy contributions are generally inaccurate. These limitations are typical for most coarse-grained models [74].

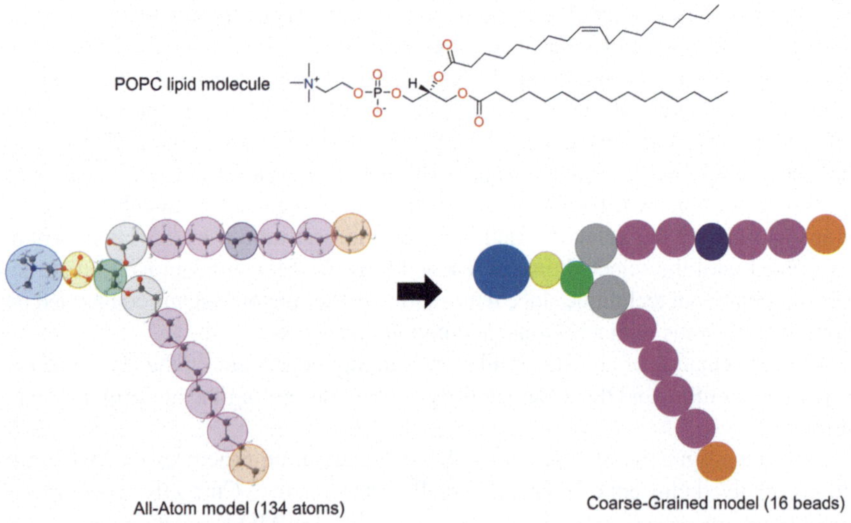

Fig. 11.6 All-atom and coarse-grained representations of palmitoyl oleyl phosphatidylcholine (POPC) lipid molecule. For POPC, a fully-atomistic representation requires to explicitly treat 134 atoms, whereas a CG model only deals with 16 pseudoatoms or beads [75]

The accuracy of a CG model depends mainly on: (i) the model resolution (how many atoms or monomers per bead), (ii) the mapping procedure (how the bead positions are defined as a function of the atoms or monomers coordinates), and (iii) the potential energy function (i.e., the coarse-grained force field) [72]. In general, MD simulations using coarse-grained force field (CG-FF) can be described by formulas similar to an all-atom force field (AA-FF). Nevertheless, CG-FFs use a wide variety of additional expressions that go beyond the classical formula to describe the energy. The final CG-FF model therefore depends not only on the specific CG pattern but also on the method chosen to transfer atomistic to CG potentials.

Some good CG-FF models for the proper description of the supramolecular polymerization process have successfully been designed over the last years [2, 4, 13, 28]. On the one hand, the group of Balasubramanian developed an ad hoc CG model for the BTA monomer to study the aggregation of different BTA-based oligomers in an organic solvent (nonane) [28]. On the other hand, G. Pavan and coworkers employed a modified version of the popular MARTINI coarse-grained force field [73, 76] to simulate BTA-based supramolecular polymers in water [2, 13].

11.2.5 Solvent Models

Solvent effects play a fundamental role in supramolecular chemistry [77] and need to be incorporated for a more realistic theoretical description. In computational molecular sciences, there exist different solvent models [78–80] to capture the fundamental solute–solvent and solvent–solvent interactions. In general, these theoretical solvent models can be classified into two classes: explicit and implicit solvent models [81, 82].

Implicit or continuous solvent models are computationally efficient and can provide a reasonable description of solute–solvent behaviour, but do not account for local fluctuations in solvent density around a solute molecule. In this type of models, the solvent is described as an homogeneous dielectric around the solute, which is encapsulated in a cavity (Fig. 11.7). The charge distribution of the solute polarizes the dielectric continuum, which in turn polarizes the solute charge distribution. In quantum chemistry, where charge distributions are computed from electronic structure calculations (e.g., HF, Post-HF, or DFT calculations), the implicit solvent models describe the solute–solvent interactions as a perturbation to the solute Hamiltonian [78, 83]. Amongst the continuum solvent models, the polarizable continuum model (PCM) [84], the conductor-like screening solvation model (COSMO) [85], and the solvation model based on density (SMD) [86] are the most popular approaches to include solvent effects in quantum-mechanics calculations.

In the context of MM/MD and SQM computations, the common implicit solvent models used are those that combine the Poisson-Boltzmann (PB) or generalized Born (GB) approximations for electrostatic contributions [87] with the solvent-accessible surface area (SASA) method for the non-polar interactions [88]. These implicit solvent models directly evaluate the free energy of solvation (ΔG_{solv}), which can be

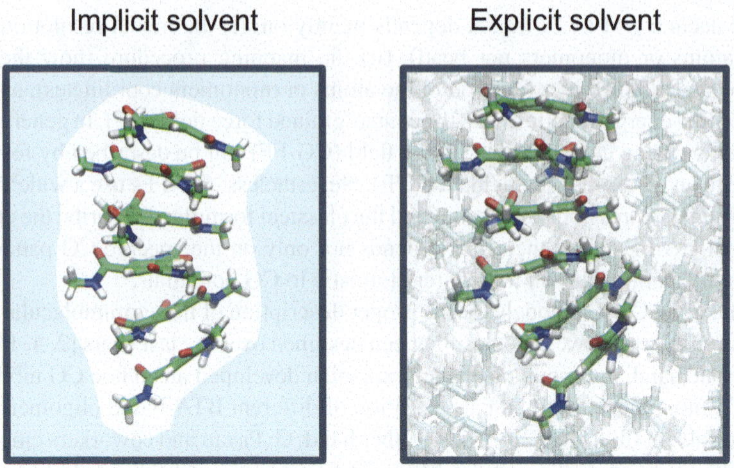

Fig. 11.7 Illustration of implicit (left) and explicit (right) solvation models. In the implicit model, the methylcyclohexane molecules are treated by a continuum (blue background) and the supramolecular aggregate is surrounded by a cavity. In the explicit model, a supramolecular aggregate (shown as a ball-and-stick model) is surrounded by explicit methylcyclohexane molecules (shown as transparent stick models in grey)

written as [89]

$$\Delta G_{\text{solv}} = \Delta G_{\text{polar}} + \Delta G_{\text{apolar}} \tag{11.14}$$

where (ΔG_{solv}) is evaluated as the sum of the electrostatic solvation energy ΔG_{polar} (polar contribution) and the non-polar contribution ΔG_{apolar} between the solute and the continuum solvent.

On the other hand, explicit solvent models are those methods in which the solvent molecules around the solute (Fig. 11.7) are theoretically and explicitly described. In principle, these solvent models offer a more realistic picture by considering the solute–solvent and solvent–solvent interactions. Unfortunately, these approaches are computationally very demanding, especially when the solvent is treated in a quantum mechanical manner. For this reason, explicit solvent models are generally used in classical MD simulations or in hybrid schemes (known as QM/MM), where the solute is treated at quantum mechanical level whereas the solvent is classically described with a force field [81, 82]. The QM/MM hybrid schemes are common in biomolecular simulations [90].

11.2.6 Theoretical Aspects of Supramolecular Polymerization

One of the key aspects in the field of SPs is to understand the polymerization mechanism. Two mechanisms for the supramolecular polymerization growth are generally proposed (cooperative and isodesmic, Fig. 11.8) [10]. In the cooperative mechanism, a few molecules initially come together to form a small aggregate (nucleus) that subsequently grows (elongation) to give rise to the final polymer. The cooperative mechanism is thermodynamically characterized by two equilibrium constants; one associated with the formation of the nucleus (K_{nuc}) and the other one associated with the elongation process (K_{elo}), as illustrated in Fig. 11.8a. In the isodesmic mechanism, the association strength is nearly independent of the oligomer length and, consequently, the polymerization mechanism is only characterized by one equilibrium constant (Fig. 11.8b).

From a theoretical point of view, a property that can be computed to determine the mechanism of supramolecular polymerization is the binding energy per interacting pair ($\Delta E_{\text{bind},n-1}$) as a function of the aggregate size. $\Delta E_{\text{bind},n-1}$ can be calculated according to [91]

$$\Delta E_{\text{bind},n-1} = \frac{E_{\text{stack}} - nE_{\text{monomer}}}{n-1} \tag{11.15}$$

where E_{stack} is the total energy of the stacked aggregate with n monomeric units and E_{monomer} is the energy calculated for the monomer. The more negative $\Delta E_{\text{bind},n-1}$ is found, the more favourable the self-assembly process is. $\Delta E_{\text{bind},n-1}$ can be evaluated in the gas phase or including solvent effects. The evolution of

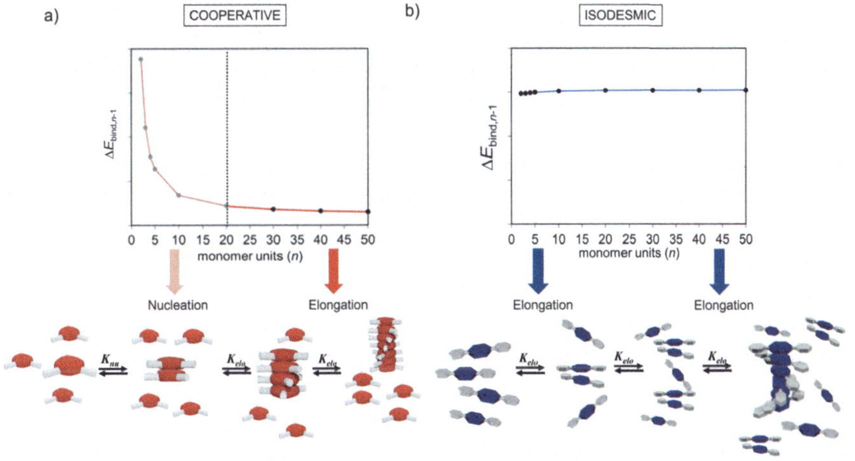

Fig. 11.8 Variation of the binding energy per interacting pair as a function of the oligomer size for a cooperative (**a**) and isodesmic (**b**) supramolecular polymerization mechanism

$\Delta E_{\text{bind},n-1}$ for aggregates of increasing size can be used to suggest the mechanism of the supramolecular polymerization growth. For a cooperative mechanism (Fig. 11.8a), $\Delta E_{\text{bind},n-1}$ decays (i.e., becomes more negative) with the oligomer size until a plateau is reached, where $\Delta E_{\text{bind},n-1}$ barely changes with the number of monomer units. These two behaviours, which depend on the selected system, are associated to the nucleation and elongation processes, respectively. For an isodesmic mechanism, $\Delta E_{\text{bind},n-1}$ hardly changes with the oligomer size as shown in Fig. 11.8b.

A cooperative mechanism is generally more attractive since it provides polymers with a smaller degree of polydispersity. Although it is not easy a priori to predict the polymerization mechanism by chemical design, the scientific community has learned some lessons from different studies [12, 28, 92]. Monomers that can interact by complementary long-range interactions (such as hydrogen bonds, permanent dipole, or electrostatic interactions together with π–π interactions) generally favour the cooperative self-assembly. For example, all atom MD simulations in an explicit apolar solvent revealed that the supramolecular polymerization mechanism of perylene bisimides derivatives took place by a cooperative mechanism owing to the complementary of the dipole–dipole interactions between carbonate groups and the π-π interactions between the conjugated perylene bisimide cores [92]. On the other hand, the self-assembly of hexaperi-hexabenzocoronenes (HBC), driven primarily by π-stacking interactions between the monomeric units, occurred through a isodesmic mechanism [12].

11.3 Structural Models of Supramolecular Polymers and Polymerization Mechanisms

Some of the main goals of computational modelling in the field of SPs are to provide structural models at atomistic resolution, shed light on the nature and strength of the non-covalent interactions governing the supramolecular self-assembly process, and unveil the possible polymerization mechanism. Over the last years, computational chemistry simulations have already contributed to expand the understanding of the supramolecular polymerization process in numerous examples [2, 4, 27, 29, 93, 94]. In this section, we have selected two representative molecular families in the field of supramolecular polymers to highlight the potential of computational modelling in this scientific area. First, we showcase how a quantum–mechanical approximation that combines DFT and semi-empirical calculations has successfully allowed providing structural atomistic models for the different possibilities that N-annulated perylene bisimide (N-PBI) derivatives exhibit during the supramolecular polymerization. In this case, the study of the self-assembly process was carried out considering several pre-stacked monomers in a reasonable initial configuration that is then relaxed to obtain information on the optimal geometry of the supramolecular polymer and the associated energy. Second, we illustrate, by using the archetypical benzene-1,3,5-tricarboxamide (BTA) system, how classical molecular simulations at either atomistic or coarse-grained level can be also applied to get a better comprehension of the molecular aggregation at different spatial and time scales inaccessible for quantum mechanical approaches.

11.3.1 Supramolecular Polymerization at Quantum-Chemical Level

Recently, L. Sánchez and coworkers reported the synthesis of two series of *N*-PBI derivatives (compounds **1** and **2**, Fig. 11.9) [14]. The *N*-PBI compounds are composed of a planar perylene π-conjugated core with two imide groups and are prone to self-assembly due to the optimal π-π interactions between the π-conjugated platforms. Both *N*-PBI derivatives share the same π-conjugated skeleton but hold different peripheral substituents attached to the imide positions (ethyl trialkoxy-benzoate groups for **1** and trialkoxybenzamide fragments for **2**). The presence of benzamide fragments in **2** allowed the formation of intramolecular hydrogen bonds and, thus, the coexistence of two possible molecular conformations for **2** (M and M*, Fig. 11.9). This situation suggested a more complex self-assembly scenario for compound **2** compared to *N*-PBI **1**.

By controlling the experimental conditions, several aggregated species were experimentally detected for **2**. In MCH, *N*-PBI **2** formed a stable (thermodynamic) aggregate labelled as $\mathbf{2J_2}$. Nonetheless, compound **2** was also able to generate metastable kinetic aggregates due to the M* conformer. When the M* monomer was in an ice bath, a kinetic aggregate ($\mathbf{2J_1}$) was formed. This $\mathbf{2J_1}$ aggregate transformed into the $\mathbf{2J_2}$ assemble in a consecutive process. In contrast to **2**, only a

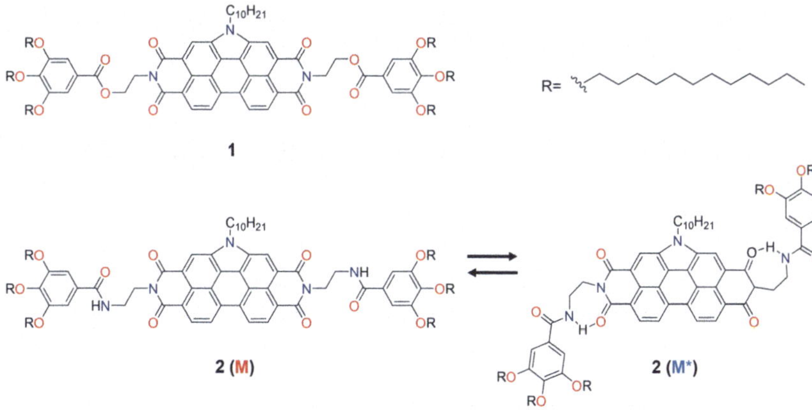

Fig. 11.9 Chemical structures of the *N*-PBI compounds **1** and **2**. For *N*-PBI **2**, two relevant molecular dispositions (M and M*) are shown. Adapted with permission from Ref. [14]. Copyright 2020 WILEY-VCH Verlag GmbH & Co. KGaA, Weinheim

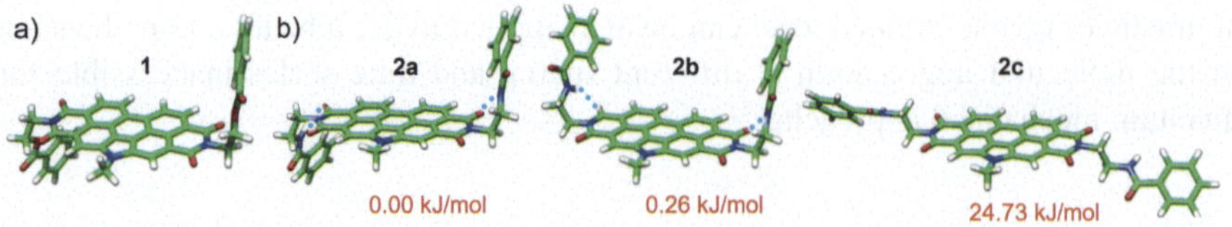

Fig. 11.10 Minimum-energy structures computed at the B3LYP-D3/6-31G** level for the most stable structures of *N*-PBI **1** and **2**. Adapted with permission from Ref. [14]. Copyright 2020 WILEY-VCH Verlag GmbH & Co. KGaA, Weinheim

single supramolecular aggregate was detected for *N*-PBI **1**, irrespective of the experimental conditions. In this context, computational multiscale modelling was key to obtain atomistic models of the supramolecular aggregates of compounds **1** and **2**, and understand the effect of peripheral groups and experimental conditions in the supramolecular self-assembly [14].

Prior to addressing the self-assembly process, computational screenings of the different conformers accessible to compounds **1** and **2** were carried out under solvent conditions by DFT calculations at the B3LYP-D3/6-31G** level. Note that the experimental knowledge about the conformational space of **1** and **2** was scarce. For **1**, only one stable conformer was predicted with the peripheral benzoate groups situated out of the plane of PBI and in an *anti*-disposition (Fig. 11.10a). The planar geometry of the *N*-PBI core revealed that optimal π-π interactions between the *N*-annulated perylene platforms could take place in a supramolecular growth. Apart from the π-stacking between the *N*-PBI units, extra non-covalent interactions between the external and flexible benzoate groups could act contributing to the stabilization of a supramolecular polymer. For **2**, two stable conformers according to the disposition of the peripheral benzamide groups with respect to the *N*-PBI plane (*anti* **2a** and *syn* **2b**) were calculated and found to be practically degenerate in energy. Both **2a** and **2b** conformers gave rise to seven-membered pseudocycles with short intramolecular H-bonds of around 2.1 Å (Fig. 11.10b). Furthermore, the π-conjugated *N*-PBI scaffold was preserved to be planar, which would favour π-π interactions between the *N*-PBI units for an optimal supramolecular growth. Finally, a conformer with extended peripheral arms and without intramolecular H-bonds (**2c**) was also optimized (Fig. 11.10b). However, conformer **2c** turned out to be less stable compared to **2a** by 24.73 kJ/mol.

To theoretically explore the supramolecular polymerization of compound **1** and **2**, oligomers of different size were modelled and optimized using the GFN2-xTB method. Note that the size of the supramolecular oligomers hinders the use of DFT techniques and, therefore, other less-demanding computational approaches needed to be employed. As mentioned above (Sect. 11.2.2), the modern semi-empirical GFN2-xTB method is a cost-effective but reasonably accurate approximation to deal with weakly-bounded supramolecular systems of large molecular size. For compound **1**, four possible assemblies were obtained with high stability, either with a growing stack slipped along the long molecular axis or with a shift along the short axis. For the most stable assembly (Fig. 11.11), the monomer presents an *anti*-disposition of

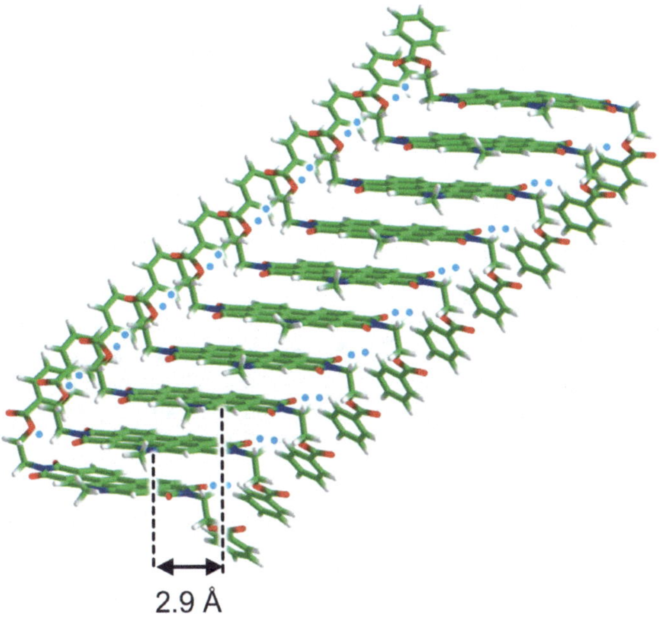

Fig. 11.11 Minimum-energy structure calculated at the GFN2-xTB level for the most stable decamer of *N*-PBI **1**. In the *N*-PBI aggregate representation, carbon atoms are in green, hydrogen in white, nitrogen in dark blue, oxygen in red, and H-bonds are represented by dots in light blue. Adapted with permission from Ref. [14]. Copyright 2020 WILEY-VCH Verlag GmbH & Co. KGaA, Weinheim

the benzoate groups and grows up with a shift along the long *N*-PBI axis. In this aggregate, several stabilizing non-covalent interactions are present; π-π interactions between the *N*-PBI cores and the peripheral benzoate groups, weak CH···O bonds, and dipole–dipole electrostatic forces between vicinal benzoate groups.

Apart from structural aspects, the type of polymerization mechanism was also theoretically investigated. The binding energy per interacting pair $\Delta E_{\text{bind},n-1}$ was calculated for ideal oligomers of increasing size ($n = 1$ to 50 monomers) built up from the central monomer of the most stable decamer of **1** at the GFN2-xTB level. Note that upon increasing the oligomer size, $\Delta E_{\text{bind},n-1}$ presented a rapid decay from dimer to decamer with a significant energy difference (>15 kJ/mol). After $n = 10$, the decrease in $\Delta E_{\text{bind},n-1}$ was attenuated and $\Delta E_{\text{bind},n-1}$ converged to a value of *ca.* –94 kJ/mol when $n = \infty$ (Fig. 11.12). This $\Delta E_{\text{bind},n-1}$ trend suggested a cooperative supramolecular polymerization, in good agreement with the experimental results [14].

As mentioned above, the incorporation of amide groups for compound **2** opened the door to a complex supramolecular pathway achieving J-type aggregates ($\mathbf{2J_1}$ and $\mathbf{2J_2}$) and one H-type aggregate (**2H**) depending on the experimental conditions (solvent and the application of sonication). To gain insights into the self-assembly complexity of compound **2**, different supramolecular oligomer models

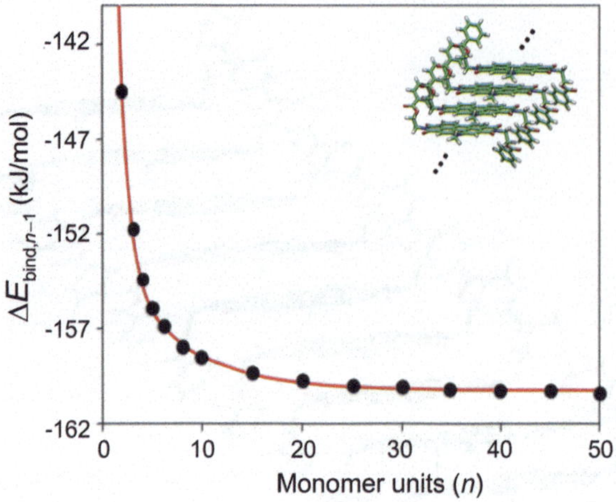

Fig. 11.12 Evolution of the binding energy per interacting pair ($\Delta E_{\text{bind},n-1}$) as a function of the oligomer size n for **1**. Adapted with permission from Ref. [14]. Copyright 2020 WILEY–VCH Verlag GmbH & Co. KGaA, Weinheim

were constructed and fully optimized at the GFN2-xTB level. Note that in this case, where **2** exhibited pathway complexity, the experimental knowledge was important to narrow the assembling possibilities. For instance, variable temperature ^{1}H NMR spectra indicated that the supramolecular polymer **2J$_1$** resulted from the self-assembly of the M* monomer with intramolecular H-bonds. Based on that experimental information, two different decamers were built from the **2a** (*anti*) and **2b** (*syn*) conformers (Fig. 11.10b) owing to the small energy difference between them (0.26 kJ/mol). The most stable decamer model for the aggregate **2J$_1$** (Fig. 11.13) consisted of a columnar arrangement, in which the monomeric units were π-stacked (~3.3 Å between the *N*-PBI cores) and shifted along the long *N*-PBI axis by *ca.* 2.8 Å. The benzamide groups were oriented in *anti* with significant intermolecular contacts (benzene–benzene distance of 4.0 Å). The intramolecular H-bonds were preserved (2.2 Å) in the final supramolecular assemble.

To propose an atomistic model for the aggregate **2J$_2$**, an initial decamer based on the previous **2J$_1$** model but manually distorted to promote intermolecular H-bonds instead of intramolecular ones was constructed and fully optimized at GFN2-xTB (Fig. 11.14). The GFN2-xTB method predicted a well-formed π-stack with relatively strong intermolecular H-bonding interactions, where the monomeric units were slightly rotated. The stacked moieties were alternatively shifted by 3.5 and 4.5 Å along the long molecular axis, which implied a significant increase with respect to **2J$_1$**. The *N*-PBI cores established an effective π-stacking with distances in the 3.2–3.5 Å range. Finally, strong intermolecular H-bonds (1.8–2.1 Å) between the amide groups were computed.

Finally, the self-assembly of the aggregate **2H** was modelled by considering highly directional and very strong H-bond intermolecular interactions, and with minimal

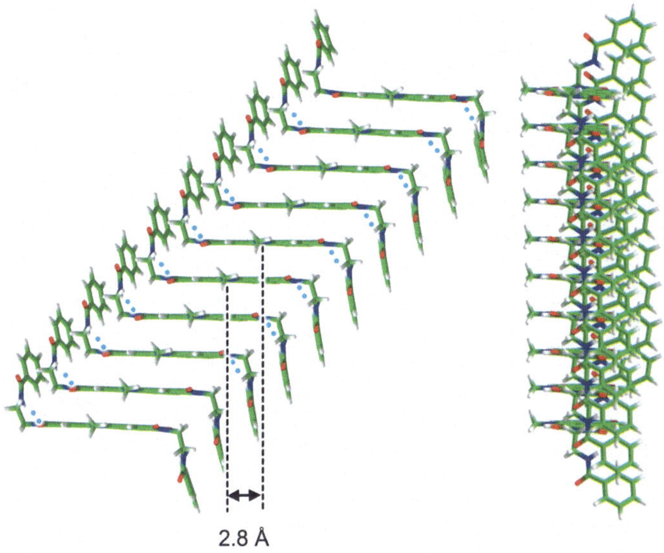

Fig. 11.13 Minimum-energy geometry for aggregate $\mathbf{2J_1}$ calculated at GFN2-xTB. In the *N*-PBI aggregate representation, carbon atoms are in green, hydrogen in white, nitrogen in dark blue, oxygen in red, and H-bonds are represented by dots in light blue. Adapted with permission from Ref. [14]. Copyright 2020 WILEY–VCH Verlag GmbH & Co. KGaA, Weinheim

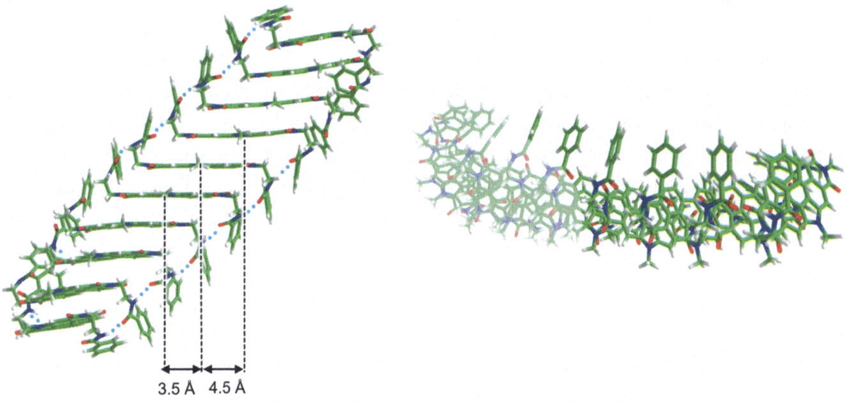

Fig. 11.14 Minimum-energy geometry for aggregate $\mathbf{2J_2}$ calculated at GFN2-xTB. In the *N*-PBI aggregate representation, carbon atoms are in green, hydrogen in white, nitrogen in dark blue, oxygen in red, and H-bonds are represented by dots in light blue. Adapted with permission from Ref. [14]. Copyright 2020 WILEY–VCH Verlag GmbH & Co. KGaA, Weinheim

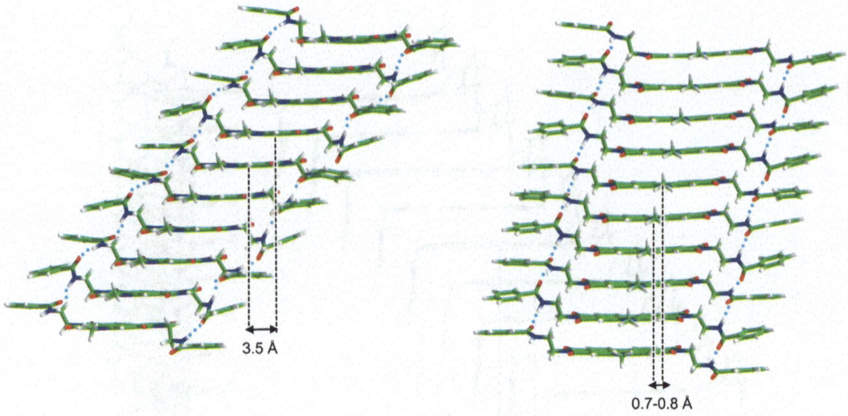

Fig. 11.15 Minimum-energy geometry for aggregate **2H** calculated at GFN2-xTB. In the *N*-PBI aggregate representation, carbon atoms are in green, hydrogen in white, nitrogen in dark blue, oxygen in red, and H-bonds are represented by dots in light blue. Adapted with permission from Ref. [14]. Copyright 2020 WILEY–VCH Verlag GmbH & Co. KGaA, Weinheim

displacement upon growth along the long molecular axis of the *N*-PBI as shown in Fig. 11.15. The supramolecular polymer grew in a stair-like arrangement, shifted along the short molecular axis by 3.5 Å and with a small displacement along the long axis (less than 0.8 Å). In this H-type aggregate, the monomeric units interacted with short π-stacking interactions of *ca.* 3.3 Å and short intermolecular amide H-bonds with distances of 1.7–1.8 Å.

The above example has clearly shown that a quantum-chemical approximation based on DFT and semi-empirical (GFN2-xTB) calculations can be extremely helpful to provide atomistic models that account for the different supramolecular polymers formed by a family of *N*-annulated perylene bisimides (*N*-PBIs). This computational approximation is especially relevant when pathway complexity emerges, and the typical puzzling experimental findings are difficult to be interpreted without further theoretical assistance.

11.3.2 Supramolecular Polymerization by Classical Simulations

Although quantum chemical calculations provide important insights into the complex supramolecular polymerization process, these are restricted to small-size systems and static properties. To address dynamical aspects of polymerization processes, it is generally necessary to resort to classical molecular modelling techniques due to their mathematically simplicity and computational efficiency. Amongst the different classical approximations, coarse-grained models have gained interest in the field

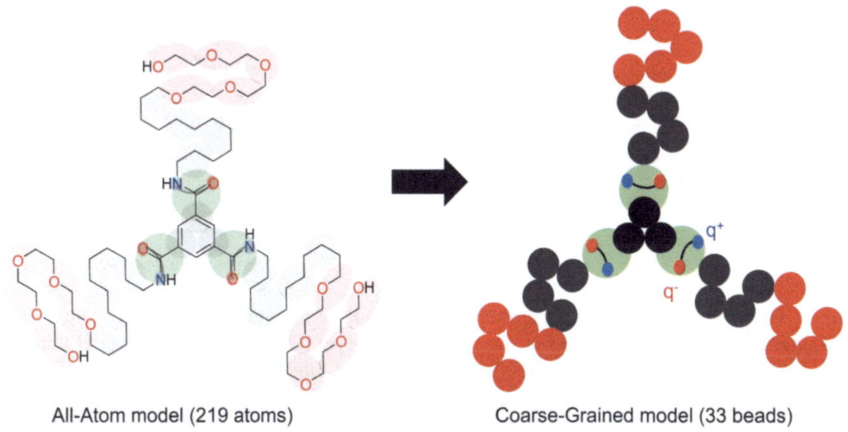

Fig. 11.16 Chemical structure of the BTA used to derive the MARTINI-based coarse-grained model (left). The simplified system based on beads (coloured spheres) is shown in the right [13]

of supramolecular polymers since they enable to monitor the time evolution of supramolecular polymerization at submolecular resolution.

An interesting CG model was that proposed by G. Pavan et al. and specially designed for a polar BTA in water. The model, which was based on the popular MARTINI force field [13], allowed studying the supramolecular polymerization of BTAs in water over long-time scales, but maintaining a remarkable consistency with all-atom force field models. As briefly explained above, the basic idea behind a CG approximation in classical molecular simulations is to transform a fully atomistic system in a fictitious system with a reduced dimensionality but retaining the essence of the realistic system. Then, the fictitious system is theoretically described by a classical force field with a smaller number of interactions to deal with. Figure 11.16 displays the targeted BTA employed for the development of the CG model, for which a bead was utilized for every 3 heavy atoms. The directional nature of the H-bonds between the amide groups in BTAs posed a relevant challenge for the CG MARTINI force field, where all interactions were typically represented by a non-directional LJ potential. For this case, an explicit treatment of the H-bonds was included by a permanent dipole in the amide beads mimicking the rigid orientation of the amides and the directionality of the H-bonds in BTA-based supramolecular polymers.

The CG model was positively validated by reproducing all the key features of the BTA-based supramolecular polymer: (a) the behaviour of the monomer in solution, (b) the strength of monomer–monomer interaction, and (c) the cooperativity of the self-assembly. Regarding the monomer behaviour, the CG force field was able to reproduce the folding of the BTA molecules in water (especially the side chains) due to hydrophobic effects showing a decrease of the solvent-accessible surface area to the BTA monomer. The intermolecular forces between BTA monomers were satisfactorily described by giving a free-energy profile for the dimerization similar to that obtained by fully-atomistic models via metadynamics simulations. With respect to the

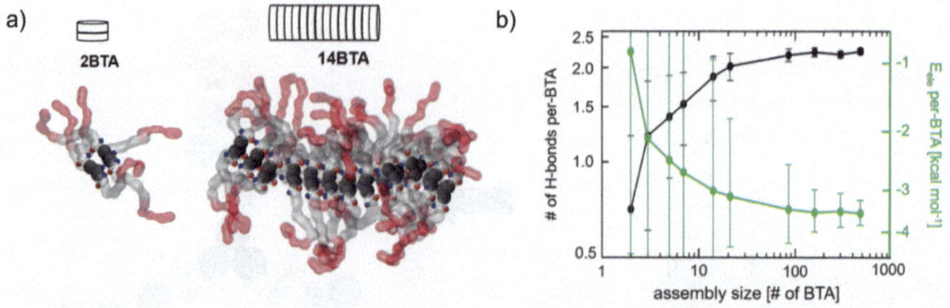

Fig. 11.17 **a)** Two representative snapshots of BTA stacks of different size extracted from the CG MD simulation. **b)** Average number of equivalent H-bonds per BTA (black) and average H-bonding energy per BTA (green) as a function of the assembly size. Image adapted with permission from [13]. Copyright 2017 American Chemical Society

cooperativity, the CG simulations for oligomers of different size revealed an increase of the H-bonds per BTA from ~0.7 (for dimers) to ~2.2 for the largest oligomer (Fig. 11.17b). Consequently, the average H-bonding energy per BTA underwent a rapid decay for small BTA clusters and an attenuated region for larger oligomers (Fig. 11.17b) in line with a cooperative mechanism.

After the complete validation of the CG model, MD simulations with the CG force field were carried out for 160 dissolved (disassembled) monomers (Fig. 11.18a). The simulations revealed that BTA monomers self-assembled quickly (the first ∼0–20 ns) in small-size clusters which evolve slowly to give supramolecular fibres in the time scale of μs (Fig. 11.18a). This trend was also supported by the abrupt decrease in the number of BTA clusters present in the solution along the simulation (Fig. 11.18b, line dotted red). The degree of order of the BTA aggregates along the MD run was also investigated (Fig. 11.18b, solid red line) by means of the parameter Φ (an average coordination number between BTA cores ranging from 0 for totally dissolved BTA to a maximum of 2 for an ideal highly-ordered polymer). The evolution of the parameter Φ indicated that, up to ∼30 ns, the BTA aggregates

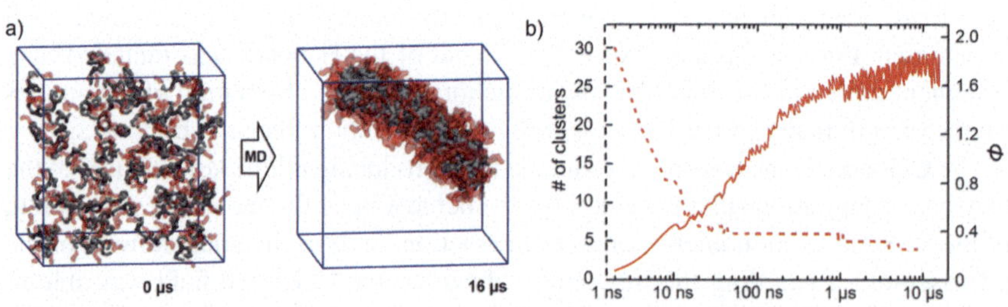

Fig. 11.18 **a)** Initial and equilibrated (final) snapshots of the CG-based MD simulation for the BTA self-assembly. **b)** Number of BTA clusters and order parameter Φ (core−core coordination) as a function of the simulation time. Image adapted with permission from [13]. Copyright 2017 American Chemical Society

were mainly disordered ($\Phi < 1$). Only at long-time scales, the index Φ reached values close to 2 ($\sim$1.8 after 16 μs) generating a long and ordered supramolecular fibre of *ca.* 85 monomers (Fig. 11.18a). Therefore, two phases with different time scales during the polymerization process were clearly identified; the BTA monomers quickly self-assembled forming disordered aggregates that, upon reaching a certain size ($\sim$ 20–30 BTAs), then underwent a structural reorganization and evolved into more ordered BTA oligomers. Finally, the growth into the supramolecular fibres proceeded through the fusion of these ordered oligomers.

This subsection has showed that classical coarse-grained force field models can be very attractive to unravel the dynamical aspects of the supramolecular polymerization process. The CG technique has allowed for a water-soluble BTA to identify and monitor the nucleation and elongation steps during the polymerization at submolecular resolution, which is a detailed knowledge unattainable from current experimental techniques.

11.4 Transfer and Amplification of Chirality

Nature has developed a multitude of hierarchical biological components with fascinating and complex superstructures (e.g., proteins, sugars, DNA, etc.) to perform information storage and transport, pharmaceutical and toxicological activities, and localized chemical transformations [95]. Amongst these macromolecular components, chirality is an essential and basic characteristic [96]. Supramolecular polymers can give rise to helical columnar aggregates that exhibit chirality [93]. In fact, these polymers turn out to be excellent models for understanding fundamental issues such as transfer and amplification of chirality. Computational modelling has already helped to get a deeper understanding of the chiral transfer and amplification phenomena at molecular level. In this section, we discuss how quantum-chemical calculations (at the DFT and GFN2-xTB level) have successfully assisted to rationalize the processes related to the transfer and amplification of chirality in supramolecular polymers based on tricarboxamide derivatives.

Recently, L. Sánchez and coworkers reported the synthesis of a complete family of chiral and achiral oligo(phenylene ethynylene) tricarboxamide (OPE-TA) derivatives (compounds **3–5**, Fig. 11.19a) [19, 97] to understand in more detail the chiral transfer and amplification phenomena in the OPE-TA-based supramolecular polymers compared to other well-known SPs (e.g., BTA-based polymers) [19]. In particular, the research was focused on the effect that the number of stereogenic centres in the monomer exerted on the final helical (chiral) supramolecular polymers.

Experimentally, it was demonstrated by circular dichroism (CD) that the presence of stereogenic centres of absolute *S* configuration in the side chains of OPE-TAs, irrespective of the number of stereogenic centres (**3a–c**), gave rise to supramolecular columnar *P*-type helices. On the contrary, those OPE-TAs with chiral chains with absolute *R* configurations (**4a–c**) resulted in *M*-type helices (Fig. 11.20) [19, 97, 98].

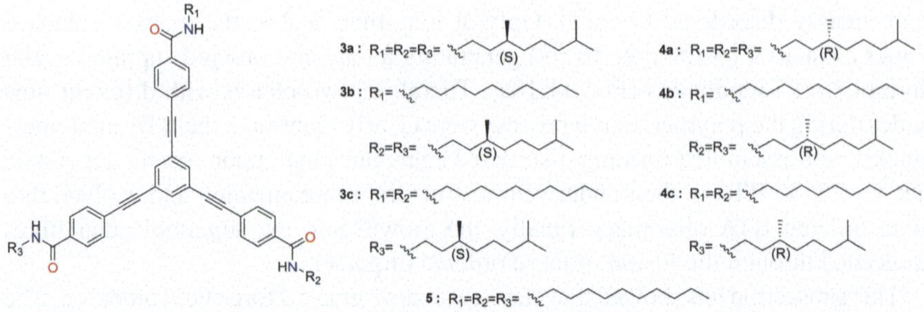

Fig. 11.19 Chemical structure of the symmetrically and asymmetrically substituted OPE-TA **3**, **4**, and **5**. Image adapted with permission from [19]. Copyright 2019 American Chemical Society

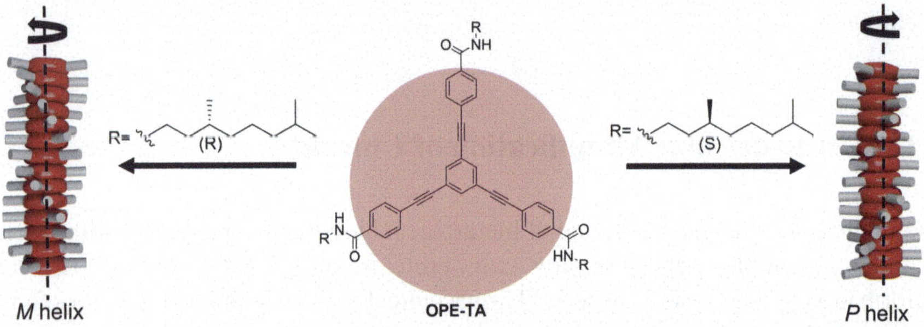

Fig. 11.20 Schematic illustration of the relationship between the point chirality embedded in the side chains of **3a** and **4a** and the helical outcome upon supramolecular polymerization

This chiral transfer phenomenon was further studied and rationalized by computational modelling with the GFN2-xTB method. In particular, the tendency of these OPE-TA derivatives to polymerize into *M*-type helical assemblies was analysed. To do so, supramolecular helical aggregates of 20 monomers of *M*- and *P*-type for compounds **4a–c** were built and fully optimized. Compound **4a**, which only had chiral side chains, favoured the formation of an *M*-type helix by 6.3 kJ/mol compared to the *P*-type helix (the energy difference was calculated per monomeric unit). In contrast to compound **4a**, the presence of achiral and chiral chains in **4b** and **4c** gave rise to two different supramolecular arrangements, eclipsed and staggered, according to the distribution of chains along the helical structure. In an eclipsed-type arrangement, the achiral and chiral chains are located in different arms along the helical structure (Fig. 11.21a, c), whereas in the staggered-type arrangement achiral and chiral are intercalated along the helical axis (Fig. 11.21b, d).GFN2-xTB predicted that the eclipsed arrangement is more favourable than staggered dispositions for aggregates of **4b.** Furthermore, the eclipsed *M*-type structure was calculated more stable than its analogous *P*-type helix by an energy difference per monomeric unit of 7.0 kJ/mol. However, the staggered helical arrangement of **4c** is more stable than the

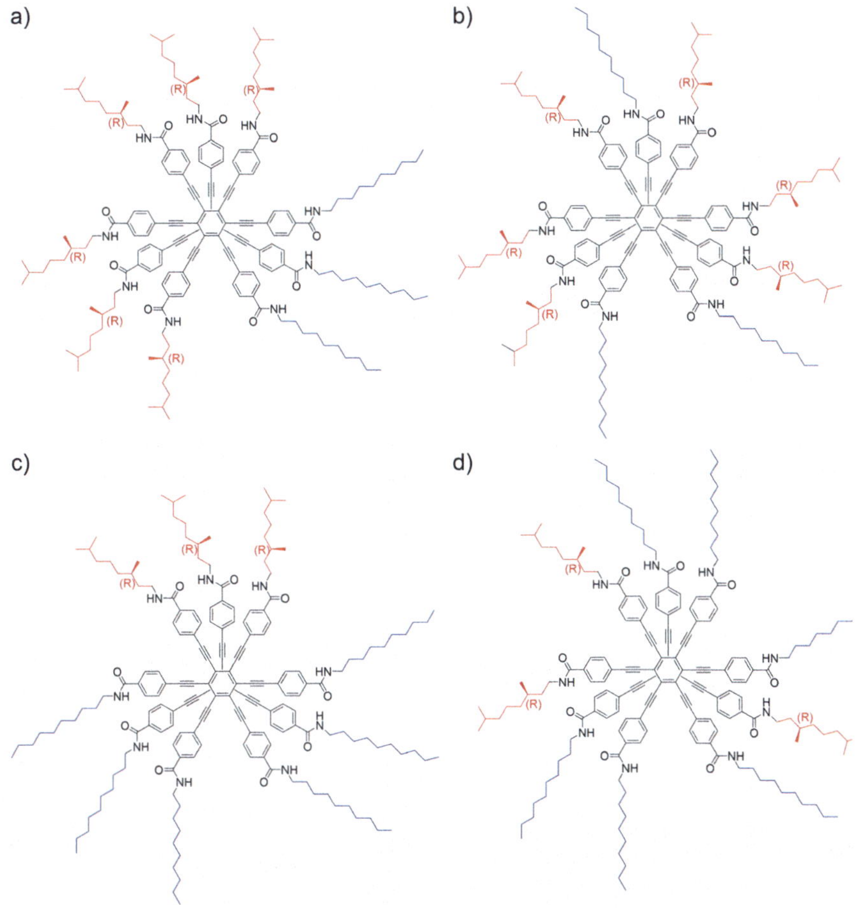

Fig. 11.21 Eclipsed and staggered disposition, respectively, of chiral chains along the helical stack of OPE-TAs **4b** (**a** and **b**) and **4c** (**c** and **d**). Image adapted with permission from [19]. Copyright 2019 American Chemical Society

eclipsed structure. Within the staggered disposition, the *M*-type helical supramolecular polymer was found to be 5.0 kJ/mol per monomeric unit more stable than *P*-type helix.

GFN2-xTB calculations were therefore able to predict and support that, regardless of the number of chiral aliphatic chains, OPE-TAs **4a–c** self-assembled in preferential *M*-type helical stacks whereas derivatives **3a–c** polymerized in *P*-type helical arrangements in line with the experimental findings. In other words, GFN2-xTB confirmed that only one stereogenic centre in the side chains of OPE-TAs is sufficient to effectively transfer chirality to the entire supramolecular polymer.

Regarding the chiral amplification phenomenon, sergeants and soldiers (SaS) experiments indicated that the decrease in the number of stereogenic centres in the side chains was accompanied by a drastic decrease in the ability to amplify chirality at supramolecular level. The influence of the number of stereogenic centres on the amplification of chirality was also analysed by performing majority rules (MR) experiments. These MR experiments revealed that cutting down the chiral information per monomeric unit was accompanied by a reduced ability of the system to achieve a complete amplified state. For mixtures of **3a** + **4a** only a 48% enantiomeric excess (ee) was required to obtain a homochiral mixture, whereas higher values of ee were necessary (57% for **3b** + **4b** and >80% for **3c** + **4c**) to generate a clear chiral response. These experimental data were in line with those drawn from the SaS findings, but were in contrast to those reported for BTAs, for which reducing the number of stereogenic centres per monomeric unit caused a more efficient chiral amplification.

To shed light on the controversial trends in OPE-TAs and BTAs related to the chiral amplification process, minimum-energy geometries at GFN2-xTB for oligomers of **4a–c** and BTAs with the same side chains were carefully analysed. The most interesting intermolecular parameter to explain the different trends was the rotation angle between adjacent monomer units along the growth axis (θ, Fig. 11.22a). For the oligomers of **4a–c**, the rotation angle θ systematically decreased upon reducing the number of chiral chains, going from an average value of 18.51° in **4a** to 18.38° in **4b** and 18.31° **4c**. Larger values of the stacking rotation would imply smaller steric hindrance between neighbouring chains and, consequently, a smaller energetic cost for the incorporation in the helical stack of an extra monomer with the "wrong" chiral configuration to keep a homochiral polymer. For analogous BTA-based supramolecular oligomers, GFN2-xTB calculations showed that the number of chiral side chains barely had an impact on the stacking rotation with a value

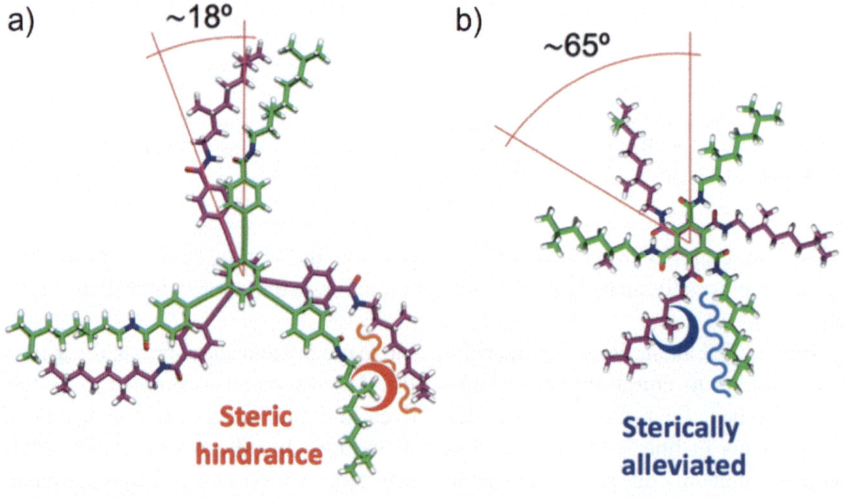

Fig. 11.22 Schematic illustration of the steric hindrance effect exerted by the chiral, aliphatic chains of vicinal units for the OPE-TA **4a** (**a**) and the analogous BTA (**b**). Image adapted with permission from [19]. Copyright 2019 American Chemical Society

of *ca.* 65° (Fig. 11.22b). With this helical organization in BTAs, the chiral side chains of neighbouring molecular units do not interact as they are spatially distant. As θ remains practically constant in BTAs, increasing the number of chiral centres certainly increased the steric volume in the column. Hence, the energy penalty for introducing an additional chiral monomer in an unpreferred helical stack would rise with the number of the chiral side chains in the monomeric unit, which is opposed to the behaviour observed and predicted for OPE-TAs.

In this section, we have shown that the GFN2-xTB method can be seen as an affordable computational technique to understand fundamental questions related to phenomena of transfer and amplification of chirality in supramolecular polymers. First, GFN2-xTB allowed to explain that only one stereogenic centre in the side chains of OPE-TAs is sufficient for an effective transfer of chirality at the supramolecular level. Second, the structural models proposed for the OPE-TA and BTA-based supramolecular polymers have helped to rationalize the controversial experimental findings about the chiral amplification process according to the rotation angle of the columnar helices.

11.5 Solvent Effects

Solvent effects are extremely important in supramolecular chemistry. The supramolecular polymerization process starts from a good solvent in which the molecular units are totally dissolved. Subsequently, self-assembly is induced by adding a poor solvent (i.e., increasing the number of unfavourable solute–solvent interactions) [99]. The selected solvent conditions (ratio between good and poor solvents) can therefore determine the stability of the aggregated structures [100] and even the morphology [77]. Therefore, computational modelling including solvent effects is generally needed to provide a more complete comprehension of the supramolecular polymerization process. In this section, we showcase some recent examples where different solvent models (implicit and explicit) were employed in the field of SPs. First, we show how the semi-empirical GFN2-xTB method in combination with the implicit solvent GBSA model was able to successfully provide different structural atomistic models of two *N*-annulated PBI derivates that exhibit pathway complexity. Second, we illustrate that AA-MD simulation techniques, including the solvent explicitly, can assist to monitor the self-assembly of small-size BTA oligomers in *n*-nonane. Finally, we also discuss the supramolecular polymerization process of polar BTAs by means of either fully atomistic or coarse-grained force fields for the solute (BTAs) and implicit models for the solvent (water).

Recently, a combined experimental-theoretical study of the self-assembly properties of a new *N*-PBI compound (**6**, Fig. 11.23) compared to its *N*-PBI **1** analogue (previously discussed, Fig. 11.9) was reported [101]. Interestingly and in contrast to **1**, a solvent-driven supramolecular polymerization for compound **6** took place; i.e., aggregates with different optical and chiroptical properties were experimentally detected in methylcyclohexane (MCH) and toluene. To provide atomistic models for

6

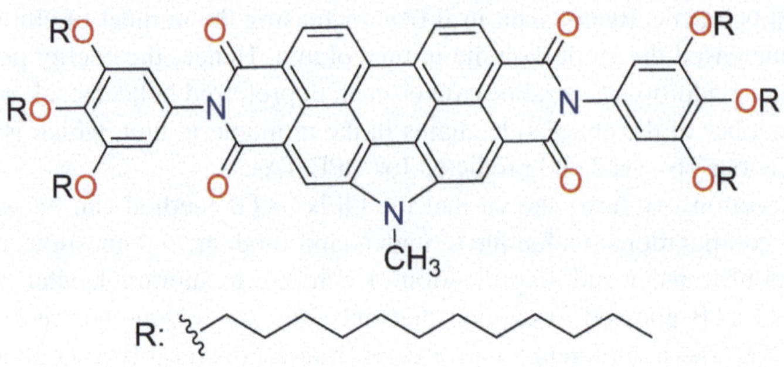

Fig. 11.23 Chemical structure of the N-PBI derivative **6**. Image adapted with permission from [101]. Copyright 2021 American Chemical Society

the different aggregates and explain the chiroptical features, a quantum-chemical approach with an implicit solvent model (GBSA) was adopted.

Compound **6** and its supramolecular polymers were theoretically studied at the GFN2-xTB level in gas phase. The structure of the monomer was simplified by substituting the peripheral—$OC_{12}H_{26}$ chains attached to the terminal phenyl rings by hydrogen atoms. At monomeric level, only two conformations, differing in the eclipsed (**6a**) or staggered (**6b**) conformation of the peripheral benzene rings with respect to the long *N*-PBI axis, were found due to the rigid structure of the terminal phenyl groups. Both conformers were predicted to be almost degenerate in energy, and were used to build up supramolecular aggregates of 5 monomeric units. The pentamer **6A$_5$** was made up of monomers **6a** arranged in staggered disposition, whereas **6B$_5$** was formed by monomeric units of **6b** self-assembled in a helical orientation (Fig. 11.24). In both structures, a number of stabilizing non-covalent interactions are predicted; π-stacking between the central *N*-PBI cores and the peripheral phenyl groups together with CH(phenyl)···O interactions (distances of 2.02 Å in **6A$_5$**

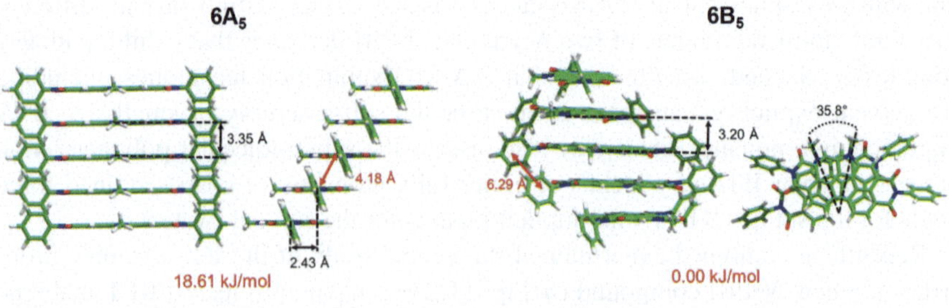

Fig. 11.24 Minimum-energy structures (with their relative energy indicated) calculated at the GFN2-xTB level for the most stable supramolecular pentamers of N-PBI **6**. Image adapted with permission from [101]. Copyright 2021 American Chemical Society

Table 11.1 Self-assembly free energy ($\Delta G^{solv}_{bind,n-1}$, in kJ/mol) estimated for the two most-stable oligomers of **6** in toluene and n-hexane solution at the GFN2-xTB/GBSA level

Oligomer	$\Delta G^{n\text{-hexane}}_{bind,n-1}$	$\Delta G^{toluene}_{bind,n-1}$
$6A_5$	−19.63	−24.94
$6B_5$	−27.93	−24.35

and 2.25–2.48 Å in **6B$_5$**). Although the oligomer **6B$_5$** was found to be 18.61 kJ/mol more stable than oligomer **6A$_5$**, both pentamers were initially considered for the study in different solvent conditions.

To further investigate the role of the environment on the supramolecular polymerization, the **6A$_5$** and **6B$_5$** pentamers previously optimized in gas-phase were recomputed at the same level (GFN2-xTB) but including solvent effects by means of the GBSA method in *n*-hexane and toluene as solvents. Note that *n*-hexane was used as a representative solvent that emulates the MCH nature since MCH was not parameterized for the GBSA model. The self-assembly free energy per interacting pair ($\Delta G^{solv}_{bind,n-1}$) in toluene and *n*-hexane were calculated (Table 11.1). In *n*-hexane, the relative stability computed in gas phase for **6A$_5$** and **6B$_5$** was preserved, and the helical aggregate **6B$_5$** turned out to be more stable than the pentamer **6A$_5$** by 8.30 kJ/mol. However, the stability was reversed in toluene and the linear assemble (**6A$_5$**) became slightly more stable by 0.59 kJ/mol. The relative stability of the supramolecular oligomers of *N*-PBI **6** was therefore predicted to be dependent on the solvent nature in line with the experimental evidences. Additionally, the most stable aggregates theoretically proposed to be generated in *n*-hexane (**6B$_5$**) and toluene (**6A$_5$**) were successfully used to simulate the absorption and CD spectra in concordance with the experimental ones [101]. Therefore, this is a good example in which a quantum-chemical approach (GFN2-xTB) combined with an implicit solvent model (GBSA) has enabled to rationalize the solvent-controlled supramolecular polymerization process of *N*-PBI **6**.

Explicit solvent effects during the supramolecular polymerization process have been also considered by means of classical MD simulations. A good example is that reported by S. Balasubramanian and coworkers [102], in which the supramolecular polymerization under ambient conditions of a small-size BTA decorated with peripheral alkyl chains in *n*-nonane was studied by all-atom MD simulations with the DREIDING force field (Fig. 11.25a) [28]. The MD simulations of pre-stacked oligomers (Fig. 11.25b) of different size allowed to estimate a potential energy per monomeric unit. The evolution of the potential energy per monomer as a function of oligomer size (Fig. 11.25c) revealed cooperativity in line with the outcomes experimentally obtained for BTA-based supramolecular polymers in apolar solvents. To understand in more detail the dynamical aspects of the self-assembly in BTAs, MD simulations starting from disassembled BTA molecules in a box of *n*-nonane molecules were also carried out. The BTA monomers were able to self-assemble into small oligomers that eventually formed a columnar helical aggregate of 6 monomers (Fig. 11.25d) on a

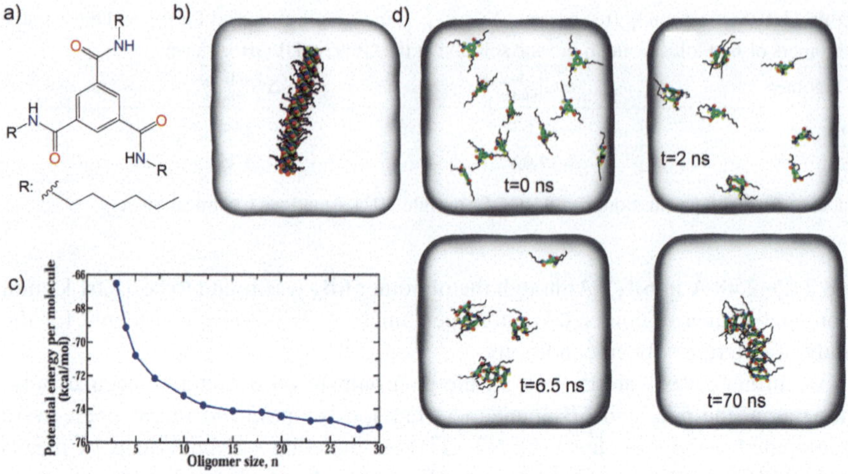

Fig. 11.25 **a)** Chemical structure of the BTA used with the DREIDING force field. **b)** Snapshot of a 20-mer of BTA solvated in n-nonane. **c)** The evolution of the potential energy per monomer as a function of oligomer size. **d)** Snapshots exhibiting the progress of the supramolecular self-assembly of BTA in n-nonane starting from a random configuration at 0, 2, 6.5, and 70 ns. Image adapted with permission from [102]. Copyright 2014 American Chemical Society

time scale of nanoseconds. Another interesting aspect demonstrated by the MD simulations was that the BTA molecules were found to supramolecularly organize with two amide NH groups and one carbonyl oxygen pointing out in the same direction. Stabilization of that supramolecular structure was due to the favourable antiparallel alignment of the dipole moments of the hydrogen bonds.

This example illustrates how explicit solvent effects can be accounted for in a classical manner to theoretically describe the supramolecular polymerization process. All-atom MD simulations including solvent molecules can assist to reveal dynamical aspects of supramolecular polymerization unable to be captured by quantum-chemical approaches and implicit solvent models.

Although AA-MD simulations including explicit solvent are very attractive computational techniques to monitor the self-assembly process in SPs, this technique is limited to supramolecular polymers of small size or to reduced oligomer models. When the size of the system under study becomes huge, computationally less demanding techniques are highly desired. In this regard, different possibilities that combine implicit and explicit solvent models and coarse-grained force fields have been proposed [4, 13, 28]. Amongst the different affordable computational protocols to describe SPs including solvent effects, it is necessary to stress two computational protocols developed by Pavan et al. [4, 13].

The first computational approximation initially employed all-atom MD simulations (explicit solvent) to model the self-assembly of a BTA-based supramolecular polymer in water (Fig. 11.16, system previously used in Sect. 11.3) [13]. The popular General Amber Force Field (GAFF) [69] was parameterized to provide a reliable

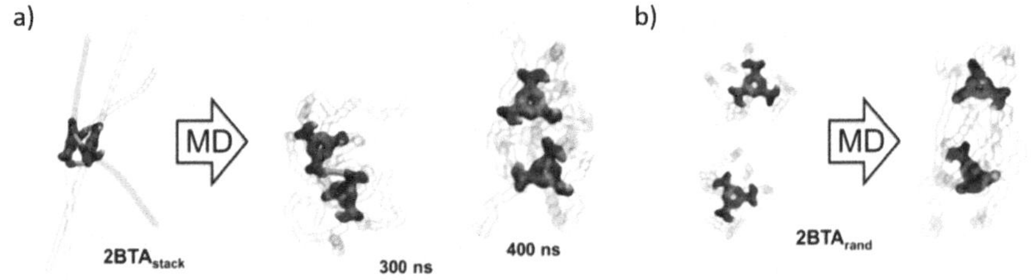

Fig. 11.26 Initial (extended) and equilibrated structures of pre-ordered **a)** and randomly-ordered **b)** BTA dimers extracted from the MD simulation. Image adapted with permission from [71]. Copyright 2016 American Chemical Society

picture of the supramolecular growth in BTAs governed by H-bonding. Interestingly, the authors started the MD simulations from pre- and randomly-ordered stacks of BTA aggregates of increasing oligomer size. Their outcomes revealed interesting aspects of the supramolecular polymerization mechanism. For instance, the MD simulations showed that small aggregates (e.g., dimers or trimers) were found to be quite unstable (Fig. 11.26) and easily disassembled during the early steps of the MD trajectory. Only oligomers of larger size (e.g., BTA pentamers) remained assembled and stable owing to the effective π-π interactions and would act as nuclear species for the supramolecular elongation process. Nevertheless, up to oligomers of 14 units, BTA aggregates were still found to be non-uniform and discontinuous. Long, uniform, and persistent BTA polymers were obtained from assemblies with size ≥14 BTA units. These computational findings seemed to be in controversy with the general assumption that dimeric species were those which acted as initial nuclear agents for the supramolecular polymerization process. That aspect was revisited with large-scale MD simulations using coarse-grained models (see below).

Apart from the time evolution of the self-assembly process visualized from all-atom MD simulations, thermodynamic aspects to unveil the polymerization mechanism were also extracted from MD simulations but with a slightly different protocol. In particular, G. Pavan and coworkers evaluated the self-assembly free-energies (ΔG) per monomeric unit of each supramolecular oligomer by using a cost-effective computational approach. All self-assembly ΔG values were calculated as,

$$\Delta G = \Delta E_{\text{gas}} + \Delta G_{\text{solv}} - T\Delta S \tag{11.16}$$

where ΔE_{gas} was the gas-phase molecular mechanics energy provided by the force fields, ΔG_{solv} was the total solvation energy and $T\Delta S$ is the entropic term of the free-energy. In the efficient computational protocol, the oligomer structures of BTA aggregates were extracted from the equilibrated MD simulations (the last 100 ns) to calculate an average ΔE_{gas}, whereas the solvent effects (ΔG_{solv}) were now treated with an implicit solvent model (GBSA). That combined computational approach provided a pattern for the evolution of the self-assembly energies as a function of the oligomer size (Fig. 11.27), characteristic for a cooperative mechanism of polar

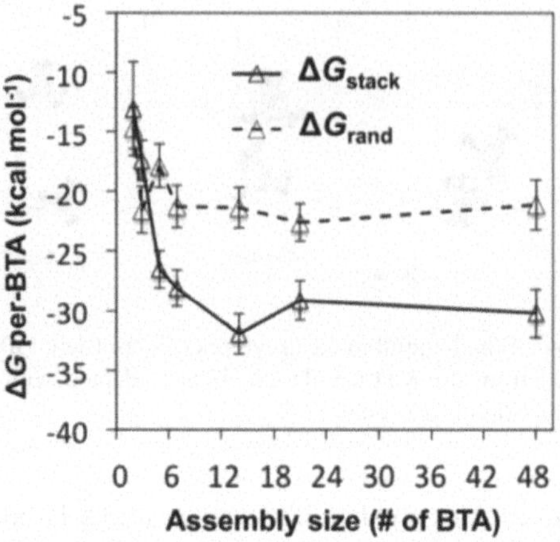

Fig. 11.27 Self-assembly free energies per BTA unit for pre-ordered and randomly-ordered BTA assemblies of increasing size. Image adapted with permission from [71]. Copyright 2016 American Chemical Society

BTAs in water, which was consistent with the experimental observations and other theoretical studies [13]. Note that a cooperative behaviour was predicted for both pre- and randomly-ordered stacks but being more significant for the pre-ordered aggregates. The computational protocol proposed by G. Pavan and coworkers is an excellent example that highlights that combining explicit and implicit solvent models allows to visualize the supramolecular polymerization process at atomistic resolution and also to evaluate thermodynamic aspects to determine the potential polymerization mechanism [71].

The second computational approximation consisted of MD simulations using a CG force field for solute molecules combined with implicit solvents to explore the self-assembly of BTAs in water [4]. The main advantage of this computational approach was its efficiency. This approximation enabled the simulation of the self-assembly of thousands of BTA molecules in large solvent boxes reaching concentration levels in the simulations similar to those generally used in supramolecular polymerization experiments.

The implicit-solvent CG model for a water-soluble BTA supramolecular polymer was opportunely tuned to obtain a dimerization free-energy profile consistent with that obtained using an explicit solvent and CG FFs for the BTA monomers [13, 71]. Therefore, the CG model was able to correctly capture the behaviour of the individual monomer, monomer-monomer interactions in solution and, more challenging, the cooperative self-assembly. After the complete validation of the model, a BTA-based system composed by 1000 monomers randomly dissolved in a cubic box ($30 \times 30 \times 30$ nm) was built to start the study. That box represented a concentration of *ca.* 48 mM. After 7.5 μs of simulation, a spontaneously long assemble (>70 nm, Fig. 11.28) was generated. In line with the previous findings using a CG model for BTAs and explicit solvent molecules (water) [13], the BTA units self-assembled in small and disordered

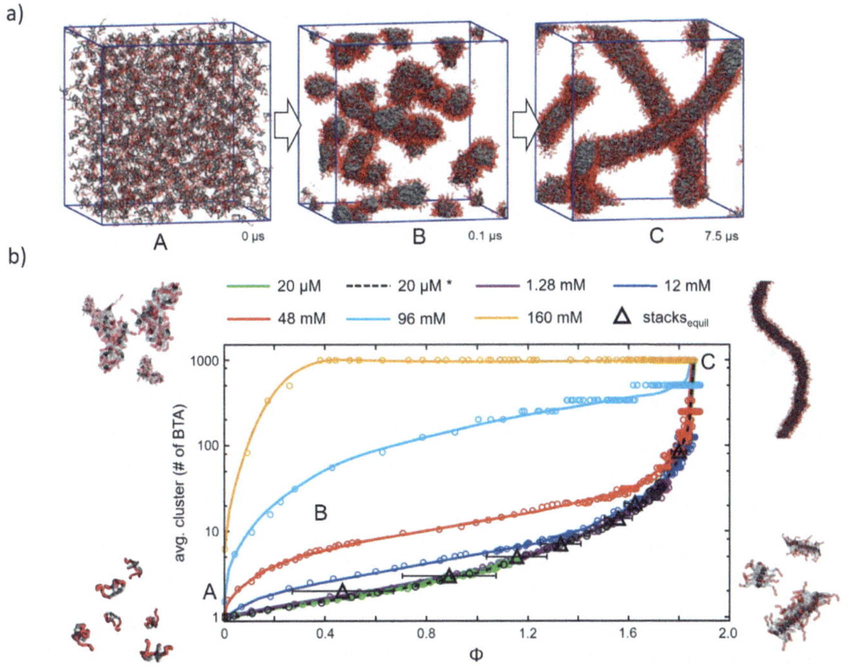

Fig. 11.28 **a)** Polymerization of BTA supramolecular fibres obtained from the implicit-solvent CG model. At 48 mM, randomly dispersed monomers (A) rapidly aggregate into small, disordered clusters (B), that on longer-time scales evolve into supramolecular fibres (C). **b)** Average size of the BTA clusters as a function of Φ for different concentrations along the MD trajectory using the implicit-solvent CG model Image adapted with permission from [4]. Copyright 2017 American Chemical Society

oligomers in a fast time scale whilst the ordering of the aggregates proceeded on a slower time scale. That conclusion was drawn by monitoring the evolution of the average oligomer size and the stacking order (measured by the order parameter Φ) during the supramolecular polymerization process (Fig. 11.28).

In a further step, the effect of the monomer concentration on the supramolecular polymerization was analysed by generating different concentration conditions. Several solvent boxes of different size but keeping constant the number of monomers in the system (solute) were systematically built from low to high concentrations (20 µM, 1.28 mM, 12 mM, 48 mM, 96 mM, and 160 mM). As shown in Fig. 11.28b, the polymerization process at high concentrations followed a curve tending to an "L" shape, meaning that a fast aggregation into large disordered clusters occurs before the conversion into highly-ordered supramolecular fibres. This is consistent with the experimental evidence that, at high concentrations, water-soluble BTAs can result in hydrogels [103]. On the other hand, decreasing the concentration gives rise to curves with a reverse "L" pattern, in line with the formation of small-ordered clusters for long-time scales before proceeding towards a large-ordered supramolecular growth [4].

The above discussion has highlighted that the combination of coarse-grained force fields and implicit solvent models can be seen as an interesting and computationally-efficient approach to theoretically describe systems prone to supramolecularly polymerize. For the first time, this computational technique opened the door to simulations in "realistic experimental conditions" (dilute concentrations) to explore a wide range of potential supramolecular polymers.

11.6 Concluding Remarks

In this chapter, we show how computational modelling at quantum and classical level is particularly helpful to gain insights into the complex supramolecular polymerization process at molecular scale. In particular, we have displayed using a few selected examples that a quantum-chemical approach (combining DFT with augmented dispersion-corrected terms and the cost-effective semi-empirical GFN2-xTB method) is able to provide: (i) structural models for the supramolecular polymers at atomistic level, (ii) energetics of the polymerization mechanism, and (iii) basis for chiral transfer and amplification phenomena. On the other hand, we have illustrated that classical molecular dynamics simulations are also very useful techniques to explore molecular self-assembling phenomena intractable at quantum mechanical level. For instance, we have shown that molecular dynamics simulations based on specifically-adaptive coarse-grained force fields can be used to characterize the steps during the dynamic supramolecular polymerization process even at dilute (realistic) concentrations.

A bright future for computational modelling in the field of supramolecular polymers is therefore anticipated for the following years. Efficient computational techniques but accurate enough to capture cooperativity effects (e.g., the implicit-solvent coarse-grained approximation) and pathway complexity are expected to bring an atomistic detail of the polymerization process difficult to achieve at experimental level. Nevertheless, some important challenges still need to be addressed. One important theoretical challenge is to rationalize at molecular level why a particular solvent or a mixture of solvents can determine the supramolecular organization of the final polymers from a dynamic perspective.

Although this chapter has mainly focused, from a computational perspective, on the supramolecular polymerization growth mostly dominated by non-covalent π-stacking and H-bond interactions, the computational techniques discussed along this chapter can be also used in other contexts of supramolecular chemistry. In particular, these techniques would be appropriate for those supramolecular systems that self-assemble governed by intermolecular electrostatic interactions. In that context, the quantum-chemical techniques (e.g., DFT and semi-empirical) would capture these electrostatic interactions reasonably well without further dispersion corrections.

References

1. Li P, Xu G, Wang N, Guan B, Zhu S, Chen P, Liu M (2021) 0D, 1D, and 2D Supramolecular nanoassemblies of a porphyrin: controllable assembly, and dimensionality-dependent catalytic performances. Adv Funct Mater 31:2100367. https://doi.org/10.1002/adfm.202100367
2. Bochicchio D, Pavan GM (2018) Molecular modelling of supramolecular polymers. Adv Phys X 3:316–338. https://doi.org/10.1080/23746149.2018.1436408
3. Varela-Aramburu S, Morgese G, Su L, Schoenmakers SMC, Perrone M, Leanza L, Perego C, Pavan GM, Palmans ARA, Meijer EW (2020) Exploring the potential of Benzene-1,3,5-tricarboxamide supramolecular polymers as biomaterials. Biomacromol 21:4105–4115. https://doi.org/10.1021/acs.biomac.0c00904
4. Bochicchio D, Pavan GM (2017) Effect of concentration on the supramolecular polymerization mechanism via implicit-solvent coarse-grained simulations of water-soluble 1,3,5-Benzenetricarboxamide. J Phys Chem Lett 8:3813–3819. https://doi.org/10.1021/acs.jpclett.7b01649
5. Grabicki N, Dumele O, Sai H, Powers-Riggs NE, Phelan BT, Sangji MH, Chapman CT, Passarelli JV, Dannenhoffer AJ, Wasielewski MR, Stupp SI (2021) Polymorphism and optoelectronic properties in crystalline supramolecular polymers. Chem Mater 33:706–718. https://doi.org/10.1021/acs.chemmater.0c04123
6. Bakker MH, Kieltyka RE, Albertazzi L, Dankers PYW (2016) Modular supramolecular ureidopyrimidinone polymer carriers for intracellular delivery. RSC Adv 6:110600–110603. https://doi.org/10.1039/C6RA22490C
7. Haedler AT, Kreger K, Issac A, Wittmann B, Kivala M, Hammer N, Köhler J, Schmidt HW, Hildner R (2015) Long-range energy transport in single supramolecular nanofibres at room temperature. Nature 523:196–199. https://doi.org/10.1038/nature14570
8. Wittmann B, Wenzel FA, Wiesneth S, Haedler AT, Drechsler M, Kreger K, Köhler J, Meijer EW, Schmidt HW, Hildner R (2020) Enhancing long-range energy transport in supramolecular architectures by tailoring coherence properties. J Am Chem Soc 142:8323–8330. https://doi.org/10.1021/jacs.0c01392
9. Fouquey C, Lehn JM, Levelut AM (1990) Molecular recognition directed self-assembly of supramolecular liquid crystalline polymers from complementary chiral components. Adv Mater 2:254–257. https://doi.org/10.1002/adma.19900020506
10. De Greef TFA, Smulders MMJ, Wolffs M, Schenning APHJ, Sijbesma RP, Meijer EW (2009) Supramolecular polymerization. Chem Rev 109:5687–5754. https://doi.org/10.1021/cr900181u
11. Aida T, Meijer EW, Stupp SI (2012) Functional supramolecular polymers. Science (80-.). 335:813–817. https://doi.org/10.1126/science.1205962
12. Kulkarni C, Balasubramanian S, George SJ (2013) What molecular features govern the mechanism of supramolecular polymerization? ChemPhysChem 14:661–673. https://doi.org/10.1002/cphc.201200801
13. Bochicchio D, Pavan GM (2017) From cooperative self-assembly to water-soluble supramolecular polymers using coarse-grained simulations. ACS Nano 11:1000–1011. https://doi.org/10.1021/acsnano.6b07628
14. Greciano EE, Calbo J, Ortí E, Sánchez L (2020) N-Annulated perylene bisimides to bias the differentiation of metastable supramolecular assemblies into J- and H-Aggregates. Angew Chemie Int Ed 59:17517–17524. https://doi.org/10.1002/anie.202005837
15. Swathi K, Sissa C, Painelli A, Thomas KG (2020) Supramolecular chirality: a caveat in assigning the handedness of chiral aggregates. Chem Commun 56:8281–8284. https://doi.org/10.1039/d0cc01922d
16. Orvay F, Cerdá J, Rotger C, Ortí E, Aragó J, Costa A, Soberats B (2021) Influence of the Z/E isomerism on the pathway complexity of a squaramide-based macrocycle. Small 17:1–9. https://doi.org/10.1002/smll.202006133

17. Garrido M, Martínez-Periñán E, Calbo J, Rodríguez-Pérez L, Aragó J, Lorenzo E, Ortí E, Martín N, Herranz MÁ (2021) Supramolecular assembly of pyrene-tetrathiafulvalene hybrids on graphene: structure–property relationships and biosensing activity. J Mater Chem C 9:10944–10951. https://doi.org/10.1039/d1tc01442k
18. Dorca Y, Valera JS, Cerdá J, Aragó J, Gómez R, Ortí E, Sánchez L (2018) Synergy of axial and point chirality to construct helical N-heterotriangulene-based supramolecular polymers. ChemNanoMat. 4:781–784. https://doi.org/10.1002/cnma.201800186
19. Greciano EE, Calbo J, Buendía J, Cerdá J, Aragó J, Ortí E, Sánchez L (2019) Decoding the consequences of increasing the size of self-assembling tricarboxamides on chiral amplification. J Am Chem Soc 141:7463–7472. https://doi.org/10.1021/jacs.9b02045
20. Wagner W, Wehner M, Stepanenko V, Ogi S, Würthner F (2017) Living supramolecular polymerization of a perylene bisimide dye into fluorescent J-aggregates. Angew. Chemie - Int. Ed. 56:16008–16012. https://doi.org/10.1002/anie.201709307
21. Cantekin S, De Greef TFA, Palmans ARA (2012) Benzene-1,3,5-tricarboxamide: a versatile ordering moiety for supramolecular chemistry. Chem Soc Rev 41:6125–6137. https://doi.org/10.1039/c2cs35156k
22. You F, Zhou X, Huang H, Liu Y, Liu S, Shao J, Zhao B, Qin T, Huang W (2018) N-Annulated perylene diimide derivatives as non-fullerene acceptors for solution-processed solar cells with an open-circuit voltage of up to 1.14 v. New J Chem 42:15079–15087. https://doi.org/10.1039/c8nj02566e
23. Li J, Loh XJ (2008) Cyclodextrin-based supramolecular architectures: Syntheses, structures, and applications for drug and gene delivery. Adv Drug Deliv Rev 60:1000–1017. https://doi.org/10.1016/j.addr.2008.02.011
24. Mayoral MJ, Guilleme J, Calbo J, Aragó J, Aparicio F, Ortí E, Torres T, González-Rodríguez D (2020) Dual-Mode chiral self-assembly of cone-shaped subphthalocyanine aromatics. J Am Chem Soc 142:21017–21031. https://doi.org/10.1021/jacs.0c07291
25. Cui Q, Elstner M (2014) Density functional tight binding: values of semi-empirical methods in an ab initio era. Phys Chem Chem Phys 16:14368–14377. https://doi.org/10.1039/c4cp00908h
26. Bannwarth C, Ehlert S, Grimme S (2019) GFN2-xTB—An accurate and broadly parametrized self-consistent tight-binding quantum chemical method with multipole electrostatics and density-dependent dispersion contributions. J Chem Theory Comput 15:1652–1671. https://doi.org/10.1021/acs.jctc.8b01176
27. Kulkarni C, Reddy SK, George SJ, Balasubramanian S (2011) Cooperativity in the stacking of benzene-1,3,5-tricarboxamide: the role of dispersion. Chem Phys Lett 515:226–230. https://doi.org/10.1016/j.cplett.2011.09.028
28. Bejagam KK, Balasubramanian S (2015) Supramolecular polymerization: a coarse grained molecular dynamics study. J Phys Chem B 119:5738–5746. https://doi.org/10.1021/acs.jpcb.5b01655
29. Korlepara DB, Bejagam KK, Balasubramanian S (2017) Supramolecular Polymerization of N, N′, N″, N‴-tetra-(Tetradecyl)-1,3,6,8-pyrenetetracarboxamide: a computational study. J Phys Chem B 121:11492–11503. https://doi.org/10.1021/acs.jpcb.7b10171
30. Hohenberg P, Kohn W (1964) Inhomogeneous electron gas. Phys Rev 136:B864–B871. https://doi.org/10.1103/PhysRev.136.B864
31. Born M, Oppenheimer R (1927) Zur Quantentheorie der Molekeln. Ann Phys 389:457–484. https://doi.org/10.1002/andp.19273892002
32. Venanzi, C.A.: Reviews in Computational Chemistry, Volume 8. (1996)
33. Perdew JP, Burke K, Ernzerhof M (1996) Generalized gradient approximation made simple. Phys Rev Lett 77:3865–3868. https://doi.org/10.1103/PhysRevLett.77.3865
34. Tao J, Perdew JP, Staroverov VN, Scuseria GE (2003) Climbing the density functional ladder: nonempirical meta–generalized gradient approximation designed for molecules and solids. Phys Rev Lett 91:3–6. https://doi.org/10.1103/PhysRevLett.91.146401
35. Calbo J, Ortí E, Sancho-García JC, Aragó J (2015) The nonlocal correlation density functional VV10: a successful attempt to accurately capture noncovalent interactions. In: Dixon DA (ed.) Annual Reports in Computational Chemistry. pp 37–102. Elsevier

36. Tainer JA, Getzoff ED, Beem KM, Richardson JS, Richardson DC (1982) Determination and analysis of the 2 Å structure of copper, zinc superoxide dismutase. J Mol Biol 160:181–217. https://doi.org/10.1016/0022-2836(82)90174-7
37. Yanai T, Tew DP, Handy NC (2004) A new hybrid exchange-correlation functional using the Coulomb-attenuating method (CAM-B3LYP). Chem Phys Lett 393:51–57. https://doi.org/10.1016/j.cplett.2004.06.011
38. Chai JD, Head-Gordon M (2008) Systematic optimization of long-range corrected hybrid density functionals. J Chem Phys 128:1–15. https://doi.org/10.1063/1.2834918
39. Brémond E, Ciofini I, Sancho-García JC, Adamo C (2016) Nonempirical double-hybrid functionals: an effective tool for chemists. Acc Chem Res 49:1503–1513. https://doi.org/10.1021/acs.accounts.6b00232
40. Grimme S (2006) Semiempirical hybrid density functional with perturbative second-order correlation. J Chem Phys 124:1–16. https://doi.org/10.1063/1.2148954
41. Steinmann SN, Corminboeuf C (2011) Comprehensive benchmarking of a density-dependent dispersion correction. J Chem Theory Comput 7:3567–3577. https://doi.org/10.1021/ct200602x
42. Tkatchenko A, Scheffler M (2009) Accurate molecular van der Waals interactions from ground-state electron density and free-atom reference data. Phys Rev Lett 102:6–9. https://doi.org/10.1103/PhysRevLett.102.073005
43. Becke AD, Johnson ER (2007) Exchange-hole dipole moment and the dispersion interaction revisited. J Chem Phys 127:1–6. https://doi.org/10.1063/1.2795701
44. Dion M, Rydberg H, Schröder E, Langreth DC, Lundqvist BI (2004) Van der Waals density functional for general geometries. Phys Rev Lett 92:22–25. https://doi.org/10.1103/PhysRevLett.92.246401
45. Vydrov OA, Van Voorhis T (2010) Nonlocal van der Waals density functional: the simpler the better. J Chem Phys 133:244103.https://doi.org/10.1063/1.3521275
46. Calbo J, Ortí E, Sancho-García JC, Aragó J (2015) Accurate treatment of large supramolecular complexes by double-hybrid density functionals coupled with nonlocal van der waals corrections. J Chem Theory Comput 11:932–939. https://doi.org/10.1021/acs.jctc.5b00002
47. Grimme S (2006) Semiempirical GGA-type density functional constructed with a long-range dispersion correction. J Comput Chem 27:1787–1799. https://doi.org/10.1002/jcc.20495
48. Grimme S, Bannwarth C (2016) Ultra-fast computation of electronic spectra for large systems by tight-binding based simplified Tamm-Dancoff approximation (sTDA-xTB). J Chem Phys 145:054103.https://doi.org/10.1063/1.4959605
49. Antony J, Grimme S (2006) Density functional theory including dispersion corrections for intermolecular interactions in a large benchmark set of biologically relevant molecules. Phys Chem Chem Phys 8:5287–5293. https://doi.org/10.1039/b612585a
50. Grimme S, Antony J, Ehrlich S, Krieg H (2010) A consistent and accurate ab initio parametrization of density functional dispersion correction (DFT-D) for the 94 elements H-Pu. J Chem Phys 132:154104.https://doi.org/10.1063/1.3382344
51. Caldeweyher E, Ehlert S, Hansen A, Neugebauer H, Spicher S, Bannwarth C, Grimme S (2019) A generally applicable atomic-charge dependent London dispersion correction. J Chem Phys 150:154122.https://doi.org/10.1063/1.5090222
52. Grimme S, Ehrlich S, Goerigk L (2011) Effect of the damping function in dispersion corrected density functional theory. J Comput Chem 32:1456–1465. https://doi.org/10.1002/jcc.21759
53. Caldeweyher E, Mewes JM, Ehlert S, Grimme S (2020) Extension and evaluation of the D4 London-dispersion model for periodic systems. Phys Chem Chem Phys 22:8499–8512. https://doi.org/10.1039/d0cp00502a
54. Yilmazer ND, Korth M (2015) Enhanced semiempirical QM methods for biomolecular interactions. Comput Struct Biotechnol J 13:169–175. https://doi.org/10.1016/j.csbj.2015.02.004
55. Bannwarth C, Caldeweyher E, Ehlert S, Hansen A, Pracht P, Seibert J, Spicher S, Grimme S (2021) Extended tight-binding quantum chemistry methods. Wiley Interdiscip Rev Comput Mol Sci 11:1–49. https://doi.org/10.1002/wcms.1493

56. Hehre WJ, Stewart RF, Pople JA (1969) Self-consistent molecular-orbital methods. I. Use of gaussian expansions of slater-type atomic orbitals. J Chem Phys 51:2657–2664. https://doi.org/10.1063/1.1672392
57. Seifert G, Joswig JO (2012) Density-functional tight binding-an approximate density-functional theory method. Wiley Interdiscip Rev Comput Mol Sci 2:456–465. https://doi.org/10.1002/wcms.1094
58. Klopman G (1964) A semiempirical treatment of molecular structures. II. Molecular terms and application to diatomic molecules. J Am Chem Soc 86:4550–4557. https://doi.org/10.1021/ja01075a008
59. Hoffmann R (1963) An extended Hückel theory. I Hydrocarbons J Chem Phys 39:1397–1412. https://doi.org/10.1063/1.1734456
60. Dewar MJS, Thiel W (1977) Ground states of molecules. 38. The MNDO method. Approximations and parameters. J Am Chem Soc 99:4899–4907
61. Dewar MJS, Zoebisch EG, Healy EF, Stewart JJP (1985) AM1: A new general purpose quantum mechanical molecular model1. J Am Chem Soc 107:3902–3909. https://doi.org/10.1021/ja00299a024
62. Stewart JJP (1989) Optimization of parameters for semiempirical methods I. J Comput Chem 10:221–264. https://doi.org/10.1002/jcc.540100209
63. Christensen AS, Kubař T, Cui Q, Elstner M (2016) Semiempirical quantum mechanical methods for noncovalent interactions for chemical and biochemical applications. Chem Rev 116:5301–5337. https://doi.org/10.1021/acs.chemrev.5b00584
64. Pople JA, Beveridge DL (1970) Approximate molecular orbital theory. McGraw Hill Higher Education
65. Stewart JJP (2013) Optimization of parameters for semiempirical methods VI: More modifications to the NDDO approximations and re-optimization of parameters. J Mol Model 19:1–32. https://doi.org/10.1007/s00894-012-1667-x
66. Hospital A, Goñi JR, Orozco M, Gelpí JL (2015) Molecular dynamics simulations: advances and applications. Adv Appl Bioinforma Chem 8:37–47. https://doi.org/10.2147/AABC.S70333
67. Decherchi S, Cavalli A (2020) Thermodynamics and kinetics of drug-target binding by molecular simulation. Chem Rev 120:12788–12833. https://doi.org/10.1021/acs.chemrev.0c00534
68. Brocorens P, Linares M, Guyard-Duhayon C, Guillot R, Andrioletti B, Suhr D, Isare B, Lazzaroni R, Bouteiller L (2013) Conformational plasticity of hydrogen bonded bis-urea supramolecular polymers. J Phys Chem B 117:5379–5386. https://doi.org/10.1021/jp401915y
69. Wang J, Wolf RM, Caldwell JW, Kollman PA, Case DA (2004) Development and testing of a general Amber force field. J Comput Chem 25:1157–1174. https://doi.org/10.1002/jcc.20035
70. Frenkel, D., Smit, B.: Understanding Molecular Simulation. Academic Press (2002)
71. Garzoni M, Baker MB, Leenders CMA, Voets IK, Albertazzi L, Palmans ARA, Meijer EW, Pavan GM (2016) Effect of H-Bonding on order amplification in the growth of a supramolecular polymer in water. J Am Chem Soc 138:13985–13995. https://doi.org/10.1021/jacs.6b07530
72. Baron R, Trzesniak D, De Vries AH, Elsener A, Marrink SJ, Van Gunsteren WF (2007) Comparison of thermodynamic properties of coarse-grained and atomic-level simulation models. Chem Phys Chem 8:452–461. https://doi.org/10.1002/cphc.200600658
73. Marrink SJ, De Vries AH, Mark AE (2004) Coarse grained model for semiquantitative lipid simulations. J Phys Chem B 108:750–760. https://doi.org/10.1021/jp036508g
74. Kmiecik S, Gront D, Kolinski M, Wieteska L, Dawid AE, Kolinski A (2016) Coarse-Grained protein models and their applications. Chem Rev 116:7898–7936. https://doi.org/10.1021/acs.chemrev.6b00163
75. Wang ZJ, Deserno M (2010) A systematically coarse-grained solvent-free model for quantitative phospholipid bilayer simulations. J Phys Chem B 114:11207–11220. https://doi.org/10.1021/jp102543j

76. Marrink SJ, Risselada HJ, Yefimov S, Tieleman DP, De Vries AH (2007) The MARTINI force field: coarse grained model for biomolecular simulations. J Phys Chem B 111:7812–7824. https://doi.org/10.1021/jp071097f
77. Mabesoone MFJ, Palmans ARA, Meijer EW (2020) Solute-Solvent interactions in modern physical organic chemistry: supramolecular polymers as a muse. J Am Chem Soc 142:19781–19798. https://doi.org/10.1021/jacs.0c09293
78. Cramer, C.J (1999) Implicit solvation models: equilibria, structure, spectra, and dynamics—chemical reviews (ACS Publications). Chem Rev
79. Tomasi J, Mennucci B, Cammi R (2005) Quantum mechanical continuum solvation models. Chem Rev 105:2999–3093. https://doi.org/10.1021/cr9904009
80. Zhang LY, Gallicchio E, Friesner RA, Levy RM (2001) Solvent models for protein-ligand binding: comparison of implicit solvent poisson and surface generalized born models with explicit solvent simulations. J Comput Chem 22:591–607. https://doi.org/10.1002/jcc.1031
81. Zhang J, Zhang H, Wu T, Wang Q, Van Der Spoel D (2017) Comparison of implicit and explicit solvent models for the calculation of solvation free energy in organic solvents. J Chem Theory Comput 13:1034–1043. https://doi.org/10.1021/acs.jctc.7b00169
82. Steinmann SN, Sautet P, Michel C (2016) Solvation free energies for periodic surfaces: comparison of implicit and explicit solvation models. Phys Chem Chem Phys 18:31850–31861. https://doi.org/10.1039/c6cp04094b
83. Scalmani G, Frisch MJ (2010) Continuous surface charge polarizable continuum models of solvation. I. General formalism. J Chem Phys132:114110-1–114110-15. https://doi.org/10.1063/1.3359469
84. Benedetta Mennucci RC (2007) Continuum solvation models in chemical physics: from theory to applications. Wiley
85. Klamt A, Schüürmann G (1993) COSMO: a new approach to dielectric screening in solvents with explicit expressions for the screening energy and its gradient. J Chem Soc Perkin Trans 2:799–805. https://doi.org/10.1039/P29930000799
86. Marenich AV, Cramer CJ, Truhlar DG (2009) Universal solvation model based on solute electron density and on a continuum model of the solvent defined by the bulk dielectric constant and atomic surface tensions. J Phys Chem B 113:6378–6396. https://doi.org/10.1021/jp810292n
87. Wang E, Sun H, Wang J, Wang Z, Liu H, Zhang JZH, Hou T (2019) End-Point binding free energy calculation with MM/PBSA and MM/GBSA: strategies and applications in drug design. Chem Rev 119:9478–9508. https://doi.org/10.1021/acs.chemrev.9b00055
88. Kollman PA, Massova I, Reyes C, Kuhn B, Huo S, Chong L, Lee M, Lee T, Duan Y, Wang W, Donini O, Cieplak P, Srinivasan J, Case DA, Cheatham TE (2000) Calculating structures and free energies of complex molecules: combining molecular mechanics and continuum models. Acc Chem Res 33:889–897. https://doi.org/10.1021/ar000033j
89. Sun H, Duan L, Chen F, Liu H, Wang Z, Pan P, Zhu F, Zhang JZH, Hou T (2018) Assessing the performance of MM/PBSA and MM/GBSA methods. 7. Entropy effects on the performance of end-point binding free energy calculation approaches. Phys Chem Chem Phys 20:14450–14460. https://doi.org/10.1039/c7cp07623a
90. Senn HM, Thiel W (2009) QM/MM methods for biomolecular systems. Angew. Chemie - Int. Ed. 48:1198–1229. https://doi.org/10.1002/anie.200802019
91. Filot IAW, Palmans ARA, Hilbers PAJ, Van Santen RA, Pidko EA, De Greef TFA (2010) Understanding cooperativity in hydrogen-bond-induced supramolecular polymerization: a density functional theory study. J Phys Chem B 114:13667–13674. https://doi.org/10.1021/jp1072928
92. Kulkarni C, Bejagam KK, Senanayak SP, Narayan KS, Balasubramanian S, George SJ (2015) Dipole-Moment-Driven cooperative supramolecular polymerization. J Am Chem Soc 137:3924–3932. https://doi.org/10.1021/jacs.5b00504
93. Nieto-Ortega B, García F, Longhi G, Castiglioni E, Calbo J, Abbate S, López Navarrete JT, Ramírez FJ, Ortí E, Sánchez L, Casado J (2015) On the handedness of helical aggregates of C3 tricarboxamides: a multichiroptical characterization. Chem Commun 51:9781–9784. https://doi.org/10.1039/c5cc03054d

94. Das A, Vantomme G, Markvoort AJ, Ten Eikelder HMM, Garcia-Iglesias M, Palmans ARA, Meijer EW (2017) Supramolecular copolymers: structure and composition revealed by theoretical modeling. J Am Chem Soc 139:7036–7044. https://doi.org/10.1021/jacs.7b02835
95. Liu M, Zhang L, Wang T (2015) Supramolecular chirality in self-assembled systems. Chem Rev 115:7304–7397. https://doi.org/10.1021/cr500671p
96. Li SL, Xiao T, Lin C, Wang L (2012) Advanced supramolecular polymers constructed by orthogonal self-assembly. Chem Soc Rev 41:5950–5968. https://doi.org/10.1039/c2cs35099h
97. García F, Viruela PM, Matesanz E, Ortí E, Sánchez L (2011) Cooperative supramolecular polymerization and amplification of chirality in C3-symmetrical OPE-based trisamides. Chem A Eur J 17:7755–7759. https://doi.org/10.1002/chem.201100898
98. García F, Sánchez L (2012) Structural rules for the chiral supramolecular organization of OPE-based discotics: induction of helicity and amplification of chirality. J Am Chem Soc 134:734–742. https://doi.org/10.1021/ja210443m
99. Zang L, Che Y, Moore JS (2008) One-dimensional self-assembly of planar π-conjugated molecules: Adaptable building blocks for organic nanodevices. Acc Chem Res 41:1596–1608. https://doi.org/10.1021/ar800030w
100. Mes T, Smulders MMJ, Palmans ARA, Meijer EW (2010) Hydrogen-bond engineering in supramolecular polymers: polarity influence on the self-assembly of benzene-1,3,5-tricarboxamides. Macromolecules 43:1981–1991. https://doi.org/10.1021/ma9026096
101. Martínez MA, Doncel-Giménez A, Cerdá J, Calbo J, Rodríguez R, Aragó J, Crassous J, Ortí E, Sánchez L (2021) Distance matters: biasing mechanism, transfer of asymmetry, and stereomutation in N-Annulated perylene bisimide supramolecular polymers. J Am Chem Soc 143:13281–13291. https://doi.org/10.1021/jacs.1c06125
102. Bejagam KK, Fiorin G, Klein ML, Balasubramanian S (2014) Supramolecular polymerization of benzene-1,3,5-tricarboxamide: a molecular dynamics simulation study. J Phys Chem B 118:5218–5228. https://doi.org/10.1021/jp502779z
103. Goor OJGM, Hendrikse SIS, Dankers PYW, Meijer EW (2017) From supramolecular polymers to multi-component biomaterials. Chem Soc Rev 46:6621–6637. https://doi.org/10.1039/c7cs00564d

Index

M. A. Aboudzadeh and A. Frontera (eds.), *Supramolecular Assemblies Based on Electrostatic Interactions*,
https://doi.org/10.1007/978-3-031-00657-9

R

S

The manufacturer's authorised representative in the EU is Springer Nature Customer Service Centre GmbH, Europaplatz 3, 69115 Heidelberg, Germany. If you have any concerns regarding our products, please contact ProductSafety@springernature.com

Printed and bound by CPI Group (UK) Ltd, Croydon, CR0 4YY
01/10/2026
02225893-0011